THE OXFORD HANDBOOK OF

INTERACTIVE AUDIO

THE OXFORD HANDBOOK OF

INTERACTIVE AUDIO

Edited by

KAREN COLLINS, BILL KAPRALOS,

and

HOLLY TESSLER

OXFORD
UNIVERSITY PRESS

OXFORD
UNIVERSITY PRESS

Oxford University Press is a department of the University of Oxford.
It furthers the University's objective of excellence in research, scholarship,
and education by publishing worldwide.

Oxford New York
Auckland Cape Town Dar es Salaam Hong Kong Karachi
Kuala Lumpur Madrid Melbourne Mexico City Nairobi
New Delhi Shanghai Taipei Toronto

With offices in
Argentina Austria Brazil Chile Czech Republic France Greece
Guatemala Hungary Italy Japan Poland Portugal Singapore
South Korea Switzerland Thailand Turkey Ukraine Vietnam

Published in the United States of America by
Oxford University Press
198 Madison Avenue, New York, NY 10016

First issued as an Oxford University Press paperback, 2017

Library of Congress Cataloging-in-Publication Data
The Oxford handbook of interactive audio / edited by Karen Collins, Bill Kapralos, and Holly Tessler.
pages cm
Includes bibliographical references and index.
ISBN 978-0-19-979722-6 (hardcover : alk. paper); 978-0-19-065105-3 (paperback : alk. paper)
1. Interactive multimedia. 2. Video game music—Analysis, appreciation. 3. Computer game music—
Analysis, appreciation. I. Collins, Karen, 1973-, editor. II. Kapralos, Bill, editor.
III. Tessler, Holly, editor IV. Title: Handbook of interactive audio.
QA76.76.I59O94 2014
006.7—dc23
2013029241

Contents

SECTION 4 PERFORMANCE AND INTERACTIVE INSTRUMENTS

SECTION 5 TOOLS AND TECHNIQUES

SECTION 6 THE PRACTITIONER'S POINT OF VIEW

List of Common Acronyms Found in the *Handbook*

AI: Artificial Intelligence, referring to machine learning ability.

API: Application Programming Interface, a specification designed to interface between software.

DAW: Digital Audio Workstation, a home computer recording studio.

DLC: DownLoadable Content, content that is commonly an add-on to games or other software that can be downloaded by the user.

DSP: Digital Signal Processing, in referring to sound, is the various effects used to enhance or change a sound wave.

FM: Frequency Modulation: in the context found here, FM is used in regard to an early form of sound synthesis (as opposed to a radio broadcast mechanism).

FPS: First-Person Shooter, a genre of game in which the player is in first-person perspective, commonly holding a gun.

GANG: Game Audio Network Guild, an organization of game sound designers and composers. http://www.audiogang.org.

GUI: Graphical User Interface, an image- or icon-based interface.

HCI: Human–Computer Interaction, a branch of computer science that focuses on the interaction between humans and computers (hardware and software).

HRTF: Head-Related Transfer Function describes the location- and distance-dependent filtering of a sound by the listener's head, shoulders, upper torso, and most notably, the pinna of each ear.

IASIG: Interactive Audio Special Interest Group, an industry-led organization that creates specifications and standards, and research reports on audio-related topics. http://www.iasig.org

MIDI: Musical Instrument Digital Interface, a music industry specification for interfacing between instruments and software.

MIR: Music Information Retrieval, a branch of computer science that focuses on our ability to search and retrieve music files.

MMO/MMORPG: Massively Multiplayer Online Game/Online Role-Playing Game, online games in which there are multiple simultaneous players over a network.

NIME: New Interfaces for Musical Expression, an annual conference that brings together work on new and emerging musical interfaces and instruments.

NPC: Non-Player Characters, those characters in a game over which the player has no control.

RPG: Role-Playing Game, a genre of games in which the player undertakes a series of quests or solves puzzles, usually in a vast virtual world.

SID: Sonic Interaction Design, the study of sound in interaction design, and the focus of a major European Science Foundation project called COST-SID: http://sid.soundobject.org/.

VST/VSTi: Virtual Studio Technology, a software interface that brings together audio software (synthesizers, editors, effects, plug-ins) and often used to refer to audio effects plug-ins: VSTis are instruments.

List of Software Found in the *Handbook*

Acid, Sony: http://www.sonycreativesoftware.com/acidsoftware
Animoog, Moog: http://www.moogmusic.com/products/apps/animoog-0
Band-in-a-Box: PGMusic: http://www.pgmusic.com/
ChucK, Ge Wang and Perry Cook: http://chuck.cs.princeton.edu/
Composer's Desktop Project: http://www.composersdesktop.com/
C64 Digi, Robin Harbron, Levente Harsfalvi, and Stephen Judd: http://www.ffd2.com/fridge/chacking/c=hacking20.txt
CryEngine, Crytek: http://mycryengine.com/
Cubase, Steinberg: http://www.steinberg.net/en/products/cubase/start.html
Curtis, The Strange Agency: https://itunes.apple.com/app/megacurtis-free/id317498757?mt=8
Digital Performer, Motu: http://www.motu.com/products/software/dp/
FamiTracker, jsr: http://famitracker.com/
Fmod, Firelight Technologies: http://www.fmod.org/
Garage Band, Apple: http://www.apple.com/ilife/garageband/
GoatTracker, Lasse Öörni: http://www.sidmusic.org/goattracker/mac/
Instant Heart Rate, Azumio: http://www.azumio.com/apps/heart-rate/
It2nsf, mukunda: http://battleofthebits.org/lyceum/View/it2nsf/
iTunes, Apple: http://www.apple.com/itunes/
Jitter, Cycling '74: http://cycling74.com/
Little Sound DJ, Johan Kotlinski: http://littlesounddj.com/
Live, Ableton: http://www.ableton.com/
Logic, Apple: http://www.apple.com/ca/logicpro/
Max/MSP, Cycling '74: http://cycling74.com/
MediaPlayer, Google Android: https://play.google.com/store/apps/details?id=com.codeaddictsofcseku.androidmediaplayer&hl=en
MiniMoog, Arturia: http://www.arturia.com/evolution/en/products/minimoogv/intro.html
Music Macro Language. Jikoo. http://woolyss.com/chipmusic-mml.php.
MySong, Microsoft Research: http://research.microsoft.com/en-us/um/people/dan/mysong/
Nanoloop, Oliver Wittchow: http://www.nanoloop.de/advance/index.html
Nerdtracker II, Michel Iwaniec: http://nesdev.com/nt2/

OpenMPT, Olivier Lapicque: http://openmpt.org/
Plogue Bidule, Plogue: http://www.plogue.com/
Plogue Chipsounds, Plogue: http://www.plogue.com/products/chipsounds/
ProTools, Avid: http://www.avid.com/US/products/family/Pro-Tools
Pure Data (Pd): http://puredata.info/
Reactable, Reactable systems: http://www.reactable.com
Recognizr, The Astonishing Tribe: http://www.tat.se/videos/
Sibelius, Avid: http://www.sibelius.com/home/index_flash.html
SID DUZZ' IT, Gallefoss, Glen Rune, and Geir Tjelta: http://home.eunet.no/~ggallefo/sdi/index.html
Songsmith, Microsoft Research: http://research.microsoft.com/en-us/um/redmond/projects/songsmith/
SonicNotify: http://sonicnotify.com/
SoundPool, Google Android: http://developer.android.com/reference/android/media/SoundPool.html
Supercollider: http://supercollider.sourceforge.net/
Unity3D, Unity Technologies: http://unity3d.com/
Wavelab, Steinberg: http://www.steinberg.net/en/products/wavelab.html
Weka, Mark Hall, Eibe Frank, Geoffrey Holmes, Bernhard Pfahringer, Peter Reutemann, and Ian H. Witten: http://www.cs.waikato.ac.nz/ml/weka/
Wwise, Audiokinetic: http://www.audiokinetic.com/en/products/208-wwise

List of Games Found in the *Handbook*

Adventures of Batman and Robin. 1995. Sega Genesis.
Alone in the Dark. 1992. Infogrames.
Anarchy Online. 2001. Funcom 2001.
Arkanoid. 1986. Taito.
Asheron's Call 2: Fallen Kings. 2002. Turbine.
Astro Marine Corps. 1989. Creepsoft
BioShock. 2007. 2K Boston.
Bitesize. 2012. BBC.
Braid. 2008. Number None Inc.
Broken Sword: Angel of Death. 2006. Sumo Digital.
Brütal Legend. 2009. Double Fine Productions.
Call of Duty. 2003–2012. Activision.
Civilization IV. 2005. Firaxis Games.
Contra 3: The Alien Wars. 1992. Konami.
Child of Eden. 2011. Ubisoft.
Crysis. 2007. Crytek.
Dance Central. 2010. Harmonix.
Dead Space. 2008. EA.
Demor. 2004. Blendid.
Desert Falcon. 1982. Atari.
Digger. 1983. Windmill.
Dimensions. 2011. RealityJockey.
Donkey Kong. 1981. Nintendo.
Donkey Konga. 2003. Namco.
Dungeons and Dragons. 1974. TSR, Wizards of the Coast. Gygax, Gary and Dave Arneson.
Earthbound. 1994. Nintendo.
Elder Scrolls: Oblivion. 2006. Bethesda Game Studios.
Electroplankton. 2005. Indieszero.
Epoch. 2012. Uppercut Games.
Eternal Sonata. 2007. Namco-Bandai.
Fallout : New Vegas. 2010. Bethesda Game Studios.
Flower. 2009. thatgamecompany.

Frogger. 1981. Konami.
Fruit Farmer. 2007. Locamatrix http://www.locomatrix.com/
Guitar Hero. 2005. Harmonix.
Half-Life 2. Electronic Arts. Valve. 2004.
Halo: Combat Evolved. 2001. Bungie Studios.
Kane and Lynch: Dead Men. 2007. IO Interactive.
Kirby's Dream Land 2. 1995. Nintendo.
L.A. Noire. 2011. Rockstar.
Legend of Zelda. 1986. Nintendo.
Legend of Zelda : Skyward Sword. Nintendo. 2011.
Mega Man Zero. 2002. Inti Creates.
Metal Gear 2: Solid Snake. 1987. Konami.
Mi Vida Loca, Spanish for beginners. 2012. BBC Languages.
Michael Jackson—The Experience. 2010. Ubisoft.
Minesweeper. 1990. Microsoft.
Minute War. GPSGames.org http://minutewar.gpsgames.org
Monty on the Run. 1985. Peter Harrup.
MUD. 1978. Roy Trubshaw and Richard Bartle.
Myst. 1993. Cyan.
NBA Live 95. 1994. EA.
Need for Speed. 2005. EA.
NHL '12. 2011. EA.
Operation Flashpoint. 2011. Codemasters.
Operative: No One Lives Forever. 2000. Fox Interactive.
Pac-Man. 1980. Namco.
Panel De Pon. 1995. Intelligent Systems.
Papa Sangre. 2011. Somethin Else.
Patapon. 2008. Pyramid.
Pong. 1972. Atari.
Portal 2. 2011. Valve Corporation.
Project Zero. 2001. Tecmo.
Quake. 1996. id Software.
Quake III Arena. 1999. id Software.
Raw Recruit. 1988. Mastertronic Group Ltd.
Resident Evil 4. 2005. Capcom.
Retro City Rampage. 2012. Vblank Entertainment.
Rock Band. 2007. Harmonix.
Secret of Monkey Island. 1990. LucasArts.
Shatterhand. 1991. Natsume.
Silent Hill III. 2003. Konami.
SingStar. 2004. London Studio.
Sonic Advance 3. 2004. Sonic Team.
Space Manbow. 1989. Konami.

Splinter Cell. 2005. Ubisoft.
Spore. 2009. EA.
SSX. 2012. EA.
Star Wars: The Old Republic. 2011. BioWare.
Starcraft II. 1999. Blizzard.
Super Mario 64. 1996. Nintendo.
Super Mario Bros. 1984. Nintendo.
Super Mario World. 1990. Nintendo.
Tetris. 1984. Alexey Pajitnov.
Thief: Deadly Shadows. 2004. Ion Storm.
Tom Clancy's EndWar. 2008. Ubisoft.
Uncharted 2: Among Thieves. 2008. Naughty Dog.
Uncharted 3: Drake's Deception. 2011. Naughty Dog.
Unreal Tournament 2004. 2004. Epic.
Urban Terror. 2005. Silicon Ice.
Vib-Ribbon. 2000. Nana On-Sha.
World of Warcraft. 2004. Blizzard.
Xenon 2—Megablast. 1989. The Assembly Line.

List of Contributors

Marc Ainger is a sound artist who works in the area of computer and electronic music, often in combination with other media such as film, dance, and theater. His works have been performed throughout the world, including at the American Film Institute, the KlangArts Festival, Gageego New Music Ensemble, Guangdong Modern Dance, the Royal Danish Ballet, Streb, the New Circus, and *Late Night with David Letterman*. As a sound designer he has worked with IRCAM, the Los Angeles Philharmonic, the Olympic Arts Festival, and Waveframe, among others. He is currently head of the theory and composition program at the Ohio State University.

Anders-Petter Andersson is a sound designer, Ph.D. in Musicology and currently holds a position as Postdoctoral researcher at Kristianstad University in Sweden. Since 1999 he has worked within the group MusicalFieldsForever together with Birgitta Cappelen and Fredrik Olofsson, creating interactive musical Art installations. The group MusicalFieldsForever has since 1999 created interactive art installations that explore new forms of expression and the democratic potential of interactive media, by creating open, audio-tactile Art installations – musical fields. A musical field is open for co-creation on many levels. The group exhibits their installations internationally. Since 2006 they have worked with tangible musical interaction for people with disabilities, in a health context. Currently in the project RHYME.no (RHYME.no) at the Oslo School of Architecture and Design (AHO) in Norway.

David Bessell has been active in the field of popular music for many years. He also studied classical composition and orchestration at the Royal College of Music, London, and jazz guitar with John Etheridge. He holds a Doctorate in Music and currently teaches Music and Music Technology at Plymouth University. He can be found performing on guitar or electronics from time to time in a variety of styles. https://sites.google.com/site/davebessellmusic/home.

M.J. Bishop is inaugural director of the University System of Maryland's Center for Innovation and Excellence in Learning and Teaching, which was established in 2013 to enhance and promote USM's position as a national leader in higher education academic innovations. The Center conducts research on best practices, disseminates findings, offers professional development opportunities for institutional faculty and administrators, and supports the 12 public institutions that are part of the system as they continue to expand innovative academic practices. Prior to coming to USM, Dr. Bishop was Associate Professor and Director of the Lehigh University College of Education's

Teaching, Learning, and Technology Program where, in addition to being responsible for the institution's graduate programs in teacher education and instructional technology, she also played a leadership role in several campus-wide university initiatives. Author of numerous national and international articles, her research interests include exploring how various instructional media and delivery systems might be designed and used more effectively to improve learning. Dr. Bishop taught courses in instructional design, interface design, and Website and resource development at Lehigh.

Alan F. Blackwell is Reader in Interdisciplinary Design at the University of Cambridge Computer Laboratory. He is an authority on visual representation and notation, especially with regard to the usability of programming languages. He collaborates regularly with music researchers, especially through Cambridge's Centre for Music and Science, and has a specific research interest in notations for artistic production and performance, working with a wide range of contemporary choreographers and composers. Together with his students and collaborators, he has a long-standing interest in the tools and practices of Live Coding.

Niels Böttcher graduated from Aalborg University in Copenhagen, at the Institute of Architecture, Design and Media Technology. His PhD was on the topic of procedural audio in computer games with a special focus on motion controllers. Niels has an ongoing interest in the relationship between gesture and sound in musical controllers, computer games, and related applications. He has been very active in building DIY music instruments and has been performing all over Europe in various electronic music groups. In 2002 he founded the record label JenkaMusic, which has more than sixteen international releases.

Birgitta Cappelen is an industrial designer, interaction designer and Associate Professor at the Oslo School of Architecture and Design (AHO) in Norway. She has worked within the field of screen based interactive media since 1985, and with Art and research within Tangible Interaction and Smart Textile since 1999 in the group MusicalFieldsForever. The group MusicalFieldsForever has since 1999 created interactive art installations that explore new forms of expression and the democratic potential of interactive media, by creating open, audio-tactile Art installations – musical fields. A musical field is open for co-creation on many levels. The group exhibits their installations internationally. Since 2006 they have worked with tangible musical interaction for people with disabilities, in a health context. Currently in the project RHYME.no (RHYME.no) at the Oslo School of Architecture and Design (AHO) in Norway.

Nick Collins is a composer, performer, and researcher who lectures at the University of Sussex. His research interests include machine listening, interactive and generative music, and musical creativity. He coedited *The Cambridge Companion to Electronic Music* (Cambridge University Press, 2007) and *The SuperCollider Book* (MIT Press, 2011), and wrote the *Introduction to Computer Music* (Wiley, 2009). Sometimes he writes in the third person about himself, but is trying to give it up. Further details, including publications, music, code and more, are available from http://www.sussex.ac.uk/Users/nc81/index.html.

Fionnuala Conway is a musician, composer, and multimedia artist. She has been lecturing on the MPhil in Music and Media Technologies course at Trinity College, Dublin, since 2002 and was appointed Course Director in 2006. With a background in music and music technology, she has worked as composer and performer and produced work in a wide variety of forms, from traditional materials to interactive digital media, wearable technology, installations and theatre presentation, including Art of Decision and Urban Chameleon.

Andrew Dolphin is a composer and digital artist currently working as a lecturer in Music, Sound and Performance at Leeds Metropolitan University, UK. He recently completed a PhD at SARC (Sonic Arts Research Centre), Queen's University in Belfast, Northern Ireland. He completed his MMus at Goldsmiths, University of London, and BA Hons in Sonic Art at Middlesex University. His recent projects have focused upon the exploration and practical application of computer game and physics engine technologies in the creation of creative works in the fields of sound art and music composition. Themes of play, allocation of compositional control to players, user accessibility, and symbolic representations of sound, synthesis, and music control parameters are often key themes in the game engine projects. http://www.dysdar.org.uk.

Inger Ekman earned her MSc in computer science from the University of Tampere, Finland, in 2003. Since then, she has worked at teaching and researching the experiential aspects of gaming and interactive media at the University of Tampere and Aalto University. Currently, she is pursuing a doctoral degree on game sound. Her research interests combine design practice and UX research with theoretic approaches grounded in psychoacoustics and emotion theory. She has published on game experience in journals such as *Gaming and Virtual Worlds, Simulation & Gaming, Computer & Graphics*, in books, and in numerous conference proceedings.

Andy Farnell is a computer scientist from the United Kingdom, specializing in audio DSP and synthesis. Pioneer of procedural audio and the author of MIT textbook *Designing Sound*, Andy is visiting professor at several European institutions and consultant to game and audio technology companies. He is also an enthusiastic advocate and hacker of free open-source software, who believes in educational opportunities and access to enabling tools and knowledge for all.

Mike Frengel holds BA, MA, and PhD degrees in electroacoustic music composition from San Jose State University, Dartmouth College, and City University, London, respectively. He has had the great fortune to study with Jon Appleton, Charles Dodge, Larry Polansky, Denis Smalley, Allen Strange, and Christian Wolff. His works have won international prizes and have been included on the Sonic Circuits VII, ICMC '95, CDCM vol. 26, 2000 Luigi Russolo and ICMC 2009 compact discs. Mike serves on the faculty of the music departments at Northeastern University and Boston Conservatory, where he teaches courses in music technology and composition.

Melanie Fritsch works as research assistant at the Forschungsinstitut für Musiktheater since October 2008, teaches in the Music Theatre Studies department at the University of Bayreuth, and is also PhD candidate. She studied Performance Studies, Contemporary German Literature and Musicology in Berlin (Freie and Humboldt Universität) and Rome. During this time, she also freelanced for various theater and music theater productions, and has worked at various German and Italian cultural institutions both within Germany and Italy. Currently she is finishing her doctoral dissertation in the research area of video games and music. Her other research focuses are Performance Studies (Music as Performance), Liveness, Virtual Worlds Research, and Theatre and Dance History and Aesthetics. See also: http://uni-bayreuth.academia.edu/mfritsch

Tom Garner conducted his PhD and subsequent postdoctoral research at the University of Aalborg, with interests focusing upon auditory perception, most prominently concerning the affective potential of sound within virtual environments. He has since lectured in computer science at the University of Kent and is currently a Research Fellow in Interactive Technologies at the University of Portsmouth where he is working on a range of projects concerning the perceptual quality of sound within virtual reality applications.

Tim van Geelen is a Dutch interactive sound specialist, and a teacher at one of Holland's highest-standing colleges. In 2008, he graduated in adaptive audio for games, and has since employed his specialty in the fields of, among others, serious games, education, and live performance. Apart from a passion for innovative audio, he also plays bass guitar and practices Kundalini yoga. He is always looking for collaboration on and innovation of interactive and adaptive sound. He can be contacted through www.timvangeelen.com.

Mark Grimshaw is the Obel Professor of Music at Aalborg University, Denmark. He has published widely across subjects as diverse as sound, biofeedback in computer games, virtuality, the Uncanny Valley, and IT systems and also writes free, open source software for virtual research environments (WIKINDX). His books include the anthologies *Game Sound Technology & Player Interaction* (IGI Global 2011) and *The Oxford Handbook of Virtuality* (Oxford University Press 2014), and, with co-author Tom Garner, a monograph entitled *Sonic Virtuality* (Oxford University Press 2015). A two-volume anthology, *The Oxford Handbook of Sound & Imagination*, is due in 2018 from Oxford University Press.

Michael Gurevich is Assistant Professor of Performing Arts Technology at the University of Michigan, where he teaches media art, physical computing, and electronic chamber music. Framed through the interdisciplinary lens of Interaction Design, his research explores new aesthetic and interactional possibilities that emerge through performance with real-time computer systems. He holds a PhD in computer music from Stanford, and has worked at the Sonic Arts Research Centre at Queen's University, Belfast, and Singapore's Institute for Infocomm Research. He has published in the New Interfaces for

Musical Expression, computer music and HCI communities, and served as Music Chair for NIME 2012.

Norbert Herber is a musician and a sound artist. His work explores the relationship between people and sound within mediated environments—spaces created by software, sensors, speakers, and other mediating technologies. This music is more likely to be heard on a personal computer, mobile device, or installation space than on CD or vinyl. Field recordings, live instruments, and electronics are brought together in an ever-changing, generative mix of texture and tone that leverages the processing capabilities of contemporary technology to create music specific to a place and time. Using this approach Norbert is focused on creating sound and music in digital environments for art, entertainment, and communications applications. His works have been performed and exhibited in Europe, Asia, South America, and in the United States.

Jan Paul Herzer studied Audio Engineering at the SAE Hamburg, and Sound Studies—Acoustic Communication at the Berlin University of the Arts. He works as sound designer, musician and programmer in the spectrum between acoustic scenography, functional sound design, and installation art. He is one founder of Hands on Sound, an artist collective and design agency that specializes in sound design for architectural space and makes extensive use of interactive and generative audio concepts. Jan Paul Herzer currently lives and works in Berlin.

Maia Hoeberechts served as project manager on the AMEE research project at the University of Western Ontario with the goal of developing an emotionally adaptive computer music composition engine. Dr. Hoeberechts worked in many different capacities at Western including as a lecturer, lab manager, and research associate prior to her assuming a new position in the NEPTUNE Canada science team based at the University of Victoria, where she currently serves as Research Theme Integrator for Engineering and Computational Research.

Damian Kastbauer is a freelance technical sound designer working to help bridge the gap between sound designers, composers, and game developers. Utilizing the functionality of game-audio-specific implementation authoring tools, his goal is to create dynamic sound interactions that leverage interactive techniques to make good sound content sound great. In addition to working remotely and onsite helping games make glorious noises, he can be found scribing the Aural Fixations column in *Game Developer Magazine* and pontificating on sound at http://www.LostChocolateLab.com.

Michael Katchabaw is an Associate Professor in the Department of Computer Science at the University of Western Ontario. His research focuses on various issues in game development and virtual worlds, with dozens of publications and numerous funded projects in the area, supported by various government and industry partners. At Western, Dr. Katchabaw played a key role in establishing its program in game development as one of the first in Canada, as well as the Digital Recreation, Entertainment, Art, and Media (DREAM) research group.

Tom Langhorst is a lecturer in game sound at Fontys University of Applied Sciences, Netherlands. His work and research focusses on the cross over between design, perception and technology (such as game AI). He was educated as musician, music theorist and composer and worked in the game, entertainment and advertisement industry and as interaction designer for product innovation. More recently Tom is also involved in research and development of games for healthcare and he is advisor of the Games4Health Europe conference.

Victor Lazzarini is a Senior Lecturer in Music at the National University of Ireland, Maynooth. His research work includes over 100 articles in the areas of musical signal processing, computer music languages, and audio programming. He is the co-editor of Audio Programming (MIT Press, 2010), which is a key reference volume in Computer Music. Victor is also an active composer of instrumental and electronic music, and one of the developers of Csound.

Jon Inge Lomeland holds an MA in ethnomusicology and musicology from the University of Bergen, Norway, where he studied music and emotions in the game *World of Warcraft*. He teaches music in addition to composing music for games.

Kiri Miller is Associate Professor of Music at Brown University. Her research focuses on interactive digital media, communities of practice, amateur musicianship, and popular music. Miller is the author of *Traveling Home: Sacred Harp Singing and American Pluralism* (University of Illinois Press, 2008) and *Playing Along: Digital Games, YouTube, and Virtual Performance* (Oxford University Press, 2012). She has published articles in *Ethnomusicology, American Music, 19th-Century Music*, the *Journal of American Folklore, Game Studies*, and the *Journal of the Society for American Music*. Her work has been supported by fellowships from the Radcliffe Institute for Advanced Study and the American Council of Learned Societies.

Helen Mitchell read music at Edinburgh University, gaining the Fraser Scholarship upon graduation. After completing a diploma from the London Guildhall School of Music and Drama, she spent a further year specializing in solo performance and repertoire at Liverpool University. She studied the flute with Roger Rostrun (Hallé Orchestra), Richard Chester (Royal Scottish National Orchestra) and Colin Chambers (Royal Liverpool Philharmonic Orchestra). In 1992 she was appointed Professor of Flute and Saxophone at the Royal Marines School of Music, Deal, Kent, and in 1998 embarked on further postgraduate studies in music technology at York University. She currently lectures in Creative Music Technology at the University of Hull.

Chris Nash (chris@nashnet.co.uk) is a professional programmer and composer, and currently Senior Lecturer in Music Technology (Software Development for Audio, Sound, and Music) at the University of the West of England (UWE Bristol, UK). He completed his PhD on music HCI at the University of Cambridge, looking at theoretical and analytical methods for modeling and designing interfaces for composition, supported by a longitudinal study of over 1,000 DAW users, empirically investigating user experience with respect to flow, learning, virtuosity, creativity, and liveness. His current research projects focus on digitally-supported amateur musicianship and learning, and end-user programming for music. Around his research, he is the developer of the award-winning reViSiT composition tool, and has written music for TV and radio, including the BBC.

Niels C. Nilsson holds a Master's degree in Medialogy from AAU Copenhagen and is currently a Ph.D. fellow at Aalborg University Copenhagen under Rolf Nordahl. His Ph.D. does in general terms revolve around an investigation of the factors influencing the perceived naturalness of Walking-In-Place locomotion within technologically immersive virtual reality. Moreover, his research interests include presence research, user experience evaluation and consumer virtual reality systems.

Rolf Nordahl is Associate Professor at Aalborg University Copenhagen. His research lies within VR, (Tele)-Presence, Sonic Interaction Design, audio-haptics, multimodal perception and developing novel methods and evaluation techniques for VR, Presence and Games. He is principal investigator for several research and commercial projects including the EU funded project Natural Interactive Walking, and has done seminal work in the EU-project BENOGO. He is member of IEEE and likewise is recognized as an expert for the Danish Evaluation Institute, responsible for national accreditation of educations. He has performed series of invited lectures on his research areas at recognized universities, such as Yale University (Connecticut, US).

Nye Parry is a sound artist, composer, and Research Fellow at CRiSAP, University of the Arts, London. He has made numerous sound installations for museums including the National Maritime Museum, the British Museum, and the Science Museum in London, as well as creating concert works, gallery installations, and over twenty scores for contemporary dance. He has a PhD in electroacoustic composition from City University and teaches at the Guildhall School of Music and Drama and Trinity Laban Conservatoire. Between 2003 and 2011 he ran the MA in Sonic Arts at Middlesex University, where he also did research on locative media.

Natasa Paterson is a Dublin-based composer and performer. Natasa completed her MPhil in Music and Media Technologies at Trinity College, Dublin, and is currently studying for a PhD, exploring composition for location-aware audio applications. Natasa was project manager of the Irish Composers' Collective, the 2012 Ad Astra Composition Competition winner, and is a Fulbright Scholar. Her compositional work include pieces for choir, piano, string, and brass quartet, and the use of electroacoustic processes with performances at the National Concert Hall, Samuel Beckett Theater, Cake Contemporary Center, and Center for Creative Practices. www.natasapaulberg.com.

Leonard J. Paul has worked in the games industry since 1994 and has a history in composing, sound design and coding for major game titles at companies which include

Electronic Arts, Backbone Entertainment and Radical Entertainment. His titles have sold over 9.7 million units and include Need for Speed, NBA Jam and Retro City Rampage. He has over ten years of experience teaching video game audio at institutions such as the Vancouver Film School and the Arts Institute and is the co-founder of the School of Video Game Audio. Leonard has spoken at many industry conferences such as the Game Developers Conference at locations in the USA, Brazil, UK, Canada, Switzerland, Colombia, Germany and other countries worldwide. He is a well-known documentary film composer, having scored the original music for multi-awarding winning documentary The Corporation which remains the highest-grossing Canadian documentary in history to date. His website is: http://VideoGameAudio.com.

Dave Raybould is a Senior Lecturer at Leeds Metropolitan University where he teaches game audio, sound design and synthesis. A regular contributor conferences in the field he is also a member of the papers review committee for the Audio Engineering Society 'Audio for Games' conferences and co-authored "*The Game Audio Tutorial: A Practical Guide to Sound and Music for Interactive Games*" (Focal Press).

Holly Rogers is Senior Lecturer in Music at the University of Liverpool. Recent fellowships have included a postdoctoral position at University College Dublin, a senior research post at Trinity College Dublin and a Fulbright scholarship at the DocFilm Institute in San Francisco. She has published on a variety of audiovisual topics including music and experimental cinema, visual music, video art-music and composer biopics and is author of *Visualising Music: Audiovisual Relationships in Avant-Garde Film and Video Art* (Verlag, 2010), *Sounding the Gallery: Video and the Rise of Art-Music* (OUP, 2013) and editor of *Music and Sound in Documentary Film* (Routledge, 2014).

Guillaume Roux-Girard is a PhD student in film studies at the University of Montreal. His current research focuses on the sound aesthetics of videogames. His recent publications include entries about sound and the *Metal Gear* series in the *Encyclopedia of Video Games* (ABC-Clio Press, 2012), and a chapter about sound in horror videogames in the anthology *Game Sound Technology and Player Interaction: Concepts and Developments* (IGI Global 2011).

Benjamin Schroeder is a researcher, artist, and engineer living in Brooklyn, New York. Benjamin's interests span several different time-based media, including animation, sound, and physical interaction. His work investigates the power, promise, and beauty of computational media, asking questions about how computation and interaction extend our creative reach. Benjamin has presented his research work at such venues as SIGGRAPH, SMC, NIME, and the ICMC. Benjamin works as a software engineer at Google and is a PhD candidate in computer science at the Ohio State University.

Stefania Serafin is Professor with special responsibilities in sound for multimodal environments in the Medialogy section at Aalborg University in Copenhagen. She teaches and researches on sound models and sound design for interactive media and multimodal interfaces.

Jeff Shantz is a PhD candidate in the Department of Computer Science at the University of Western Ontario. While his doctoral research involves the study of graph algorithms, he has served a valuable role as research associate for the AMEE research project at Western, involved in both the development of the core engines and the Pop Tones game.

Richard Stevens is a Senior Lecturer and Teacher Fellow at Leeds Metropolitan University, UK, where he leads the MSc in Sound and Music for Interactive Games. He is a leading evangelist for game audio education, chairing the Education Working Group of the Interactive Audio Special Interest Group (IASIG) through to the publication of their "Game Audio Curriculum Guideline" document, and promoting the subject through regular conference talks, panels, and workshops. In 2011, he coauthored the first practical textbook in the field, *The Game Audio Tutorial.*

Rafał Zapała is a composer and a faculty member at the Academy of Music in Poznań. http://www.zapala.com.pl/. He also works at Studio Muzyki Elektroakustycznej Akademii Muzycznej w Poznaniu (SMEAMuz Poznan). He graduated composition (MA, PhD) and choir conducting (MA); participant of K. Stockhausen Concerts and Courses (Kurten 2008), Acanthes Courses (Metz, 2010 with IRCAM, T. Murail and B. Furrer), and others; founder and head of ARCHE New Music Foundation and many ensembles (contemporary, improvised, electronic music). Zapała does not recognize any boundaries between music acquired through academic education, experience of the counterculture, and collaborating with artists from other fields of art (http://www.zapala.com.pl).

About the Companion Website

www.oup.com/us/ohia

The *Oxford Handbook of Interactive Audio* is a collection of articles on interactivity in music and sound whose primary purpose is to offer a new set of analytical tools for the growing field of interactive audio. Since interactive audio is inherently a multimedia experience, we have assembled a series of links, sounds and videos collected from the *Handbook*'s authors, with the aim of providing additional reading and audiovisual material to support the ideas and artists introduced here. The book begins with the premise that interacting with sound differs from just listening to sound in terms of the audience and creator's experience. Like the book itself, the companion website is intended to be a helpful resource to researchers, practitioners, theorists and students across a range of disciplines. The website includes links to a range of websites, projects, blogs, tutorials, experiments, artistic, creative and musical works, and ongoing research about and involving interactive audio.

THE OXFORD HANDBOOK OF

INTERACTIVE AUDIO

INTRODUCTION

KAREN COLLINS, HOLLY TESSLER, AND BILL KAPRALOS

THE *Oxford Handbook of Interactive Audio* is a collection of chapters on interactivity in music and sound whose primary purpose is to offer a new set of analytical tools for the growing field of interactive audio. We began with the premise that interacting with sound is different from just listening to sound in terms of the audience's and creator's experience. Physical agency and control through interactivity add a level of involvement with sound that alters the ways in which sound is experienced in games, interfaces, products, toys, environments (virtual and real), and art. A series of related questions drive the *Handbook*: What makes interactive audio different from noninteractive audio? Where does *interacting with* audio fit into our understanding of sound and music? What are the future directions of interactive audio? And, How do we begin to approach interactive audio from a theoretical perspective?

We began the *Oxford Handbook of Interactive Audio* by approaching authors who work with interactive audio across a wide spectrum, hoping that, together, we may begin to answer these questions. What we received in return was an incredible array of approaches to the idea of interacting with sound. Contributors to the *Handbook* approach the ontological and philosophical question of, "What is interactive audio, and what can it do?" from a number of different perspectives. For some, an understanding of sound emerges through developments and advancements in technology, in writing software programs and codes, or building original hardware and equipment to create new types of sound. For others, interactive audio is more of an aesthetic consideration of how its inherent power can be used in creative projects and art installations. For still others, new perspectives on audio emerge through exploration of its communicative power: how audio works as a link between not only the human–machine interface, but also—and increasingly—between human beings.

From the outset, our goal was to put together a volume of work that was both inclusive and dialectical in nature, a volume that would be humanities-driven, but that would also take into account approaches from practitioners and those within the natural sciences and engineering disciplines. Rather than direct contributors to write to a specific brief, we instead encouraged them to interrogate, interpret, and challenge current theories and understandings of interactive audio, in whatever forms and contexts were meaningful to them. What has emerged from this type of open-ended mandate demonstrates not only a remarkable range of scholarship but also the inherent importance of interactive audio to so many different areas.

However, beneath the seemingly wide disparity between the approach and subject matter of the chapters, a series of themes began to clearly surface and recur across disciplines. It was these themes that eventually led to the overall structure of the *Oxford Handbook of Interactive Audio* and its separation into six sections: (1) Interactive Sound in Practice; (2) Videogames and Virtual Worlds; (3) The Psychology and Emotional Impact of Interactive Audio; (4) Performance and Interactive Instruments; (5) Tools and Techniques; and (6) The Practitioner's Point of View. These sections are to some extent driven by the overarching themes that tie them together, although as will be made apparent upon reading, there is considerable overlap between sections, making our organizational structure just one of any number of ways of presenting and making sense of so many diverse and diffuse ideas.

Interactive Sound in Practice

The first section, *Interactive Sound in Practice*, presents research drawn from an arts perspective, with a particular focus on interactive audio as a component of art practice (where "art" is defined broadly). What is clear from the chapters in this section is the idea that interactivity in the arts arose as a defining element of the twentieth-century avant-garde. Interactivity facilitated (and was facilitated by) a new relationship between audience and creator, a relationship that broke down the "fourth wall" of artistic practice. The fourth wall is a term borrowed from performance theory that considers the theatrical stage as having three walls (the rear and two sides) and an invisible fourth wall between the actors and audience. "Breaking" the fourth wall has become an expression for eliminating the divide between performer or creator and audience. Alongside this creator–audience dissolution is the new emphasis on art as an experience and practice, rather than a text or object. The shift in the arts in the twentieth century from object-based work to practice-based work is a shift that has been referred to as a change of focus on *doing*: a shift to an aesthetics of relationships (Bourriaud 2002; Green 2010, 2). Gell for instance suggests a redefinition of art as the "social relations in the vicinity of objects mediating social agency... between persons and things, and persons and persons via things" (Gell 1998, 5).

One of the challenges of thinking of interactivity in these terms—that is, as an ongoing social construct—is that it brings up difficult questions about the nature of texts as finished products (Saltz 1997, 117). Tied closely to the concept of the open work (an idea of "unfinishedness" that was made famous by John Cage, although the idea certainly existed much earlier), interactivity presents work that is always evolving, always different, and never finished. Interactive texts are inherently unfinished because they require a participant with whom to interact before they can be realized in their myriad forms: a player is needed for a game, and an audience is required for an interactive play. The structures that are inherent in interactive media encourage a greater affordance for, and a greater interest on the part of, the audience toward coauthorship. In this way, notions of interactivity both feed into and draw from postmodern aesthetics, shifting away from "art" and "play" as cogent and unproblematic terms, moving toward a system that defines interactivity as a necessarily individualized and interpretive process.

From a technological–industrial perspective, it becomes evident that interactivity has been, in no small measure, influenced by advances in digital machines and media. Marshall McLuhan and Barrington Nevitt predicted as early as 1972 that the consumer–producer dichotomy would blur with new technologies. Rob Cover argues that "the rise of media technologies which not only avail themselves to certain forms of interactivity with the text, but also to the ways in which the pleasure of engagement with the text is sold under the signifier of interactivity is that which puts into question the functionality of authorship and opens the possibility for a variety of mediums no longer predicated on the name of the author" (Cover 2006, 146).

The dissolution of creator–audience divide and the rise of the audience-creator is explored in a variety of forms in this section of the book. Holly Rogers takes on this history in video art in "Spatial Reconfiguration in Interactive Video Art," drawing on Frances Dyson's conceptualization of the change as going from "looking at" to "being in" art (Dyson 2009, 2). It is further interrogated in Nye Parry's "Navigating Sound: Locative and Translocational Approaches to Interactive Audio," which explores the influence of the avant-garde on site-specific and environmental sound.

In each of the chapters in this section, it is clear that the role of the audience has gone from one of *listening* to one of *sound-making*. The audience is no longer disconnected from the sounds produced in the environment, but is actively involved in adding to, shaping, and altering the sonic environment around them. This activity is made explicit in Andrew Dolphin's chapter on sound toys, "Defining Sound Toys: Play as Composition." Dolphin questions the role of the composer as a kind of auteur, suggesting instead that interactive audio leads to a democratization of sound-making practice in the form of affordable, user-friendly interactive toys.

The new means to interact with sound may lead to potentially new ways to enhance learning, an idea explored by M. J. Bishop in her chapter, "Thinking More Dynamically About Using Sound to Enhance Learning from Instructional Technologies." Finally, Jan Paul Herzer explores the concept of an audience's participation in an interactive environment, an environment where audio becomes a component of a functional interactive ecosystem, in "Acoustic Scenography and Interactive Audio: Sound Design for Built Environments."

Videogames and Virtual Worlds

Perhaps one of the most influential drivers of interactive audio technology today is that of videogames and virtual worlds. For those who have grown up playing videogames, interacting with audio (and video) is an almost instinctive process. Our physical interaction with sound, coupled with the meaning derived from these sounds (and our interaction with them), directly informs the ways in which videogames and game franchises are created. Publishers and online companies rely on audio to communicate key ideas about the game and gameplay through sound and music. Videogames have offered a uniquely commercial avenue for the exploration and exploitation of interactive audio concepts, from generative and procedural content to nonlinear open-form composition. The nonlinear nature inherent in videogames, along with the different relationship the audio has with its audience, poses interesting theoretical problems and issues. One of the most significant aspects has been the influence of games on sound's structure, particularly the highly repetitive character of game audio and the desire for variability.

The chapters in the *Videogames and Virtual Worlds* section explore the influence of interactivity on sound's meanings and structures. Inherent in all of the chapters in this section is the idea that games are fundamentally different from film, and that interactivity drives this difference. In "The Unanswered Question of Musical Meaning: A Cross-domain Approach," Tom Langhorst draws on elements of psychoacoustics, linguistics, and semiotics to explore the meaning behind seemingly simple sounds of early 8-bit games such as *Pong* and *Pac-Man*, suggesting that new methods must be developed to explore interactive sound in media.

Jon Inge Lomeland takes a different approach to meaning in "*How Can Interactive Music be Used in Virtual Worlds Like *World of Warcraft* Lomeland approaches the meaning of game music for the audience in terms of the nostalgia that builds around the highly repetitive music tied to hours of enjoyment with a game. As games evolve over time, what changes should be made to the music, without altering the attachments that players develop to that music, and what response does new music get from its audience?

Guillaume Roux-Girard further explores the listening practices of game players in "Sound and the Videoludic Experience." Roux-Girard suggests methods that scholars can employ in analyzing interactive music, focusing on the experiential aspects of play. Roux-Girard, Lomeland, and Langhorst all focus on the idea that interactivity alters the relationship that players have with music, and suggest that game music cannot be analyzed outside the context of the game, but that there is a fundamental necessity to include the player's experience in any analysis.

Just as games can influence music's structure, the final two chapters of the section suggest how music can influence the structure of games. In "Designing a Game for Music: Integrated Design Approaches for Ludic Music and Interactivity," Richard Stevens and Dave Raybould take a cue from famed sound designer Randy Thom's well-known article "Designing a Movie for Sound" (1999). In this article, Thom argues that sound can be a driving force for film if the film is written to consider sound right from the beginning. The idea was later explored by game sound director Rob Bridgett in his *Gamasutra* article, "Designing a Next-gen Game for Sound" (2007), where he argues that it is necessary to design games with "sound moments" in order to entice the audience. Stevens and Raybould offer their own take on this important concept, suggesting that previous definitions of interactivity have focused merely on the idea of *re*activity, and by reconceptualizing the notion of interactivity itself, we may begin to think about new ways of developing games around audio, rather than developing the audio around the game, as is commonly done. Melanie Fritsch offers us some insight into music-based games in her chapter, "Worlds of Music: Strategies for Creating Music-based Experiences in Videogames." By presenting three case studies of musically interactive games, Fritsch brings forth the notion that games are *activities*, driven by our physical, embodied interaction.

The Psychology and Emotional Impact of Interactive Audio

Historically, researchers into human cognition believed thinking and problem-solving to be exclusively mental phenomena (Clancey 1997, in Gee 2008). But more contemporary research, specifically that of embodied cognition theory, holds that our understanding of the world is shaped by our ability to physically interact with it. According to embodied cognition theory, our knowledge is tied to the original state that occurred in the brain when information was first acquired. Therefore, cognition is considered "embodied" because it is inextricably tied to our sensorimotor experience; our perception is always coupled with a mental reenactment of our physical, embodied experience (Collins 2011).

In the third section of the *Handbook, The Psychological and Emotional Impact of Interactive Audio*, embodiment through sound technology is explored by taking an embodied cognition approach, as is done in the two chapters that focus on videogames; Mark Grimshaw and Tom Garner's "Embodied Virtual Acoustic Ecologies of Computer Games" and Inger Ekman's "A Cognitive Approach to the Emotional Function of Game Sound." The importance of the role that our body plays in experiencing interactive sound—not only through the direct physical interaction with sound, but also through the multimodal act of listening—is explored in the following two chapters, Rolf Nordahl and Niels C. Nilsson's "The Sound of Being There: Presence and Interactive Audio in Immersive Virtual Reality" and Stefania Serafin's "Sonic Interactions in Multimodal Environments: An Overview."

Nordahl and Nilsson explore the importance of sound to the concept of immersion and presence. The theory of immersion most currently in favor within the game studies and virtual reality community is related to Csíkszentmihályi's (1990) concept of "optimal experience" or "flow." Csíkszentmihályi describes flow as follows: "The key element of an optimal experience is that it is an end in itself. Even if initially undertaken for other reasons, the activity that consumes us becomes intrinsically rewarding" (Csikszentmihalyi 1990, 67). He outlines eight criteria for the flow experience: (1) definable tasks; (2) ability to concentrate; (3) clear goals; (4) immediate feedback; (5) "deep but effortless involvement that removes from awareness the worries and frustrations of everyday life"; (6) sense of control over their actions; (7) disappeared concern for self; and (8) altered sense of the duration of time.

Several attempts have been made to identify the elements of virtual environments or games that lead to or contribute to immersion. One of the least explored areas of immersion is the influence of sound. Nordahl and Nilsson attempt to define presence and immersion in the context of interactive virtual environments, exploring the auditory influence as well as specific auditory techniques on immersive experiences. Serafin expands on this argument by focusing specifically on sound as one component within a multimodal system.

The interactions that occur between our sensory modalities can vary depending on the context they are operating in. Our perception of one modality can be significantly affected by the information that we receive in another modality. Some researchers have studied the interactions among modalities in general (Marks 1978). Others have focused on the interactions of two specific sensory modalities, such as vision and touch (Martino and Marks 2000), sound and touch (Zampini and Spence 2004), sound and taste (Simner, Cuskley, and Kirby 2010), and sound and odor (Tomasik-Krótki and Strojny 2008). Serafin interrogates these cross-modal interactions with sound, examining how an understanding of our perceptual system may improve our ability to design and create technologies.

Indeed, an understanding of the emotional and cognitive aspects of sound can potentially lead to much greater engagement with a variety of media. Anders-Petter Andersson and Birgitta Cappelen even show in "Musical Interaction for Health Improvement" that sound (specifically, music) can influence and improve our health. Natasa Paterson and Fionnuala Conway's "Engagement, Immersion and Presence: The Role of Audio Interactivity in Location-aware Sound Design" specifically focuses on the role of sound in the design of location-aware games and activities, arguing for greater engagement and immersion through sound design.

Performance and Interactive Instruments

The fourth section of the *Handbook, Performance and Interactive Instruments*, brings together emerging ideas about how we physically interact with audio: through what devices, media, and technologies? New generations of game consoles manifest the idea that we physically interact with audio: through devices shaped like guitars and light sabers, through hand-held controllers and other gestural interaction devices. However, what are the constraints of these systems? How are designers and engineers working to overcome current technical and industrial limitations? In addition, how does the increasingly important role of social and online media influence the ways in which people interact with audio? In seeking solutions to these and other questions, the work of authors in this section challenge traditional thinking about audio and the environment, about performer and audience, about skill and virtuosity, about perception and reality.

Each author presents a different perspective on what interactive sound means in terms of digital sound production and consumption: exploring liveness, instrument creation, and embodiedness. Kiri Miller explores interactivity through dance in "Multisensory Musicality in Dance Central." Miller argues that through the performative practice of dance, and the social interactions that take place around games like *Dance Central*, audiences may develop a new relationship to music and sound.

Mike Frengel and Michael Gurevich each explore interactivity in the performing arts from the perspective of the composer and performer, rather than audience. This is not to say that an audience isn't a component of that performance. Indeed, Frengel argues that "Interactivity in the performing arts is distinctive because there is a third party involved—the spectator. In concert music performances, the interaction typically occurs between a performer and a system, but it is done for an audience that remains, in most cases, outside the interactive discourse." Both Frengel's "Interactivity and Liveness in Electroacoustic Concert Music" and Gurevich's "Skill in Interactive Digital Music Systems" examine the relationship between the performer and the audience in electronic (and particularly digital) interactive music, exploring what it means to perform with technology.

Research has shown that we can recognize and feel the emotion conveyed by a performer when we listen to music (Bresin and Friberg 2001). An embodied cognition approach as to why this occurs suggests that we understand human-made sounds (including those generated by playing a musical instrument) in terms of our own experience of making similar sounds and movements. We therefore give meaning to sound in terms of emulated actions, or corporeal articulations (Leman 2008). More specifically, we mentally and sometimes physically imitate the expressiveness of the action behind the sound, based on our "prior embodied experience of sound production" (Cox 2001). Winters describes, "The mimetic hypothesis might also provide an explanation for why we might find ourselves unconsciously "imitating" the emotion seemingly being expressed, in addition to any willing participation in a game of make-believe" (Winters 2008). Electronically generated or synthesized sounds and music remove this corporeal connection to causality. Issues of liveness frequently arise in the discussions of electronic music. What is made clear in Frengel's and Gurevich's chapters is that digital electronic instruments can disguise some of the important performative aspects of music. Marc Ainger and Benjamin Schroeder's "Gesture in the Design of Interactive Sound Models" focuses on this role of gesture in the relationship between performer, instrument, and listener, suggesting some means to overcome the lack of gesture in some types of digital music performance. Nick Collins suggests that the machine can become a performer in its own right, an intelligent responsive instrument that can listen and learn, in "Virtual Musicians and Machine Learning." This idea is further expanded upon by Norbert Herber in "Musical Behavior and Amergence in Technoetic and Media Arts." Herber suggests generative music systems can offer one means to enhance the live experience, as variation and difference can be brought into performance.

Tools and Techniques

The concept of machine learning, and how the machine "talks" back to us and interacts with us, brings us to the section on *Tools and Techniques*, focusing on the enabling nature of new tools, technologies, and techniques in interactive audio. Within *Tools and Techniques*, ontological implications of questions regarding the evolving, ongoing, and often contested relationship between human and machine are explored. The essence of the interactivity lies within the medium of interaction and therefore, unsurprisingly, computers, hardware, and software are the media integral to the production of digital audio. New technologies such as digital sensors have enabled interactivity to thrive in the arts but how, specifically, can these media influence interaction with sound? In some instances, such as in music for film and television, audio is transmitted in one direction: from creator to listener, with little or no interactivity involved; and in others, sound can and indeed must be interactive, as is the case with videogames. Despite this difference, implicit in all of these cases is the understanding that technology is simply a tool—true creativity is an inherently human trait. But is such a statement necessarily the case? The research presented in this section questions the essential elements of interactivity by linking these findings to wider questions about creativity and creative work. Is creativity, by definition, something that can be produced only by human beings? Can machines produce output that evokes emotion?

Chris Nash and Alan F. Blackwell begin the section in "Flow of Creative Interaction with Digital Music Notations" by exploring the relationship between digital music notation and creation, examining the software at the heart of digital music production, from sequencer or tracker-based systems such as Pro Tools to graphic programming software such as Max/MSP. They present a series of design heuristics based on their research into the influence that software has on creativity. David Bessell's "Blurring Boundaries: Trends and Implications in Audio Production Software Developments" provides a useful corollary to Nash and Blackwell by providing us with a historical overview of the digital audio workstation, or DAW, focusing on the development of this musical software.

The next two chapters focus on generative and procedural production systems for videogames. Procedural music soundtracks may offer some interesting possibilities that may solve some of these complications with respect to composing for games. On the other hand, procedural music composers are faced with a particular difficulty when creating for videogames: the sound in a game must accompany an image as part of a narrative, implying that sound must fulfill particular functions in games. Cues need to relate to each other, to the gameplay level, to the narrative, to the game's run-time parameters, and even to other games in the case of episodic games or those that are part of a larger series. Procedural music and sound in (most) games, therefore, must be bound by quite strict control logics (the commands or rules that control playback), in order to function adequately (see Collins 2009). In particular, music must still drive the emotion of the game, a fact explored by Maia Hoeberechts, Jeff Shantz, and Michael Katchabaw in "Delivering Interactive Experiences through the Emotional Adaptation of Automatically Composed Music." Niels Böttcher and Stefania Serafin specifically focus on the question of how procedural sound relates to the gestural interactions of the player in "A Review of Interactive Sound in Computer Games: Can Sound Affect the Motoric Behavior of a Player?."

The *Tools and Techniques* section of the *Handbook* is rounded out by Victor Lazzarini's "Interactive Spectral Processing of Musical Audio," which explores emerging ideas in interactive spatial sound and interactive spectral processing. Although such tools and techniques often occur "behind the scenes" of the creative and experiential aspects of sound production and listening, the ideas and concepts are driving new tools and technologies that are sure to become familiar to us in the future.

The Practitioner's Point of View

The final section of the book, *The Practitioner's Point of View*, steps back from some of the academically inspired issues and questions to consider interactive audio from the point of view of some of its practitioners. The collection of chapters presented in this section coalesce around considerations of the past, present, and future of interactive audio. "Let's Mix it up: Interviews Exploring the Practical and Technical Challenges of Interactive Mixing in Games" by Helen Mitchell presents interview material with game sound designers, outlining some of the creative and technical challenges of designing interactive sound. Damian Kastbauer, an audio implementation specialist for games, explores what "Our Interactive Audio Future" might look like, introducing some of the technical work that is being undertaken through a narrative of sound synthesis in the future. Leonard J. Paul's "For the Love of Chiptune" explores what it means to compose with game sound tools, and how practitioners can develop their own aesthetic within a community of composers.

Andy Farnell, one of the leading proponents of procedural audio, introduces us to his take on "Procedural Audio Theory and Practice," providing a useful complementary chapter to some of the theoretical work presented in other chapters. Likewise, complementing the chapters in *Performance and Interactive Instruments*, composer Rafał Zapała gives his theory on and techniques for live electronic and digital performance with "Live Electronic Preparation: Interactive Timbral Practice." Finally, game composer and sound designer Tim van Geelen introduces us to "New Tools for Interactive Audio, and What Good They Do," suggesting how new hardware, software, and techniques may lead us forward in our production and understanding of interactive audio.

A Series of Lists . . .

The crossover between chapters has meant that there are common references to products and concepts that recur throughout the Handbook. In order to facilitate an ease of referencing common software, games, and acronyms, we have compiled three lists following this Introduction: (a) a list of acronyms; (2) a list of software; and (3) a list of games. It is our hope that by presenting the information collated in this fashion, readers will be more easily able to follow up on references. Likewise, we have presented a list of further references for those readers who wish to seek out videos, images, sound files, and other content beyond what we could include in this text. This latter list was compiled by the authors of the chapters included here, and is presented as a kind of "recommended reading, viewing, and listening list."

References

Bourriaud, N. 2002. *Esthétique relationnelle*. Dijon: Les presses du réel.

Bresin, Roberto, and Anders Friberg. 2001. Expressive Musical Icons. In *Proceedings of the 2001 International Conference on Auditory Display*, ed. J. Hiipakka, N. Zakarov. and T. Takala, 141–143. Espoo, Finland: Helsinki University of Technology.

Bridgett, Rob. 2007. Designing a Next-gen Game for Sound. *Gamasutra*, November 22. http://www.gamasutra.com/view/feature/130733/designing_a_nextgen_game_for_sound.php.

Clancey, W. J. 1997. *Situated Cognition: On Human Knowledge and Computer Representations*. Cambridge, UK: Cambridge University Press.

Collins, Karen. 2009. An Introduction to Procedural Audio in Video Games. *Contemporary Music Review*, 28(1): 5–15

——. 2011. Making Gamers Cry: Mirror Neurons and Embodied Interaction with Game Sound. *ACM AudioMostly 2011: 6th Conference on Interaction with Sound*. Coimbra, Portugal, September 2011, 39–46.

——. 2013. *Playing With Sound: A Theory of Interacting with Sound and Music in Video Games*. Cambridge, MA: MIT Press.

Cover, Rob. 2006. Audience Inter/active: Interactive Media, Narrative Control and Reconceiving Audience History. *New Media and Society* 8(1): 139–158.

Cox, Arnie. 2001. The Mimetic Hypothesis and Embodied Musical Meaning. *Musicae Scientiae* 5(2): 195–212.

Csikszentmihalyi, Mihaly. 1990. *Flow: The Psychology of Optimal Experience*. New York: Harper-Perennial.

Dyson, Frances. 2009. *Sounding New Media: Immersion and Embodiment in the Arts and Culture*. Berkeley: University of California Press.

Gee, James Paul. 2008. Video Games and Embodiment. *Games and Culture* 3(3–4): 253–263.

Gell, Alfred. 1998. *Art and Agency: An Anthropological Theory of Art*. Oxford: Oxford University Press.

Green, Jo-Anne. 2010. Interactivity and Agency in Real Time Systems. *Soft Borders Conference and Festival Proceedings*, 1–5. São Paulo, Brazil.

Leman, Marc. 2008. *Embodied Music Cognition and Mediation Technology*. Cambridge, MA: MIT Press.

Marks, Lawrence E. 1978. *The Unity of the Senses: Interrelations among the Modalities*. New York: Academic Press.

Martino, Gail, and Lawrence E. Marks. 2000. Cross-modal Interaction between Vision and Touch: The role of Synesthetic Correspondence. *Perception* 29(6): 745–754.

McLuhan, Marshall, and Barrington Nevitt. 1972. *Take Today: The Executive as Dropout*. New York: Harcourt, Brace and Jovanovich.

Saltz, David Z. 1997. The Art of Interaction: Interactivity, Performativity, and Computers. *Journal of Aesthetics and Art Criticism* 55(2): 117–127.

Simner, J., C. Cuskley, and S. Kirby. 2010. What Sound Does That Taste? Cross-modal Mappings across Gustation and Audition. *Perception* 39(4): 553–569.

Thom, Randy. 1999. Designing a Movie for Sound. *Film Sound*. http://filmsound.org/articles/designing_for_sound.htm.

Tomasik-Krótki, Jagna, and Jacek Strojny. 2008. Scaling of Sensory Impressions. *Journal of Sensory Studies* 23(2): 251–266.

Winters, Ben. 2008. Corporeality, musical heartbeats, and cinematic emotion. *Music, Sound, and the Moving Image* 2(1): 3–25.

Zampini, Massimiliano, and Charles Spence. 2004. The role of auditory cues in modulating the perceived crispness and staleness of potato chips. *Journal of Sensory Studies* 19(5): 347–363.

SECTION 1

INTERACTIVE SOUND IN PRACTICE

CHAPTER 1

SPATIAL RECONFIGURATION IN INTERACTIVE VIDEO ART

HOLLY ROGERS

VIDEO art has always been immersive: but it can also be performative and interactive. New forms of technology and easy-to-use audiovisual interfaces have enabled artists to hand the compositional control of their sounds and images to visitors. However, in order to physically participate in video work, audiences must cross a sacred divide that has, until relatively recently, been a fundamental component of music performance and art exhibition. Once in the heart of the video work, visitors are able to dissolve the boundaries that separate performers from audience, and artwork from viewers. But they are also given the chance to draw together different disciplines; to combine music and image to form new intermedial structures. Although New York City–based video artist Gabriel Barcia-Colombo describes his audiovisual work as "video sculpture," for instance, he encourages interactive, spatial audiovisuality through the use of knobs, sensors, and software such as Jitter, a visual language program for Max/MSP that enables users to process video in real time. In order to take "cinematic experiences and mak[e] them into real-world interactions," many of his pieces feature tiny projected people, often trapped inside everyday objects, such as blenders, suitcases or glass utensils. In *Jitterbox* (2007), a piece described by Barcia-Colombo as an "interactive video jukebox," a small dancer appeared trapped in a glass dome atop a 1940s radio (see Figure 1.1). The visitor was able to change the channel of the radio, choosing between several songs from the 1940s: as the music changed, the dancer responded to the new beat, adjusting style and time according to the will of the user.

Canadian dancer and artist Marie Chouinard explored a different route to audio-visual interactivity in her 2004 participatory video installation, *Cantique 3*. Installed as part of the Monaco Dance Forum, the piece consisted of two large monitors, each linked to a flat-screen interface. On one screen, a man's face was seen in close profile; he looked toward the other screen, on which a woman's profile peered back at him. The touch-screen panels showed five lines resembling a musical stave: a small, frozen image of the man sat on one stave; and a snapshot of the woman occupied the other.

FIGURE 1.1 Gabriel Barcia-Colombo, *Jitterbox* (2007).

Two "players" were invited to interact with the touch-screen "mixing boards" by moving the frozen images along, and up and down, the lines. When the face of a character was touched by their player, the corresponding large image was activated so that it burst into motion and guttural, abrasive vocalizations that ranged from hoarse whispers to frenzied, onomatopoeic shrieks: "We are in the presence of the birth of language...and its critique," explained Chouinard. The two players composed with their images simultaneously, initiating an audiovisual counterpoint whose responsive, process-driven structures were controlled entirely by the composerly desires of the visitors.

Invited to set the *Jitterbox* in motion and to create an audiovisual composition for *Cantique* 3, visitors became physically and aesthetically integrated into the artwork. With this in mind, interactive video can be understood as a facilitator for spatial merging. But what happens when visitors are asked to participate in—or even control—an intermedial discourse? Can internal and external spaces really be combined? And what occurs when a traditional musical, artistic, or "cinematic experience," is turned into a "real-world interaction," subject to constant reconfiguration?

The crossing of physical and aesthetic borders enabled by video technology when it arrived on the commercial market in 1965 accelerated several strands of creative experimentation that had already begun to blossom during the twentieth century. Speaking of the interpersonal actions between people operating within the segregated performance space of drama (and by extension, the music concert), Richard Schechner (1968,

44) identified three "primary transactions": the communication between performers, between performers and audience members, and between individual members of the audience. While everyone present at a dramatic or musical event takes part in at least one of Schechner's transactions, the nature of each interaction differs between cultures, ideologies, and eras. Since the nineteenth century, for instance, the modern concert hall has developed a physical and conceptual segregation between a "performing space and a listening space" (Blesser and Salter 2007, 130). Remaining physically separated from the creative sonic hub, listeners sit in silence, thoroughly immersed and emotionally engaged in the music, yet unable to affect the flow of sound. The concert auditorium's design, Christopher Small (1998, 26–7) argues, not only "discourages communication among members of the audience," it is also planned "on the assumption that a musical performance is a system of one-way communication, from composer to listener through the medium of the performers." While it is important to note that listening is rarely a passive experience, the physical separation can prevent a concert from becoming performative. Modern gallery spaces—or what Brian O'Doherty calls "the white cube"— are often organized in a similar way: with walls painted white and noise kept to a minimum, visitors to the "neutral void" are asked to look but not touch; as in Small's concert hall, they remain separated, at least physically, from the artwork presented (O'Doherty 1976, 15).

Although there are examples of earlier interactive, performative music and art, it was during the twentieth century that a sustained attack on the rigidity of viewing and listening conventions was launched from many quarters. At the heart of the dissolution of "one-way communication" lay the promotion of unrepeatable, inclusive music performance, the embrace of unique audio configurations found in John Cage's chance-determined pieces, Berio's graphically notated works (which give performers a great deal of interpretative input), Stockhausen's use of broadcast radio (which is different for every performance), and Terry Riley's fragment-controlled improvisations among others. Despite operating according to different aesthetics, the result of such experimentation was music that was structurally different in each performance and musical progressions that could be determined to a greater or lesser extent by performers or the audience.

As composers began to loosen control in order to give performers and audience members a sonic, structural control over their music, visual artists began to reconfigure traditional exhibition spaces by pulling visitors into the physical heart of their work. Although forms of reciprocal communication can be found in many schools of visual practice, it is most clearly articulated in installation art, an impermanent sculptural practice defined by Erika Suderburg (1996, 4) in terms of spatial activation: " 'installation' is the art form that takes note of the perimeters of that space and reconfigures it." The reconfiguration of space can be found in the earliest examples of installation art in France, such as Yves Klein's completely empty gallery space, *Le Vide*; and Arman's response, *Le Plein*, in which the same gallery was so full of found objects that visitors were unable to get in (Galerie Iris Clert, Paris: 1958, 1960). A similar aesthetic developed in America, where Claes Oldenburg, in *The Street*, and Jim Dine, in *The House*, assembled artifacts found discarded on the streets of New York in the city's Judson Gallery in

1960. In December of the following year, Oldenburg rented a New York shop for two months, where he installed *The Store*, an exhibit that functioned at once as studio, commercial gallery, and shop. Oldenburg and Dine sought to merge public and art spaces by bringing the street indoors, while simultaneously encouraging the audience to enter the installation's environment rather than to view and objectify it: to integrate interior and exterior spaces.

In her critical history of installation art (Bishop 2005), Claire Bishop explains that the genre not only reconfigures the "white cube," it also initiates an "activation" of viewers who, confronted with assembled fragments, must decide where to stand in order to interpret, or complete the piece. As the century progressed toward the late 1960s, the philosophical shifts in art aesthetic, as in music, prompted a fundamental relocation of focus from the fixed object to a process that could include, to varying extents, the viewer.

Emerging together with video art in the mid 1960s, performance art—in the form of "Happenings," "Events," "Actions," and so on—dealt another strong blow to traditional methods of art and music consumption. Writing in 1979, RoseLee Goldberg noted that artists often invited performance into their work "as a way of breaking down categories and indicating new directions" when a creative progression had "reached an impasse": "Live gestures have constantly been used as a weapon against the conventions of established art" (Goldberg 2001, 7). In this way, the inclusion of live performance and theatricality into artwork contributed to the devaluation of the commodity value of art, as pieces created were often not repeatable (at least not exactly) and could not be collected or sold: "performance was the surest means of disrupting a complacent public" (154). At the time, however, performance and video artist Vito Acconci expressed a hatred for the designation "performance" because it evoked the theater, a space divided into two areas separated by a "mystic gulf" (Wagner) that kept apart actors and audience: the word, Acconci explained, suggested a "point you went toward," an "enclosure" that could provide only "abstractions of the world and not the messy world itself" (Kaye 2007, 74). The lure of a "messy" potential in performance was explored by Cage and Allan Kaprow, orchestrator of the Happening, among others, who encouraged spontaneous participation from their audience members in order to better integrate the segregated spaces of traditional performance and exhibition environments. Writing about the reasons behind his recourse to the live gesture, Kaprow explained that his inspiration came from the public arena rather than from the artworld; live performance work was not only an attack on "the conventions of established art," but also on those responsible for maintaining its sanctified edifices (Reiss 2001, 15).

Many of Kaprow's Environments, for instance, were located outside the gallery space, functioning in lofts, courtyards, and other public spaces where it was easy for anyone to get involved: "There are no clear distinctions between . . . art of any kind (Happenings) and life," he explained (Kaprow 2003, 73). However, he also worked in traditional spaces, where the aesthetic of inclusion assumed an even more radical edge. Visitors to his exhibition at the Hansa Gallery, New York, for instance, did not "come to look *at* things," but rather were placed at the center of a dynamic and malleable event and given the option to interact according to their "talents for 'engagement' " (11): "there are freedoms for the viewer... but they are revealed only within the limits dictated by the art work's immediate as well as underlying themes." Although there were restraints, these boundaries did not provide a prior meaning, or "finite object," but rather encouraged participation in a continually changing process. In order to do this, Kaprow reasoned, the artist must possess a "disregard for security," a willingness to fail (20).

As music was expanding out of its traditional spatial parameters into the audience's space during the 1960s, and as art reached out toward the spectators, inviting them to cross the normal threshold between work and receiver, the two disciplines began to come together. The introduction of portable, relatively cheap and easy to handle video equipment in the middle of the decade provided the final nudge toward a truly intermedial fusion of music and art.

Early on, video was used as part of audiovisual multimedia performances, installations, and Happenings in order to re-mediate and enlarge preexistent practice. The video format was unique in its ability to record and transmit sound and image at the same time in a cheap and convenient manner. For this reason, artists found that they could easily sound their visual experimentation while musicians could visualize their music with little or no training. Because of video's potential for audiovisuality, many key players during the medium's earliest years were trained musicians: Nam June Paik, Steina Vasulka, and Robert Cahen, for instance; others, such as Tony Conrad, Bill Viola, and Bruce Nauman, although not musically trained, were nevertheless heavily involved in music as performers or composers. Video intermediality had a particularly profound effect on the visual arts that, unlike music, do not traditionally require realization through performance. As video introduced a temporal element into the static arts, allowing images to unfold though time like music, a shift from art-as-object to art-as-process was initiated, a transition that contributed to the "dematerialisation of the art object" during the twentieth century (Lucy Lippard, in Oliveira et al. 1994, 28).

Performance art fed fluently into early video practice, partly because many practitioners, such as Paik, Joan Jonas, Carolee Schneemann, Ulrike Rosenbach, and VALIE EXPORT, were involved with both disciplines. Kaprow's desire to include the public in his work by making the gallery space part of normal life was a sentiment that lay at the heart of early video work: "As a medium that is economically accessible and requires minimal technical skills to master, video is ideally suited as a vehicle for the close integration of art and life," explains Tamblyn (1996, 14). Emerging from within this discourse, early video artists and composers treated the new audiovisual technology like a performer, a technological presence able to improvise audiovisually and to be reactive to its changing environment via a closed-circuit feed rather than exhibiting prerecorded or preedited footage and sound.

Of course, not all video includes sound; nor is all video work installational or sculptural. As an artistic tool, video has been used to create single-channel works, guerrilla-style documentary, and work for broadcast television. Yet in its earliest years, the video format required separate technologies for recording and playback: as a result, the easiest and most revolutionary way to make use of the medium was as a live component of multimedia events. And it was here, in the real-time, experiential mobilization of a live audience, that video's audiovisuality most clearly arose.

1.1 Immersion

Through the use of a closed-circuit feed, or by taking over an entire room, video work can immerse its visitors completely. Moreover, once across the normally forbidden threshold that separates work and life, visitors become the material of the piece, able to assume varying levels of compositional control by pulling together all three of Schechner's primary transactions. With reference to new media, Frances Dyson identifies a change in engagement toward "'being in,' rather than 'looking at,' virtual environments," a perceptual relocation that enables the visitor to occupy real and fictional spaces at the same time (Dyson 2009, 2). As a result, Dyson explains that immersion becomes:

> a process or condition whereby the viewer becomes totally enveloped within and transformed by the "virtual environment." Space acts as a pivotal element in this rhetorical architecture, since it provides a bridge between real and mythic spaces, such as the space of the screen, the space of the imagination, cosmic space, and literal, three-dimensional physical space. (1)

Immersive environments that remap spectatorial habits from one-way communication to two-way activity help to bind spectator to spectacle by removing the barriers of passivity and the physical space between viewer and art exhibition; listener and music recital. Neuropsychology has articulated the spatial reconfigurations that immersive, or interactive, environments can enable by identifying three different spatial interfaces: personal space, which is inhabited by the body; peripersonal space, which "is the region within easy reach of the hands"; and extrapersonal space, which includes "whatever lies beyond peripersonal space":

> Although the brain uses different representations and approaches to interacting in different spaces, there are ways to "bridge the gap" between spaces, allowing the brain to work in one space using the same approach that it uses in another. It has been found that the brain can naturally bind personal and peripersonal space, but binding extrapersonal space is more difficult. (Shoemarker and Booth 2011, 91)

The use of tactile interfaces in *Jitterbox* and *Cantique 3* helps to bind personal and peripersonal space with the extrapersonal by transporting the user into the virtual worlds of Barcia-Colombo's singing radio and Chouinard's gesticulating faces; but by "bridging the gap" between the two physical locations, the extrapersonal becomes synonymous with the mythic space identified by Dyson. The result can be unnerving.

The invitation to step into a mythic space is most clearly articulated in works that not only defamiliarize the traditional gallery area, but also replace it by asking visitors to step into a separate arena. Tony Oursler's video environments, for instance, transport visitors into a brand new world where they are immersed on all sides by videoed images in the same way as a listener is immersed in music at a concert. In *System for Dramatic Feedback* (1994), visitors walking into a darkened room are greeted by a rag doll, its face animated by a video projection that shouts "No! No!" If they dare to enter after this warning, they find themselves in a complete environment in which a pile of ragdolls with animated faces twitch and jitter and a large screen shows rows of cinema-goers eating popcorn with inert faces, a trope on the passivity of cinematic, and by extension, art consumption. Once in this environment, explains the artist, "the division between media and real world has dissolved" (Oursler 1995). Bodily immersion also lies at the heart of much of Bill Viola's work, with audiovisual environments such as *Five Angels of the Millennium* (2001) and *Ocean without a Shore* (2007) dissolving awareness of the original surroundings and transporting visitors straight into an extrapersonal, communal space. For the visitor, the result is akin to participating in a music recital, jumping through the frame and into a painting, or dissolving into the fictional diegesis of a film.

In her exhibition *Eyeball Massage* at London's Hayward Gallery, Swiss video artist Pipilotti Rist presented numerous versions of spatial merging within a single gallery space, by asking the viewers constantly to oscillate between different modes of engagement. In *Lungenflügel* ("Lobe of the Lung," 2009), visitors were invited into an area set off from the rest of the gallery space by four video walls and hanging layers of material and encouraged to sit, lie or stand on a bed of cushions. Once across the threshold of the whole-room installation, visitors could choose where to sit, where to look, and for how long to stay. Immersed in a continuous, atmospheric wash of sound (by Anders Guggisberg) that evoked "the sounds of the moving fluids inside of our bodies that we don't pay much attention to normally; a melody of heartbeats, things moving inside your stomach" (Rist 2011, 15), color-saturated images roamed across the main articulated projection frame, while visual counterpoints flicked across the screens to the side and back. The form of immersion demanded by *I'm Not the Girl who Misses Much* (1986) was less relaxing; in order to see the videoed artist singing and miming to the Beatles' song "Happiness is a Warm Gun" (1968), visitors had to stick their heads through small holes in a suspended box; once inside they were able not only to watch and listen to a video of Rist dancing to the Lennon track, but also to witness at close proximity the heads of other visitors who had happened upon the installation at the same time.

1.2 Interactivity

But while an audience is invited into the spatial heart of immersive video environments, they are not always able to contribute to the structure, content, or flow of a work. Here we can articulate a distinction between immersion and interactivity. As we have seen, music is immersive: and yet the performance of art music is not traditionally performative. Listeners are immersed in sound, which is able to move through their space and to surround them entirely. They may also be transported into the sound*world* where they are able to perform a personal dialog with the music. But they nevertheless remain unable to change the course of the performance itself. The same is often true of immersive video environments, such as Oursler's *System for Dramatic Feedback* and Rist's *Lungenflügel*.

Other artists have pushed through the immersive barrier to enable visitors to assume a hands-on creative role. The possible levels of video interactivity, which have characterized video work from the beginning, are manifold: a work can interact with a space and initiate a dialog with the visitors within it; sound and image can be manipulated by visitors in order to create individual audiovisual pathways; or visitors in different locations can be drawn together via technological intervention.

We saw above that early video enabled performative intermedial spaces by inviting visitors into the realm of the projected image and amplified sound in order to better probe issues of public and private spaces, democratic decisions, and interpersonal connections. Once a fundamental element of a piece, audience members could introduce "flexibility, changeability, fluency" into the creative formula (Cage, in Goldberg 2001, 124). Tracing the etymology of "inter" to the Latin for "among," Margaret Morse explains that the suffix to interactive "suggests a linking or meshing function that connects separate entities": interactivity, she continues "allows associative rather than linear and casual links to be made between heterogeneous elements" (Morse 1990, 18; 22). In interactive video work, the "meshing function" operates not only between media (in the form of intermedia), but also between a work's components and those who choose to engage with it. The significance of the visitor to this mesh was explained early on by video artists Steina and Woody Vasulka, who described the Kitchen Videotape Theatre, which they founded at the Mercer Arts Center, New York, as "a theatre utilizing an audio, video, and electronic interface between performers (including actors, musicians, composers, and kinetic visual artists) and audience": within this theater, video work was considered as an "activity" rather than an "art a priori" (Steina and Woody Vasulka, in Salter 2010, 120).

One form of video interactivity relies on a visual or audio contribution from visitors. The illusion of bodily transference into the mythic space of the work, for instance, can be achieved by presenting visitors with their own videoed images. Early video work in particular achieved an interactive component largely through exploration of the closed-circuit feed, which could use images and sounds from the audience to produce a responsive, site-specific form of mimesis and transformation, a process of inclusion that lay at the heart of Les Levine's early work *Iris* (1968). Installed in Levine's studio, *Iris* was a closed-circuit feed that promoted an interplay between three video cameras—which recorded visitors as they moved around the performance space—and a stack of six television monitors. With their images presented through the monitors in real time, visitors were able to change how the installation looked and the speed with which the images progressed, initiating a performative interplay between mediated space and the "real" space of the work, as the artist explained:

> I don't tend to think of my work purely in psychological terms, but one must assume some psychological effect of seeing oneself on TV all the time. Through my systems the viewer sees himself as an image, the way other people would see him were he on television. In seeing himself this way he becomes more aware of what he looks like. All of television, even broadcast television, is to some degree showing the human race to itself as a working model. It's a reflection of society, and it shows society what society looks like. It renders the social and psychological condition of the environment visible to that environment. (Youngblood 1970, 339)

Iris remained in a constant state of flux, with each moment of its existence utterly unique. At first, visitors reported unsettling psychological fissures when included as a key component to a video work. Recalling *Iris*, for example, theorist Gene Youngblood suggests that visitors are made to feel self-conscious because the work

> turns the viewer into information. The viewer has to reconsider what he thought about himself before. He must think about himself in terms of information. You notice people in front of *Iris* begin to adjust their appearance. They adjust their hair, tie, spectacles. They become aware of aspects of themselves which do not conform to the image they previously had of themselves. (339)

By drawing together the processes of videoing and experiencing, creator and receiver, video existed in, and moved through, the transient time and space of the visitor by displacing them into their own unsettling extrapersonal space. This method of visual transportation lay at the heart of many of Paik's musical works, such as the *TV Cello* (1971), an instrument constructed from three television monitors linked to a closed-circuit feed of the audience; when a cellist (Charlotte Moorman) played the sculpture, not only were electronic sounds produced, but the images underwent associated forms of distortion and manipulation.

The increasing institutional support for video work by major galleries and museums from the mid-1980s onward, the increasing availability of funding for moving image and audio art, and accelerated technological innovation have provided increased opportunities for artists to use the meshing function of video interactivity in a variety of ways. While early pieces such as *Iris* and *TV Cello* physically repositioned the visitor into the heart of the installation, the first interactive video disc that enabled viewers to determine their own course through a work was Lynn Hershman Leeson's *LORNA* (1983–4), a work that acted as "a natural progression from time-based sculptural strategies" (Leeson 2005, 77). In *Iris*, visitors became visual material whether or not they acted for the camera; as explained in relation to the Happening, the spectator became an "important physical component of the art environment" regardless of their will to participate (Kaprow 2003, 93). Leeson (2005, 78) differentiated between these works and her new form of engagement, explaining that true "Interactive systems *require* users to react":

> A (pre)condition of a video dialog is that *it does not talk back*. Rather, it exists as a moving stasis; a one-sided discourse; like a trick mirror that absorbs instead of reflects. Perhaps it was nostalgia that led me to search for an interactive video fantasy—a craving for control, a longing for liveness, a drive toward direct action. This total, cumulative, and chronic condition I suffered from is reputedly a side effect... of watching television. (Leeson 1990, 267)

In order to give her visitors the opportunity to react to *LORNA*, Leeson provided them with a remote control similar to the one her videoed agoraphobic protagonist used to change her television channels. Lorna appeared unable to make her own decisions and sat staring at her TV monitor, overwhelmed (we are told) by alienation and loneliness. Juxtaposed against her inability to act was the heightened free will of the user, who was able to alight on various objects in Lorna's virtual room in order to release a sound or video module. Depending on the objects selected, or the choices provided (there were three options for the phone, for instance), the user released a different narrative for Leeson's character, which resulted in one of three possible endings (Lorna either shot herself, shot her TV, or decided to move to Los Angeles). Despite this interactive freedom, however, Leeson points out that "these systems only appear to talk back. That they are alive or independent is an illusion. They depend upon the architectural strategy of the program. However, there is a space between the system and player in which a link, fusion, or transplant occurs. Content is codified. Truth and fiction blur..." (Leeson 1990, 271). Evoking ideas of a spatial interaction—or "transplant"—Leeson's description of her work is predicated on the ability of visitors to step across the threshold of the white cube and assume control over her work's structure.

While *LORNA* only "appear[s] to talk back," more recent work makes use of technological advances in order to allow visitors a truly influential role over an installation's progression. Mary Lucier's *Oblique House (Valdez)* (1993), installed in an abandoned car dealership in Rochester, New York, asked visitors to step into a house haunted by the sounds and images of people who had encountered a loss as the result of a natural disaster (the 1964 earthquake), or a man-made catastrophe (the 1989 oil spill) in the city of Valdez, Alaska. At first, monitors situated in the corners of the room were silent, showing only facial close-ups of three women and one man. As visitors entered the space, sensors near each monitor picked up their movements, prompting the images to lurch into slow-motion life and embark on their testimonials in highly resonant, processed timbres. Via movement, visitors were able to set off several recollections at once, resulting in duets, trios, and quartets for the departed. As the stories combined, a common thread of pain and solace emerged from the cacophony, an ever-changing soundscape composed by the visitors.

David Small's and Tom White's video installation *Stream of Consciousness* (1997–8; later retitled *An Interactive Poetic Garden*) gave a different form of control to the user. Here, a rock garden housed several linked pools. Water flowed down through the pools before coming to rest in a large, glowing basin onto which words were projected from above. Described by its creators as an exploration into the "open-ended active and passive modes of interaction," the installation invited visitors to manipulate a hand interface in order to direct the text, halt its flow, or "change the content of the words themselves" in order "to evoke the fluid contents of consciousness" (Small and White, n.d.). By interacting with a word through the interface, visitors could create a blue aura around the text: when a word was pressed directly, it appeared larger until additional words began to form. As water in the pool moved, older words were discarded as the liquid drained from the basin to leave, eventually, only the words chosen by the user. Site-specific, *An Interactive Poetic Garden* had to be performed in real time, existing only at the moment of interaction; moreover, the work was performative—embracing the "flexibility, changeability, fluency" enabled by the creative vitality of each user.

Video performer and sound artist Camille Utterback explored similar ideas of creative interactivity in her 1998 installation *Vicissitudes*, a work that made use of specialized, yet user-friendly, technology in order to embrace "the messy world itself" (Acconci, in Kaye 2007, 74). Like *Cantique 3*, Utterback's work explored the nature of language and linguistic constructs, but this time operated through sensor-based apparatus that allowed the installation to respond to the movement of visitors as they moved in and through the gallery's space. The work comprised two audio-tracked interviews, which were linked to physical props located in the exhibition area: in one recording, people recollected the moments in their lives in which they felt happy, or up; in the other, they recalled situations that made them feel unhappy, or low. Visitors were invited to make use of the props: when the ladder was scaled, for instance, the volume of the first audio-track increased; when a visitor lay on the chalked outline, the second soundtrack became more audible. "Many of our linguistic constructs rely on physical metaphor, though they have become transparent to us due to their common usage," explained the artist: "Through its interface, this piece explores the embodiedness of language itself" (Utterback 2004, 224). Asked to navigate through the piece according to their "talents for 'engagement,'" visitors were given responsibility over the sound of the piece, able to compose with the available material to produce a soundtrack with a large amount of variability, a control that replaced the autonomy of the artist-composer with the impermanent nature of audience-controlled process art.

Forms of bodily engagement also form the basis of Christa Erickson's work. The artist has asked visitors to sit on a swing (*Invertigo*, 1997) or play on a seesaw (*MNEMONIC DEVICES: See/Saw*, 2000, 2007). In *Whirl* (2007), an installation in which "memory and nostalgia is revealed as a warped phenomena," the bodily interaction apparent in Erickson's earlier work became even more personal (see Figure 1.2). A pinwheel was linked to a record player and a video projector: when a visitor blew the wheel, the installation burst into life, flinging a group of children wildly around a circle swing and sending warped life into a vinyl recording of nursery rhymes. As the visitor ran out of breath, the sound and images slowed to a standstill, awaiting reactivation by another gust of life-giving breath. While *Whirl* occupied a similar aesthetic position—to fuse art and life—to that of Kaprow and others, it also

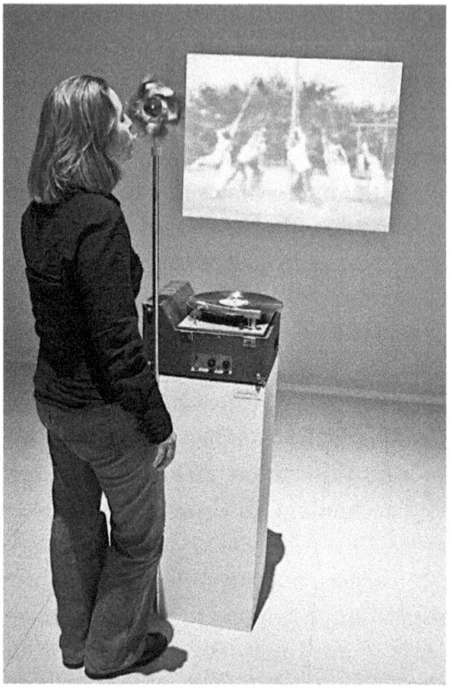

FIGURE 1.2 Christa Erickson, *Whirl* (2007). Installation Still Image.

activated a form of fragile memory by highlighting not only the content and its relationship to the viewer, but also the technology, as old and new forms of audiovisual equipment interacted with one another.

Tropes of nostalgia and reminiscence also lie at the heart of Erickson's motion-tracking, interactive installation *Search* (2005–7), a work that poeticized the movement of people across spatial environments by posing oncoming visitors with a silent, frozen picture of a globe and a hand; as they approached, however, the image burst into motion as the hand began to spin the globe: "Today's global culture has accelerated the creation of many diasporas. People move, travel, flee, and are displaced for personal, economic, environmental, and political reasons. Many long for home, family, culture, and moments of respite in a busy world," explains Erickson (2005–7). In order to evoke the nostalgia for an absent home, the visitors' movements generated streams of words, which emerged then weakened; at the same time, sound began to materialize, becoming increasingly melancholy and noticeable when a visitor stood still:

> There are two categories of words. One set relates to wandering, including active words like drift, roam, flee, migrate, seek, etc. The other set are what one might desire when they stop moving, including words like home, refuge, respite, family, shelter, etc. These words mix and merge on screen as traces of bodies in motion. (Erickson 2005–7)

Whereas earlier works, such as Levine's *Iris* and Paik's *TV Cello*, contributed to the "dematerialization of the art object" by using the visitor as compositional material, thus ensuring a continually different audiovisual progression, more recent work, such as *Whirl* and *Search*, required the visitor actively to participate; to "react," as Leeson would say.

1.3 From the Miniature to the Communal

While the degree of participation in reactive work continues to grow, the type of spatial interaction required remains highly variable. The examples above have been informed by a desire to use the meshing function of video; to obscure or dissolve the boundaries that can separate work from visitor, art from life. However, they have all operated from within the white cube. There are many examples of early video work that was performed outside of the gallery, a move beyond the institution particularly favored by Paik. But the first video equipment was large and cumbersome, and such events could be difficult to achieve. Recent technological innovation has enabled video artists to produce work that can intervene more easily in real life. Such interventions can occur in one of two ways: either through a miniaturization of experience, or by operating in enlarged, communal arenas. The use of touch-screen technology, for instance, has promoted a variety of interiorized interactive audiovisual experiences accessible from beyond the gallery environment, ranging from Brian Eno's generative iPhone and iPod touch app *Bloom*, which invites the user to create ambient musical phrases and a variety of colored shapes simply by tapping the screen, to the interactive iPad component to Björk's recent *Biophilia* project, in which apps accompanying songs allow the user not only to access musical analyses and information, but also to assume compositional control over a song's structure: "each app isn't just a music video or even an instrument: it's something in between," explains interactive media artist and *Biophilia* designer Scott Snibbe (Björk n.d., Tour App Tutorial). This "something in between" thrums most clearly in the "Crystalline" app, an interactive journey that enables the user to tilt the iPad in order to construct her own unique structure for the song. Given control over a set of crystals, the user navigates a system of tunnels: upon reaching a crossroads, she must choose her direction. Each choice leads not only to a new visual experience, but also determines the structure of the song, of which there are numerous possible versions.

Other new forms of audiovisuality can encourage interactivity not only between user and machine, but also between participants: a form of social interrelation that can expose the personal listening spaces promoted by Eno and Björk to a peripersonal—even extrapersonal—audience. Andrew Schneider explains his *Prolixus* (2007), part of a

FIGURE 1.3 Andrew Schneider, *Prolixus* (2007).

series of wearable devices, as a contraption that "makes it possible to say things to yourself *as* other people" (Figure 1.3). It is

> a matching set of interactive video mouths to be worn over the users own mouths by way of the helmets to which they are affixed. They each consist of a five-inch LCD screen attached by metal rod to a bike helmet. The LCD displays either the wearer's mouth or the matching wearer's mouth. Switching between the two mouths requires the users to either slam their heads into something hard, or slap their own or each other's helmets.... The signal from the wired camera behind each screen is fed into the helmet's DPDT relay, acting as an A/B switcher. The other feed for the switcher comes from the wireless receiver also mounted on the back of each helmet. The wireless transceiver on one helmet is tuned to receive the wireless signal from the matching helmet and vice versa. This means that each helmet's LCD screen has the potential to display either wearer's mouth at any time.... A wearer can only discern what is on his or her own screen by looking into a mirror, or judging the reaction they are receiving from their surroundings. (Schneider, n.d.)

Schneider's interactive video mouth probes the boundaries between different levels of personal space, allowing the user at once to recede into their personal world, but also to see this safe interiority exposed for close scrutiny by the other user and those experiencing the contraption. The *Prolixus*, then, not only initiates, but also highlights the osmosis-like flow between work and receiver, between inside and out.

Other artists have sought to create more communal, large-scale audiovisual interventions. Those involved with the London-based group Greyworld, an audiovisual collective whose work focuses on public, environmental interventions that are often temporary in nature and can be installed without permission, have created particularly musical installations. For *Railings* (1996), for instance, a Parisian street balustrade was tuned to sound "The Girl from Ipanema" when an object such as an umbrella was passed along it. Here, viewers become direct participants, with the sole ability to sound the installation and thus bring it to life: as Kaprow said about his earlier work, "Art and life are not simply commingled; the identity of each is uncertain" (2003, 82). However, the desire to merge the identities of real and videoed environments has not just led to the situation of audiovisual work outside the white cube; it has also resulted in a direct interaction with it. Lee Wells, for instance, has installed interactive video pieces along airport terminal tunnels (*Video Forest*, Kimpo Airport, Seoul, 2009) and across bridges (*Bright Nights*, Manhattan Bridge, New York, 2009). The 2- and 3D video mapping created by the Netherlands-based company Nuformer works toward a similar transformational end. Moving images roam across public buildings such as theaters and government buildings. Accompanied by sound effects and music, the images convert familiar structures into medieval cathedrals, jungle scenes, or underwater worlds, or make them appear to burst or shatter entirely.

As we have become increasingly familiar with the audiovisual forms that now fill our world, the mythic spaces of video art have expanded into our everyday lives. Yet those working with video have embraced the messy potential of the medium to create immersive and/or interactive audiovisual environments from the outset: to promote art and music practices as an activity, not an art form a priori. In order to achieve this, artists and musicians have had to loosen their creative control by offering to an audience a set of parameters that are open to varying levels of manipulation: they have to embrace a willingness to fail. Born into an arena of intense musical and artistic experimentation, the video format enabled an enlargement of creative ideas that were already being articulated in other genres. But its methods of delivery and ability to transport images and sounds across spaces—to move visitors into their extrapersonal space—lent itself particularly well to dissolving traditional forms of one-way communication to form real-world interactions that were, and are, subject to continual reimagination.

References

Barcia-Colombo, Gabriel. 2007. *Jitterbox*. http://www.gabebc.com/#Jitterbox.

Bishop, Claire. 2005. *Installation Art: A Critical History*. London: Tate Publishing.

Björk. n.d. *Biophilia*: Tour App Tutorial. http://www.youtube.com/watch?v=n8cox6dO2bg

——. n.d. *Biophilia*: Crystalline App Tutorial. http://www.youtube.com/watch?v=EzfzXNssNn sandfeature=relmfu

Blesser, Barry, and Linda-Ruth Salter. 2007. *Spaces Speak, Are you Listening? Experiencing Aural Architecture*. Cambridge, MA: MIT Press.

Chouinard, Marie. 2004. *Cantique 3*. http://www.mariechouinard.com/cantique-no-3-189.html.

Dyson, Frances. 2009. *Sounding New Media: Immersion and Embodiment in the Arts and Culture*. Berkeley: University of California Press.

Erickson, Christa. 2005–7. *Search* http://emedia.art.sunysb.edu/christa/search.html.

Goldberg, RoseLee. 2001. *Performance Art: From Futurism to the Present*. London: Thames and Hudson.

Leeson, Lynn Hershman. 1990. The Fantasy beyond Control. In *Illuminating Video: An Essential Guide to Video Art*, ed. Doug Hall and Sally Jo Fifer, 267–74. San Francisco: Aperture/Bay Area Video Coalition.

——. 2005. Private 1: An Investigator's Time-Line. In *The Art and Films of Lynn Hershman Leeson: Secret Agents, Private I*, ed. Meredith Tromble, 13–104. Berkley: California University Press.

Kaprow, Allan. 2003. *Essays on the Blurring of Art and Life, ed.* Jeff Kelley. Berkeley: University of California Press.

Kaye, Nick. 2007. *Multi-media: Video, Installation, Performance*. Oxford: Routledge.

Morse, Margaret. 1990. Video Installation Art: The Body, the Image, and the Space-in-Between. In *Illuminating Video: An Essential Guide to Video Art*, ed. Doug Hall and Sally Jo Fifer, 153–67. New York: Aperture/Bay Area Video Coalition.

O'Doherty, Brian.1976, reprinted in 1986. *Inside the White Cube: The Ideology of the Gallery Space*. Berkeley: University of California Press.

Oliveira, Nicholas de, Nicola Oxley, and Michael Petry. 1994. *Installation Art*. London: Thames and Hudson.

Oursler, Tony. 1995. System for Dramatic Feedback. http://www.moma.org/interactives/exhibitions/1995/videospaces/oursler.html.

Reiss, Julie. 2001. *From Margin to Center: The Spaces of Installation Art*. Cambridge, MA: MIT Press.

Rist, Pipilotti. 2011. Lobe of the Lung. In *Pipilotti Rist: Eyeball Massage*. London: Hayward Gallery Leaflet.

Salter, Chris. 2010. *Entangled: Technology and the Transformation of Performance*. Cambridge, MA: MIT Press.

Schechner, Richard. 1968. 6 Axioms for Environmental Theatre. *Drama Review* 12 (3): 41–64.

Schneider, Andrew. n.d. *Prolixus* http://experimentaldevicesforperformance.com/

Shoemarker, Garth, and Kellogg S. Booth. 2011. Whole Body Large Display Interfaces for Users and Designers. In *Whole Body Interaction*, ed. David England, 87–100. London: Springer-Verlag.

Small, Christopher. 1998. *Musicking: The Meanings of Performing and Listening*. Middletown, CT: Wesleyan University Press.

Small, David, and Tom White. n.d. An Interactive Poetic Garden. http://acg.media.mit.edu/projects/stream/InteractivePoeticGarden.pdf.

Suderburg, Erika. 1996. Introduction: On Installation and Site Specificity. In *Space, Site, Intervention: Situating Installation Art*, ed. Erika Suderburg, 1–22. Minneapolis: University of Minnesota Press.

Tamblyn, Christine. 1996. Qualifying the Quotidian: Artist's Video and the Production of Social Space. In *Resolutions: Contemporary Video Practices*, ed. Michael Renov and Erika Suderburg, 13–28. Minneapolis: University of Minnesota Press.

Utterback, Camille. 2004. Unusual Positions: Embodied Interaction with Symbolic Spaces. In *First Person: New Media as Story, Performance and Game*, ed. Noah Wardrip-Fruin and Pat Harrigan, 218–26. Cambridge, MA: MIT Press.

Youngblood, Gene. 1970. *Expanded Cinema*. Boston: Dutton.

CHAPTER 2

NAVIGATING SOUND

Locative and Translocational Approaches to Interactive Audio

NYE PARRY

The emergence of recorded sound in the twentieth century saw an unprecedented shift in the way music and spoken word were integrated into our cultural lives. The ability to take recordings home in the form of records and later CDs gave us access to both musical and narrative audio experiences in our domestic environments, making them part of our daily routine, away from the communal and ritualized settings of the concert hall, church, or theatre. Music, particularly, became to a large extent a private experience, a direct engagement with organized sounds at a time and place of our choice. Just as the printing press transferred literature from the public realm—the reading of scriptures in church or the communal performance of mystery plays—to the private domain of the individual reader, recording had the dual effect of democratizing access to a huge number of musical performances, and turning the consumption of those performances into a largely private affair, under the control of the individual listener. This new domestic experience of music is inherently non-linear. As Jonathan Kramer observed:

> Recording has not only brought distant and ancient musics into the here and now, it has also made the home and the car environments just as viable for music listening as the concert hall. The removal of music from the ritualized behavior that surrounds concertgoing struck a blow to the internal ordering of the listening experience. Furthermore, radio, records, and, more recently, tapes allow the listener to enter and exit a composition at will. (Kramer 1981, 531)

Around the time these words were written, Sony released the first personal stereos, in the form of the Walkman, further narrowing the focus of musical listening to the individual and in particular to the internalized experience of headphone listening. This headphone use could be viewed as isolating the listener from the environment, however as Chambers observes, the relationship between Walkman users and their surroundings is more complex: "the Walkman offers the possibility of a micro-narrative, a customized story and soundtrack, not merely a space but a place, a site of dwelling. Our listening acts as an escape from our lived environment while also intersecting with this environment forming an accidental soundtrack to our real lives" (Chambers 2004, 100). The recording, in this interpretation, interacts with and even augments the experience of the outside world. Musical experiences are situated within the spaces we inhabit and we may come to associate certain pieces of music with particular times and places, experiencing our own library of recorded sound in direct relation to the landscape we inhabit.

In this scenario, the relationship between the musical artwork and our physical surroundings is still somewhat arbitrary: we may choose a particular piece of music to accompany a particular landscape or activity, but beyond this the relationships that arise at each moment are determined by chance and happy coincidence. However, technology has once again moved on and we can now design audio experiences that know, and respond directly to, the listener's location. This opens a fertile space for the sonic artist to explore, in which the individual experience of the acoustically augmented environment can become meaningful.

The ubiquity of powerful mobile computing devices, in the form of smartphones that combine the ability to store and play back significant amounts of data in response to a range of inputs, from accelerometers, video, compass data, and global positioning system (GPS), presents a vast range of possibilities to artists and experience designers, to directly engage with the situated and nonlinear nature of recorded sound. The listener's movement though the listening environment may be monitored and used to directly influence what is heard. In particular, sound can be triggered or manipulated in response to the listener's location as reported by sensors such as a smartphone's GPS chip or compass, allowing the specific location or direction of travel to influence the temporal flow of sounds, whether they be prerecorded or synthesized in real time. This is the domain of locative media.

In this chapter, I investigate the unique potential of locative media to address fundamental issues in sonic interaction. I show how the use of physical movement in space as an interface may allow users to engage directly with the underlying spatial metaphors of interaction design and musical structure. I draw on an earlier wave of experimentation with nonlinear structures in the musical avant-garde of the 1950s to elucidate the relationship between spatially conceived compositional structures and the emergent temporal forms experienced by the user, and introduce the notion of a *translocational* approach to locative media, in which portable, non-site-specific applications allow users to explore the intrinsic structural relationships of the work through direct engagement with a location of their own choice.

2.1 Site-specific and Translocational Media

On the whole, the most prominent use of location data for interaction has taken an absolute or site-specific form. In mobile applications designed to find a nearby restaurant or bank, or give an accurate weather report, the ability to tie GPS data to maps and "points of interest" databases is a fundamental feature of this mode of interaction. More creative applications of technology such as geo-caching games and interactive audioguides take a similar approach, drawing on the machine's apparent awareness of specific features of the environment to add a layer of realism or topicality to the virtual experience. *Augmented reality* applications map the virtual space onto the real with the desire to "break the frame" and create artworks that engage directly with the specifics of the real world. The notion of an *augmented aurality*, a term used by the website http://www.notours.org to describe its dramatic locative audio tours, highlights the unique potential of locative technologies to extend and enhance our experience of place through listening. This augmentation has been one of the attractions of mobile technologies for performance companies such as Blast Theory (Benford et al. 2006) or the Danish Company Katapult (Hansen, Kortbek, and Grønbæk 2012) whose Mobile Urban Drama project integrates location-specific audio clips on cellphones with live actors. One obvious drawback to this approach is that audiences are limited by their access to the location for which an application has been created or are even expected to attend at particular performance times, so pieces such as these tend to remain tied to traditional performance models dominated by a focus on the public event. This may be viewed as being at odds with the way mobile and ubiquitous media are increasingly used to extend and enhance day-to-day activities, and the disjuncture between a public event and a personalized experience is often apparent (see also Chapter 16 in this volume).

An alternative approach to location-based media experiences, which I shall term *translocational*, can be experienced privately by individual users in their own time and in any location. Currently, more familiar in the world of GPS games such as *MinuteWar* or Locomatrix' *Fruit Farmer* than in art-orientated projects, translocational media experiences use location-sensing technologies such as GPS to build virtual spaces that users explore without explicit reference to the actual environment in which they are situated. Instead, the virtual space, while still overlaid onto the environment, is self-contained and constructed by the software, relative to the user's starting location. Locative music experiences such as Ben Mawson's Android-based *Take Me by the Hand* or Strijbos and Van Rijswijk's *Walk with Me* also fall into this category because the authors will build a version for any proposed location and the materials do not directly reference the particular location in which the work is experienced. *On a Theme of Hermes* by Satsymph (http://www.satsymph.co.uk) allows users to define an area by walking around its boundaries if they are not in one of the specified locations. In such an audio piece users may, for example, move between different sound environments, cross auditory boundaries, or approach sounds originally heard in the distance. Importantly, they may also return to places already visited to find sounds that may be identical with, or may have evolved from, those experienced in that location before.

It may at first seem counterproductive to exclude one of the unique features of locative technology, the awareness of absolute location, in this way, seemingly reducing the GPS device to a mere motion sensor. However, it is important to emphasize that the translocational approach does not render the physical listening environment irrelevant simply because it is not directly addressed in the content. On the contrary, the experienced environment, whatever it may be, acts as a frame of reference, a ground on which the experience takes place. An awareness of the environment is vital to the experience, allowing the listeners to build a structural map, a cognitive representation of the *virtual* space, through reference to the features of the *actual* space around them. I will argue that this ability to map the structures that underpin the layout of the virtual space adds significant value to the experience of interaction, allowing us to intuitively interrogate the artwork through our physical exploration of space and our natural ability to associate ideas with locations. In place of the *extrinsic* associations of site-specific media, locations become *intrinsically* associated with particular audio events, such as particular musical motives, timbral combinations, or perhaps characters in a story, wherever they were first encountered (for example, "this is where I heard the gardener"). This kind of exploration draws on our fundamental aptitude for spatial organization of memories. As Meredith Gattis points out, "The experience of re-visiting a place demonstrates that sometimes space can be a more powerful organizer of memory than time" (Gattis 2001, 3). This insight lies behind the use of spatial mnemonics, which dates back to Aristotle: "Remembering really depends on the potential existence of the stimulating cause... For this reason some use places for the purposes of recollecting" (*De Oratore*, trans. E. W. Sutton, quoted in Yates 1992, 48). The use of spatial mnemonics developed through the Middle Ages and Renaissance, culminating in the construction of complex conceptual memory theaters or memory palaces, such as those of Camillo and Bruno (Yates 1992). These spatial mnemonics allowed practitioners of the art of memory to organize complex ideas into logical structures by linking them with familiar or learnt spatial locations.

Similarly, the use of location sensing in translocational audio experiences allows listeners to associate sonic events with actual locations, so that potentially they can return to them to find familiar sounds or to discover new material that may have been developed from, or have associations with, material that was heard in that location before. The act of returning can reinforce the listeners' understanding of the virtual space, allowing them to construct mental maps, associating sounds with the environmental cues around them. In this way, translocational media offer unique one-time experiences based on the exploration of an underlying spatial structure that may be revealed through intuitive interaction. Translocational media can therefore be seen to address the fundamental disjuncture between nonlinear (spatial) structures created by composers, authors, or designers, and the inherently linear nature of the user experience, in which the results of interactions cannot but unfold as a neatly ordered sequence of events in time.

In the following paragraphs I investigate this relationship between nonlinear structure and linear form, drawing on ideas that emerged in the musical avant-garde of the 1950s at least partially in response to the historical shifts brought about by the recording technologies discussed above. These ideas may be interpreted as a shift from a linear to a nonlinear or spatial conception of musical structure. Experiments in open form composition, arising from the conceptual separation of musical structures and the emergent forms governed by them, offer an insight into the design of interactive audio applications. The emphasis on spatial conceptions of structure among the composers involved in these developments can be seen to echo similar models in interaction design, which in turn derive from our fundamental understanding of structural relationships in spatial terms. In this context, locative and translocational modes of interaction are of particular interest as they may reveal and elucidate the underlying spatial metaphors of sonic interaction design, drawing on users' embodied understanding of spatial relationships and bodily navigation.

2.2 Structure and Form

In a number of writings from the 1940s and 50s, John Cage proposed a fourfold division of music (Prichett 1993, 38) into: Structure, Method, Material, and Form (Cage 1978). His clearest definitions of these categories, which as Jenkins (2002) has pointed out were not static but developed considerably during the course of his writings, are found in the 1949 essay "Forerunners of Modern Music," in which he states: "Structure in music is its divisibility into successive parts from phrases to long sections. Form is content, the continuity. Method is the means of controlling the continuity from note to note. The Material of music is sound and silence" (Cage 1978, 62). The separation of the concepts of structure and form may come as a surprise to readers used to classical notions of sonata form, rondo form, and so on, in which form and structure are essentially interchangeable concepts. For Cage, structure and form are not only separate categories but have quite different points of origin in the process of composition. Structure is "mind-controlled," while form "wants only freedom to be" (Cage 1978, 62; cf. Boulez and Cage 1993, 39) and concerns the "morphological line of the sound continuity" (Kostelanetz 1971, 79). This distinction reflects Cage's method of composition at the time, in which a priori structures based on proportional subdivisions of time were essentially "filled" with material (sounds and silences) according to various methods, gradually developing from his Gamut technique, through the use of charts to his eventual use of the I Ching (Pritchett 1993). In other words, the differentiation between form and structure reflects a growing sense that in Cage's music, "structure and sound material can be composed separately" (van Emmerik 2002, 234).

As his compositional style developed, Cage's underlying structures became increasingly abstracted from their rhythmic roots to become conceived of as atemporal frameworks, necessary for the production of the work but not necessarily perceived in the emergent musical form. Cage initially defined structure in temporal terms, famously insisting, in the lecture "Defense of Satie" (Kostelanetz 1971, 77–84), on the importance of duration as the only valid basis for musical structure. However, by the time he came to compose *Music of Changes* (1951), the temporal basis of his "rhythmic" structure had become purely theoretical as he began to employ chance procedures at each structural subdivision, not only to determine pitches and durations of individual notes, but also to decide at what tempo the next unit of structure was to be played. In other words, the carefully proportioned "temporal" subdivisions would be executed in varying lengths of time, determined by chance, making the carefully planned proportions impossible to perceive in the sounding result. Cage states: "My recent work... is structurally similar to my earlier work: based on a number of measures having a square root, so that the large lengths have the same relation within the whole that the small lengths have within a unit of it. Formerly, however, these lengths were time-lengths, whereas in the recent work the lengths exist only in space, the speed of travel through this space being unpredictable" (Cage 1978, 57).

Cage's reimagining of structure as atemporal—indeed as essentially spatial—seems to have a liberating effect. In a letter to Boulez, Cage wrote, "The rhythmic structure is now magnificent because it allows for different tempi: accellerandos, ritards etc." (Boulez and Cage 1993, 95) and the importance of precompositional decision making is common to both composers. Van Emmerik draws parallels between Cage's a priori conception of structure and the strategies of Boulez's total serialism, both in their historical context and their methodology, noting that "Cage's notion lends composition using rhythmic structures a highly abstract nature, and frequently results in certain discrepancies between the musical continuity as it was composed and as it is perceived" (Van Emmerik 2002, 234). This a priori notion of structure may help to clarify the difficult distinction that Cage makes between structure and form. *Structure* in this interpretation has become atemporal, an essentially spatial framework underpinning the musical *form* which is inherently temporal, pertaining to the continuity of sounds as perceived by the listener. As his compositional style developed, Cage increasingly obscured the underlying structure by various means (eventually by adopting the random choice of tempo at structural boundaries mentioned above). In terms of the fourfold division, his choice of methods and materials could determine whether the underlying structure of the work was hidden from the listener or whether it was elucidated or revealed by the perceived form. It is clear that the extent to which form and structure elucidate each other is a compositional choice.

At one extreme, we may celebrate the disjuncture between structure and emergent form, reveling in the way complexity arises out of an ordered system, as Henry Flynt states: "The audience receives an experience which simply sounds like chaos but in fact what they are hearing is not chaos but a hidden structure which is so hidden that it cannot be reconstructed from the performed sound" (in Piekut 2011, 76). On the other hand we may join with Steve Reich in calling for complete clarity, as he does in his seminal essay "Music as a Gradual Process": "John Cage has used processes and has certainly accepted their results, but the processes he used were compositional ones that could not be heard when the piece was performed... What I am interested in is a compositional process and a sounding music that are one and the same thing" (Reich 2004, 305).

"Process" in Reich's conception perhaps comes closer to Cage's notion of "method" than his original definition of "structure" as the subdivision of the whole into parts. However the very essence of Reich's conception of process seems to be to integrate

Cage's categories of structure, method, form, and even material into a single perceived unity. His assertion that "Material may suggest what sort of process it should be run through . . . , and processes may suggest what sort of material should be run through them" (Reich 2004, 305) stands in contrast to Cage's assertion in 1958, in relation to the *Sonatas and Interludes*, that "nothing about the structure was determined by the materials which were to occur in it" (Kostelanetz 1971, 19). Reich recognizes that, in Cage's terms, the extent to which the perceived form reveals the underlying structure is determined by the method by which the structure is articulated, and the materials that embody this articulation. As Pritchett observes, "Form is actually the result of method acting on materials" (1993, 39). I would go a step further and suggest that form results from method acting on materials in a given structure. Even when it is impossible to perceive the structural subdivisions in the sounding flow of the music, as in the *Music of Changes* (1951), the structure still determines important aspects of form, such as changes in the density of material over time. In this sense, structure is at least partly generative, governing aspects of the perceived form. The presentation of Cage's categories in diagrammatic form running from left to right: Structure—Method—Material—Form, as it appears in Boulez and Cage (1993, 39) (with the left of the diagram labeled "Mind" and the right labeled "Heart") reveals a set of causal connections in which structure ultimately determines form through the application of methods to materials.

What is emerging is a vision of structure and form in which a "consciously controlled," frequently spatially conceived, structure has the potential to produce a variety of forms that are "unconsciously allowed to be." In Cage's work, this division leads directly to the open forms of his indeterminate works of the later 1950s. In these works each performance manifests an entirely different musical form from a single structure laid down by the composer.

This situation is of course familiar to the interaction designer. The nature of interactive sound work in particular is such that many possible temporal realizations may be generated from one underlying structure according to the actions of the user, just like in Cage's early indeterminate works, fixed structures may produce numerous temporal forms. However, each individual realization of the interactive artwork uniquely unfolds in linear time and is experienced as a simple succession of events, just as a traditional piece of music might be. The linearity of the resulting experience could lead to the criticism that the interaction is rendered meaningless unless the piece is experienced repeatedly (a criticism often leveled against open-form compositions, see below). However, as I have argued above, it is at least theoretically possible to create a work in which the temporal unfolding of the experienced form may reveal aspects of the underlying structure to the listener even on one hearing. The extent to which structure is revealed by the experienced form is a compositional or design choice, dependent on the utilization of particular methods and materials by the author of the experience. That is to say, the sounds themselves and the way they are mapped on to the structure have a direct bearing on the extent to which the structure is revealed by emergent forms. The materials may be sound files or notes and sounds generated by synthesis engines of various types, the methods may simply be triggering and cross-fading files or may extend to sophisticated generative techniques.

We might be tempted to equate structure with the program code which, when executed, produces a variety of results. This interpretation is in fact quite a good fit with Cage's indeterminate works in which the score often consists of graphical tools and a set of instructions used to create a performance. However, it is perhaps more useful to think of structure as an inherently spatial representation, a conceptual metaphor that imparts order to the materials based on our innate ability to navigate and understand our real-world environment. The importance of spatial reasoning in understanding conceptual structures has been recognized in interaction design and data representation. The desktop metaphor, which has become the primary means of structuring home and office computing applications, is a familiar example. Hypermedia too are largely conceived of spatially, as David Saltz points out:

> Notice that spatial metaphors govern the rhetoric of hypermedia: people move along paths from link to link, traveling through cyberspace. Rather than functioning either as performers or as authors, hypermedia audiences function as explorers. They are like tourists, rushing through the areas that do not interest them, lingering when they find something that strikes their fancy, meandering down an intriguing alley way, perhaps getting lost for a while before finding their way back to a familiar landmark. All the while, the interactors keep their eyes on the road. Their object of attention is the work, not themselves in the work. (Saltz 1997, 118)

Many Hypermedia structures, of course, do not entirely obey the laws of two-dimensional Euclidian space and this can be advantageous. Penny discusses how: "the interfaces to virtual worlds are seldom mapped one to one. Generally a small movement in the real world produces a large movement in the virtual world," suggesting that, "This tendency replicates the paradigm of the labour saving machine" (Penny 1996). The Internet is another good example, where hyperlinks allow a user to jump between non-adjacent pages that might be hard to reach through a linear route of successive "next" buttons. However, the underlying spatial metaphor is clearly reflected in the language we use to describe the experience: We "surf" the net, using "forward" and "back" buttons to "navigate" pages.

This conception of a metaphorical spatial structure underpinning the interactive artwork is of particular interest in the sphere of locative and translocational audio, since the physical means of interaction with locative media draw on precisely those faculties of spatial awareness and navigation that underpin the structural metaphors used in the interaction design. The direct engagement with our physical environment demanded of the user is potentially a highly intuitive means of interaction with media content. For this to be the case, the underlying structure must remain largely consistent with our expectations of the behavior of ordinary space and should be mapped onto physical space in a coherent manner.

2.3 Embodied Conceptual Metaphors for Time and Music

That an essentially spatial concept of structure should emerge, both in the field of musical composition and in interaction design should come as no surprise. Indeed, it is hard to conceive of any kind of "structure" without recourse to spatial reasoning. In *Philosophy in the Flesh*, Lakoff and Johnson suggest a reason for this: "Reason is not disembodied, as the tradition has largely held, but arises from the nature of our brains, bodies, and bodily experience. This is not just the innocuous and obvious claim that we need a body to reason; rather, it is the striking claim that the very structure of reason itself comes from the details of our embodiment. The same neural and cognitive mechanisms that allow us to perceive and move around also create our conceptual system and modes of reason" (Lakoff and Johnson 1999, 4). The use of spatial reasoning to understand temporal concepts is particularly prevalent: "Every day we take part in 'motion-situations'—that is, we move relative to others and others move relative to us. We automatically correlate that motion... with those events that provide us with our sense of time" (151). Of particular interest are two fundamental metaphors for time, which Dedre Gentner (Gentner 2003, 203) identifies as the *ego-moving* metaphor and the *time-moving* metaphor. The former, in which time is considered stationary and the observer moves through it, is characterized by such statements as "I am going to do that" or "We are fast approaching the holidays." The latter, in which time moves past a stationary observer is reflected in the expressions such as "the years to come" or "night follows day" (204). Lakoff and Johnson characterize these as "The Moving Time Metaphor" and the "Time's Landscape Metaphor" (Lakoff and Johnson 1999, 141, 145)

As Johnson and Larson (2003) have pointed out, our conception of music as events in time is also structured by these two fundamentally opposed perspectives. Their position is summarized by Spitzer: "Given that we typically conceptualize time either as 'motion through space' ('The Moving Times Metaphor') or as a 'landscape' through which we ourselves move ('The Time's Landscape Metaphor'), we can imagine music either as moving past us or as a structure we navigate (audiences prefer the former, letting the piece flow past; analysts choose the latter, moving 'through' or 'across' a score)" (Spitzer 2004, 63). In the first case, when the music is perceived as moving past the stationary listener, the musical material itself appears to have agency, the developing phrases of the composition being the subject of development and transformation. The moving music metaphor relates strongly to the idea of musical narrative that characterizes musical thought throughout the eighteenth and nineteenth centuries. Grounded in the operatic form, this conception encourages the listener to identify with fictive musical characters in an unfolding drama. As Christopher Small suggests, "A work in the western concert tradition is a pattern of sounds that is always performed in the same combinations... Those sound combinations are metaphorically invested with meaning through the operation of a semiology of sound relationships that has been developed over the past four centuries or so, and the way in which they are put together tells a story that presents us with certain paradigms and models of human relationships" (Small 1998, 187–8). As listeners, we identify with an imaginary musical character immersed in the relentless flow of the music, locked into its fate. We observe the music flowing past and in a particularly convincing performance may even get "swept along" by it. This narrative conception of music is in turn intimately tied to the development of the tonal system, in which hierarchies of cadential patterns drive the music forward in waves of tension and resolution.

Adopting the music as a landscape metaphor, on the other hand, it is possible to conceive of music as stationary, a landscape to be explored by a moving listener, able to make reference to musical landmarks and memories of places encountered along the way. This perspective emerged strongly in the writings of composers seeking to find new musical languages that rejected the teleological structures of tonal music. Iannis Xenakis, for example "felt that by almost exclusively emphasizing music's forward direction in the temporal sphere European musicians had enervated music by too little attention to static, non-temporal aspects of musical architecture" (Gann 1996, 153).

We may observe this shift in perspective particularly clearly in composers allied to serialism who start to discuss their works in decidedly topographical terms. Pierre Boulez uses moving-listener metaphors to express the nonlinear nature of serial music: "I want the musical work not to be that series of compartments which one must inevitably visit one after the other; I try to think of it as a domain in which, in some manner, one can choose one's own direction" (Boulez 1968, 26). It should come as no surprise, then, that many serial composers started to investigate open-form composition in which the score, rather than specifying the performance unambiguously, offers elements of choice to the performer, leading to alternative readings in each performance. M. J. Grant (2005) suggests that open-form conceptions of music arise as a direct consequence of serial thought, and indeed the exploration of modular and reconfigurable approaches to musical composition seems to emerge naturally from a notion of music in which the listener is an active participant, an explorer of a musical landscape, rather than the stationary observer of a musical journey undertaken by an unacknowledged protagonist embodied in the musical material itself. She draws on Pousseur's description of music as a "field of relations," itself a spatial metaphor, stating that

> Serial music is not linear, that is, there is not a logical process of events, rather a field of relations. But neither is it an undifferentiated field—it is not white noise. The important point is the statistical nature of this process, the tendency against the foreseeability of events. It is in this sense that the "contradiction" of serial and open form is invalid: serial form *per se* is open form, and I would go so far as to say that in this sense it is only a more extreme situation than in much new music—openness not necessarily from the standpoint of production, but perception: the openness of perceived form (Grant 2005, 158–9).

Thus, for Grant, the idea of the open-form work, a single score with many possible realizations in which, in the words of Eco, "every performance *explains* the composition, but does not *exhaust* it" (Eco 2004, 171), is inherent in the serial aesthetic. I would suggest that the emergence of the open-work concept among serial composers is directly linked to the shift in perspective from a "static listener, moving music" metaphor to a "moving listener, static music" metaphor. This conception encourages exploration as a primary mode of listener behavior and raises the possibility of alternative paths through a two-dimensional musical landscape, as Boulez describes: "I have often compared a work with the street map of a town: you don't change the map, you perceive the town as it is, but there are different ways of going through it, different ways of visiting it" (Boulez and Deliège 1976, 82). Boulez is eager to emphasize that the multitude of realizations of such a work in no way diminishes either the integrity of the work or the role of the composer as author of the experience (taking pains to distance himself from the chance procedures of Cage): "I have often heard it said that the introduction of free elements in music is an abdication on the part of the composer. I believe, however that the introduction of a dimension of freedom rather entails an increase in the composer's powers, since it is far more difficult to build a town than to build a street: A street leads from one point to another, whereas a town has lots of streets and presents many different directions for building" (Boulez and Deliège 1976, 85).

It is easy to see why, as this multidimensional spatial conception of musical structure took hold, many composers started to concern themselves with space as a compositional parameter. Brant (1998), Stockhausen (1964, 105), and Berio (2000, 154), to name but a few of the most prominent exponents, began to see physical space as integral to the music's structure, a means of clarifying complex polyphonies (Brant) or elucidating the interchange of material between discrete timbral groups (Stockhausen, Berio). Clearly growing out of these concerns, and of particular relevance to the argument presented here, the 1960s saw the first spatial sound installations, in which the geographical layout of sounds replaced their temporal succession as a structuring principle.

Max Neuhaus's *Drive-in Music* (1967), which has been cited as the first real sound installation (Tittel 2009, 57) clearly demonstrates the division between a spatial structure defined by the composer and its resultant temporal forms as experienced by the listener. The piece consists of a number of sine-tone mixtures, each broadcast from an individual short-range transmitter along the side of a road, so that listeners driving along the road enter the broadcast range of each transmitter in succession. The alternation of sounds heard therefore depends on the speed of the car (tempo) and its direction (order of succession). La Monte Young's *Sine Wave* installations also deserve consideration here. These installations consist of carefully tuned sine-wave chords, which interact with the natural acoustics of the space they are sited in, as Gann describes: "Because each pitch has a different Wavelength, each is reinforced at some points in a room by bouncing back on itself in phase and canceled out at other points in the same room where the bounce-back is 180 degrees out of phase... Thus every point in the space has its own pattern of reinforced and canceled frequencies" (Gann 1996, 188). In both of these examples a spatial structure is revealed as temporal form through the listener's actual movement in space. The listener is cast in the role of an explorer, discovering musical material and decoding the structure of the composition (Neuhaus) or the space it inhabits (Young).

2.4 Conclusions

The open works of the 1950s and 60s are presented to a listening audience by performers who, through their choices in performance, may be considered to be interacting with and completing the work. The fact that the listener is ignorant of the alternatives and presented with only one (or occasionally two) realizations of the work in a given concert, has been cited by Nattiez as highly problematic. In his view, the linear temporal character of the work's reception renders the alternative versions obsolete from the listener's perspective: "If in order to be understood, the *poietic* phenomenon of 'openness' must be explained before or during a performance then this 'openness' is not perceptible on the *esthesic* level" (Nattiez 1987, 86).

The sound installations that emerged in the sixties, and the interactive works of the present age, which allow the listener to replace the performer, directly engaging in the interactions that produce the experienced temporal forms of the work, can be thought of as addressing this criticism by handing the element of choice directly to the listener. However, a single experience is still essentially linear, a one-time revelation of one possible solution (form) to the structural puzzle created by the artist. As such, the alternative possibilities may yet remain obscure and the act of interacting may often be considered a mere gimmick, giving users an illusion of control without any real understanding of the consequences of their actions.

I have argued that the extent to which the experienced form of an interactive experience reveals the underlying structure is a matter of compositional choice. Locative and translocational media can allow users to directly interrogate the underlying, spatially conceived structures of sonic interaction design through exploration of their topography, mapped onto a real environment, and may, if well designed, greatly increase the clarity of the interactive experience. By memorizing the locations of particular sonic events in a locative audio work, we may return to test out our cognitive maps, gaining an insight into the architecture of the work. This practice in turn allows us to further categorize and conceptualize the sound we are hearing in relation to the structure we discover. An understanding of the spatial metaphors of the underlying structure may elucidate the emergent form, just as the form may elucidate the structure. Furthermore, the use of physical movement in real space to gain access to the spatial metaphors of the interaction design draws on the embodied nature of structural understanding, allowing users to intuitively navigate the architecture of the work. The understanding gained in this way may even give users of locative and translocational media the possibility of projecting alternative possible realizations onto their one-time experience, revealing the "openness" of the interactive artwork in a single interaction.

Acknowledgments

The ideas in this paper are informed by the Locating Drama project undertaken by Parry, Bendon, Boyd Davis, and Moar at the Lansdown Centre for Electronic Arts at Middlesex University in collaboration with the BBC in 2007 (Parry et al. 2008), as well as the author's translocational iPhone composition *Triptych*.

References

Benford, Steve, Andy Crabtree, Martin Flintham, Adam Drozd, Rob Anastasi, Mark Paxton, Nick Tandavanitj, Matt Adams, and Ju Row-Farr. 2006. Can You See Me Now? *ACM Transactions on Computer-Human Interaction* 13 (1): 100–133.

Berio, Luciano. 2000. *Luciano Berio: Two Interviews*. London: Marion Boyars.

Boulez, Pierre. 1968. *Notes of an Apprenticeship*. New York: A. A. Knopf.

Boulez, Pierre, and John Cage. 1993. *The Boulez–Cage Correspondence*, ed. Jean-Jacques Nattiez. Cambridge: Cambridge University Press.

Boulez, Pierre, and Célestin Deliège. 1976. *Pierre Boulez: Conversations with Célestin Deliege*. Translated by B. Hopkins. London: Eulenburg.

Brant, Henry. 1998. Space as an Essential Aspect of Musical Composition. In *Contemporary Composers on Contemporary Music*, ed. Elliott Schwartz and Barney Childs, 221–242. Cambridge, MA: Da Capo.

Cage, John. 1978. *Silence: Lectures and Writings*. London: Marion Boyars.

Chambers, Iain. 2004. The Aural Walk. In *Audio Culture: Readings in Modern Music*, ed. Christoph Cox and Daniel Warner, 98–102. New York: Continuum.

Eco, Umberto. 2004. The Poetics of the Open Work. In *Audio Culture: Readings in Modern Music*, ed. Christoph Cox and Daniel Warner, 167–175. New York: Continuum.

van Emmerik, Paul. 2002. An Imaginary Grid: Rhythmic Structure in Cage's Music up to Circa 1950. In *John Cage: Music, Philosophy, and Intention, 1933–1950*, ed. David W. Patterson, 217–238. New York: Routledge.

Gann, Kyle. 1996. The Outer Edge of Consonance. In *Sound and Light: La Monte Young and Marion Zazeela*, ed. Richard Fleming, William Duckworth, and Richard Fleming, 153–194. Lewisburg, PA: Bucknell University Press.

Gattis, Merideth. 2001. Space as a Basis for Abstract Thought. In *Spatial Schemas and Abstract Thought*, ed. Merideth Gattis, 1–12. Cambridge, MA: MIT Press.

Gentner, Dedre. 2003. Spatial Metaphors in Temporal Reasoning. In *Spatial Schemas and Abstract Thought*, ed. Merideth Gattis, 203–222. Cambridge, MA: MIT Press.

Grant, M. J. 2005. *Serial Music, Serial Aesthetics: Compositional Theory in Post-war Europe*. Cambridge, UK: Cambridge University Press.

Hansen, Frank, Karen Allan, Johanne Kortbek, and Kaj Grønbæk. 2012. Mobile Urban Drama: Interactive Storytelling in Real World Environments. *New Review of Hypermedia and Multimedia* 18 (1–2): 63–89.

Jenkins, Chadwick. 2002. Structure vs. Form in *The Sonatas and Interludes for Prepared Piano*. In *John Cage: Music, Philosophy, and Intention, 1933–1950*, ed. David Patterson, 239–262. New York: Routledge.

Johnson, Mark, and Steve Larson. 2003. "Something in the Way She Moves": Metaphors of Musical Motion. *Metaphor and Symbol* 18 (2): 63–84.

Kostelanetz, Richard, ed. 1971. *John Cage*. London: Allen Lane.

Kramer, Jonathan. 1981. New Temporalities in Music. *Critical Inquiry* 7 (3): 539–556.

Lakoff, George, and Mark Johnson. 1999. *Philosophy in the Flesh: The Embodied Mind and Its Challenge to Western Thought*. New York: Basic Books.

Mawson, Ben. 2012. *Take Me by the Hand*. http://www.benmawson.com/music/TMBTH.htm.

Nattiez, Jean-Jacques. 1987. *Music and Discourse: Toward a Semiology of Music*. Translated by Carolyn Abbate. Princeton, NJ: Princeton University Press.

Parry, Nye, Helen Bendon, Stephen Boyd Davis, and Magnus Moar. "Locating Drama: A Demonstration of Location-Aware Audio Drama." In *Interactive Storytelling*, edited by Ulrike Spierling and Nicolas Szilas, 41–43. Lecture Notes in Computer Science 5334. Springer Berlin Heidelberg, 2008. http://link.springer.com/chapter/10.1007/978-3-540-89454-4_6.

Penny, Simon. 1996. From A to D and Back Again: The Emerging Aesthetics of Interactive Art. *Leonardo Electronic Almanac*. http://sophia.smith.edu/course/csc106/readings/Penny_interaction.pdf.

Piekut, Benjamin. 2011. *Experimentalism Otherwise: The New York Avant-Garde and its Limits*. Berkeley: University of California Press.

Pritchett, James. 1993. *The Music of John Cage*. Cambridge, UK: Cambridge University Press.

Reich, Steve. 2004. Music as a Gradual Process. In *Audio Culture: Readings in Modern Music*, ed. Christoph Cox and Daniel Warner, 304–306. New York: Continuum.

Saltz, David Z. 1997. The Art of Interaction: Interactivity, Performativity, and Computers. *Journal of Aesthetics and Art Criticism* 55 (2): 117–127.

Small, Christopher. 1998. *Musicking: The Meanings of Performing and Listening*. Middleton, CT: Wesleyan University Press.

Spitzer, Michael. 2004. *Metaphor and Musical Thought*. Chicago: University of Chicago Press.

Stockhausen, Karlheinz. 1964. *Texte 2: Aufsätze 1952–1962 Zur Musikalischen Praxis* ed. Dieter Schnebel. Cologne: Verlag M. DuMont Schauberg.

Strijbos and Van Rijswijk. 2011. *Walk with Me*. http://itunes.apple.com/us/app/walk-with-me/id461519712.

Tittel, Claudia. 2009. Sound Art as Sonification, and the Artistic Treatment of Features in Our Surroundings. *Organised Sound* 14 (1): 57–64.

Yates, Frances A. 1992. *The Art of Memory*. London: Pimlico.

CHAPTER 3

DEFINING SOUND TOYS

Play as Composition

ANDREW DOLPHIN

In this chapter, sound toys are examined and discussed as a playful medium for composition as they offer access to music composition and sound creation. Sound toys can be considered as interactive, sonic-centric systems in which the end user may trigger, generate, modify, or transform sound. Playful approaches to composition offered by sound toys provide a novice user access to composition through symbolic representation of often complex underlying systems. The visual domain becomes a dynamic and artful animated user interface for player exploration of sound and/or music.

Sound toys could be considered as compositional systems that allow players access to parameters of composition, various types of musical experiences, and sound worlds. "Sound toys" is considered an appropriate term to describe many playful, accessible, and exploratory sonic-centric audiovisual interactive composition systems and software applications.[1] The term "toy" suggests playful interactions, whether these are ludic or exploratory, and implies a level of accessibility for the end user(s) or player(s). Sound toys may be designed as open-form compositions, compositional tools, or instruments that may be influenced by a number of fields relating to electroacoustic and electronic music, sound art, and contemporary computer music, also exhibiting interdisciplinary approaches relevant to many other artistic and technological fields. Whether sound toys can be considered to be instruments, compositions, or tools for composition depends upon the nature of the sound toy, the level of control offered to the player, and the type or styles of player interactions with the computing system.

Sound toys offer options and choice for the player, providing scope for varied interactions and sonic output. The range of interaction approaches is also potentially broad and may range from a more linear approach with some degree of openness to a system which offers a multitude of possible pathways, providing varied and more extensive possible sonic experiences. Where more options are offered, these types of sound toys become less predictable in terms of sonic outcome, with many diverging branches of possible outcome. This approach can be aligned with the notion of a "field of possibilities" (Eco 1959, 170).[2] The player may be offered a significant range of potential possible experiences that may be quite diverse, yet these may still exist within a specific prepared framework for interaction that provides a designed (or composed) play space for sonic exploration and discovery, with Toshio Iwai's *Electroplankton* (2005) being a relevant example.

When considering the realm of sound toys, issues of definition and classification arise: what terminology is most appropriate to describe or categorize these works in a way that effectively communicates their creative interests? The term *computer game* is somewhat misrepresentative of many software sound toys' creative concerns, as this term is loaded with social expectations of what constitutes a computer game and it could potentially become a barrier to an audience understanding a sound toy's themes and intended interactions. Many sound toys avoid an intentionally competitive framework. There are frequently no defined characters, no winners, no violence, and many cannot be completed as such. Other common computer game characteristics, such as rigid rules, specific objectives, and resulting rewards when objectives are achieved, competition and scoring, need not be incorporated, with the player instead offered exploratory audiovisual experiences that are primarily concerned with sound. It is therefore suggested that the term best suited to these works is *sound toys*. This term is used as it conveys that the works are predominantly sound-centric, and the term *toy* implies an intended playful experience for the user, with further implications of casual or recreational experience. Sound toys may be explored for seconds or hours at the will of the user or player, and often cannot be completed. There are many examples of computer applications that could be termed sound toys, or sound-centric applications available for mobile devices. Relevant examples include: RjDj (2008), Bloom (2008), Biophilia (2012), Aura Flux (2010), Sonic Wire Sculptor (2010), Soundrop (2010), SoundyThingie (2010), Daisyphone (2009).

Sound toys offer a playful framework for composition in which sound is organized over time. More broadly, they provide scope for developing existing modes of artistic presentation and dissemination of playful composition artifacts influenced by the fields of sound art, computer music, electroacoustic and electronic music. Sound toys frequently provide a platform for interaction, with allocation of some degree of improvised compositional control to the player, often a nonexperienced user.

In this chapter, sound toys are considered as frameworks for composition and as composition tools, and their relationships to the "open work" as defined by Umberto Eco in "The Poetics of the Open Work" (1959) are explored. Approaches in applying real-time sound and synthesis systems for composition in a sound toy context are discussed alongside proposed system models in which sound is a central feature, or a defining artistic style. While some techniques and technologies are discussed, artistic and aesthetic issues are given primary attention. Examples such as Toshio Iwai's *Electroplankton*, Brian Eno and Peter Chilvers' *Bloom*, and Björk's *Biophilia* provide a starting point for discussion, introducing ideas of accessible symbolic control of generative music parameters.

The sound toy medium offers improved access to music making, with mobile devices offering opportunities for composing and participating in composition to be more widely experienced. Computer game-related technologies and techniques offer opportunities for developing existing modes of artistic presentation and dissemination within the realms of sound art, contemporary computer music, electroacoustic and electronic music composition. Sound toy systems may be developed using tools which are perhaps less often associated with the field of composition, but their development is frequently influenced by diverse aspects of compositional practice and related techniques, processes, themes, and aesthetic concerns.

3.1 Is this Really Composition? The Open Work and Play as Composition

While originally intended for a very different musical context, the theories of Umberto Eco expressed in "The Poetics of the Open Work" are relevant to the creative pursuits of interactive non-linear sound toys in which the user is invited to exercise choice and interact. As a result, this interactivity affects or influences the music or sonic output produced, thereby making the process inherently compositional. Eco describes instrumental musical works in which the performer, or performers, may use their "judgement on the form of the piece" (Eco 1959, 167), for example deciding the length of a note, instigating the next musical phrase, changing dynamics, to influence or control the overall structure and form of the piece. This process extends beyond variations of a work based on a musician's or performer's interpretation of a score (an accepted part of Western traditional scored music) to the performers input amounting "to an act of improvised creation" (Eco 195, 167). In an open work the performer's role becomes extended to exist within the realms of composition.

Despite differences in context when applying this theory to the sound toy medium, the notion of the open work has definite resonances within many sonic-centric interactive systems, such as *Electroplankton* and *Biophilia*.[3] Eco's reference to the comments of Henri Pousseur when describing his work *Scambi* further illustrate the relevance to interactive sound toys, as they provide the user with a "field of possibilities" (Eco 1959, 170), and invite the player to exercise choice. Eco's theories were evidently originally intended for a different musical context, these theories being conceived and published at a time predating ubiquitous computer based gaming and interactive technologies, yet their significance to the field of sound toys is apparent. Eco discusses the complex interplay of motive forces and a revised vision of cause and effect that moves away from "a rigid, one-directional system: now a complex interplay of motive forces is envisaged, a configuration of possible events, a complete dynamism of structure" (170). These ideas can be related to some of the creative concerns of many sound toys, particularly when generative processes or simulated physics systems are employed. One simple event, at one specific moment in time, has the potential for complex knock-on effects on the resulting cascading field of dynamic possibilities. It is suggested that relating Eco's theories and definitions to the field of sound toys allows many of these to be appropriately described and defined as open works, or indeed open-form compositions.

3.2 Who is the Composer?

Sound toys provide the user with varied degrees of compositional input and control. Compositional input is multidimensional, with a number of different converging sources of compositional input. The importance or significance of each input as an element of composition is somewhat open to interpretation. Compositional input contributing to the final sound output, or performance of a sound toy such as Toshio Iwai's *Electroplankton*, can be attributed to three primary forces or agents, each dictating or influencing characteristics of the piece. *Electroplankton* is a notable example of game technologies being applied within an algorithmic music composition context. In this "game," the symbolic and playful representations of the algorithmic musical processes allow easy access for a novice and a reasonable level of compositional control for the player. In this example no sound parameter names are included within the visual play space. The three compositional forces are:

Composer/designer	offline
User/player	real time
Simulated physics	real time

A basic sound toy model of compositional input is presented Figure 3.1, in which different areas of composition, interaction, influence and control are represented. The composer or designer is responsible for designing and creating the framework for composition, making compositional decisions during the construction and development of the sound toy work. Modes of interaction, sound materials, transformation processes, compositional options and constrictions, and modes of presentation and representation are all dictated by the composer–designer.

The user–player engages with the system in real time, responding to both visual and aural feedback from the system. Where some form of physics engine is employed, there is frequently a codependency between the human player and the simulated physics system, which acts as a third compositional agent, adding an algorithmic or generative component to the system. Menzies' *Phya* (2008) is a relevant example of research exploring the use of physics systems in a sound-centric context. In systems in which simulated physics systems are implemented, the algorithmic component is accompanied by symbolic representations of the algorithmic processes in the virtual visual space. These visual representations provide the user/player with some insight into this aspect of the system, which is enhanced through play, exploration, and learning.

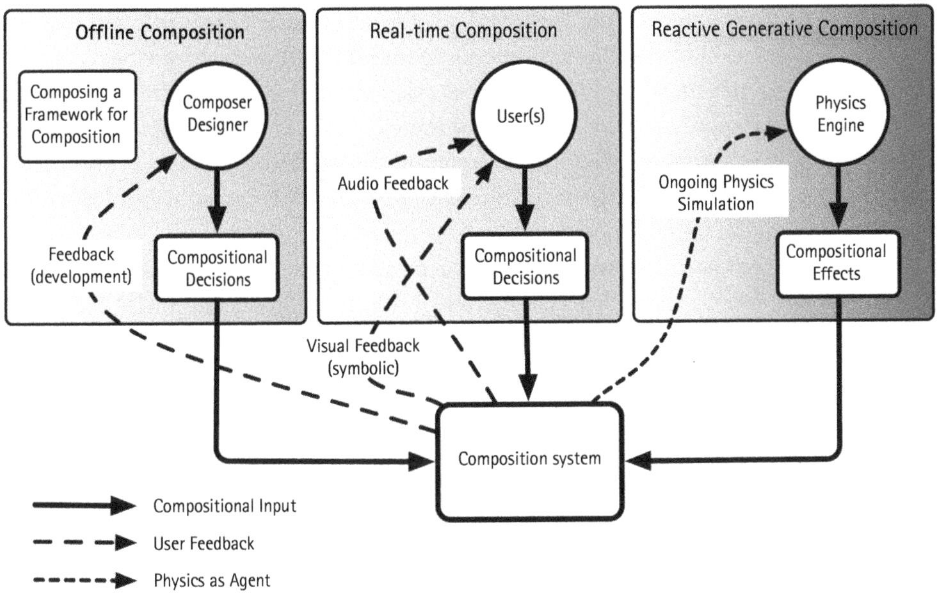

FIGURE 3.1 Designing for Composition: Three Compositional Forces.

Symbolic representation of the simulated physics system allows real-time interaction between the user/player and the system in both visual and aural domains, also allowing anticipatory responses that enable the user/player to react to forthcoming events. *Hanenbow* in *Electroplankton* is a relevant example here. These two compositional forces (simulated physics system and user/player) influence each other throughout play. The user/player interacts with virtual objects and a physics/generative system to control or influence aural and visual domains, shaping the structure of the piece within the framework prepared and "composed" by the composer/designer. The user/player is therefore not merely a passive listener but instead plays an active and significant compositional role. The context of this activity may vary, and ranges from a player's recreational activity using a software application on a personal mobile device (as in *Electroplankton*), to a visitor or participant in an interactive sonic art installation space (see the work of Julian Oliver, for instance). The term *sound toy* is therefore potentially applicable and relevant to different artistic contexts and technologies for dissemination and delivery.

Many sound toys could be considered open-form compositions, as the prepared framework often imposes musical or sonic restrictions, which could be considered a compositional act by the composer/designer. Also, no final fixed version exists, with each player having their own individual experience of the work, with the sonic output being dependent on the nature of the interactions. Eno's and Chilvers' *Bloom* (2008) is a relevant example here. Sound toys frequently use an open form that provides the user/player with scope for compositional input within a confined symbolic sonic play space. User/player input may determine form and structure on a macro level; or they may control the microstructure or microevent level of sound. In some cases this may even be on a spectral frame-by-frame basis. While some sound toys may explore aspects of contemporary computer music's and electronic music's aesthetics and sensibilities, integral features of many sound toys are also relevant to fields such as multidisciplinary art and composition, interactivity, audiovisual interfaces, and audiovisual composition incorporating transdomain mappings, algorithmic or generative composition, and real-time synthesis and digital signal processing.

A definition that best fits a sound toy will depend on the interactive approach used, but classification is to some degree a matter of opinion, often with much overlap of possible definitions. However, as their primary creative concern is the shaping and structuring of sound over time, on either a micro or macro level (or time scale), it is suggested that they can be considered as being inherently compositional, albeit within a defined compositional framework that is to some degree precomposed by the composer–designer. Allocation of compositional parameters to external real-time "agents" results in the works being fundamentally open and therefore of no fixed duration. The interaction of the user/player with often quasigenerative systems provide scope for varied and sometimes unexpected results, some of which may not have been anticipated by the composer/designer. Eno's (1975) perspective is relevant here as he states that he tends "towards the roles of planner and programmer, and then become[s] an audience to the results." Sound toys can therefore be considered as interactive or reactive systems, that may also be generative, or semigenerative, that implement a form of dialog or exchange between the player and the symbolically represented system.

3.3 External Agency for Composition

External agency as a compositional device is familiar in contemporary music composition and sound art. Whalley (2009) discusses a number of perspectives, and contextualizes artistic application of agents and agency primarily in a software-based context. The external agency could also be an object (*musique concrète*), environment (soundscape), mathematical equation (algorithmic composition), data (sonification), or an end user (interactivity). Sound toys frequently touch on a number of these categories of external compositional influence. However, it should be noted that the external agent could be considered as a component part of the composition process. Human organization, intervention, and interaction with the materials and structures is often also a fundamental component of composition, with many sound toys exploring the coexistence and interplay between external agency and human reaction, creative intention and control.

Many sound toys introduce two key external agents for composition: the end user interacting with the work, and a simulated physics or semigenerative system, potentially with both influencing the resulting sonic structures. In many sound toys, codependencies and interrelationships exist between these two agents that determine the end result, or sonic output of the work.

3.4 Interaction Approach

The interaction approach adopted in different sound toys varies greatly, and is often influenced by the type of sound toy and intended level of player interaction. There are, however, frequent key commonalities in the types of interaction approaches implemented. These can be to some degree aligned with theories outlined by Paine (2002). Of particular relevance is Paine's discussion of interactions that do "not include any pre-defined pathways" (2002, 295). This is a characteristic of sound toys such as *Luminaria* in *Electroplankton*. While it may be argued that defined pathways may result in greater musical and structural coherence, with further composed elements enforced by the composer/designer, the decision to avoid a single structured pathway through a work encourages a range of potential sonic outcomes or experiences for the player.

In his discussion of interactivity, Paine (2002) also introduces Wishart's theories of *dynamic morphology* (Wishart 1996). Wishart states that, "an object will be said to have a dynamic morphology if all, or most, of its properties are in a state of change" (Wishart 1996, 93). Paine views this idea as, "a conceptual framework for dealing with streamed data that facilitates an exploration of dynamic timbre in interactive, responsive music systems, and more broadly as a conceptual framework for the design of truly interactive systems, covering human–computer interface and sound synthesis applications" (Paine 2002, 295). Time scales relating to interaction therefore become significant.

3.5 Time Scales in Interactive Sound Toy Systems

Levels of interaction can be directly related to the time scales of the control systems implemented, and the level of granularity of control. Roads (2001, 11) suggests the macro level of musical time corresponds "to the notion of form, and encompasses the overall architecture of a composition." The micro timescale is described by Roads as being, "a broad class of sounds that extends from the threshold of timbre perception . . . up to the duration of short sound objects" (Roads 2001, 20–21). While in the context of sound toys we may not necessarily be dealing with control of sound in microseconds, a higher-input resolution could be considered within a micro timescale level, as opposed to a macro timescale level (which may, for example, use lengthy triggered prepared audio samples as the sound output). In sound toys, where there is a moment-by-moment time resolution for control, this can be considered as being on the micro level. Microlevel control can also be aligned with what Farnell (2008, 318) terms *procedural audio*. Farnell describes procedural audio as being "highly dynamic and flexible, it defers many decisions until runtime" (301).

It is suggested that sound toys that offer microlevel control of sound over time, with control of timbre or spectrum provide the player with more significant options for composition, influence and variations within a constrained framework. Therefore this approach has the potential to provide the player with a more flexible and rewarding interaction experience.

3.6 Opportunities at a Micro Level

While sound toys offer opportunities for "casual" sonic experiences, historically, limited processing power has to some degree restricted the types of real-time sound generation processes that can be implemented, with this limitation becoming less and less of an issue as technologies continue to develop. These limitations seemingly resulted in a predominantly fixed audio sample-based approach being adopted in many sound toy and gaming systems, making microlevel control less likely. The implementation of complex and intensive real-time sound generation or transformation systems is now achievable on a relatively small-form factor. This issue is significant as microlevel control and spectral-level processing become more achievable and accessible, providing further options for real-time control of sound and increased levels of interactivity.

Spectral analysis and resynthesis techniques are familiar in electroacoustic music composition practice, with spectral transformation techniques being frequently used for sound object metamorphosis and abstraction. Examples are composers such as Dennis Smalley (1997), accompanied by his writings on "spectromorphology," and Trevor Wishart (1987) and his *Composers' Desktop Project* software, which offers spectral processing features.

Smalley defines a particular approach to music that is primarily concerned with sound spectrum, which he terms "spectromorphological thinking." In Smalley's view this is "applicable to a wide variety of electroacoustic music, cutting across national boundaries and individual styles" (Smalley 1997, 109). He defines its relevance being "more concerned with spectral qualities than actual notes, more concerned with varieties of motion and flexible fluctuations in time rather than metrical time, more concerned to account for sounds whose sources and causes are relatively mysterious or ambiguous rather than blatantly obvious" (109). This approach is considered relevant to emerging approaches in sound toy applications that use microlevel control as their foundation, but yet also focus less on metrical time, notes, and traditional harmony. There is scope for playful and symbolic interpretation of Smalley's definition of "spectromorphological thinking" in the medium of sound toys, in which spectral motion, traversal, and transformation may be central themes. Virtual-object motion in the visual domain may be intimately linked to spectral motion and progression in the aural domain, with sound controlled on a micro level, with each spectral frame determined by the position and motion of objects in a virtual visual play space. Navigation in the virtual visual space represents navigation of spectral sound space, with a degree of compositional control being allocated to the player. It is suggested that this type of approach offers significant scope for sonic variation and nuance, that might engage the player more fully, and for increased periods of time.

3.7 Definitions and Classifications of Sound Toys

Interactions that occur on a micro level introduce some further issues of definition, and it is suggested that it is appropriate to consider some sound toys as potentially exhibiting behaviors of an "instrument." The term "instrument" is considered as also being relevant in an open-work context. In Björk's *Biophilia* many of the individual pieces have the option to be used as an "instrument," and could therefore be classified accordingly. The different modes offered in *Biophilia* suggest each piece may be experienced as a song (or composition), but also as an instrument. In this case, different classifications within a single piece of work exist, which is presented as a form of album, or collection of works. Is this sound toy therefore best defined as an open-form composition, composition tool, or as an instrument? It is suggested that issues of definition can be considered as a classification continuum between these three areas or definitions (see Figure 3.2).

At what point may an open work be also classified as exhibiting behaviors of an instrument, in the sense of an instrument offering particular sonorities and timbral qualities, with infinite possibilities from a compositional perspective? Perhaps sound toys become easier to classify as open-form compositions where more significant amounts of materials within the framework for composition are predetermined? Where increased player

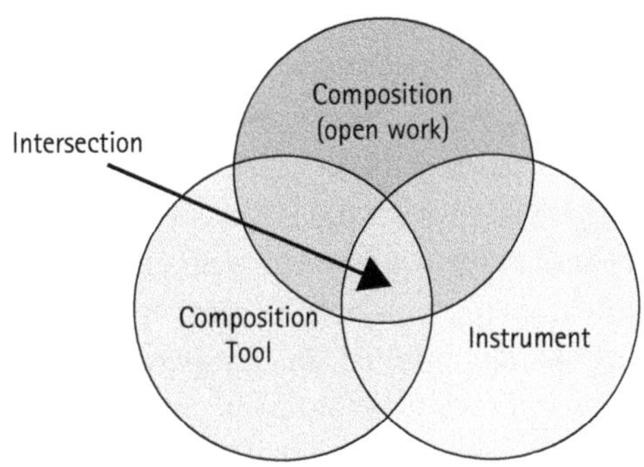

FIGURE 3.2 Intersection of terms of classification.

options for compositional input and influence are provided, classification as an open form composition is still relevant, but classification as an instrument, of some form, is also to some degree appropriate. As these terms of possible classification (as open-form composition, composition tool, or instrument) are frequently relevant to many sound toys, a continuum of definitions acknowledges that these three distinct classification areas are often intrinsically interrelated and may overlap. Absolute classification can be considered as being somewhat open to interpretation and may exist at a point of intersection. Where a sound toy may be placed within this area of intersection will likely be different for each system.

Definitions of the role of the player interacting with the work are also subject to similar issues of classification: Is the end user a player, composer, or participant? The roles of the player experiencing the work are multifaceted, with tensions between concepts of composition and intention sometimes evident. Equally, the role of the artist creating the work also becomes open to issues of definition, as the framework is composed. However, the eventual outcome cannot be fully determined due to individual interaction styles and any stochastic processes implemented. Modes of interaction may exist within the areas where these classification terms intersect. Where the framework for interaction and audiovisual elements are to some degree designed or composed, boundaries of definition and classification are often unclear, and it is the intersection between these possible boundaries that is particularly intriguing.

It is significant to note that there are many cases of sound-based applications that emulate traditional studio equipment or synthesis tools. Moog's *Animoog* application is a relevant example here. Many of these types of software application are perhaps easier to classify as instruments, as they often closely emulate an original instrument, often using control paradigms that imitate traditional synthesizer interfaces. While these types of application are relevant to some degree, it is suggested that they are primarily designed to be instrument-like, and are perhaps not best defined as sound toys.

3.8 Classification

There are frequent gradients of definition when attempting to classify sound toys, with classification or positioning along a continuum of definitions being open to interpretation. Sound toys that are perhaps more clearly definable as an open work often deal with larger prepared sonic structures (or samples), so definition as an instrument is deemed less appropriate. In sound toys in which the frame-by-frame interactions of the player result in sonic behaviors or outcomes, a more instrument-like experience is more likely. While the term "instrument" is perhaps not the most appropriate term, there are some relevant interaction relationships implied, in that player exploration and learning of the methods of interaction and resulting sonic outcome provide scope for recreating or reperforming sonic materials in an instrument-like fashion. Here the player has significantly increased options for choice and variation regarding output.

Where sound materials are interchangeable or replaceable, there is some movement away from classification as an open work, as the player may exercise further choice, and may introduce sound materials that the composer/designer did not anticipate. Where control is on a macro level with larger sections of prepared audiovisual materials, or where there are greater aspects of constraint, these sound toys can perhaps be more clearly classified as being an open work, moving away from both instrument and compositional tool classifications. In this example, it is worth noting that the structure of the player's experience may often still be open, where overall form, structure and duration may be flexible, resulting in a wide variety of possible experiences of the work, with no predefined pathway enforced or suggested for the player.

3.9 A Sound Toy Structural Model

A generalized sound toy model is presented to highlight the potential for complex and dynamic interactions in this medium, both internal to the system, and externally with the participating player. This model is derived from analysis of existing sound toys, such as *Electroplankton* and *Biophilia*, as well as some distillation of more specific models developed for personal practice, resulting in the works *SpiralSet, MagNular, Dioxide Dissolves, Cyclical Flow*, and *ResOscope* (Dolphin 2008–11) and *Urbicolous Disport* (Ash and Dolphin 2012). This model addresses the three compositional forces introduced earlier (see Figure 3.3). The model begins with the player that interacts with the system, or composes, using an input device. Player input may determine virtual-object and environment behaviors within the virtual visual symbolic play space. Player input may also determine user-interface component settings, which are not contained within the virtual space. Player input may be mapped directly to sound properties or processes. When the player input explicitly controls virtual objects, the results of this may be direct, for example where an object's coordinate data (position) is mapped to a sound parameter, such as amplitude. Alternatively, player input may introduce indirect results that occur once an algorithmic or generative process is set in motion, as a result of player interaction. In this model a physics engine serves this generative function where complex nondeterministic processes may be set into motion by the player, in which multiple virtual objects may continue to interact within the system without any further direct input from the player. This process is iterative, in that the simulated physics system determines subsequent conditions that determine the next iteration, dependent on how the physics system has been implemented. This approach can be aligned with Eno's idea, "that it's possible to think of a system or a set of rules which once set into motion will create music for you" (1996). Although in this case the system is only partially generative, as often the player will continue to interact, dynamically updating variables, conditions, or rules.

Implementation of a semigenerative simulated physics system need not be "realistic" or exhibit behaviors familiar to the real world. Wishart acknowledges that, "we are not

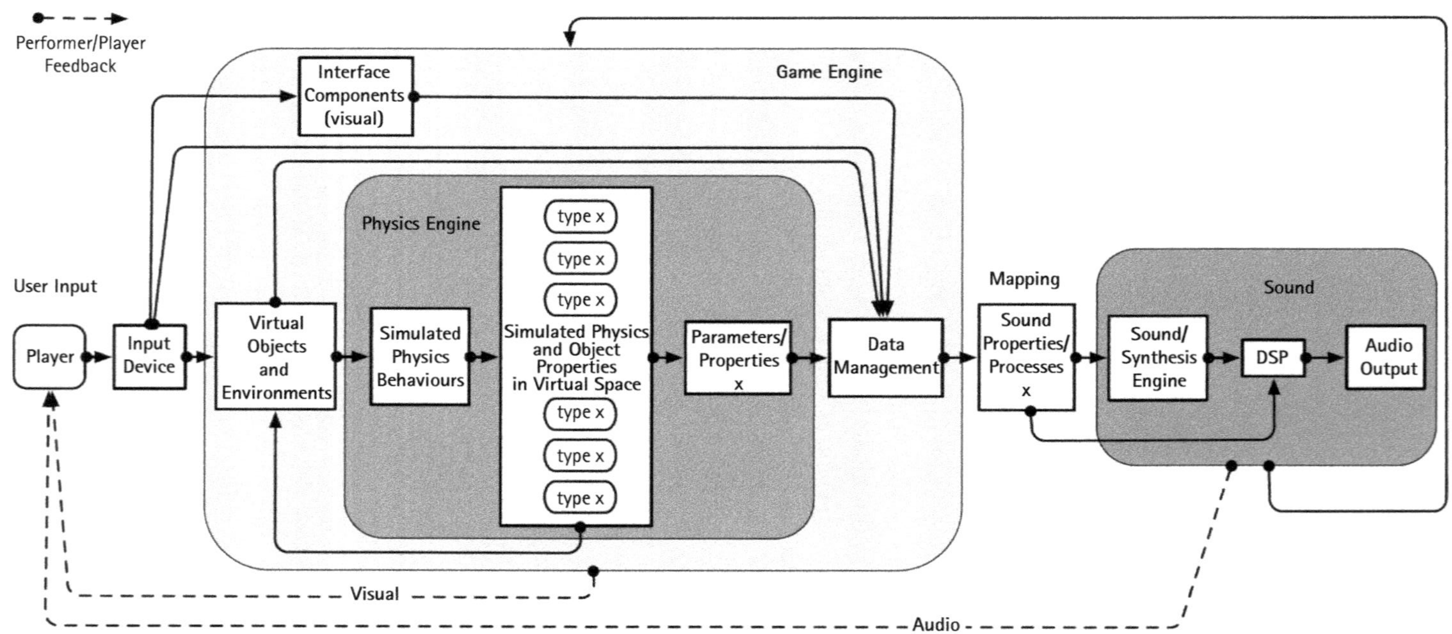

FIGURE 3.3 Sound toy model.

confined to basing our sound-models on existing physical objects or systems. We may build a model of a technologically (or even physically) impossible object" (Wishart 1996, 327). This is also true of a simulated physics based system in a sound toy context. These types of algorithmic processes may result in the player influencing but not fully controlling the system, as in *Hanenbow* in *Electroplankton*. Output properties of this stage may be for example object collisions, collision magnitude, object speed, force, direction, position, size, distance, state, and so on. Any required data is then managed, scaled, coupled, or filtered appropriately so that it may be used to determine sound properties, parameters, or processes. This is usefully thought of as the mapping stage. The mappings determine the types and ranges of dynamic control of the sound or synthesis engine, and/or any digital signal processing that occurs. Mappings may be simple (one-to-one) or complex (one-to-many, many-to-one, many-to-many). Their symbolic representation in the visual domain may be transparent, in which resulting behaviors of player interactions are apparent for the player, or oblique, in that interactions are more difficult for the player to decipher. Where oblique, the audible result of interactions can be understood only through player interaction, and there are often limited direct indications of sonic outcome in the visual interface. Where interaction and resulting outcomes are unclear for the player, they may have no way of determining what the sonic outcomes of their interactions might be without play. This is true of examples such as *Electroplankton*.

With continuing input from the performer and the physics (or generative) system, complex dynamic streams of data can be used to determine, control, or influence sonic or musical results. There is then the option of extending interactions within the system by using aspects of sonic or musical output to further control or influence the state of the physics or generative system. Basic parameters such as pitch, amplitude, and onset may be used for example. More sophisticated techniques could also be implemented, such as pattern or gesture recognition. In this model, player feedback is continuous, with output from both visual and audio domains influencing future interactions. These techniques combined—dynamic systems, transdomain mappings, generative processes, and player feedback in multiple domains—provide significant scope for variations in outcome.

3.10 Sound Toy Technologies for the Composer

The application and integration of technologies commonly associated with the field of contemporary computer music and composition are becoming increasingly integrated in game-like environments. Relevant to this are the sonic experiences offered by RjDj, which uses an implementation of Pure Data. Digital artist and composer explorations of software such as Pure Data, Max/MSP, SuperCollider, and Chuck for the development of innovative sound toys, or games in which sound is treated as a primary component frequently use a range of technical and artistic approaches familiar to contemporary computer music and electronic music.

Computer game-related technologies are viable tools for the creation and delivery of sound toys and interactive sound-centric works. From a technical perspective, the robustness of game engine technologies used alongside flexible sound technologies such as libpd, Wwise and FMOD offer flexible options for sound toy design. Networking technologies such as Open Sound Control (OSC) may also be used within a game engine, allowing flexible audio technologies to be implemented, such as Max/MSP/Jitter and SuperCollider, with one-way or bidirectional data communication between the applications. This flexibility allows sound (or music) to be generated, synthesized, or processed outside of the game engine, using external software for sound and synthesis, allowing computer music artists and researchers to explore game-related tools to realize their interactive works and prototype new technologies. The game engine's graphical capabilities may then be used to create a virtual environment for the symbolic control of sound, a game-engine component that could be considered to be an *animated user interface* (AUI), rather than a *graphical user interface* (GUI), generating real-time control data for the external sound system. It should be noted that in many sound toys, the visual component's function is not simply a GUI, as this is frequently an integral, functional, and artistic component of the sound toy, with specific aesthetic, stylistic, and interaction features. In sounds toys such as *Electroplankton*, aspects of the control systems, symbolic representations, and artistic style are to some extent familiar to the field of computer games. It should be noted that while sound toys are considered to be games by some, with *Electroplankton* being a relevant example in this respect, common computer game characteristics, such as competition as a motivation for interaction, are often avoided, encouraging the participant to solely focus on audiovisual experience and sonic-centric interaction. Playful composition and a sound-centric approach are considered to be integral themes and attributes of a sound toy.

Integrated physics engine technologies also offer opportunities to develop and implement systems that adhere to the *three compositional forces* model previously outlined. The game engine's integrated physics engine has the capacity for complex virtual object interactions, which is particularly enticing from a sonic perspective, see Mullan (2011) for discussion on physics engine integration with physical modeling synthesis techniques in a virtual environment. A physics engine may also be used as a form of generative composition agent within a sound toy. Varied artistic and sonic design options are available and interactive functionalities are flexible when using these types of tools. Working with computer game technologies for sonic purposes provides the opportunity for the composer/designer to draw on existing experiences of a possible audience and their understanding and appreciation of increasing levels of complexity and interactivity now found in modern computer games.

Game engine software offers many possibilities for the creation and delivery of interactive sound or music works that allow the player control over compositional and sound parameters. Sound artists frequently explore ideas and experiment with techniques that allow the visual domain to intimately coexist with, or directly control, sound parameters using transdomain mapping techniques (for instance, *Audiovisual Environment Suite* by Golan Levin). Network technologies allow communication between a game engine and external sound and synthesis software, providing varied creative possibilities for a composer or sound artist (who may or may not be a game programmer), who may then use familiar tools for the development of audio systems. Integrating external, flexible, and open-ended sound software enables sound artists and/or composers to work with specialist tools and techniques to explore interdisciplinary approaches for creating new repertoire, which may be informed by varied perspectives relevant to music and sound.

3.11 Conclusions

The term *sound toy* can be applied to many current and emerging interactive and/or reactive applications and systems in which some aspects or elements of composition are made available for the player. The player may be in personal a recreational situation, or may be a participant or visitor in other artistic presentation contexts, such as an art installation or performance.

Sound toys provide the player with scope for "musicking" (Small 1998), offering varying degrees of compositional input, control, influence, or decisions within a defined framework. While sound toys are sound-centric, the medium is not solely concerned with sound, and this evidently has certain aesthetic implications. It is very much a matter of personal perspective as to what may constitute meaningful composition, and any conclusive viewpoint on what makes for meaningful composition in a sound toy context is left for the reader to decide. While sound toys draw on a number of different technological and cultural reference points, it is suggested that sound toys are not simply pieces of software, and many could be considered as interdisciplinary compositional repertoire in an open form. Media and methods for composition will evidently continue to develop, offering further opportunities for interdisciplinary interactive practices to emerge and grow, in a period in time in which the democratization of media is becoming increasingly prevalent. Sound toys provide an inclusive platform for composition, and participation in the experience of open-form compositions. In the words of Wishart, "The era of a new and more universal sonic art is only just beginning" (1996, 331).

Notes

1. The term "sound toys" is directly relevant to many of the types of audio-visual artworks presented on the website repository soundtoys.net, which was originally established in 1998. The term is predominantly used in a computing context here, which is the primary focus of this chapter.
2. All citations from Umberto Eco's "The Poetics of the Open Work" refer to the paper's republication in Christoph Cox and Daniel Warner, eds. (2004), *Audio Culture: Readings in Modern Music*, 167–175, New York: Continuum.

3. *Electroplankton* offers the player a series of "games" in which musical sequences and patterns are generated according to player interactions using a stylus on the touch screen interface of the *Nintendo DS*. In *Biophilia*, a series of musical pieces are presented with varied options for player intervention, influence and control of resulting sound events, determining the overall musical structure.

References

Aura Flux. n.d. http://www.higefive.com/apps/flux/.
Dolphin, Andrew. 2009. Compositional Applications of a Game Engine. In *Proceedings of the Games Innovations Conference, 2009 (ICE-GIC 2009), International IEEE Consumer Electronics Society*, 213–222. London: IEEE.
——. 2009. SpiralSet: A Sound Toy Utilizing Game Engine Technologies. In *Proceedings of the 2009 International Conference on New Interfaces for Musical Expression (NIME)*, 56–57. Pittsburgh. http://www.nime.org/proceedings/2008/nime2008_087.pdf.
Eco, Umberto. 1989. *The Poetics of the Open Work. In The Open Work, translated by Anna Cancogni*, 1–23. Cambridge, MA: Harvard University Press.
Eno, Brian. 1975. *Discreet Music*, CD-ROM, UK: EG Records.
——. 1996. Evolving Metaphors, in my Opinion, is what Artists Do. Paper presented at the Imagination Conference in San Francisco, June 8, 1996. http://www.inmotionmagazine.com/eno1.html.
Eno, Brian, and Peter Chilvers. 2008. *Bloom*, http://www.generativemusic.com/.
Farnell, Andy. 2008. *Designing Sound*. London: Applied Scientific Press.
Iwai, Toshio. *Electroplankton*. http://electroplankton.com/.
Levin, Golan. 2000a. *Painterly Interfaces for Audiovisual Performance*. Master's thesis, Massachusetts Institute of Technology, Program in Media Arts and Sciences.
——. 2000b. *An Audiovisual Environment Suite*. http://acg.media.mit.edu/people/golan/aves/.
Menzies, Dylan. 2009. Phya and VFoley: Physically Motivated Audio for Virtual Environments. In *Proceedings of the 35th AES International Conference on Audio for Games*. New York: Audio Engineering Society.
Mullan, E. 2009. Driving Sound Synthesis from a Physics Engine. In *Proceedings of the Games Innovations Conference, 2009 (ICE-GIC 2009), International IEEE Consumer Electronics Society*, London, 1–9.
Nimoy, Joshua. n.d. *BallDroppings*. http://www.balldroppings.com/.
Oliver, Julian, and Stephen Pickles. 2007. Fijuu2: A Game-based Audio-visual Performance and Composition Engine. In *Proceedings of the 2007 International Conference on New Interfaces for Musical Expression (NIME)*, 430. New York.
Paine, Garth. 2002. Interactivity, Where to from here? *Organised Sound* 7 (3): 295–304.
——. 2007. Sonic Immersion: Interactive Engagement in Real-time Immersive Environments. *SCAN Journal of Media Arts and Culture*, 4(1).
Raber, Hansi. n.d. *SoundyThingie*. http://www.soundythingie.net/.
Reality Jockey. n.d. *RjDj*. http://rjdj.me/.
Roads, Curtis. 2002. *Microsound*, Cambridge, MA: MIT Press.
Small, Christopher. 1998. *Musicking: The Meanings of Performing and Listening*. Hanover, NH: Wesleyan University Press.

Smalley, Denis. 1997. Spectromorphology: Explaining Sound-shapes. *Organised Sound* 2 (2): 107–126.
Soundtoys. http://www.soundtoys.net/.
Whalley, Ian. 2009. Software Agents in Music and Sound Art Research/Creative Work: Current State and a Possible Direction. *Organised Sound* 14 (2): 156–167.
Wishart, Trevor. 1996. *On Sonic Art*. New York: Routledge.

CHAPTER 4

THINKING MORE DYNAMICALLY ABOUT USING SOUND TO ENHANCE LEARNING FROM INSTRUCTIONAL TECHNOLOGIES

M. J. BISHOP

For those who are not hearing impaired, real-world sounds are extremely useful for communicating information about things like when to shift gears in our cars or stop pouring liquids, the weight and material of a slammed door, the proximity of an impending thunderstorm, or the true level of our spouse's irritation (Bregman 1993; Deutsch 1986; McAdams 1993). The education field has, therefore, speculated for some time now on sound's potential to increase the "bandwidth" of learning. According to Hannafin and Hooper (1993), incorporating sound with other instructional modalities capitalizes on the additive effects of learners' coding mechanisms by compelling learners to act on information from multiple sources. Paivio (1986) called this strategy "dual-coding," maintaining that seeing an object and hearing its accompanying sound will result in better memory performance than either seeing or hearing it would by itself. But Dunn, Dunn, and Price (1979) argued that the need to incorporate sound into instruction is even more fundamental; it is a matter of accommodating some individuals' auditory learning styles. The Dunns are not alone in this contention. While the terminology varies—learning styles, learner aptitude, multiple intelligences, and modality strengths, to name a few—many authors have concluded that some individuals learn better auditorially than they do visually (see, for example, Armstrong 1994; Barbe and Swassing 1979; Gardner 1983, 1993; Keefe 1979; Snow 1997). These theorists agree that the extent to which educators can incorporate multiple modalities into their instruction is the extent to which that instruction will be suited to the specific needs of various learners.

Instructional designers, those involved in the design and development of learning resources, have therefore sought ways to use sound in instructional computer programs for years. In the early 1960s, for example, student terminals connected to the mainframe-based IBM 1500 Tutorial System included specialized reel-to-reel tape players that played sounds to accompany the instruction (Bennion and Schneider 1975). Lengthy fast-forwarding and rewinding delays caused by a tape player's linearity, however, relegated sound's use to self-contained primary examples or very specific and brief, attention-getting narrative cues (Dale 1969). In the mid-1980s, videodisc players that provided "random access" to audio and video recordings became fairly widely available in schools (Technology Milestones 1997). While this meant that desired audio or video segments could be played back with only a small time delay, the analog signal format that was used isolated the presentation to a separate television monitor, leaving the audio and video signals physically "removed" from the interactivity of the computer interface. Digital overlay boards developed in the late-1980s to translate videodisc signals from analog to digital formats only partially solved the problem; audio and video segments still often were operated using "player" software that was separate from the instructional software. By the late 1980s and early 1990s, computerized instruction written for computer-driven multimedia configurations typically involved a lot of reading on the computer screen that was supplemented, if the user chose, by clicking to view a separate visual or audio presentation (see for example, *The Adventures of Jasper Woodbury*, 1988–1992; *The Great Solar System Rescue*, 1992; *Interactive Nova*, 1990; *Introduction to Economics*, 1986; *The Living Textbook*, 1990). These applications often relied heavily upon the user's ability and desire to explore the available media, not upon the software's own dynamic presentation of integrated information types (Gygi 1990; Mayes 1992). More highly integrated interface sounds were not technologically possible, in fact, until the early 1990s, when Creative Labs introduced their relatively inexpensive SoundBlaster sound card for the PC and Macintosh released the Mac LC with standard integrated sound-recording capabilities.

Clearly, digital sound production techniques and reproduction technologies have improved dramatically over the last twenty years since these technological developments. But while the film and gaming industries have been exploring sound's role to enhance the end-user experience for some time (see Bishop 2000; Bishop and Sonnenschein 2012), instructional software programs do not appear to use sound very extensively. For example, a recent content analysis (Bishop, Amankwatia, and Cates 2008) of twelve award-winning instructional products by found that the use of sound was still relegated primarily to error messages, self-contained examples (recording of a historical speech), or screen-text narration. Despite the fact new audio technologies have made it possible to incorporate sound as a highly integrated part of the interface, it appears designers of instructional technologies are not thinking very creatively about how sound might be used more systematically or artfully to enhance learning (Calandra, Barron, and Thompson-Sellers 2008). Why is that the case? Does sound have a more prominent role to play to enhance learning from instructional technologies? How might interactive audio technologies change the way we think about designing instruction with sound?

To find the answers to those questions, this chapter reviews the traditional theoretical foundations and existing research on sound's use in instructional technologies. It then explores some new ways of thinking about the role sound might play as designers consider how increasingly interactive technologies alter the way learners can and should experience sound in instructional technologies to enhance learning.

4.1 Traditional Ways of Thinking about Sound's Use

Explorations into the design and evaluation of instructional materials over the years have been grounded at the intersection of learning and communications theories (see Bishop 2013, for a full review). Specifically, in order to optimize learning from instructional materials, instructional designers have traditionally sought to balance what we know about the capacities and limitations of learners' cognitive information processing against what we know about message design for effective and efficient communications. Each of these theories and its implications for the design of instruction is discussed below.

4.1.1 Cognitive information-processing theory

Cognitive information-processing theory posits that humans learn in much the same way computers process information (Atkinson and Shiffrin 1968). Models that have evolved from this perspective typically represent human information processing as a system made up of three stages or "stores": sensory memory, short-term memory, and long-term memory. Information from all five senses (sights, sounds, smells, tastes, and haptics) enters the system in parallel at the sensory memory stage (Broadbent 1958). Because sensory memory can process incoming stimuli only in serial, however, the system must make preperceptual, split-second decisions (either consciously or unconsciously) about what information to attend to and what to ignore. Individuals remain essentially unaware of information not selected for attention (Treisman and Gelade 1980). Information that is chosen, however, then passes to the short-term memory stage for further processing.

Short-term memory is the point in the system at which one first becomes conscious of the information being processed (Driscoll 2005). Here, individuals work to prepare information for long-term storage through a process called *encoding*. Processing at this stage requires that effort be applied as the individual actively tries to make sense of incoming stimuli by organizing, categorizing, grouping, and comparing the new information against prior learnings retrieved from long-term memory. Short-term memory is limited both in terms of its duration (estimated to be only about 20–30 seconds without further processing; Peterson and Peterson 1959) and its overall capacity (estimated to be about seven plus or minus two "chunks" of information at once, Miller 1956). Thus, there is a limit to the amount of information, or maximal *cognitive load*, that an individual can process in short-term memory at any given time (see Clark, Nguyen, and Sweller 2006; Mayer and Moreno 2003; Paas, Renkl, and Sweller 2003; Sweller, Ayers, and Kalyuga 2011). Although it may be that cognitive load varies somewhat depending upon the nature of the input stimuli (Craik 1979), our capacity for processing incoming data is certainly limited to some finite quantity. Information that exceeds cognitive processing capacity is dropped from short-term memory without being further processed.

The final destination in the information-processing model is long-term memory. Here, memories are stored either as *episodic* (your memory of what you had for dinner last night) or *semantic* (your abstracted memory of what a hamburger is) (Tulving 1972, 1983). Research to date indicates that, while information stored here can eventually become irretrievable, long-term memory is of virtually limitless duration and capacity. Control or "metacognitive" processes oversee the entire cognitive system by regulating the exchange of information between sensory memory and short-term memory, determining which search-and-retrieval strategies should be used to access information from long-term memory, and deciding when sufficient information has been retrieved (Flavell 1976).

Important as the cognitive information-processing model has been for explaining and consolidating much of the existing data on human cognition, the model is not without its shortcomings. Several information-processing theorists contend that one particularly troublesome deficiency is the model's unitary short-term store, which implies that input from each of the senses, or *modalities*, is processed along exactly the same route and in exactly the same way (see Bregman 1990; Humphreys and Bruce 1989; Marr 1982; Moore 1982; Pinker 1985; Warren 1982). If this were true, they argue, it would not be possible for people to process multiple input and output modalities simultaneously as they do. Studies over the last thirty years by Baddeley (2003) and his colleagues indicate that there may be many different short-term stores—at least one per modality—each with its own strengths and weaknesses (see also Baddeley 2000, 2001, 2002; Baddeley and Andrade 2000). This multistore *working memory* concept may explain more accurately how each of the modalities, including sound, can have its own "specialty" and can be uniquely suited to its specific role in information processing (Alten 1999).

4.1.2 Communication theory

In 1949 Shannon and Weaver proposed that all communication processes begin when a *source*, desiring to produce some outcome, chooses a *message* to be communicated. A transmitter then *encodes* the message to produce a signal appropriate for transmission over the *channel* that will be used. After the message has been transmitted, a *receiver* then *decodes* the message from the signal transmitted and passes it on to the destination.

In person-to-person communication, where one individual performs both the message-creation and encoding functions and another individual performs both the message-decoding and receiving functions, it may be useful to refer to only a source and a receiver (see, for example, Hankersson, Harris, and Johnson 1998; Newcomb 1953). Further, while Shannon and Weaver defined a channel generally as any physical means by which a signal is transmitted, some theorists prefer to distinguish between the artificial technical channels of more mechanistic communication (such as telephones, films, and newspapers) and the natural sensory channels typical of human communication (such as seeing, hearing, touching, smelling, and tasting) (see Moles 1966; Travers 1964a, 1964b). According to the Shannon–Weaver model, however, whether technical or natural, all channels have limited capacity. In humans, *channel capacity* generally refers to the physiological and psychological limitations on the number of symbols or stimuli that individuals can process (Severin and Tankard 1979). When more symbols are transmitted than a channel can handle, some information is lost. This loss is called *equivocation*.

While the Shannon–Weaver model was primarily intended to explain mechanistic communication over telephone channels, the researchers' 1949 publication did also discuss communication more broadly in terms of the semantic meaning of a message and its pragmatic effects on the listener as well. At "Level A," they suggested, message designers concern themselves primarily with technical noise that affects how accurately signals can be transmitted. At "Level B" message designers focus on the semantic noise that prevents the receiver from accuracy interpreting the signal sent. At "Level C" message designers seek to overcome conceptual noise that arises when connotative mismatches between the sender and receiver cause the message to fail to have the desired effect.

Regardless of the level, the Shannon–Weaver model suggested overcoming all types of noise in the system involves increasing the *redundancy* in messages. Redundancy between and among the cues of a message consists of the relationships and dependencies among those cues (Attneave 1959). Redundancy is the information that cues share: the parts that "overlap." In fact, while the word "redundancy" is commonly defined as something that is superfluous or unnecessary, in communication systems the surplus may not necessarily be uncalled for. Redundancy that helps a receiver separate transmitted information from system noise increases understanding and is, therefore, desirable. That said, redundancy not needed by the receiver or that fails to increase understanding can be a burden on the system. Leonard (1955) suggested that channel limits mean unnecessary redundancy may actually impede the flow of new information and, consequently, decrease communication effectiveness. It appears that the trick to effective and efficient message design for communication is in knowing how much and which sort of between-cue message redundancy to include in order to counteract noise (Krendl et al. 1996).

4.2 Instructional Implications of Information-processing and Communication Theories

Traditionally in the field of instructional technology, learning theory and communications theory have been viewed as two sides of the same coin: learning theory explores the ways in which receivers decode messages sent, and communications theory explores how senders should encode those messages to assure they achieve the desired outcomes (Berlo 1960). Table 4.1 demonstrates this orthogonal relationship, depicting the ways in which information-processing limitations within each of the three stages affects learning outcomes at each level of communication. The rows in Table 4.1 illustrate each level of potential communication problem while the columns represent the information-processing limitations, all three stages of which are active to varying degrees at each level of communication. So, at Level A, learner difficulties in directing attention, isolating relevant information, and retrieving existing schemas cause technical difficulties that prevent the instructional message from being selected at all.

Table 4.1 Problems in instructional communication (adapted from Bishop 2000, Bishop and Cates 2001).

	Sensory memory: acquisition noise	Working memory: processing noise	Long-term memory: retrieval noise	Outcome
Level A. Technical difficulties cause message-*transmission* problems.	Learner has trouble directing attention to the instructional message.	Learner cannot isolate and disambiguate relevant information contained in the instructional message.	Learner's existing schemas are not activated by the instructional message.	LEARNER FAILS TO **SELECT** MESSAGE
Level B. Semantic difficulties cause message-*interpretation* problems.	Learner has trouble focusing attention on the instructional message.	Learner cannot organize the information contained in the instructional message.	Learner does not use the information contained in the instructional message to build upon existing knowledge.	LEARNER FAILS TO **ANALYZE** MESSAGE
Level C. Conceptual difficulties cause message-*effectiveness* problems.	Learner has trouble sustaining attention on the instructional message over time.	Learner cannot elaborate upon the information contained in the instructional message.	Learner does not use the information contained in the instructional message to construct transferable knowledge structures.	LEARNER FAILS TO **SYNTHESIZE** MESSAGE

At Level B, the learner's problems focusing attention, organizing the information, and building on existing knowledge mean the message does not get adequately analyzed. And, at Level C, the learner's trouble in sustaining attention, elaborating on the new information, and constructing transferable knowledge structures means the message will not be well synthesized for long-term storage and easy retrieval when needed later.

From this perspective at the intersection of cognitive information processing and communication theories, then, sound is among the modalities or "cues" available to designers for use within the instructional communications system. The goal is to use sound, often in combination with other modalities, to "front load" instructional messages with the redundancy needed in order to overcome acquisition, processing, and retrieval information-processing limitations at each level of potential communication problems and optimize learning within the system. Unfortunately, findings from recent research on the use of sound to enhance learning have been somewhat mixed.

4.3 Recent Research Evidence for sound's Use to Enhance Learning

Research over the last fifteen years on "multimedia learning" by Mayer and his colleagues seems to indicate that, while students may learn better from graphics or animations combined with narration than from graphics or animations combined with onscreen text (*modality principle*, see Mayer and Moreno 1998; Moreno and Mayer 1999), the addition of nonspeech sounds to multimedia instruction appears to show less potential than hoped and, in some cases, may even be detrimental to learning (*coherence principle*). In two experiments by Moreno and Mayer (2000), for example, participants viewed a short (180 second), narrated animation on either how lightning storms develop (Experiment 1) or how hydraulic braking systems work (Experiment 2). In each experiment, one group received only the narrated animation (N), one group received the narrated animation with the addition of environmental sounds (NS), one received narrated animation with music (NM), and one group received the narrated animation with the addition of both sounds and music (NSM).

Findings indicated that adding sound effects and music to a narrated animation presentation significantly reduced learners' retention and transfer scores in both lessons and that adding only sound effects also harmed learning in the braking lesson (Experiment 2). The authors suggested these results were consistent with the idea that auditory adjuncts can overload the learner's auditory working memory and concluded "in multimedia learning environments, students achieve better transfer and retention when extraneous sounds are excluded rather than included" (Moreno and Mayer 2000, 124).

A cursory read of the findings from these studies and the *coherence principle* guidelines derived from them might lead one to conclude that the addition of any sounds other than screen narration are "extraneous material" that should be eliminated from instructional presentations—which may help to explain why so few instructional software programs currently make much use of sound. However, it is important to note that Mayer and his colleagues qualify "extraneous" sounds as those that are *interesting but irrelevant* to the material under study (Mayer 2001, 123). What might the findings have been if the music and sound effects chosen had not been just "bells and whistles" (Moreno and Mayer 2000, 117), but rather were germane to the material under study and implemented in a way that made clearer how they were related to lesson concepts? Is it possible for sound effects and music to be incorporated into interactive multimedia presentations in ways that might enhance learning from them without overloading working memory? Stated differently, what would make a sound *relevant* to an instructional presentation?

4.4 New Ways of Thinking about Sound's Use

In her book on sound design for games, Collins (2008, 3) defined *dynamic audio* as an umbrella term encompassing both *interactive audio*—sound events that react to the user's direct input—and *adaptive audio*—sound events that react to the state of the user's progress in the activity (the game, in this case). The author observed further that dynamic audio shifts the user's role from the passive "receiver" of a sound signal to (at least partly) the "transmitter of that signal, playing an active role in the triggering and timing of these audio events." The author went on to observe that this represents a rather significant paradigm shift in our thinking about sound's role in the interface: "Existing studies and theories of audience reception and musical meaning have focused primarily on linear texts" (Collins 2008, 3). Leman (2007) agreed, arguing further that interactive audio shifts sound's role from passive content delivery mechanism to interactive mediator of "perception-action" loops. This section explores new ways of thinking about how sounds might be incorporated more *dynamically* to facilitate cognition, improve motivation, and support knowledge construction.

4.5 Designing with Dynamic Sound to Facilitate Learner Cognition

Harrison (1972) proposed that in order to classify as communication, a stimulus really must be a *sign* that can be used to represent other potential stimuli, the way a flag stands for patriotism. Further, this sign must clearly be part of a larger *code* or set of signs that has been firmly established in advance, with procedures for combining the signs meaningfully (or *syntax*) and meanings common to the members of some group. This matches well with Fiske (2011), who suggested that messages might be generally categorized in terms of their *representational* or *presentational* codes. Representational codes—such as languages, musical notations, and other symbolic figures—typically are used to produce *works* of communication. Once transmitted, a cue built from representational code exists independently, standing for something apart from itself and its source (like the word "door"). On the other hand, presentational codes—like gestures, musicality, and other forms of expressiveness—typically are used to produce *acts* of communication. A cue built from presentational code both echoes information contained in some existing representational cue and supplies additional information. Secondary presentational cues appear to be interpretable only within the context of a primary representational cue. In the absence of a primary cue, the receiver may supply his or her own derived cue based on information acquired from other environmental stimuli or retrieved from existing schemas. For example, a waving, raised hand is a presentational cue that often accompanies a friendly verbal greeting. When no words are exchanged, understanding the message requires the receiver to infer the primary cue from the context of the situation and from his previous experience. Coming from an old friend, the receiver might supply a "hello" primary cue. Coming from a uniformed police officer, however, the receiver might instead supply a "stop" primary cue. Thus, it appears that in order for secondary cues, like music and sound effects, to have meaning for the learner in instructional messages, a presentational code for them must first be clearly established.

Turning back to the two Moreno and Mayer (2000, 124) experiments discussed earlier, the researchers reported that in the braking lesson only two mechanical sounds (pistons moving, brakes grinding) were repeated several times throughout the animation—apparently at fairly random spots throughout the presentation that in no way made clear their correspondence to the underlying concepts they might represent. According to the authors, these sounds "may have been too intrusive, arbitrary, and ambiguous to associate with the other materials in the lesson." And, while the sound effects used in the lightning lesson were more carefully matched to their respective events in the animation, *seven* different natural (realistic) sounds were used and played only once during the presentation: (1) a gentle wind for the start of the process; (2) water condensing in a pot for cloud formation; (3) clinking of ice crystals forming; (4) a stronger wind indicating downdrafts; (5) a static sound for the development of electrical charges; (6) a crackling sound for the charges moving between cloud and ground; and (7) thunder for the final lightning flash. While, presumably, these seven sounds fall within Miller's "seven plus or minus two" rule for predicting cognitive load (1956), this seems like a lot of primary and secondary cue connections to make in a very short period of time (180 seconds), making it unlikely the inclusion of these sounds could help students recall the related concepts after the presentation.[1] Additionally, it may be some of these realistic sounds were not sufficiently distinctive to be particularly meaningful in relation to their underlying constructs—such as the sound of water condensing in a pot or the difference between the sound of ice crystals forming versus the sound of static.

What if, instead of presenting each sound only once in a linear animation, the representational syntax for each sound was more clearly established throughout the lesson through repeated simultaneous presentation and one-to-one correspondence, so that the auditory secondary cues might eventually be used to communicate information without supplying a primary cue? Additionally, while the experimenters did draw on some fairly well-established presentational codes—like gentler and stronger wind sounds to accompany air movement—why not first establish an initial "wind sound" that is then built upon more adaptively as the learner progresses through the lesson's content on this concept? Further, given that new codes easily can be established through convention and/or context, why restrict the sound effects used to realistic representations only? Would more "metaphorical" sounds like ice clinking in a glass and the whistle of a teapot have established stronger prior associations from long-term memory and created clearer distinctions between the lightning lesson's concepts of ice crystals forming and water condensing? If an auditory syntax had been more clearly established in this way, might sound effects have helped to supply the redundancy needed to overcome communication noise without adding appreciably to the cognitive load for participants in these experiments?

4.6 Designing with Dynamic Sound to Improve Learner Motivation

While learner motivation is a construct that has been researched for many years, educational theorists interested in motivation have recently begun taking a more holistic view of the learning *experience* that has cognition at its core but that also embraces affect in learning as well (Wilson 2005). Sousa (2006) and others have observed that, although emotion is largely misunderstood in education as something that is "unscholarly," it is nonetheless a powerful force in learning and memory (see also Craig et al. 2004; Kort, Reilly, and Picard 2001). "Students are more likely to remember curriculum content in which they have made an emotional investment," Sousa concluded (2006, 84). According to Dewey (1987), creating the environment for this kind of emotional investment in a learning experience requires that we consider the aesthetics of that experience in addition to the factors that enhance cognitive processing.

Drawing largely from the arts—particularly literary criticism—Parrish (2005, 2008, 2009, 2010) has been exploring the aesthetics of learning experiences and has come up with a set of principles and guidelines for thinking about message design, some of which suggest alternative approaches to the problems of cognitive load and avoiding split attention. Aesthetic considerations of the learning experience go beyond the traditional instructional system components of subject matter, instructional method, learner, instructor, and context to include also "the way the learner feels about, engages with, responds to, influences, and draws from the instructional situation (Parrish 2005, 512). The idea is provide much more than "an attractive frame or surface to instructional events" but rather to "show strong connections to valued instructional theories derived from traditional sources" which, like aesthetic experience, are also aimed at helping learners construct meaning (Parrish 2005, 525). Thus, a necessary ingredient for aesthetic experience is the learner's active participation and contribution as well. According to Parrish, "the opposites of an aesthetic experience are boredom; mindless routine; scattered, dispersed activity; or meaningless, imposed labor" (2009, 514). Parrish concluded that " 'Experience' in this sense describes more than a passive event. It is a transaction with the environment in which learning is an outcome (witness the saying, 'experience is the best teacher')" (2007, 512).

Turning back once again to the Moreno and Mayer (2000, 119) studies then, the researchers reported that the twenty-second instrumental music loops used in both lessons were chosen specifically because they were unrelated to the presentation and were characterized by the authors as being "synthesized and bland." While the generic music might intuitively have seemed like the best choice to serve as an irrelevant message cue in this experiment, learners' irritation with the lack of harmony (pun intended) between the music and the rest of the lesson presentation may have actually "disrupted cognition, damaged attitudes, and dissuaded persistence" (Cates, Bishop, and Hung 2005, 448; see also Ormrod 2003).

What if, instead, musical elements had been selected more deliberately and used more adaptively throughout the lesson to evoke learners' existing and, potentially related constructs? Interestingly, Parrish drew parallels to classical music as he noted that all learning experiences have a beginning, middle, and an end. Like a classical symphony, he observed, the beginnings of learning experiences require starting out strong and developing learners' sense of anticipation for what lies ahead. The middle "movement" of a learning experience, however, should proceed "in a quieter, and more thoughtful pace than does the opening, often allegro, movement" in order to help learners process the material under study. The end of learning must bring a "profound closure" to the experience, like the energetic, final movement of a symphony that "adds emotional intensity to the feeling of consummation and restored order when it is finally complete" (Parrish 2005, 25). For example, adding an instrumental version of the Doors' dark and mysterious "Riders on the Storm," with its slowly building intensity, would certainly have been more aesthetically pleasing and might have helped learners activate prior understandings and positive associations, and make connections metaphorically between the lightning formation material under study and the learners' existing schemas.

4.7 Designing with Dynamic Sound to Support Learner Knowledge Construction

But no matter how well we know our learners, we cannot anticipate everything they will need to offset "noise" in the system, particularly as we begin to explore the role aesthetics and other, more affective elements, might play in the learning experience. For instance, depending on the age and other characteristics of the audience in the example supplied above, the intended effect of playing "Riders on the Storm" as part of a lightning-formation lesson might be completely lost on learners. What is needed to truly support *all* learners' knowledge construction will be a move away from our traditional, "transmission" view of communication to a more "transaction"-oriented perspective on message design instead (Bishop 2013).

As suggested by De La Cruz and Kearney (2008) and others, movement away from an objectivist, linear paradigm of instructional message design and delivery and toward creating technology-facilitated environments that support multiple two-way communication "transactions" will require that we find ways for participants other than the initial source to support and represent their thinking while engaged in the discourse (see also Boyd 2004; Gibbons 2009; Gibbons and Rogers 2009a, 2009b). According to Luppicini, *conversation theory* explores "how people think, learn, and interact through conversational processes" and emerged in reaction to the often reductionist view "of human thinking and learning as a set of mental structures and processes that can be analyzed separately and applied to learning and instructional applications" (Luppicini 2008, 3). But conversations need not be only among humans for learning to occur—conversations can also involve technology-based communication systems as well, particularly as the rapid growth of interactive multimodal and social-networking technologies offer opportunities not previously possible (2008). In fact, Pangaro argued that it is "inevitable" that all disciplines involved "in the crafting of systems, products, and services built on technology" will eventually incorporate constructs "that explore the role of conversation, its efficiencies and effectiveness, its failures, and its aesthetics" (Pangaro 2008, 37).

Revisiting the Moreno and Mayer (2000) studies one last time, how might their findings have been different if, after the initial lightning-formation animation with one-to-one correspondence between lesson concepts and accompanying sound effects, learners had the opportunity to interact more directly with those sounds? Unlike interactions with objects in the real world, interactions with mimetic objects in technology interfaces typically make no sound at all until one has been chosen and programmed into the system. Therefore, learner interactions with an instructional technology can make any sound the designer wishes. So, rather than use a "click" sound to confirm the action of clicking a button on the screen in an online tutorial, could this opportunity be used, instead, to reinforce the lesson's content in some way that might also enhance learning?

For example, what if in a series of embedded practice activities learners were asked to drag icons representing steps in the lightning-formation process and drop them in the proper order on an "answer space" and, when they did so, the corresponding sound effects played again? Might these sound effects then also have been used to accompany learners' responses to assessment items and, therefore, possibly provide additional, auditory pathways for retrieving these concepts from long-term memory later? Additionally, how might learning be affected if a second lesson in this series elaborated on these initial sound effects by adaptively building further on the concepts presented? For example, a higher-pitched static sound for positive charges (in the clouds) versus a lower-pitched static sound for negative charges (on the ground)? Or different sorts of thunder sounds for "cloud flashes" (lighting that stays entirely within the cloud) versus a regular lightning flash (cloud to ground)? And, perhaps even more compelling, what if learners were eventually given the opportunity to select the sounds *they* thought best represented the concepts presented?

As Mayer and his colleagues have demonstrated, without a strong theoretical cognitive foundation to focus efforts to use sound in computerized lessons, the sounds used in instructional technologies not only may not enhance learning, they might detract from learning. Designing instruction with sound is clearly more complicated than simply adding "bells and whistles" as afterthoughts. Instructional designers need a strong theoretical framework for sound's optimal use in instructional technologies.

4.8 Conclusions

In an effort to describe the "design space" for sound's potential role to enhance learning from instructional technologies, Bishop (2000; Bishop and Cates 2001) developed a framework derived from the juxtaposition of cognitive information-processing and communication theories (see Table 4. 2)

The framework seeks solutions to the instructional communication problems (noise) identified in Table 4.1 by suggesting ways narration, sound effects, and music might supply the various kinds of redundancy needed to facilitate information-processing operations (columns) at each level of learning (rows). Following the cells vertically down the information-processing columns, the framework anticipates deepening acquisition, processing, and retrieval difficulties at each subsequent phase of learning (top to bottom). When tracing the cells horizontally across the learning phases, the framework similarly anticipates waning interest, curiosity, and engagement at each deeper level of processing (left to right).

Thus, when one traces the first, selection-level row of cells horizontally across the information-processing stages, the framework suggests that learner interest may be captured by an instructional message that employs sound to gain attention with novelty (cell 1), to isolate information through increased salience (cell 2), and to tie into previous knowledge by evoking existing schemas (cell 3). Similarly, learner curiosity might be aroused using sound to focus attention by pointing out where to exert information-processing effort (cell 4), to organize information by differentiating

Table 4.2 .Application of various types of redundancy to the solution of instructional communication problems (adapted from Bishop 2000, Bishop and Cates 2001).

	Content redundancy: "Amplifies" the content for message transmission	Context redundancy: Supplies framework for message interpretation	Construct redundancy: Cues appropriate constructs for message understanding	Outcomes
Level A Encourages noise-defeating learner *selection* states.	1. Use sounds to help learners direct attention. Example: Employ novel, bizarre, and humorous auditory stimuli.	2. Use sounds to help learners isolate information. Example: Group or simplify content information conveyed to help learners auditorially isolate and disambiguate message stimuli.	3. Use sounds to help learners tie into previous knowledge. Example: Recall learner's auditory memories and evoke existing schemas for sound associations.	LEARNER IS INTERESTED
Level B Encourages noise-defeating learner *analysis* strategies.	4. Use sounds to help learners focus attention. Example: Alert learners to content points by using sound to show them where to exert information-processing effort.	5. Use sounds to help learners organize information. Example: Help learners differentiate among content points and create a systematic auditory syntax for categorizing main ideas.	6. Use sounds to help learners build upon existing knowledge. Example: Use sound to situate the new material within real-life or metaphorical scenarios from learners' experience.	LEARNER IS CURIOUS
Level C Encourages noise-defeating learner *synthesis* schemes.	7. Use sounds to help learners hold attention over time. Example: Immerse learners by using sounds that help make them feel the content is relevant and meaningful to their lives.	8. Use sounds to help learners elaborate upon information. Example: Build upon established sound syntaxes to supplement the content and supply mental models.	9. Use sounds to help learners integrate new material into overall knowledge structures and prepare for transfer to new learning contexts. Example: Help learners transfer knowledge to new learning situations by building useful auditory adjuncts to overall knowledge structures that might be more easily retrieved later.	LEARNER IS ENGAGED

between content points and main ideas (cell 5), and to build upon existing knowledge by situating the material under study within real-life or metaphorical scenarios (cell 6). Likewise, a learner's level of engagement might be increased using sounds to hold attention over time by making the lesson more relevant (cell 7), to elaborate upon information by supplying auditory images and mental models (cell 8), and to prepare knowledge for later use by providing additional auditory knowledge structures that might be useful in subsequent learning (cell 9). When designed systematically into the instruction in this way, sound might supplement instructional messages with the additional content, context, and construct support necessary to overcome many of the acquisition, processing, and retrieval problems one might encounter while learning.

This more deliberate and theory-grounded approach to the selection and use of various modalities in instructional communications might be a key to identifying auditory message cues that can facilitate learning from instructional technologies. However, in a recent review of the research literature, Bishop (2013) observed that traditional perspectives on the design of instructional messages had failed to keep up with theoretical and technological developments over the last twenty years. Consequently, research and practice in this area is still firmly rooted in a linear, "transmission" view of instructional communication that fails to capitalize on the affordances of newer learner-centered technologies or to take adequately into account the learner's active role in the process. In order for multimedia sound to evolve from "an add-on to a learn-from technology" as suggested by Mann (2008, 1169), we will need to make the shift from a transmission to a transactional view of communications theory and explore the ways in which sound can be used to facilitate cognition, improve motivation, and support knowledge construction. It is from this perspective that the development of interactive audio technologies might have its greatest impact on sound's use to enhance learning by giving us new ways to think more dynamically about the use of sound in instructional technologies.

Note

1. It should be noted here that, while the "top" scoring narration-only groups scored fairly well on the knowledge-level matching tests ($M = 7.10$ out of 8 for the lightning lesson and $M = 4.15$ out of 6 for the braking lesson), this same group scored only $M = 11.05$ out of 19 on the lightning retention test and $M = 3.95$ out of 8 on the braking retention test, calling into question whether 180 seconds with either of these scientific–mechanical concepts was really sufficient for a group of novices to learn this material very thoroughly at all, regardless of the message cues used or how they were employed.

References

Alten, Stanley R. 1999. *Audio in Media*. Belmont, CA: Wadsworth.

Armstrong, Thomas. 1994. *Multiple Intelligences in the Classroom*. Alexandria, VA: Association for Supervision and Curriculum Development.

Atkinson, R. C., and R. M. Shiffrin. 1968. Human Memory: A Proposed System and its Control Processes. In *The Psychology of Learning and Motivation: Advances in Research and Theory*, ed. Kenneth W. Spence and Janet T. Spence, 89–195. New York: Academic.

Attneave, Fred. 1959. *Applications of Information Theory to Psychology: A Summary of Basic Concepts, Methods, and Results*. New York: Holt.

Baddeley, Alan D. 2000. The Phonological Loop and the Irrelevant Speech Effect: Some Comments on Neath. *Psychonomic Bulletin and Review* 7 (3): 544–549.

——. 2001. Levels of Working Memory. In *Perspectives on Human Memory and Cognitive Aging: Essays in Honour of Fergus Craik*, ed. Moshe Naveh-Benjamin, Morris Moscovitch, and Henry L. Roediger, 111–123. New York: Psychology Press.

——. 2002. Is Working Memory still Working? *European Psychologist* 7 (2): 85–97.

——. 2003. Working Memory: Looking Back and Looking Forward. *Neuroscience* 4: 829–839.

Baddeley, Alan D., and Jackie Andrade. 2000. Working Memory and the Vividness of Imagery. *Journal of Experimental Psychology: General* 129 (1): 126–145.

Barbe, Walter B., and Raymond H. Swassing. 1979. *Teaching through Modality Strengths: Concepts and Practices*. Columbus, OH: Zaner-Bloser.

Bennion, Junius L., and Edward W. Schneider. 1975. *Interactive Video Disc Systems for Education*. Provo, UT: Instructional research, development, and evaluation, Brigham Young University.

Berlo, David K. 1960. *The Process of Communication: An Introduction to Theory and Practice*. San Francisco: Rinehart.

Bishop, M. J. 2000. *The Systematic Use of Sound in Multimedia Instruction to Enhance Learning*. Dissertation Abstracts International.

——. 2013. Instructional design: Past, present, and future relevance. In *Handbook for Research in Educational Communications and Technology*, 4th edn., ed. J. M. Spector, M. D. Merrill, J. Elen, and M. J. Bishop, 373–383. New York: Springer.

Bishop, M. J., Tonya B. Amankwatia, and Ward Mitchell Cates. 2008. Sound's Use in Instructional Software to Enhance Learning: A Theory-to-practice Content Analysis. *Educational Technology Research and Development* 56 (4): 467–486.

Bishop, M. J., and Ward Mitchell Cates. 2001. Theoretical Foundations for Sound's Use in Multimedia Instruction to Enhance Learning. *Educational Technology Research and Development* 49 (3): 5–22.

Bishop, M. J., and David Sonnenschein. 2012. Designing with Sound to Enhance Learning: Four Recommendations from the Film Industry. *Journal of Applied Instructional Design* 2 (1): 5–15

Boyd, Gary McIntyre. 2004. Conversation Theory. In *Handbook of Research on educational communications and technology*, ed. David H. Jonassen, 179–197. Mahwah, NJ: Lawrence Erlbaum.

Bregman, Al. 1990. *Auditory Scene Analysis: The Perceptual Organization of Sound*. Cambridge, MA: MIT Press.

——. 1993. Auditory Scene Analysis: Hearing in Complex Environments. In *Thinking in Sound*, ed. Stephen McAdams and Emmanuel Bigand, 10–36. New York: Oxford University Press.

Broadbent, Donald E. 1958. *Perception and Communication*. New York: Pergamon.

Calandra, Brendan, Ann E. Barron, and Ingrid Thompson-Sellers. 2008. Audio Use in E-learning: What, Why, When, and How? *International Journal on E-Learning* 7 (4): 589–601.

Cates, Ward Mitchell, M. J. Bishop, and Woei Hung. 2005. Characterization versus Narration: Drama's Role in Multimedia Instructional Software. *Journal of Educational Technology Systems* 33 (4): 437–460.

Clark, Ruth, Frank Nguyen, and John Sweller. 2006. *Efficiency in Learning: Evidence-based Guidelines to Manage Cognitive Load*. San Francisco: Pfeiffer.

Collins, Karen. 2008. *Game Sound: An Introduction to the History, Theory, and Practice of Video Game music and Sound Design*. Cambridge, MA: MIT Press.

Craig, Scotty. D., Arthur C. Graesser, Jeremiah Sullins, and Barry Gholson. 2004. Affect and Learning: An Exploratory Look into the Role of Affect in Learning with AutoTutor. *Journal of Educational Media* 29 (3): 241–250.

Craik, F. I. M. 1979. Human Memory. *Annual Review of Psychology* 30, 63–102.

Dale, Edgar. 1969. *Audiovisual Methods in Teaching*. New York: Dryden.

De La Cruz, Guadalupe, and Nick Kearney. 2008. Online Tutoring as Conversation Design. In *Handbook of Conversation Design for Instructional Applications*, ed. Rocci Luppicini, 124–143. Hershey, PA: Information Science Reference.

Deutsch, Diana. 1986. Auditory Pattern Recognition. In *Handbook of Perception and Human Performance*, ed. Kenneth R. Boff, Lloyd Kaufman, and James P. Thomas, 32.1–32.49. New York: Wiley.

Dewey, John. 1987. *Art as Experience*. Edited by Jo Ann Boydston. Carbondale: Southern Illinois University Press.

Driscoll, Marcy Perkins. 2005. *Psychology of Learning for Instruction*. Boston: Allyn and Bacon.

Dunn, Rita, Kenneth Dunn, and G. E. Price. 1979. Identifying Individual Learning Styles. *Student Learning Styles: Diagnosing and Prescribing Programs*, 39–54. Reston, VA: National Association of Secondary School Principals.

Fiske, John. 2011. *Introduction to communication studies*. New York: Routledge.

Flavell, John H. 1976. Metacognitive Aspects of Problem Solving. In *The Nature of Intelligence*, ed. L. B. Resnick, 231–236. Hillsdale, NJ: Erlbaum.

Gardner, Howard E. 1983. *Frames of Mind: The Theory of Multiple Intelligences*. New York: Basic Books.

——. 1993. *Multiple Intelligences: The Theory in Practice*. New York: Basic Books.

Gibbons, Andrew S. 2009. The Value of the Operational Principle in Instructional Design. *Educational Technology* 49 (1): 3–9.

Gibbons, Andrew S. and P. Clint Rogers. 2009a. The Architecture of Instructional Theory. In *Instructional-design Theories and Models*, vol. 3: *Building a Common Knowledge Base*, ed. Charles M. Reigeluth, and Alison A. Carr-Chellman, 305–326. New York: Routledge.

——. 2009b. Coming at Design from a Different Angle: Functional Design. In *Learning and Instructional Technologies for the 21st Century*, ed. Leslie Moller, Jason Bond Huett, and Douglas M. Harvey, 15–25. New York: Springer.

Gygi, Kathleen. 1990. Recognizing the Symptoms of Hypertext…and What to do about it. In *The Art of Human-computer Interface Design*, ed. B. Laurel, 279–287. Reading, MA: Addison-Wesley.

Hankersson, Darrel R., Greg A. Harris, and Peter D. Johnson. 1998. *Introduction to Information Theory and Data Compression*. Boca Raton, FL: CRC.

Hannafin, Michael J., and S. R. Hooper. 1993. Learning Principles. In *Instructional Message Design: Principles from the Behavioral and Cognitive Sciences*, ed. Malcolm L. Fleming, and W. Howard Levie, 191–231. Englewood Cliffs, NJ: Educational Technology Publications.

Harrison, R. P. 1972. Nonverbal Behavior: An Approach to Human Communication. In *Approaches to Human Communication*, ed. Richard W. Budd and Brent D. Ruben, 253–268. Rochelle Park, NJ: Hayden.

Humphreys, Glyn W., and Vicki Bruce. 1989. *Visual Cognition: Computational, Experimental, and Neuropsychological Perspectives*. Hillsdale, NJ: Lawrence Erlbaum.

Keefe, James W. 1979. Learning Style: An Overview. In *Student Learning Styles: Diagnosing and Prescribing Programs*, 1–17. Reston, VA: National Association of Secondary School Principals.

Kort, B., R. Reilly, and R, Picard. 2001. An Affective Model of Interplay between Emotions and Learning: Reengineering Educational Pedagogy—Building a Learning Companion. In *Proceedings of the IEEE International Conference on Advanced Learning Technology: Issues, Achievements and Challenges*, ed. T. Okamoto, R. Hartley Kinshuk, and J. P. Klus, 43–8. Madison, WI: IEEE Computer Society.

Krendl, Kathy A., William H. Ware, Kim A. Reid, and Ron Warren. 1996. Learning by any other Name: Communication Research Traditions in Learning and Media. In *Handbook of Research for Educational Communications and Technology*, ed. David H. Jonassen, 93–111. New York: Macmillan.

Leman, Marc. 2007. *Embodied Music Cognition and Mediation Technology*. Cambridge, MA: MIT Press.

Leonard, A. 1955. Factors which Influence Channel Capacity. In *Information Theory and Psychology: Problems and Methods*, ed. Henry Quastler, 306–315. Glencoe, IL: Free Press.

Luppicini, Rocci. 2008. Introducing conversation design. In *Handbook of Conversation Design for Instructional Applications*, ed. Rocci Luppicini, 1–18. Hershey, PA: Information Science Reference.

Mann, Bruce L. 2008. The Evolution of Multimedia Sound. *Computers and Education* 50: 1157–1173.

Marr, David. 1982. *Vision: A Computational Investigation into the Human Representation and Processing of Visual Information*. San Francisco: Freeman.

Mayer, Richard E. 2001. *Multimedia Learning*. Cambridge, UK: Cambridge University Press.

Mayer, Richard E., and Roxana Moreno. 1998. A Split-attention Effect in Multimedia Learning: Evidence for Dual Processing Systems in Working Memory. *Journal of Educational Psychology* 90: 312–320.

——. 2003. Nine Ways to Reduce Cognitive Load in Multimedia Learning. *Educational Psychologist* 38 (1): 43–52.

Mayes, J. Terry. 1992. Multimedia Interface Design in Education. In *Multimedia Interface Design in Education*, ed. Alistair D. N. Edwards and Simon Holland, 1–22. New York: Springer-Verlag.

McAdams, Stephen. 1993. Recognition of Sound Sources and Events. In *Thinking in Sound*, ed. Stephen McAdams and Emmanuel Bigand, 146–198. New York: Oxford University Press.

Miller, George A. 1956. The Magical Number Seven, Plus or Minus Two: Some Limits on our Capacity for Processing Information. *Psychological Review* 63 (2): 81–97.

Moles, Abraham A. 1966. *Information theory and esthetic perception*. Urbana: University of Illinois Press.

Moore, Brian C. J. 1982. *Introduction to the Psychology of Hearing*. London: Academic.

Moreno, Roxana, and Richard E. Mayer. 1999. Cognitive Principles of Multimedia Learning: The Role of Modality and Contiguity. *Journal of Educational Psychology* 91: 358–368.

——. 2000. A Coherence Effect in Multimedia Learning: The Case for Minimizing Irrelevant Sounds in the Design of Multimedia Instructional Messages. *Journal of Educational Psychology* 92: 117–125.

Newcomb, Theodore M. 1953. An Approach to the Study of Communicative Acts. *Psychological Review* 60: 393–404.

Ormrod, Jeanne Ellis. 2003. *Human Learning*. New York: Prentice Hall.

Paas, Fred, Alexander Renkl, and John Sweller. 2003. Cognitive Load Theory and Instructional Design: Recent Developments. *Educational Psychologist* 38 (1): 1–4.

Paivio, Allan. 1986. *Mental Representations: A Dual Coding Approach*. New York: Oxford University Press.

Pangaro, Paul. 2008. Instruction for Design and Designs for Conversation. In *Handbook of Conversation Design for Instructional Applications*, ed. Rocci Luppicini, 35–48. Hershey, PA: Information Science Reference.

Parrish, Patrick E. 2005. Embracing the Aesthetics of Instructional Design. *Educational Technology*. IVLA Conference, October 2005. www.unco.edu/cetl/sir/making.../Aesthetic%20Principles_Web.doc.

——. 2008. Plotting a Learning Experience. In *Handbook of Visual Languages in Instructional Design*, ed. Luca Botturi, and Todd Stubbs, 91–111. Hershey, PA: Information Science Reference.

——. 2009. Aesthetic Principles for Instructional Design. *Educational Technology Research and Development* 57: 511–528.

——. 2010. Aesthetic Decisions of Teachers and Instructional Designers. In *Transformative Learning and Online Education: Aesthetics, Dimensions and Concepts*, ed. T. Volkan Yuzer and Gulsun Kurubacak, 201–217. Hershey, PA: Information Science Reference.

Peterson, Lloyd R., and Margaret Jean Peterson. 1959. Short-term Retention of Individual Verbal Items. *Journal of Experimental Psychology* 58: 193–198.

Pinker, Steven. 1985. *Visual Cognition*. Cambridge, MA: MIT Press

Severin, Werner J., and James W. Tankard. 1979. *Communication Theories: Origins, Methods, Uses*. New York: Hastings House.

Shannon, Claude E., and Warren Weaver. 1949. *The Mathematical Theory of Communication*. Urbana, IL: University of Illinois.

Snow, Richard E. 1997. Aptitudes and Symbol Systems in Adaptive Classroom Teaching. *Phi Delta Kappan* 78 (5): 354–360.

Sousa, David A. 2006. *How the Brain Learns*. Thousand Oaks, CA: Corwin.

Sweller, John, Paul Ayers, and Slava Kalyuga. 2011. *Cognitive Load Theory*. New York: Springer.

Technology Milestones. 1997. *THE Journal* 24 (11). http://www.thejournal.com/magazine/97/jun/techmile.html.

Travers, Robert Morris William, ed. 1964a. *Research and Theory Related to Audiovisual Information Transmission*. Salt Lake City: University of Utah Press.

——. 1964b. The Transmission of Information to Human Receivers. *AV Communications Review*, 12: 373–385.

Treisman, Anne M., and Garry Gelade. 1980. A Feature Integration Theory of Attention. *Cognitive Psychology* 12: 97–136.

Tulving, Endel. 1972. Episodic and Semantic Memory. In *Organization of Memory*, ed. Endel Tulving, and Wayne Donaldson, 381–403. New York: Academic.

——. 1983. *Elements of Episodic Memory*. Oxford: Clarendon.

Warren, Richard M. 1982. *Auditory Perception: A New Synthesis*. New York: Cambridge University Press.

Wilson, Brent G. 2005. Broadening our foundation for instructional design: Four pillars of practice. *Educational Technology* 45 (2): 10–15.

CHAPTER 5

ACOUSTIC SCENOGRAPHY AND INTERACTIVE AUDIO

Sound Design for Built Environments

JAN PAUL HERZER

THE inclusion of interactive audio concepts, procedural sound and adaptive music is continually gaining importance in new multimedia applications, electronic devices, and the growing game industry. Such technologies create multiple challenges concerning the development of functional and flexible sound concepts, but offer exciting opportunities for design practices.

Some disciplines in which new sound design approaches can be used in a purposeful way are architecture, interior design, and *scenography* (performance spaces and museography). Even though the emphasis in the general discourse about the design of built environments is mostly focused on visual perception (electro)acoustic design has a deep impact on the individual experience of one's surroundings. Thus, the implementation of specialized audio concepts can augment the sensory perception of environments where sound is perceived in subtle as well as in deliberately noticeable ways. Additionally, the growing awareness of acoustic ecology evokes the need for a sensitive design approach—especially regarding situations in which visitors and employees are exposed to designed sound for a long period of time (Schafer 1994).

As a result of these technological and aesthetic requirements, an increasing number of design practices focus on the creation of audio applications for built environments. Specialized sound designers and programmers generate acoustic environments, sound sculptures, and sonic interfaces in the realm of architecture, interior and exhibition design, scenography, and contemporary art in public space. There is no clear definition for this field of work, but the term "acoustic scenography" is a term that befits theory and practice.

This chapter presents a brief overview of the technical and theoretical aspects of the design of interactive and reactive sound concepts in a wide range of applications. Furthermore, it will assemble procedures and fundamentals from a practical point of view and bring out specific approaches for the implementation of algorithms, procedural sound processing, and the use of interactive systems. I will connect and consolidate some concepts, while at the same time putting together a "toolbox" that will provide suggestions for researchers and designers with a theoretical or practical interest.

I work actively in the field of acoustic scenography and therefore, logically, the following pages will be biased at certain points. The chapter is not intended to be a universal set of categories and rules, but, rather, a helpful collection of basic approaches and techniques that should encourage further experiments, research, and practical implementations on the intersection of sound design and architecture.

5.1 Acoustic Scenography

5.1.1 Perceiving Sound in Built Environments

When perceiving one's surroundings, sound is often an undervalued part of the personal experience. While the visual appearance and structure of the space is obvious to the human individual, the acoustic surrounding may appear to play a minor role. The architect Steen Eiler Rasmussen tries to contradict this common view by comparing sound and light:

> most people would say that as architecture does not produce sound, it cannot be heard. But neither does it radiate light and yet it can be seen. We see the light it reflects and thereby gain an impression of the form and material. In the same way we hear the sounds it reflects and they, too, give us an impression of form and material. Differently shaped rooms and different materials reverberate differently. (Rasmussen 1962, 224)

Sound can provide the listener with information about the structure, distance, and shape of their surroundings. The reflected sound of clapping hands can allude to the presence of a wall at close proximity, for instance. Or, the sounds of our footsteps may illustrate the surface of the floor we are walking on. The acoustic properties of a space and the audible feedback to human movement and action influence the listener's perception in different ways.

Not only do these elemental parameters shape spatial perception; research in the last decades of the twentieth century gave birth to new disciplines that put the listener and their cultural and social environment at the center of attention. Sound Studies, the "world soundscape project" (originally initiated by R. Murray Schafer), Aural Architecture, and other individual research projects started exploring the influence a person's cultural imprint and social environment have on his or her perception of sound. The individual interpretation of a sonic event and the implications of objects in a room that fulfill a specific task or play a certain role in social life are elemental to the listener's experience on site. The individual combination of sounds a listener experiences and reacts to in a space consist of more than sensory stimuli and contain a complex framework of influences and parameters. Not only physical phenomena shape the listener's aesthetic sense of space, but also cultural and social influences, orientation, music, and voice recognition play an important role.

5.1.2 Designing Sound for Built Environments

There has been much research conducted in the fields of room acoustics and sound insulation, as well as on their effects on the broad field of architecture. Parameters such as diffusion, reverberation, and absorption help describe, design, and optimize the acoustic properties of a room. The geometry of a space, its general surfaces and building materials, strongly influence its acoustics and are now increasingly planned intentionally, even in situations where the performance and playback of music (e.g., in concert halls or cinemas) is not the only purpose of a building. Computer simulations make it possible to estimate the impact of constructional changes on the final acoustic "fingerprint" of a building. While these physical and mostly calculable parameters of room acoustics have been part of design processes for some time now, the all-embracing shaping of the aural experience merits further consideration, especially since it often includes the design of artificially generated sounds coexisting with an architectural concept and interior design process (see Blesser and Salter, 2007).

The growing number of multimedia applications in everyday life—in spaces independent from their function and original purpose—have led toward the need of a steady and distinguished confrontation with acoustic environments. The number of media messages consistently grows and sound gets intentionally designed in so many ways—though the architecture one lives and works in has to be regarded in a sensitive way.

Music that has been composed specifically for a certain space has a somewhat strange reputation and is often misunderstood. Satie's *musique d'ameublement* was one historic attempt to compose music specifically designed to be subtle and subconscious while being stimulating and comfortable at the same time. Eno's "Music for Airports" followed a comparable approach, and "ambient music" is now considered a distinct musical genre. Entire business models have grown on the idea that music played back at points of sale increase sales and change customer behavior (Behne 1999).

In contrast to the subconscious manipulation of visitors in environments primarily conceived for sales, there are other applications that demand an immersive and active experience, and thus include something other than comforting and soothing music. Interactive exhibits, the playback of movies, and complex projection mapping, as well as media art and sculptures often generate the need for specialized sound design. These developments and upcoming challenges have lead to the emergence of new professional disciplines in which acoustic and aural architects, acoustical consultants, and sound designers are jointly concerned with the process of the intentional design of aural experiences.

5.1.3 A Concept of Acoustic Scenography

Scenography teaches how to design and enrich spaces and experiences through an integrative design process that combines creative, artistic and technological parameters (Bohn and Heiner 2009, 9). While architecture as an artistic language with spatial design elements includes multiple creative aspects that are concerned with the whole shape of a built environment, scenography more strongly focuses on the individual experience in connection with actively generated experiences and artificial settings, including stage design for theaters and operas. As in architecture, the range of purposes a design concept is meant to fulfill varies widely. Spaces in which scenographic concepts are implemented may reach from urban and public space and traffic areas to rooms for living, work, entertainment, performance, and education. In many cases the concepts include multiple fields, such as corporate architecture, interior design, visual arts, light, sound, and scent design. Special emphasis on scenography can be found in spaces that are meant to communicate greater ideas like educational topics or brands. While for example the design of office spaces will primarily be subject to practical considerations, museums, science centers, and spaces of corporate communication (e.g., retail stores) often offer more creative freedom in the process. In these cases a purely functional and practical design is not substituted, but gets strongly influenced by narrative, experience-oriented, and artistic approaches and thus the possibilities of designed aural experiences are immense.

Acoustic scenography, a specialized variant of scenography, connects designers and engineers who contribute to applications in scenography with applications that involve sound. These applications cover a wide range, from small objects and interfaces to complex multichannel systems used in big exhibitions. Acoustic scenography draws its fundamental techniques from conditioning, musical socialization and dramaturgy. Music, sound, and voice can trigger and influence emotions on a personal or bioacoustic level, and designers can use this knowledge to create exciting and highly immersive sound environments. As in interface design and film music, the creation of aural experiences is enhanced by using the fact that sound and music easily evoke and communicate functions and thus support functions of objects and architectural spaces as well. By using sound, a designer can both communicate about processes and point out facts, while relying on known musical figures, clichés, and conditioned functions of sounds. Even the guidance of visitors and the accentuation of specific purposes of a space are possible, and simplified, through modern technology. Bearing that in mind, special applications like the sonification of touch-screen commands or acoustic feedback in home-automation technology can be used purposefully and not only as a "gadget." Additionally, the use of sound differs from other disciplines in architecture and interior design because of the simple fact that, in contradiction to furniture, architectural figures, and light, sound can easily be altered over time. Thus, sound can indicate highlights and focus points while constantly varying the audio content.

5.2 Interactive Audio in Built Environments

As described above, acoustic scenography involves multiple fields of work and touches many different technological and conceptual applications. The implementation of complex multimedia systems and sophisticated technology for the playback of both acoustic and visual content became easier and more common in the last decade and created a "playground" for designers of diverse multimedia applications, including highly specialized audio programming and interface situations using sound. Even working and living spaces are increasingly furnished with a technological backbone that allows wireless communication, interaction, and automation in the course of a growing market for ambient intelligence and smart homes. This steady technological evolution, awareness, and acceptance in media, as well as the availability of infrastructure and suppliers for hard- and software, has enabled designers, engineers, and artists to create highly specialized and complex installations, environments, and interfaces. In terms of acoustic scenography, this evolution particularly has an impact on interactive and nonlinear applications.

This text uses the term "interactive audio" as a description for a larger scope of techniques concerning interactive, adaptive, and reactive audio and music in different multimedia formats. The variable interpretation of the term shows both the newness of the topic and the broad spectrum of techniques, abstract concepts, and ideas associated with interactive audio. As with the terminology, research in the field of interactive audio is dominated by professionals and scholars whose activities are mainly rooted in video-game audio. This field has brought to life publications and online resources, and it shares a lot of insight and ideas that the area of scenography lacks when it comes to practical and pragmatic appliance of nonlinear and interactive audio and music concepts to the built environment.

While the interactive dimension of a concept in acoustic scenography can play a vital role, further approaches toward the design of an aural experience are fundamental in a design process and intersect with each other constantly. Thus the reflection on interactive audio and acoustic scenography must include an overview and general techniques as well.

5.2.1 Conception, Design, and Composition Techniques

Acoustic scenography as part of an integral design process may be used in early stages of architectural planning or in the short-term development of an exhibition or trade fair. Whereas the level of involvement and the possibilities of influence may differ, especially concerning concepts that involve drastic changes of architectural structure and technology, general techniques, methods, and approaches toward the integration of interactive audio concepts emerge from a set of procedures and theoretical ideas.

5.2.1.1 *Spatial Orchestration*

A sound designer who specializes in the creation of aural experiences in spaces basically follows rules and structures of architecture, interior design, and scenography. Easily comparable to lighting design, sound can help to structure rooms and differentiate between functions and purposes of certain spaces by applying different sound moods and distinguishable pieces of music. Similarly, it offers the ability to guide visitors by the use of signals or to highlight certain elements and exhibits by using directional and specifically designed sound. Acoustic elements can accentuate functions and architectural shapes and figures, and field recordings or artificially composed ambiences may build the illusion of certain countries or locations. Finally, narrative content communicated through voice recordings and audio drama strongly helps to communicate didactic content, for example in exhibitions and fairs.

The fact that sound easily passes long distances and is hard to be emitted unidirectionally—and thus it travels "around corners"—it is often considered a problem in spaces where multiple objects produce sound, such as in exhibitions where atmospheric sounds accompany movies, narrators communicate content, and field recordings simultaneously create virtual settings. Overlapping sound content can cause irritation when didactic content and music are involved, and the simultaneous playback of independent sound layers causes an uncomfortable and musically atonal result. Under the premise of the inevitable interference of multiple sound sources in a space, the purposeful placement of objects and deliberate composition of content can lead to a more pleasing result. In game audio, different layers are composed in a way that they will match harmonically in any circumstance (Geelen 2008), and the same technique can be applied to sound design in spaces. Likewise, the directed and alternating positioning of different types of sound sources (human voice, natural ambience, music) and the control of timing through an upstream control unit can avoid undesirable outcomes.

Along with the specific composition of content for multiple sources, the different technological possibilities of sound reproduction in space should be considered as well. Stereophonic two-channel-playback of linear audio can be expanded through the use of modern technology to enhance the aural experience. Both the use of specialized sound-reproduction setups (e.g., surround sound and wave-field synthesis) and different speaker types (e.g., directional ultrasonic speakers, subsonic transducers, and piezo elements) can effectively change the aural experience.

When the designer keeps control over content, positioning, timing playback technology, and the resulting crossovers between different sounds, he or she will be able to compose and orchestrate the aural experience coherently and not as a collection of simultaneously but independently operating sound sources.

5.2.1.2 *Interactive, Reactive, Adaptive, and Nonlinear Audio*

One of the commonest ways to add sound to architectural space electroacoustically is through the linear playback of recorded audio, for instance using CD or DAT players and a rudimentary combination of an amplifier and speakers. Today's technology allows for more complex setups with the use of computers, sensors, and specialized software. Two of the biggest improvements that newer solutions provide are nonlinearity and interactivity of the audio content.

Strictly speaking, the use of software that is able to directly access passages of music and sound already defines its nature as nonlinear in comparison to linear media like magnetic tape and movie film (O'Sullivan and Igoe 2004, 18). But only the randomization or interactive and intentional alteration over time makes nonlinear audio so valuable for acoustic scenography. Linear playback systems confront a visitor with audible content that follows a predefined timeline and is fixed to a sequence that existed beforehand. Interactive and nonlinear compositions in contrast offer the chance of adapting to the actions of a visitor, reacting to surrounding conditions, and repeatedly varying in form and structure. Repetition of both content and playback in unwanted situations can be avoided since specific programming and setup can directly influence the process of playback.

In the process of adding an interactive component to a playback situation, a basic question should be: Which level of interactivity and consciousness experienced by the visitor is desired? Obvious and playful interactive systems may strengthen the amount of attention and thus the level of attraction an exhibit gains. On the other hand, the process of interacting with an object can draw the focus away from the initial message or content and a potential learning effect can be overlaid by pure entertainment (Simanowski 2008, 47). In some situations the unconscious adaption of background music may be appropriate, or just the audible feedback to the interaction with a touch screen is enough to enrich an experience while not being noticed by the user.

General approaches to composition and sound design for application in interactive audio systems differ from the classic and linear composition of music. Techniques such as branching and layering and the sophisticated use of musical transitions and changes allow a flexible use of the prepared material (Kaae 2008). Musical techniques in part draw inspiration and experience from ambient and minimal music. While it is possible to reconstruct finished compositions and make them work in nonlinear applications, the creation of original material should preferred.

5.2.1.3 *Generative and Procedural Audio*

Compared to sample-based nonlinear playback, the methods of the procedural and generative creation of audio content can be much more complex, but also more sophisticated. The emancipation from predetermined audio material renders it an attractive alternative to linear playback. The generation of sound and music through mathematical rules and algorithms has a history reaching back to musical automats in the seventeenth century and to experiments with aleatoric and stochastic music in the twentieth century (Ruschkowski 1998, 261–5). Besides generative composition, where the organization of rhythmic and harmonic structure gets controlled for example by changing sequences of numbers, procedural sound design can also include simple methods of simulating physical phenomena like wind, through sound synthesis. The imitation

FIGURE. 5.1 "Pulsing around Tbilisi."

Photographer Gio Sumbadze.

of natural processes and random events can be used additionally when creating virtual nature ambiences.

The added value of procedural and generative sound-design techniques for use in acoustic scenography lies in the uniqueness of every sonic event or musical figure and the avoidance of repetition. The complexity of the programming can be immense when the goal is the creation of a broad spectrum of sounds based on synthesis. Still the randomly generated variations of music and rhythmic patterns can lead to extremely variable compositions and are helpful when it comes to playback situations where people are exposed to sound for longer periods of time. Figure 5.1 shows an installation, "Pulsing around Tbilisi," that made use of a generative rhythm composition that was played back in a public pedestrian underpass in Tbilisi, Georgia. Rhythmic fragments were generated based on a number series by a microcontroller and were altered over time. The sunlight influenced the composition through the use of photoresistors. The resulting pattern slowly evolved, and the "clicking" sounds—generated just by closing and opening a circuit through a loudspeaker—hinted at the acoustic properties of the space. The procedural aspect of the programming was added in the process when it became obvious that static repetition of patterns would disturb people working in small shops surrounding the underpass.

5.3 Technical Implementation

The above-mentioned techniques and approaches mostly rely on computer technology. The use of microcontrollers, computers, and digital audio formats applies in most of these situations. Both nonlinear playback of audio content and interactive systems need a specific set of technological elements to work. General techniques stay the same independent of specific software and hardware and thus can be roughly categorized as described below.

5.3.1 Playback and Processing

Computer technology opens up extensive possibilities for the design of interactive sound environments. Specialized software enables designers to create complex systems that synthesize sound and control its playback. Solutions include software like Max/MSP and Pure Data (Pd), which were developed for the creation and programming of sound. More abstract programming languages, such as Processing and OpenFrameworks and basic coding in C or assembly, can be used in the process as well. The evolution of digital audio workstation software like Ableton Live and its extension Max4Live even enable designers with almost no experience to program and build complex setups that produce interactive and generative sound and music. In combination with microcontrollers that help translating information gathered by sensors in the environment, this software allows for the communication with input from the physical world, and can render setups highly interactive, playful and informative. Besides the use of microcontrollers like Arduino and MakeController, there are a growing number of "plug and play" solutions, which simplify the process of programming even more. Figure 5.2 illustrates a schematic of an input–processing–output setup.

In addition to solutions that include programming, even simple combinations of components can exceed the possibilities that the linear playback of audio content provides. Consumer electronics are now highly affordable and memory capacity is barely an issue any more. Cheap DVD players provide multichannel audio, compact-flash players can be triggered and controlled through sensors and microcontrollers and even smartphones can be programmed to be interfaces for playback of nonlinear audio.

One of the core elements in the creation of nonlinear audio for spaces is the inclusion of transducers for the conversion of events and conditions of the physical world into data, which is accessible and interpretable by a computer system (O'Sullivan and Igoe 2004, 19). In essence, different physical actions and states are recognized by a sensor or interface, undergo processing, and result in an electroacoustic event or a change in the way audio gets generated. Transducers of different kinds allow the monitoring of parameters such as direction and speed of movement, temperature, and brightness, and the use of the gathered information for the control and alteration of audio playback and synthesis. Today's technology opens multiple ways of sensing of and reacting to the surrounding spaces and the actions of visitors. Examples like the Microsoft Kinect illustrate that even highly sophisticated methods are increasingly available for use even in low-budget projects.

In addition to the interpretation of data from surroundings, the communication with other multimedia systems can be essential. Especially when a designer aims for a multimodal approach when developing an environment, he or she may face new challenges regarding the communication with other computers and technical setups. Fortunately, communication standards such as OSC and Arcnet exist and evolve and help, for instance with synchronizing an illumination system with the playback of audio content.

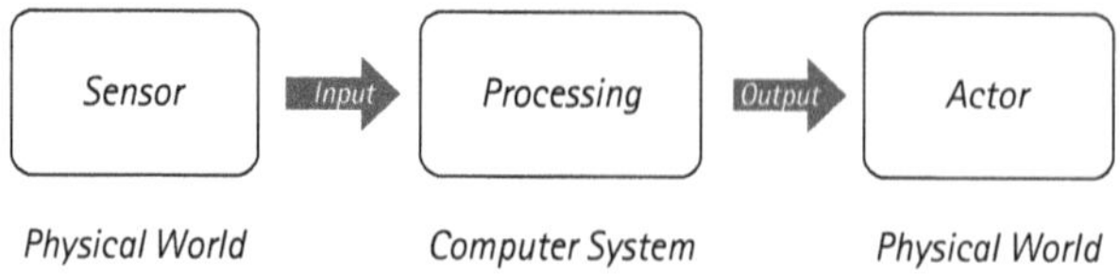

FIGURE. 5.2 Schematic Input/Processing/Output.

5.3.2 Speakers and Electroacoustic Transducers

As described above, decisions about technical components that reproduce sound are important. While in many cases the development of sound installations and exhibits focuses on programming, sound, and interaction design, the consideration of appropriate technology often gets lost en route. But the deliberate selection of speakers and transducers can strongly shape the aesthetic impression of sound.

A classic system to play back sound is a two-channel setup, which reproduces prerecorded audio using typical loudspeakers. Such a system probably will not play back all content undistorted and will imprint its frequency response on the content. In other situations, it is possible to select speakers of an adequate size to match the audio content and even intensify its acoustic attributes. There are various conventional ways to reproduce sound to simulate spatial positions through the creation of a phantom sound sourc[1], for instance through playback that attaches every sound to its very own type of speaker. While a typical surround-sound setup can produce an intense experience, the optimal listening position is limited to a small point in the room. Situations in which the visitor constantly changes his position may need other approaches. Directional playback systems that allow sound to be directly projected near one's ears through the use of ultrasonic audio have interesting possibilities.

5.4 Conclusions

The compilation of approaches, theories, and techniques given above illustrates a few noteworthy facts. First of all, a precise definition of acoustic scenography has yet to be formulated. Many different design disciplines include the creation of sound for built environments, but none really claims the applied design process as its main focus. This uniqueness is not necessarily a negative condition, but a more precise definition could increase general attention for the field of work and its importance in today's design processes. Additionally, the researchers, architects, and designers involved could benefit from a stronger exchange, particularly opening up the theoretical aspects of architecture, which could benefit from the kind of pragmatism in the design process that can be found, for example, in publications about game audio.

Today's technological evolution and its impact on the design process are immense. Specialized hard- and software, an increasing number of open-source and DIY projects, sophisticated speaker systems, and at last accessible research results open up interactive sound concepts to a broader field of interested designers and for their application in areas of different focus and with different budgets. Once a designer has surveyed and recognized the large number of tools and basic concepts, technology enables him or her to produce highly specialized, interactive, and attractive sound concepts. The use of procedural composition and interactive audio can improve the quality of the acoustic surroundings and counteract constant sensory overload as well. The steady reinvention of musical patterns through algorithms, the avoidance of repetitive figures, and especially the possibility of interacting with one's sound environment can lead to a much more pleasant auditory sensation.

On the way to accepting fields such as acoustic scenography as on a level comparable to, for instance, lighting design, aspiring designers still have to face ignorance, and they must try to educate potential customers and business partners. Sound design is still often considered to be an "add-on" or afterthought in many design situations, thus gets poorly budgeted and is rarely integrated into the planning from the start of a process. An integral design process is particularly important in fields like scenography, where the immersion of visitors often plays a vital role.

Finally, it should be pointed out that besides the need for a controlled and reviewed activity, the self-expression and artistic evolution of designers involved in the process are essential and must not be overruled by categories and academic discourse. Design disciplines like scenography may widely be seen as a more defined and structured practice than is customary in the arts, and yet the self-expression of the creator plays an important role in the process of creating a unique and immersive experience, even if it is minor compared with functional considerations. In the end, the designer is responsible for the aural experiences evoked through a sound concept, and that experience cannot be controlled by a set of rules and definitions, even though the inclusion of theoretical and systematic insights can improve and positively influence the design process.

Further Reading

Atelier Brückner, publ. Scenography: Making spaces talk. Ludwigsburg: avedition, 2011.

Collins, Nicolas. Handmade Electronic Music: The Art of Hardware Hacking. New York: Routledge, 2009.

Grueneisen, Peter. Soundspace: Architecture for Sound and Vision. Basel: Birkhäuser, 2003.

Hug, Daniel. „Ton ab, und Action! Narrative Klanggestaltung interaktiver Objekte." In *Funktionale Klänge*, edited by Georg Spehr, 143–170. Bielefeld: transcript Verlag, 2009.

Klanten, Robert and Sven Ehmann and Verena Hanschke, ed. A Touch of Code: Interactive Installations and Experiences. Berlin: Die Gestalten Verlag GmbH & Co. KG, 2011.

Sauter, Joachim and Susanne Jaschko and Jussi Ängeslevä. ART+COM: Media Spaces and Installations. Berlin: Die Gestalten Verlag GmbH & Co. KG, 2011.

Schricker, Rudolf. Kreative Raum-Akustik für Architekten und Designer. Stuttgart München: Deutsche Verlags-Anstalt GmbH, 2001.

Van Geelen, Tim. "Realizing groundbraking adaptive music." In From Pac-Man to Pop Music: Interactive Audio in Games and New Media, edited by Karen Collins, 93–102. Hampshire: Ashgate, 2008.

Note

1. The simultaneous playback of the same audio event through two loudspeakers at the same time creates the impression of one virtual sound source that is located somewhere between the two loudspeakers. See Michael Dickreiter, Handbuch der Tonstudiotechnik. Band 1: Raumakustik, Schallquellen, Schallwahrnehmung, Schallwandler, Beschallungstechnik, Aufnahmetechnik, Klanggestaltung (München: K. G. Saur Verlag KG, 1997), 124. Dickreiter, Michael. Handbuch der Tonstudiotechnik. Band 1: Raumakustik, Schallquellen, Schallwahrnehmung, Schallwandler, Beschallungstechnik, Aufnahmetechnik, Klanggestaltung. München: K. G. Saur Verlag KG, 1997.

References

Behne, Klaus-Ernst. 1999. Zu einer Theorie der Wirkungslosigkeit von (Hintergrund) Musik. In *Musikpsychologie: Bd. 14 Wahnehmung und Rezeption*, ed. Klaus-Ernst Behne, 7–23. Göttingen: Hogrefe, Verlag für Psychologie.

Blesser, Barry, and Linda-Ruth Salter. 2007. *Spaces Speak, Are You Listening? Experiencing Aural Architecture*. Cambridge: MIT Press.

Bohn, Reiner and Wilharm, Heiner. 2009. "Einführung" in Inszenierung und Ereignis: Beiträge zur Theorie und Praxis der Szenographie, edited by R. Bohn and H. Wilharm, 207–268. Bielefeld: Transcript Verlag.

Eno, Brian. 1978. *Ambient 1: Music for Airports*. Polydor AMB 001 [CD].

Geelen, Tim van. 2008. Realizing Groundbreaking Adaptive Music. In *From Pac-Man to Pop Music: Interactive Audio in Games and New Media*, ed. Karen Collins, 93–102. Aldershot, UK: Ashgate.

Kaae, Jesper. 2008. Theoretical Approaches to Composing Dynamic Music for Video Games. In *From Pac-Man to Pop Music: Interactive Audio in Games and New Media*, ed. Karen Collins, 75–92. Aldershot, UK: Ashgate.

O'Sullivan, Dan, and Tom Igoe. 2004. *Physical Computing: Sensing and Controlling the Physical World with Computers*. Mason: Corse Technology PTR.

Rasmussen, Steen Eiler. 1962. *Experiencing Architecture*. Cambridge: MIT Press.

Ruschkowski, André. 1998. *Elektronische Klänge und musikalische Entdeckungen*. Stuttgart: Reclam.

Schafer, R. Murray. 1994. *The Soundscape: Our Sonic Environment and the Tuning of the World*. Rochester, VT: Destiny.

Simanowski, Roberto. 2008. *Digitale Medien in der Erlebnisgesellschaft: Kultur—Kunst—Utopien*. Reinbek bei Hamburg: Rowohlt Taschenbuch Verlag.

SECTION 2

VIDEOGAMES AND VIRTUAL WORLDS

CHAPTER 6

THE UNANSWERED QUESTION OF MUSICAL MEANING

A Cross-domain Approach

TOM LANGHORST

THE study of musical meaning has a long tradition, from the ancient Greek philosophers to music scholars like Meyer (1956) and Bernstein (1976). Recently researchers have studied musical meaning not only from a music theory or philosophical approach but also from other scientific disciplines such as linguistics, psychology, and cognitive neuroscience. Although the issue of whether music can communicate meaning in a semantic manner is still a topic of debate (Kivy 2002), it is obvious that musical meaning is especially important in applied music, such as in videogames, where music plays an important role in the player's immersion and interaction (Collins 2008).

Designing meaningful audio for interactive applications, such as videogames or sonic user interfaces, presents designers with several challenges. One challenge is the fact that the perception of musical meaning involves cultural or learned aspects when it comes to tonality (Huron 2006). For applied music and sound design, as cross-cultural phenomenon, this cultural specificity appears to be a significant disadvantage. Nevertheless, the history of interactive music in videogames and sonic user interfaces illuminates many successful examples of meaningful musical icons in classic arcade games such as *Pac-Man, Donkey Kong, Super Mario World*, and *Pong*, and the user interface sounds of operating systems.

6.1 MEANING SUPPORTED BY PSYCHOACOUSTICS

The sounds of a user interface are especially designed to communicate meaning (Gaver 1988). If we look at one of the error sounds of the Microsoft Windows XP operating system we hear a sound that can be divided into two contrasting segments. Segment one consists mainly of higher frequencies and is followed by segment two containing not exclusively but dominant low frequencies. Figure 6.1 shows the most prominent frequencies in both segments. From this, we can conclude that the E♭ frequency is most dominant in segment one and the B♭ frequency is most dominant in segment two, although less by their loudness but more by the repetition of B♭ frequencies in the spectrum as can be seen in Figure 6.1 and Table 6.1.

Tonal hierarchic analysis of the error icon's melodic progression from E♭ to B♭, based on the theory of Lerdahl and Jackendoff (1983) and Lerdahl (2001), suggests an unresolved movement from tonic to dominant. Progressions like this, from tonic to dominant, facing the cognitive processing with an unresolved comma, question mark, or open end, are common practice in the sound design of applied music and can also be found, for instance, when inserting a USB device, in contrast to the dominant to tonic progression when the USB device is removed, or when the operating system starts up or returns from a sleeping state to its normal operating state. The tonic-to-dominant progression, as used here, shows many resemblances to an unfinished version of the

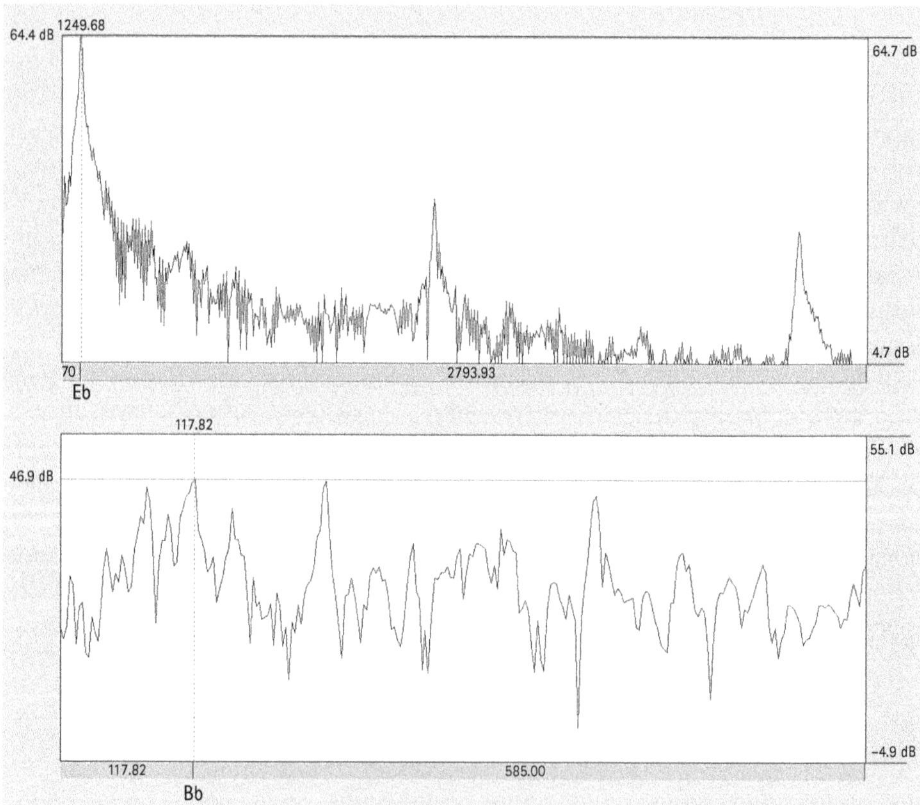

FIGURE 6.1 The most important frequencies in segment one (upper part) and segment two (lower part) show the prominence of the E♭ in the first and B♭ in the second segment.

Table 6.1 Overview of dominating frequencies in the error icon sound's spectrum, Microsoft Windows XP. The bold/bigger frequencies in the table are significantly louder than the other frequencies

Frequency (in Hz)	Pitch segment 1	Pitch segment 2
3750	B♭	
2500	E♭	
1872	B♭	B♭
1250	E♭	E♭
936		B♭
468		B♭
234		B♭
117		B♭
78		E♭

classical opening that can be found in numerous tonal compositions in the classical idiom with the purpose of establishing the tonality of the composition.

From a psychoacoustic perspective, the peak at 1250Hz, the E♭ in the first segment, is interesting. Because the frequency lies inside the range in which our sensitivity to loudness is the highest (Howard and Angus 2009, 97) and because the frequency is by far the loudest in the error icon, as a result of which the waveform of the first segment is dominated by the sine waveform at a frequency of 1250Hz, it will be the frequency that attracts all of the listener's attention. Drawing the user's attention might very well be the reason why this 1250Hz frequency is used in such a prominent way and thus contributes to the meaningful interpretation of the error icon. However, it does not account for the way in which the icon communicates the meaning of error or failure. Table 6.1 provides an overview of the dominating frequencies in the error icon spectrum. At first sight, all harmonic intervals seem to be consonants with simple frequency rations of 1:2 or 2:3.

The extent to which a harmonic interval of two sine waves can be regarded as consonant or dissonant is quantified by the amplitude modulation pattern they produce called "beatings" (Tramo et al. 2003, 138). Although the perfect fifth of E♭ and B♭ at the bottom of the second segment of the error icon can be considered as consonant in the biggest part of the range of musical pitches, in this low region it cannot, and thus will produce beatings, shown in Figure 6.2, at a rate that is perceived as dissonant. In other words, this perfect fifth interval is placed inside the critical bandwidth, the region in which the audio cannot be resolved (coded as two different frequencies), which is the basic principle of the spectral model for pitch perception (Wang and Bendor 2010), and therefore the fifth is perceived as dissonant.

The meaningful interpretation of the icon becomes clear if we regard how different aspects (factors) of music are perceived as expressions of emotion. Russell's two-dimensional Valence and Arousal model is used to describe the musical factors and

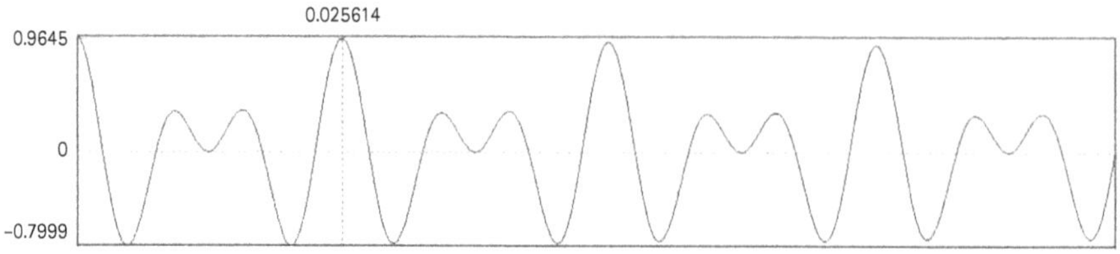

FIGURE 6.2 The autocorrelation analysis (Roads 1996, 509–11), made with Praat, of two frequencies (E♭ at 78 Hz and B♭ at 117 Hz at the lowest part of the second segment of the error icon) shows an amplitude modulation or beating pattern with 25.6 ms intervals. This pattern matches a beating frequency of 39 Hz, which lies inside the critical range of 20–200 Hz beatings that are characterizing for dissonant intervals (Tramo et al. 2003).

their perceived emotions. High pitches cause a high level of arousal, while dissonance causes a low, negative, level of valence (Gabrielsson 2010). Therefore, the error icon is designed in such a way that the first segment grabs the user's attention, and is followed by a segment that can be perceived as negative because of the negative valence. This is exactly what an error icon is supposed to do, and it does so by creating a large contrast between high and low frequencies, a concept that will be discussed later in more detail.

Dissonance of harmonic intervals of complex tones is based on the beating effect between all components of the spectrum (Howard and Angus 2009, 153–7). Therefore, there is an important difference between the dissonance in the error icon and the dissonance in more common dissonant intervals like the interval of a major second. Since the intervals within the error icon's spectrum are mostly consonant in the higher region, the perfect fifth in the error icon creates less dissonance in its higher components than, for example, a major second would. The resulting affect of negative valence is therefore of a rather subtle character.

6.2 *Pong* Success and Failure

Most people are familiar with the sound design of one of the first videogames, *Pong*. Despite their simplicity, the *Pong* sounds have become iconic examples of arcade game sounds. Furthermore, their simplicity does not rule out the fact that the designers of *Pong* had high ambitions for their game sounds. Like all game designers they faced the challenge of designing meaningful audio to provide feedback to the player's actions. Obviously semantically meaningful audio in the form of recorded or synthesized text phrases (such as "you win," "you lose") were impossible or simply too complex to implement at that time. The alternative proposed by Atari's founder, Nolan Bushnell, and others was one that included the prosodic affect of a cheering or booing crowd.

> "Once I'd gotten the game to play pretty well, Nolan said it had to have sound. And he said I want to have the sound of a crowd approving. And somebody else said I want to have hisses and boos if you lose. And I'm thinking, I have no way/idea how to make this at all. I'm already way over my budget. I've got too many chips in this thing as it is. So I simply poked around with a little audio amplifier in the circuit and found tones that sounded about right and wired them in. It was less than half a chip to put those sounds in and I said "That's it, Nolan." (Al Alcorn in BBC documentary on *Pong*, http://www.youtube.com/v/ShyRGWRcagY)"

The *Pong* sounds may have been created more or less by accident (Collins 2008, 8), but nevertheless, as Al Alcorn said, "they sounded about right." If so, this implies that the sounds do indeed communicate meaning. The question is why and how? Before answering these questions, let us take a closer look at the sounds used in *Pong*. Two of the sounds have to do with gameplay action events (*ball hits bat* and *ball hits wall*). Although sounds related to the player's actions can communicate meaning and express emotion, here I focus on sounds that evaluate the player's actions: sounds that give the player meaningful feedback. The *success* and *failure* sounds of *Pong* do exactly that and are used as communicative feedback based on the player's actions.

The most obvious way that this meaning can be established, seems to be by classical conditioning proposed by Pavlov ([1927] 2009). According to Pavlov's theory, we learn to recognize success and failure sounds because we experience the relationship between the two sounds and the gameplay actions with which they are related (*ball going out* and the *changes in the score*). It is through the recurrence of these two sounds, the learning process, that we remember the meaning of the sounds and are able to describe their meaning when we hear them, even when we aren't playing the game. Although Pavlov's theory explains how the meaningful interpretation of sounds can be learned, it does not explain *why* we perceive the two sounds as representatives of the distinct phenomena of success and failure. In Al Alcorn's idea of "right," this idea means that both sounds *sound right* and thus do represent the phenomena of success and failure in a semiotic way. Moreover, despite the fact that the failure sound is related to a negative outcome, it still can be qualified as sounding *right*. How can something bad sound right?

The answer lies in the theory of misattribution. Huron (2006) explains that the human brain is designed to predict the future successfully and therefore can evaluate a negative outcome positively if the outcome matches the prediction: "If my account is correct, then it is not the frequency per se that accounts for the experience of pleasure, but sure and accurate prediction. That is, the pleasure of the exposure effect is not a phenomenon of 'mere exposure' or 'familiarity.' It is accurate prediction that is rewarded—and then misattributed to the stimulus" (Huron 2006, 138–9).

So far, this approach explains how we can regard the success and failure sounds of *Pong* as well designed, but it still does not answer the question of why the two sounds can be regarded as meaningful (good vs. bad) in a semiotic manner. To be able to answer this question, we must examine the intrinsic characteristics of the two sounds. It is obvious that the design principle of the two sounds is based on contrast. As Figures 6.3 and

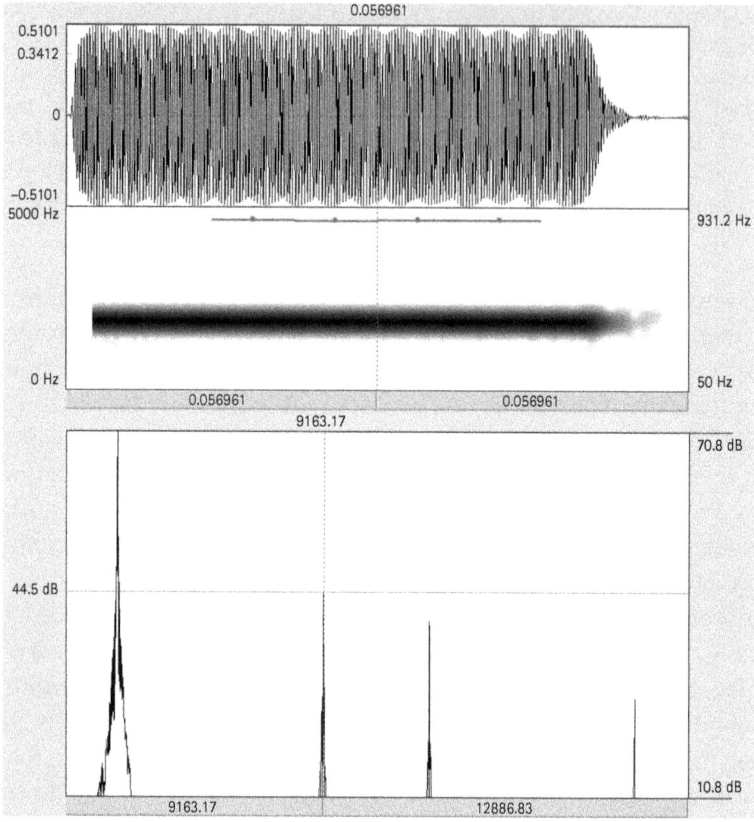

FIGURE 6.3 Praat analysis of the *Pong* success sound shows the waveform, spectrogram, pitch (931.2 Hz), and spectrum (0–20 kHz).

6.4 illustrate, success and failure differ in almost everything from pitch height to loudness, spectrum and waveform.

From Gabrielsson's study (2009) we can derive which emotions in the Valence–Arousal model of Russell may be perceived from the two contrasting sounds of *Pong*.

Because of the ambiguity of the perceived emotions (e.g., a high pitch can be perceived as anger but also as happiness) it is dangerous to conclude that the sounds derive their meaning from nothing more than their own intrinsic musical character. Gabrielsson's findings support the idea that the two sounds are based on the design principle of contrast and therefore give meaning to each other in a dialectic manner.

Although the intrinsic characterization of the two *Pong* sounds may not be sufficient or strong enough to represent their meaning, the study of Tagg and Collins (2001) shows that a higher pitch versus lower pitch can be used in a musical context to express positive versus negative. In their study, various aspects of utopian and dystopian music are described. Their analysis shows consistency in the universal design principle of contrast to communicate contrasting meanings. Their study also shows that contrast in brightness ("bright, day time, sunny" vs. "dark, night time, foggy/misty/rainy"), can be used to

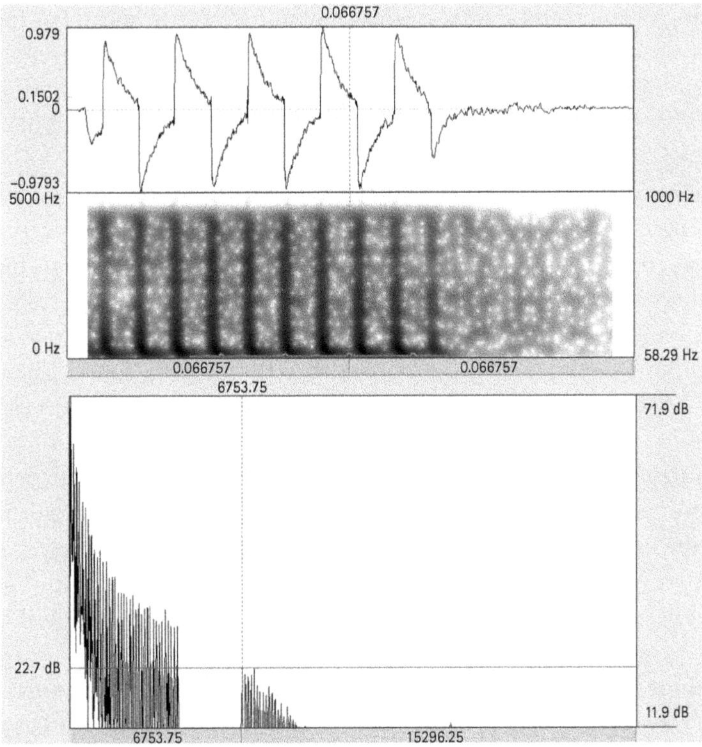

FIGURE 6.4 Praat analysis of the *Pong* failure sound shows the waveform, spectrogram, pitch (58.29 Hz), and spectrum (0–20 kHz).

describe the contrast in utopian and dystopian music. Brightness, although often used to describe musical phenomena, is in fact a vision-related description.

Several studies (e.g., Marks 1989; Collier and Hubbard 2004; Datteri and Howard 2004) describe the cross-modal relationship between the color wavelengths of light and audio frequencies and pitch. This relationship, based on neurological overlap in

Table 6.2 Overview of possible perceived emotional expressions of the *Pong* sounds. Based on Gabrielsson (2009), 143–5

Factor	Sound	Perceived emotion (Gabrielsson 2009)
High pitch	Success	Happy, graceful, serene, dreamy, exciting, surprise, potency, anger, fear, and activity
Low pitch	Failure	Sadness, dignity, solemnity, vigor, excitement, boredom, and pleasantness
Timbre	Lower/fewer harmonics (success) Complex (failure)	pleasantness anger
Loudness	Approx 6dB difference[i]	Increased loudness indicates power, intensity, excitement, tension, anger, and joy
Interval[ii]		Large intervals suggest power.

the processing of audio and vision, is consistent and inversely linear. For the two *Pong* sounds, this means that the two sounds can be perceived as meaningful, contrasting not only in the auditory but also in their cross-modal perception, where success is related to *bright* and failure is related to *dark*; or according to Tagg and Collins (2001), as *utopian* and *dystopian*.

Notice the similarity between the high–low contrast of the two *Pong* sounds and the high–low contrast in the error icon discussed earlier. Although the concept of contrast is used in both, the difference between the two sounds is caused by the fact that the error icon uses the contrasting low region for the additional psychoacoustic affect of dissonance. Dissonance can be related to a negative valence (Gabrielsson 2009) to represent the error icon's meaning, which is not embedded in the low *Pong* sound, since this sound consists of only one single tone and not of a harmonic interval and thus cannot create dissonance.

In addition to the auditory and visual domain, the meaningful perception of the *Pong* sounds can also be supported from the language (phonological) domain. Analyzing the originally intended sounds of the cheering and booing crowd, a prosodic affect providing the sounds' meaning, we can see that the most important differences between the two appear in the phonological aspects: (1) vowel timbre and (2) pitch.

The first three formants (f1, f2, and f3) are important for the difference between different vowels (Howard and Angus 2009, 220). Table 6.3, based on the studies of Peterson and Barney (1952), shows the frequencies of f1, f2, and f3 for the cheering sound's vowel *e* (ɛ[1] in *bet*) and the booing sound's vowel *oo* (ʊ in *book*) that can be compared with the vowels of cheers (ɛ) and booing (ʊ).

From Table 6. 3, it can be concluded that the difference between the two vowels is similar to the difference in the two *Pong* sounds. Since timbre can be used as a universal prosodic code element to express and communicate musical emotion during music performance (Juslin and Laukka 2003), one can conclude that the *Pong* sounds express and communicate musical emotion and thus meaning.

Altogether, the *Pong* sounds for success and failure derive their meaning from a combination of aspects from different and universal domains. Still, and perhaps because of these relationships with other domains, it is likely that the meaning of the two sounds can be learned and conditioned quickly and easily, as implied by Pavlov's theory. This idea also has to do with *Pong* gameplay, which can be considered one-dimensional in

Table 6.3 First three formants of the vowels in *bet* and *book*, based on Peterson and Barney (1952)

Vowel	Male formants (f1, f2, f3)	Female formants (f1, f2, f3)	Children's formants (f1, f2, f3)
bet	530, 1850, 2500	600, 2350, 3000	700, 2600, 3550
book	300, 850, 2250	370, 950, 2650	430, 1150, 3250

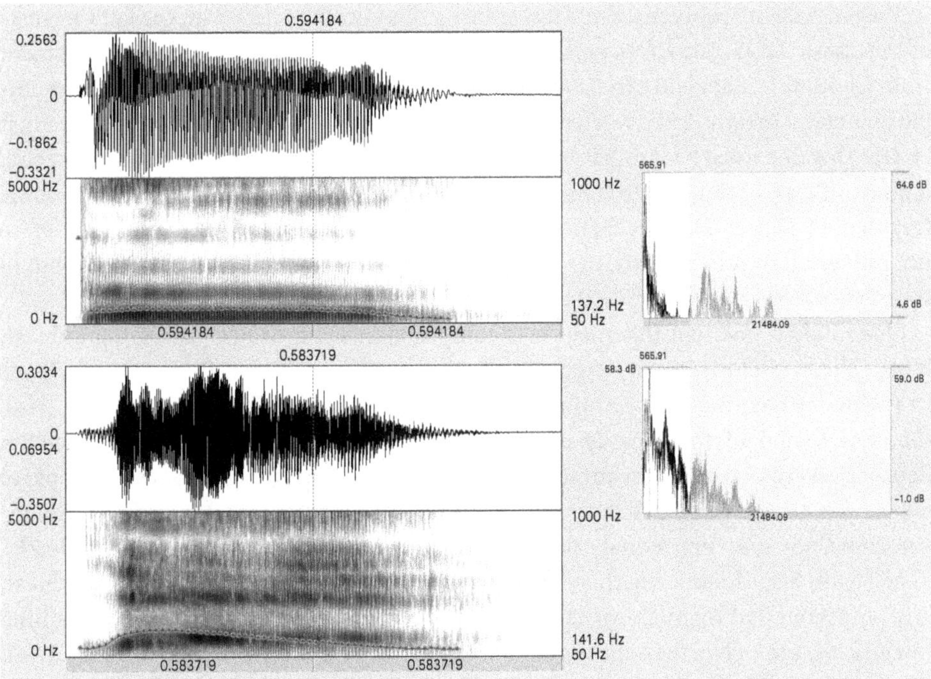

FIGURE 6.5 Praat analysis of a cheering (yeah) and a booing (boo) sound shows the waveform, spectrogram, pitch contour, and (to the right) the spectrum. In the spectrum, the vowel relevant frequency range (approx. 500 Hz to approx. 3000 Hz) is highlighted.

the sense that each gameplay challenge has only one out of two possible outcomes: the player either scores or loses a point.

6.3 Multidimensional Gameplay

Unlike *Pong*, many games have a more indirect relationship between success and failure during the gameplay of a level. In such games, the in-level gameplay challenges foresee more positive than negative outcomes, and it is more likely that sounds for success are action event related (jump, pick up, shoot, etc.). Only the accomplishment of reaching the end of a level is celebrated with a more elaborate success sound or music fragment that gives feedback on the player's actions. In-level failure is, compared to the in-level successes, less foreseen and thus more severe and often associated with *death* (losing life). In-level failure sound or music therefore needs to communicate the dramatic, deathly loss. *Donkey Kong, Super Mario World*, and *Digger* are classic games that use elaborate failure music fragments to express the loss of life.

The universality problem of musical failure icons becomes clear with the failure music of the classic DOS game *Digger. Digger*'s sound designer(s) chose a melodic fragment from Chopin's Sonata No. 2 in B♭ minor, known as the Death March; an excellent example of musical meaning in a referential way (Meyer 1956). Well chosen as it may seem, it is clear that one must be familiar with this musical piece of Chopin (to be more precise, with its title as a reference to its meaning), to understand the meaning of the melodic fragment in *Digger. Digger*'s failure icon is based on cultural knowledge and therefore not universal. In other words, the player needs to learn the meaningful relationship of the melodic icon and the fact that he has lost one life in the game.

One might argue that this melody is composed in a minor key expressing sadness, by which the melody could be perceived meaningfully, and that there is a difference in valence perception between the major and minor mode (Gabrielsson 2009). Also, the slow tempo of the melody might support this feeling of sadness (Gabrielsson 2009). However, whether the minor key is a universal identifier for sadness is a topic of debate: Bernstein (1976) describes minor as a psychoacoustic aspect in musical composition and not as an emotional category. In "'Universal' Music and the Case of Death," Tagg (1988) also doubts whether the minor key can be universally related to sadness, and points out that many European folk melodies use the minor key: "There is nothing intrinsically sad in Northern Europe about the minor key, as anyone who has sung *What Shall We Do With The Drunken Sailor?* or danced to a minor mode reel, *rull* or *polska* will willingly witness." For now, let us conclude that there is room for debate on when the minor key can be universally associated with sadness matching the loss of life in a videogame.

Analyzing the *Donkey Kong* melody, we see that the B♭ major triad forms the basic structure of the melody. The melody starts with the B♭ to F ascending perfect fifth interval confirming the chord of B♭, followed by a chromatic sequence of perfect fifths from D♭ –A♭ descending to B♭ –F. The last fifth (B♭ –F) is the same as the first fifth, only one octave lower. What follows is a broken form of the major triad of B♭ : D–F–F–B♭.

Similar to the *Donkey Kong* melody, the *Super Mario World* melody is also organized hierarchically around one major triad: C, the dominant of F major which is the main tonality of *Super Mario World*. Tonal interpretations such as the ones for the *Donkey Kong* and *Super Mario World* melodies involve a hierarchical organization of pitches, and can be found in the music theory of Schenker (1969), Lerdahl and Jackendoff (1983), and many others.

FIGURE 6.6 Transcription of the failure melodies of *Donkey Kong* and *Super Mario World*.

Although both are based on a single triad, there is an important difference in the two failure melodies, considering the tonal hierarchies in which they are used. B♭ as the tonic of *Donkey Kong*'s music does not provide the tonal structure with an unresolved urge of prolongation, whereas the C in *Super Mario World*, being the dominant of the main key of F major, indeed leaves the structural prolongation unresolved and thus can be perceived cognitively as an "open end" or "promise of continuation." This structure is consistent with *Mario*'s gameplay in which the player is given a second and even third life to play after his initial failure. The melody of *Super Mario World*, in other words, expresses failure but not eternal death and can be compared with the meaningful way in which the tonic-to-dominant progression is used in many user interface sounds, as discussed earlier.

Eternal death is presented to the player of *Super Mario World* only when he has lost his life for the fifth time and is treated to the "game over" music with a cadenza in the key of C. In this cadenza, after the broken chord of C, the C minor tones B♭ and A♭ are introduced in the melody, and are harmonically supported by the subdominant F (without third) and the tritone substitute D♭ (Lerdahl 2001, 311–12), which resolves to the C major (local) tonic chord.

Since both the game-over music and failure melody are built around C, the dominant in *Mario*'s main key F major, they can easily be followed by the "pick up" music based on a IV–V progression in F, to start a new game or next life. Nevertheless, there is an important difference between the game-over cadenza and the failure melody concerning their melodic contour. The cadenza's melody conforms much more closely to melodic conventions (Narmour 1990; Huron 2006), whereas the failure melody, as will be discussed later, has a very unconventional contour which is important in terms of its meaning. Because the cadenza combines all the harmonic functions of tonic, subdominant, and dominant and the *ritenuto* at the end of the cadenza, a code of musical expression (Juslin and Timmers 2010), it derives meaning in a much more classical way than the failure melody.

As with the *Pong* sounds, a meaningful perception of the two failure melodies might start with the concept of misattribution. In other words, the *Super Mario World* and *Donkey Kong* melodies match what we expect to hear when things go wrong, raising the question of what intrinsic characteristics cause the expression of failure or death.

Tonal, hierarchic organization also serves as an important aspect of the pattern-based (gestalt) meaning of music (Meyer 1956; Narmour 1990; Schellenberg 1997) and tonal melodies share their hierarchic structure with language intonations (Patel 2008). Based on research in language and music perception, Patel writes "despite important differences between the melodic systems in the two domains, there are numerous points of contacts between musical and linguistic melody in terms of structure and processing" (2008, 238).

The relationship between linguistic intonation and melody makes the tonal melody more universal (although to what extent is still unclear) but does not say anything about the meaningful interpretation of the *Donkey Kong* and *Super Mario World* melodies as

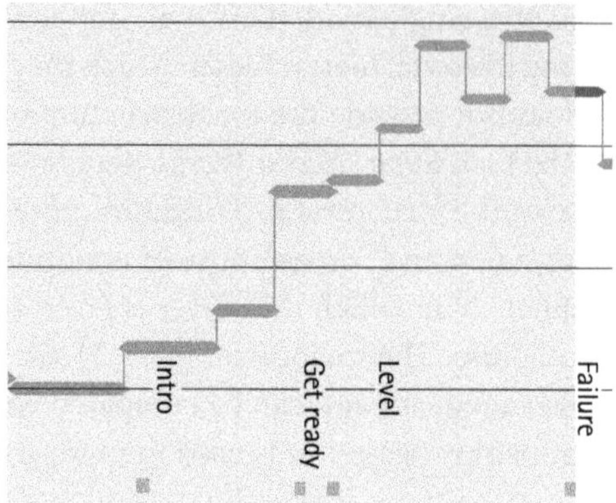

FIGURE 6.7 Breathing rate of subject listening to *Donkey Kong.*

representatives of failure and "sudden" death. For this kind of interpretation we need to return to the human expectations within tonal melodies. Both melodies have a strong closure at the end: *Donkey Kong* by the broken B♭ major chord and *Super Mario World* by the descending octave interval from C to C. Narmour (1990) explains how melodic closures contribute to a meaningful perception. Because of Narmour's complex description, Huron (2006, 157) quotes Elizabeth Margulis's characterization of closure in Narmour's theory: "The simplest way to think of Narmour's notion of closure," says Margulis, "is as an event that suppresses expectation." In other words, both melodies end with a melodic formula that brings all further listening expectations to an end; this idea can be regarded as a metaphor for death.

The first segment of the *Donkey Kong* failure icon shows the use of fast-moving chromatic intervals. Pitch expectation within a tonal context has long been studied (e.g., Krumhansl 1990). Krumhansl (1990) shows a clear hierarchy of expectancy of pitches within a tonal context. In a neurological context, expectancy of pitch is related to reaction speed and required processing time (Huron 2006, 50). This notion of expectancy means that less-expected pitches in the tonal context demand more processing effort than more-expected pitches, implying that tonal expectancy might have a physiological impact. Figure 6.7 shows the breathing rate of a subject while listening to the *Donkey Kong* sounds. The subject was given only the audio stimuli and did not play or see the game during this test. The subject was however familiar with the *Donkey Kong* game. At the last marker (the failure sound) the subject's breath rate dropped considerably. The subject was holding his breath for a short time during playback of the failure melody.

Figure 6.7 shows the subject's "ahh" reaction, for which breath-holding (sudden fall in the breathing frequency rate) is typical, when listening to the *Donkey Kong* failure melody. Huron and Margulis (2010) describe several studies concerning physiological reactions (such as heartbeat rate) and musical phenomena. Although further research is needed to support the idea of physiological responses to sudden tonal complexity in a meaningful context, this observation seems consistent with the physiological

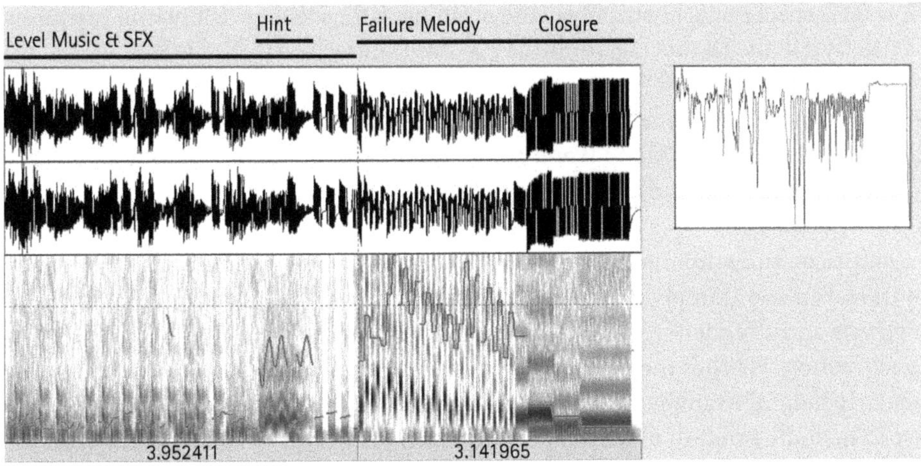

FIGURE 6.8 Praat analysis of the audio example.

reactions that occur when the expectancy is violated. Huron and Margulis (2010) show how Huron's ITPRA (Imaginative–Tension–Prediction–Reactive–Appraisal) theory (Huron 2006) can explain the physiological reactions (chills, frisson, awe, and laughter) when musical expectations are violated. They further point out that the most likely acoustic phenomenon to cause a physiological reaction is the sudden change in loudness (especially a large increase in loudness). Another acoustic phenomenon, the broadening of frequency range, and a sudden change in tempo or rhythm are less solid. In the *Donkey Kong* example, there is no sudden change in the loudness (intensity) but there is a considerable change in frequency (pitch height) between the level music and sound effects and the failure-related hint and failure melody. Also, there is a considerable increase in rhythmic density, which may be responsible for a higher level of arousal (Gabrielsson 2009).

Figure 6.8 illustrates a Praat analysis of the audio example the subject was listening to (the part where the level music and sound effects were "interrupted" by the failure icon of *Donkey Kong*). The intensity overview shows a more or less stable level of loudness, with the closure louder than the rest. The pitch analysis (the dark lines) shows the difference between the level music and the hint and first part of the failure melody. Notice how the closure of the failure melody returns to the average level of pitch heights and how the pitch is unstable in the hint part of the icon. It also shows how the rhythmic density (pulses) is increased during the failure melody (until the closure). Overall, it seems that the *Donkey Kong* failure melody derives its attention-grabbing characteristic from three different intrinsic aspects: (1) increased processing due to tonal complexity; (2) sudden broadening of frequency range (the difference between the average pitch in the level music, compared to the pitch at the beginning of the failure icon's hint and melody); and (3) the sudden increase in rhythmic density.

There is also a fourth aspect, if we take a closer look and notice that the *Donkey Kong* failure melody is preceded by a pitch-unstable sound indicated as "Hint" in Figure 6.8.

How this aspect might contribute to the meaning of the Donkey failure icon will be discussed later in the section regarding the *Pac-Man* failure icon.

So far, these aspects of the *Donkey Kong* melody can explain how the icon is able to grab our attention, and maybe even create a feeling of awe, but not how the melodies of *Donkey Kong* and *Super Mario World* can meaningfully communicate that the gameplay had an unpleasant outcome for the player. There is, however, another interesting aspect to both melodies.

Statistical rules for melodic interval succession (Narmour 1990; Schellenberg 1997; Huron 2006) imply that melodies have a downward tendency in small melodic intervals and a tendency to follow a larger interval with a smaller one in the opposite direction. Neither the *Donkey Kong* nor the *Super Mario World* melody is a particularly helpful example of these rules; each takes the listener quickly downwards over a melodic range of more than two octaves. The result is a melody that is almost impossible for nonmusicians (or probably even many experienced musicians) to sing; melodies such as these are often referred to as "instrumental" melodic progressions. Both melodies combine a relatively high pitch in a fast downward movement to a low-pitched closure. Referring to the cross-modal phenomenon of the relationship between brightness and pitch height perception, one can conclude that both failure icons derive meaning through the sudden transformation from bright to dark.

Altogether, the two failure melodies take the listener from a complex and unexpected chromatic progression (especially *Donkey Kong*) in a pitch region that can be perceived as bright, very quickly, not following the statistical rules for melodic progressions, toward a firm closure in a pitch region that can be perceived as dark. If so, it is not difficult to relate the melodies to the gameplay evaluation of losing a (game) life.

The closure of the *Super Mario World* failure melody introduces an additional aspect of meaningful perception. *Super Mario World*'s melody is harmonized and the final melodic tone is accompanied by the dominant triad C in root position in which the indicated melodic tone is the bass note. Since the region where the triad is voiced is low, this voicing causes the same dissonant affect, due to the beating amplitude modulation between the chord tones, as described earlier, in the lower part of the error icon and can therefore be perceived as negative valence (Gabrielsson 2009).

One final note can be made here regarding the fast downward movement of the melodies over a large range. Juslin and Timmers (2010, 454) envisage expression of musical performances as a multidimensional phenomenon. They describe these dimensions as five components of what they call the GERMS model (Generative rules, Emotional expression, Random fluctuations, Motion principles, and Stylistic unexpectedness). One of these components is the principle of motion "that holds that tempo change should follow natural patterns of human movement or 'biological motion' in order to obtain a pleasing shape."

If this biological or natural evaluation of perceived expression can be applied to more than just the musical movement, we would be able to conclude that the melodies of

Donkey Kong and *Super Mario World* fall outside these natural borders and thus can be perceived as an unpleasant shape instead. Their negative perception can, therefore, partly be explained by their specific melodic contours. It is clear that further research is needed to support this hypothesis but it might very well be that the relationship between positive- and negative-valenced melodic shapes or contours can be related to their more or less biological or natural presentation or form. The fact that relationships may exist between (negative) valence and the natural or biological state and appearance of the stimulus has been verified by several experiments regarding the *uncanny valley hypothesis* (Mori 1970).

A recent study at the University of California (Saygin et al. 2012) shows that there are neurological indications that humans evaluate the valence of movements more according to their motor self-image (on which they build their expectations), than on human likeness alone. In fact, humans seem not to care whether the movement originates from another human being or from a robot but, do seem to be concerned with whether the movement matches their own image of such a movement. If this idea can be applied to musical stimuli, it means that we measure the valence of audio stimuli not only by the success of our prediction, as Huron suggests, but also by the resemblance of the stimulus to our own imaginary abilities to reproduce the stimulus.

The game *Braid* is an effective example of a combination of uncanny valley or unnatural movement and music that do not match our self-imaged perception of movement and musical progression. Failure in *Braid* is related to a time-reversed moving avatar, combined with time-reversed playback of the game music, causing an unnatural, uncanny valley-like affect.

6.4 *Pac-Man* Speech

At first, the meaningful resemblance between the failure icons of *Pac-Man, Donkey Kong,* and *Super Mario World* seems to lie in the fast movement and descending contour of the melody. Analogous to the cross-modal bright-to-dark association, one expects a similar relationship in the *Pac-Man* melody. However the *Pac-Man* melodic contour has a much smaller range, from around 364Hz to 188Hz. Approximately the first part of the *Pac-Man* icon runs chromatically from F downwards to C, over no more than a perfect fourth, and is followed by a closure of two "tones." Compared to the descending contour of more than two octaves in the *Donkey Kong* and *Super Mario World* melodies, the melodic range of *Pac-Man* seems too small to justify the conclusion that the *Pac-Man* icon also derives meaning through the cross-modal perception of brightness.

In view of the biological or natural-motion hypothesis, one can say that a descending melody of less than an octave fits very well into a human self-imaged reproduction. The melody also follows the statistically expected rules of melodic

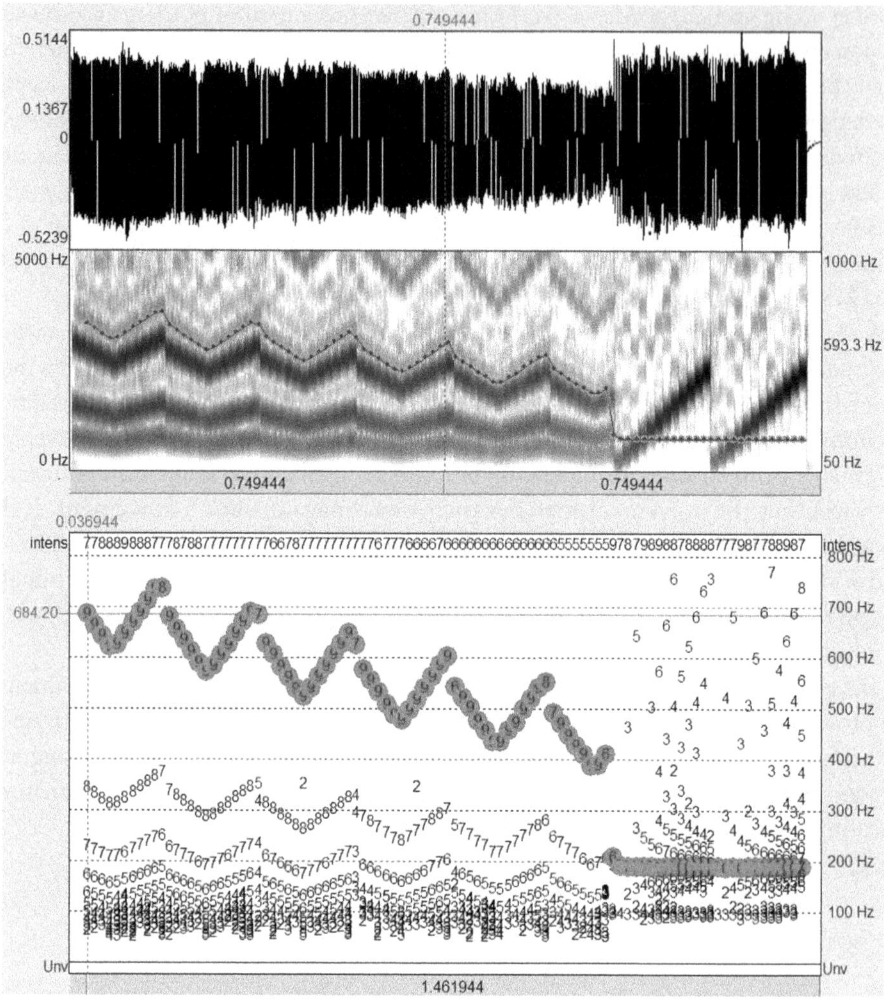

FIGURE 6.9 Praat analysis of *Pac-Man*.

progression (Narmour 1990; Schellenberg 1997; Huron 2006) for descending melodies in small steps.

In order to explain the perceived meaning of the *Pac-Man* melody in a failure-like way, similar to the *Donkey Kong* and *Super Mario World* melodies, the *Pac-Man* melody must have other intrinsic elements than the *Donkey Kong* and *Super Mario World* melodies. Typical of the *Pac-Man* melody are the pitch glides. Both the descending chromatic motif of the beginning and the two closure tones at the end have a pitch-gliding character that can be shown with the Praat analysis.

Figure 6.9 shows the Praat analysis of the waveform, spectrogram, pitch contour, and the perceived pitch probability (lower part) of the Pac-Man melody. It is notable that the chosen pathway of Praat shows a straight line at the end, implying a stable pitch.

Nevertheless, the spectrum and the complex pattern of probability numbers in this specific area show that the pitch is anything but stable. It is likely that subjects listening to this fragment will find it difficult to judge what they hear (stable or changing pitch).

Pitch glides, however, are not typical for music but are typical for speech. Patel describes the use of unstable pitch in speech intonation as follows: "Unlike musical melody, speech intonation is not built around a stable set of pitch intervals" (Patel 2008, 205). Patel asks how the absence of languages using stable pitches can be explained, and argues: "The likely reason is that spoken language mixes affective and linguistic intonation in a single acoustic channel. Affective intonation is an example of a gradient signaling system: Emotional states and the pitch cues signaling both vary in a continuous fashion" (205).

Not only can the use of unstable pitches in the melody be compared with speech but, furthermore, the *Pac-Man* melodic contour resembles the speech intonation contour in two ways. The first resemblance is the descending pitch contour of the melody, which is also a characteristic of speech. Second, the loudness of the *Pac-Man* melody fades toward the end of the first section in a decrescendo, just as the volume decreases toward the end of normal sentences. Both of these characteristics in speech can be explained by the fact that the fuel of speech is the air in our lungs and that we cannot breathe in and speak at the same time. The result is that, in speech, the intonation contour normally descends and the volume of the speech decreases toward the end of a sentence.

It seems that the *Pac-Man* melody derives its attention-capturing character from the fact that important aspects from the speech domain have been transferred to the music domain. In other words, the *Pac-Man* melody shows more resemblance to speech than to music; only the words of speech are missing. Therefore, the *Pac-Man* melody can be described as prosodic. Speech and musical melody processing share important regions in the brain (Patel 2008). If so, the exchange of categories from one domain to another does not support the idea that this is something that is likely to gain as much attention as needed for the *Pac-Man* melody to do its work as a failure icon. However crossing categories from speech to music, and vice versa, may not be as obvious as one might think. Brandt (2009) describes how the cross over from pitch gliding speech to pitch stable music involves a cross over from pragmatic and functional states into nonpragmatic and nonfunctional states of the human mind.

> The "discretization" that transforms an original glissando into a series of distinct tonal steps is crucial to the change from shouting to chanting and singing. The shared experience of articulate singing and of the song-imitating sounds of melodic and rhythmic instruments universally affects our embodied minds by creating "non-pragmatic states," i.e. states of non-functionality—of contemplation, exaltation or even trance—that are typically expected and presupposed in situations of sacredness: celebration, commemoration and invocation. (Brandt 2009, 32)

This idea means that the *Pac-Man* melody represents in itself a strong contradiction between pragmatic and functional speech and nonpragmatic and nonfunctional music.

Perhaps we can also explain the attention-capturing character of the *Pac-Man* melody from the theory of musical expectancy. Can it be that the *Pac-Man* icon derives its urgent character in a way similar to the unexpected chromatic tones in the *Donkey Kong* melody? In other words, does a musical icon that uses categories from the speech domain cause extra processing and slower reaction due to its unexpectancy? It seems that there is reason to believe it does, if we realize how successful this prosodic sound design is. Not only is it the basis for the *Pac-Man* melody but it can also be found in the failure sound of *Frogger* and at the beginning of the *Donkey Kong* failure melody (the hint part before the actual melody). It is also the sound characteristic shared by many sirens and alarms.

There is an important advantage to the prosodic melodic icons: prosody is a universal aspect of sound. Sirens are used all over the world for more or less the same purpose, in more or less the same way, with the same sound design.

As shown in Figure 6.9, the pitch melodic glide in the contour is modulated. Frequency modulation or vibrato is a typical prosodic affect used in the performance of music (Juslin and Timmers 2010). As a prosodic affect, vibrato is a universal code for the communication of expression. However, the combination of a pitch glide and vibrato, as in the *Pac-Man* melody, is rare. For many musical instrumentalists, producing such a "vibrato and glissando" is beyond their capabilities. Even for the human voice, in some respects the most flexible instrument of all, this is difficult to achieve. Therefore the affect in *Pac-Man* can be qualified as being "not reproducible" and thus unnatural or biologically impossible, as described earlier in the context of the uncanny valley hypothesis. However, besides the possible negative valence due to the unnatural state of the sound, the sound also resembles something else: laughter.

As Peretz (2010) describes, affective prosody can have two forms. One is the tone of voice, resulting in what Juslin and Timmers (2010) described as "codes for expressions"; and the other is emotional vocalization, such as laughs, cries, and screams. Peretz points out that there is not enough research to make a clear neurological distinction between the two but describes several studies that support the idea that vocal and musical emotion at least partly share the same neurological pathways, supporting the idea that prosody and vocalization can transfer emotion onto the music domain and can do so in a universal way. Peretz even points out that humans share this quality with other primates.

For a description of laughter we look at Provine (1996): "A laugh is characterized by a series of short vowel-like notes (syllables), each about 75 milliseconds long, that are repeated at regular intervals about 210 milliseconds apart. A specific vowel sound does not define laughter, but similar vowel sounds are typically used for the notes of a given laugh. For example, laughs have the structure of 'ha-ha-ha' or 'ho-ho-ho,' but not 'ha-ho-ha-ho.' There are intrinsic constraints against producing such laughs. Try to simulate a 'ha-ho-ha-ho' laugh: it should feel quite unnatural. When there are variations in the notes, they most often involve the first or last note in a sequence. Thus, 'cha-ha-ha' or 'ha-ha-ho' laughs are possible variants."

The resemblance of the *Pac-Man* melody to laughter is striking. The downwards progression at the beginning takes 1085 ms. Divided over five modulations (laughs), the duration of each laugh is 217 ms which almost equals Provine's indicated length of 210 ms. Although the tones, which are in fact large upwards glides, in the closure section are a bit shorter (190 ms) they can also be recognized as laughs, but with a vowel change. The only thing that does not exactly match Provine's description is the fact that the changed vowel is repeated one more time. *Pac-Man* laughs at you when you lose... What can be more painful than that?

6.5 Conclusions

> "While during the last 150 years linguists have developed a superb discipline of speech about speech, musicologists have done nothing at all about a discipline of speech about music. (Charles Seeger in Nattiez 1987, 150)"

Since Nattiez's criticism, music analysis has come a long way and today's music analysis of tonal hierarchy has benefited from linguistic studies (Lerdahl and Jackendoff 1983) and cross-domain studies of language and music showing neurobiological evidence for related or shared aspects in melody, rhythm, syntax, and meaning (Patel 2008). To what degree musical meaning is innate, how it may be related, for example, to psychoacoustical phenomena or learned by statistical learning (Huron 2006; Patel 2008) and can be influenced by cognitive processes like priming (Bigand and Poulin-Charronnat 2009) is still unresolved. It seems nevertheless evident that tonality learned hierarchy in music plays an important role in the meaningful perception of music. However, it seems premature to conclude that the perception of meaning in music (and more specifically in interactive applied music) is limited by cultural borders. First of all, as the examples in this chapter show, the tonal premises of interactive applied music are simple and straightforward, usually involving no more than a single triad and one-to-one tonal functions of tonic and dominant. Thus, even when it is necessary to learn this level of tonal hierarchy, implying that it does not derive from psychoacoustics, the learning process can be short and will be almost effortless.

Furthermore, due to the global and massive distribution of videogames and sonic user interfaces, these products are their own textbooks, teaching more and more people the basic rules of meaningful perception of tonality in applied music. So even though cultural differences can be observed in, for example, the relationship between musical rhythm and the mother tongue in compositions of French and English composers (Patel et al. 2006) or in Japanese culture (Patel and Daniele 2003), these differences seem less and less significant for the global lifestyle to which video games belong.

In this respect, it is remarkable that the composers of the music for *Donkey Kong* (Yukio Kaneoka), *Super Mario World* (Koji Kondo), and *Pac-Man* (Toshio Kai) all are Japanese; being part of a musical culture that differs significantly from the Western tonal tradition (Patel and Danielle 2003), they nevertheless successfully use tonal aspects when creating meaningful music. Perhaps the combination of simplicity and global distribution is the reason why tonal musical meaning can pragmatically be regarded as universal, even though theoretically it cannot.

As shown by the examples in this chapter, tonality as music theory alone is not enough to explain how musical meaning works. A meaningful perception of the interactive applied music can only be fully understood when aspects of psychoacoustics, psychology, cross-modal perception, cognitive neuroscience, linguistics, phonology, and aspects related to the biological, natural, or self-imaged perceptibility of the audio stimulus are included in music analysis. Therefore, it is to be expected that further research in cross-modality, cognition, and the shared pathways of music, language, motor, and vision will help us to better understand how universally meaningful audio can be designed, how examples of interactive applied music should be analyzed and how eventually a solid theoretical framework on musical meaning can be built.

Note

1. The International Phonetic Alphabet (IPA) symbols are used here to indicate the sounds of the vowels used by a cheering (yeah!) and booing (boo!) audience.

References

Bernstein, Leonard. 1976. *The Unanswered Question: Six Talks at Harvard*. Cambridge, MA: Harvard University Press. DVD, Kultur: 1997, 2001.

Bigand, Emmanuel, and Bénédicte Poulin-Charronnat. 2009. Tonal Cognition. *The Oxford Handbook of Music Psychology*, ed. Susan Hallam, Ian Cross, and Michael Thaut, 59–71. Oxford: Oxford University Press.

Boersma, Weenink. 1999 *Praat: Doing Phonetics by Computer*. http://www.fon.hum.uva.nl/praat/.

Brandt, Per Aage. 2009. Music and how we became human: A view from cognitive semiotics. In *Communicative Musicality: Exploring the Basis of Human Companionship* eds. Stephan Malloch and Colwyn Trevarthen, 31-44. Oxford: Oxford University Press.

Collier, William G., and Timothy L. Hubbard. 2004. Musical Scales and Brightness Evaluations: Effects of Pitch, Direction, and Scale Mode. *Musicae Scientiae* 8: 151–173.

Collins, Karen. 2008. *Game Sound: An Introduction to the History, Theory and Practice of Video Game Music and Sound Design*. Cambridge, MA: MIT Press.

Datteri, Darcee L., and Jeffrey N. Howard. 2004. The Sound of Color. *8th International Conference on Music Perception and Cognition, Evanston, IL*, ed. S. D. Lipscomb, R. Ashley, R. O. Gjerdingen, and P. Webster. Adelaide: Causal Productions.

Gabrielsson, Alf. 2009. The Relationship between Musical Structure and Perceived Expressions. *The Oxford Handbook of Music Psychology*, ed. Susan Hallam, Ian Cross, and Michael Thaut, 141–150. Oxford: Oxford University Press.

Gaver, William. 1988. *Everyday Listening and Auditory Icons*. PhD diss., San Diego, University of California, San Diego.

Howard, Daniel Martin, and Jamie Angus. 2009. *Acoustics and Psychoacoustics*. Oxford: Elsevier.

Huron, David. 2006. *Sweet Expectations: Music and the Physiology of Expectation*. Cambridge, MA: MIT Press.

Huron, David, and Elizabeth Hellmuth Margulis. 2010. Musical Expectancy and Thrills. In *The Oxford Handbook of Music and Emotion*, ed. Patrick N. Juslin and John A. Sloboda, 575–604. Oxford: Oxford University Press.

Juslin, Patrick N., and P. Laukka. 2003. Communication of Emotions in Vocal Expression and Musical Performance: Different Channels, Same Code? *Psychological Bulletin* 129: 770–814.

Juslin, Patrick N., and Renee Timmers. 2010. Expression and Communication of Emotion in Music Performance. In *The Oxford Handbook of Music and Emotion*, ed. Patrick N. Juslin and John A. Sloboda, 453–489. Oxford: Oxford University Press.

Kivy, Peter. 2002. *Introduction to a Philosophy of Music*. New York: Oxford University Press.

Krumhansl, Carol. 1990. *Cognitive Foundations of Musical Pitch*. New York: Oxford University Press.

Lerdahl, Fred. 2001. *Tonal Pitch Space*. New York: Oxford University Press.

Lerdahl, Fred, and Ray Jackendoff. 1983. *A Generative Theory of Tonal Music*. Cambridge, MA: MIT press.

Marks, Lawrence. 1989. On Cross-modal Similarity: The Perceptual Structure of Pitch, Loudness, and Brightness. *Journal of Experimental Psychology: Human Perception and Performance* 15 (3): 586–602.

Meyer, Leonard. 1956. *Emotion and Meaning in Music*. Chicago: University of Chicago Press.

Mori, Masahiro. 1970. The Uncanny Valley. *Energy*, 7 (4): 33–35.

Narmour, Eugene. 1990. *The Analysis and Cognition of Basic Melodic Structures*. Chicago: University of Chicago Press.

Nattiez, Jean-Jacques. 1987. *Music and Discourse: Toward a Semiology of Music*. Translated by Carolyn Abbate. Princeton, NJ: Princeton University Press.

Patel, Aniruddh D. 2008. *Music, Language and the Brain*. New York: Oxford University Press.

Patel, Aniruddh D., and Joseph R. Daniele. 2003. Stress-timed vs Syllable-timed Music? A Comment to Huron and Ollen (2003). *Music perception* 21: 273–276.

Patel, Aniruddh D., Iversen, John R, and Jason C. Rosenberg. 2006. Comparing the Rhythm and Melody of Speech and Music: The Case of British English and French. *Journal of the Acoustical Science of America* 119: 3034–3047.

Pavlov, Ivan Petrovich. (1927) 2009. *Conditioned Reflexes: An Investigation of the Physiological Activity of the Cerebral Cortex*. Thousand Oaks, CA: Sage.

Peretz, Isabelle. 2010. Towards a Neurobiology of Musical Emotions. *Oxford Handbook of Music and Emotion*. New York: Oxford University Press.

Peterson, G. E. and Barney, H. I. 1952. Control Methods Used in the Study of Vowels. *Journal of the Acoustical Society of America* 24: 75–184.

Provine, Robert. 1996. Laughter. *American Scientist* 84 (1): 38–47.

Roads, Curtis. 1996. *The Computer Music Tutorial*. Cambridge, MA: MIT press.

Saygin, A. P., T. Chaminade, H. Ishiguro, J. Driver, and C. Frith. 2012. The Thing that Should not Be: Predictive Coding and the Uncanny Valley in Perceiving Human and Humanoid Robot Actions. *Social Cognitive and Affective Neuroscience* 7 (4): 413–422.

Schellenberg, E. Glenn. 1997. Simplifying the Implication–Realization Model. *Music Perception* 14 (3): 293–318.

Schenker, Heinrich. 1969. *Five Graphic Music Analyses*. New York: Dover.

Seeger, Charles. 1977. *Studies in Musicology (1935–1975)*. Berkeley: University of California Press.

Tagg, Philip. 1988. "Universal" Music and the Case of Death. http://www.tagg.org/articles/deathmus.html.

Tagg, Philip, and Karen Collins. 2001. The Sonic Aesthetics of the Industrial: Re-constructing Yesterday's Soundscape for Today's Alienation and Tomorrow's Dystopia. *Sound Practice*. http://www.tagg.org/articles/dartington2001.html.

Tramo, Mark Jude, Peter A. Cariani, Bertrund Delgutte, and Louis D. Braida. 2009. Neurobiology of Harmony Perception. *The Cognitive Neuroscience of Music*, ed. Isabelle Peretz and Robert J. Zatorre. New York: Oxford University Press.

Wang, X., and D. Bendor. 2010. *Pitch. The Oxford Handbook of Auditory Science: The Auditory Brain*. New York: Oxford University Press.

CHAPTER 7

HOW CAN INTERACTIVE MUSIC BE USED IN VIRTUAL WORLDS LIKE *WORLD OF WARCRAFT*?

JON INGE LOMELAND

In November 2004, Blizzard Entertainment released the online game *World of Warcraft* (*WoW*). *WoW* is currently the most popular massively multiplayer online role-playing game (MMORPG), played by over 10 million players (Cifaldi 2011). MMORPGs are a subgenre of massively multiplayer online games (MMOs), "online games in which many players participate" (Collins 2008, 185). MMOs have roots in tabletop role-playing games like *Dungeons and Dragons*, and the first MMOs were released in the mid-1990s, although there had been online games earlier, for instance the text-based multiuser dungeons like *MUD*.

WoW is based on the previous *Warcraft* games and allows players to create characters in "Azeroth," a three-dimensional Tolkien-like fantasy world (Tolkien 1954–5). The players choose to be on the side of either the "alliance," which includes humans, dwarves, and elves, or the "horde" side that includes orcs, trolls, and goblins. They are then able to fight players of the opposite side in contested areas such as "battlegrounds." Players can also undertake various quests and fight non-player characters (NPCs) or monsters in the different lands, or "zones" as they are called, in order to "level-up" and get better equipment. The game world contains cities like "Stormwind" and "Orgrimmar," where players can meet to socialize, trade, and create groups and "guilds" with other players. Guilds are needed to face the hardest challenges in the game such as cooperative "dungeons" and "raids" against bosses that require organized tactics. The fighting is undertaken by clicking on spells or by pushing buttons that have spells assigned to them, while avoiding suffering damage.

For the first six years, *World of Warcraft* had more or less the same music, but on December 7, 2010, Blizzard Entertainment revamped large parts of the game with an expansion called *Cataclysm*. The new music for the game became available on sites like YouTube before the launch. On Blizzard Entertainment's online forums (*WoW* English forums) a player commented: "While I think this is a good change for the most part, I hope they include a classic music option as I actually enjoy many of the classic themes" (from the forum thread "Cataclysm music," August 14, 2010). As it turned out, no such option was made available. Instead, the new music kept certain themes from the original music, but used different instrumentation, minor instead of major key, and altered other musical elements like tempo.

On *Cataclysm*'s Behind the Scenes DVD, Russell Brower, Director of Audio at Blizzard Entertainment, promised that they would not change the music too much, as there were a lot of themes that people loved. Derek Duke, one of the composers of the music in *WoW*, replied that the original music had not been changed much, and was refreshed with more robust arrangements of the same themes recorded with acoustic instruments (instead of virtual or synthesized instruments). Such statements imply that there is very limited leniency for musical changes and that interactive music has some restrictions in virtual worlds like *WoW*. Offline episodic games like the *Legend of Zelda* or *Super Mario* series likewise often represent well-known themes in new ways.

In this chapter, I will use the changes to the music in *WoW* after *Cataclysm* as a point of departure for discussing how interactive music can be applied in an MMO. There are positive effects that can be gained from using interactive music, but also potential problems, such as repetition, or the loss of a sense of history and player nostalgia when the music changes. I propose ways that the music can be made more interactive while avoiding such problems.

7.1 Changes to the *WOW* Music after *Cataclysm*

The changes to the music brought about different reactions on Blizzard Entertainment's online forums. Some players were happy that the music would change: "Org[rimmar] needs new music. Or at least a remix. The drums are awesome but it gets really old after a while. There's only so long I can listen to: BUM-ba-ba-BUM-ba-ba-ba-ba-ba-BUM-ba-ba-ba." Others were skeptical: "I hope they don't replace every single theme and [that] it still exists in-game in some form. It was all great work and listening to it just takes me back (especially the Elwynn Forest theme)" (from the thread "The new music," June 16–17, 2010). Other players were positively surprised: "[I] was disappointed when [I] heard that the zone music was being changed. However now that [I] have heard it [I] am very happy with the way it was done. The melodies have been kept but the sound has been filled so much" (quote from the forum thread "New music," November 30, 2010).

To illustrate an example of the changes, the music in the zone "Elwynn Forest" was previously a three-minute song with woodwinds, horns, harps, and strings.[1] It was originally divided into three smaller parts of about one minute each (Lomeland 2009, 59). The first part, which was removed after *Cataclysm*, had woodwind and horn solos alternating with one crescendo and one diminuendo strings chord. Part two of the song was maintained in the newer version, but the instrumentation was changed. Where the original had strings and woodwinds playing chords, the new version added a choir, and the melody that was played by woodwinds in the original is now played by strings. There are also melodic differences. In the original the melody is B♭–E♭–G♭–F–B♭–E♭–A♭–G♭–F–E♭–D♭–E♭–F–E♭. This melody is altered in the new version, ending instead with F–E♭–D♭–F–E♭, dropping one of the E♭s. The third part of the original, which had a trumpet section, was also removed. The new version instead makes variations of the melody and develops the theme: E♭–F–G♭–B♭–A♭–G♭–F–E♭–D♭–F–E♭, and adds a choir singing counterpoint. Instead of dividing the song into smaller parts, the new music now has alternative versions. There is one version that has more woodwinds, another has more strings and tremolo, while a third version starts with a woodwind solo of the melody and later uses harps as *accompagnamento*. This type of variation is important for interactive music, which can otherwise be highly repetitive.

7.2 What Is Interactive Music, and Why Should it Be Used in Virtual Worlds like *WoW*?

Karen Collins defines interactive audio as "those sound events that react to the player's direct input" (2008, 4). This definition suits most of the music currently used in *WoW*, although some music is what Collins refers to as "adaptive" audio; that is "sound that occurs in the game environment, reacting to gameplay, rather than responding directly to the user" (183). When the music is interactive, it is useful because it can give hints and feedback to the players. For example, the music that starts playing when players acquire a star "power-up" in *Super Mario Bros.* lets them know that they are invincible while it plays. Interactive music is also useful in games because it can help to create variety and prevent listener fatigue. Instead of hearing the same music in every part of the game, different parts and situations can have different music, and players can hear that they are progressing through the game.

The composers of music in MMOs face more challenges than standard game composition. MMOs are not mostly linear games like *Super Mario Bros.*, where the screen allows players only to advance without being able to backtrack, or games where there are restrictions on backtracking after specific events, like in *Resident Evil 4*, where players for example cannot go back to the village after they have entered the castle. *WoW* instead allows players to go anywhere and backtrack as much they would like. The world is to some degree static: enemies that are killed are brought back to life so that players can kill them again for experience (which is needed to level-up) or for resources, currency, and equipment. Players will therefore come back to the same zone, city, raid, or battleground day after day, year after year, as long as there is something useful to be gained there. Moreover, there are no "end credits" in *WoW*, which means players spend many more hours in such a world than with other types of games, resulting in much more repetition of the music. Variation is therefore beneficial, and this can be achieved with various techniques.

7.3 Techniques for Variation

Collins (2008, 147) provides examples of commonly used game audio techniques, such as variable tempo, pitch, and volume or dynamics, and describes newer techniques such as varying musical elements like rhythm and meter, melody, harmony, (open or branching) form, mixing and timbre or digital signal processing (DSP)—"the processing of a signal (sound) digitally, including using filters and effects" (184). Layering, a form of variable mixing of music instruments, was used for instance by composer Koji Kondo in *Super Mario 64* in the level "Dire, dire docks." The music first consists of electric piano tracks while other tracks are muted, then it adds (or removes) tracks in layers depending on where Mario is, first adding strings when Mario dives under water, and then percussion when Mario reaches the other side of the water (Kaluszka 2007).

Variable open form, where sequences of song parts are put together in random order, has been used by composers since the 1700s (Collins 2008, 155). Although this technique creates variety, it can risk detaching the music from the gameplay. Composition tools have therefore been developed "where changes or branches in the performance may occur based on a condition in the game in real time" (160). One such "branching" tool was developed for the MMO *Anarchy Online* by the composers Bjørn Arve Lagim, Morten Sørlie, and Tor Linløkken. Lagim describes their approach to the challenge of musical repetition: "Our solution to this problem was to create a tool that... allowed us to create a single track built up of many smaller [samples] which were combined on the fly to create an ever-changing soundtrack" (Lagim 2002). Each sample had a number of possible transitions to other samples. Transitions are short pieces of music that bridge different parts together, for instance drum rolls or musical build-ups (Bridgett 2010, 21). These have been used in games for some time, for example in stage 6 of *Contra 3: The Alien Wars*, where a tension-building fanfare bridges the music of the first "miniboss" to that of the second miniboss. According to composer Marty O'Donnell, it is possible to skip transitions when something new happens in a game, and instead use "instantaneous transitions" to create "the surprise change" (Battino and Richards 2005, 194). However, many hard cuts may influence player immersion (see "Potential problems of using interactive music in MMOs like *WoW*" below).

Branching and transitions require composers to make many small musical pieces and a map of all possible changes between them. They also have to test that all works well together. This can be very time consuming. An alternative is to use generative music where the computer uses algorithms in order to create random variations of themes (van Geelen 2008, 96). Composers can then set "rules" for how the computer will create themes. Generative music can be created as "procedural audio," which adapts to the gameplay in code so that it can be used "in a context that makes sense" (Farnell 2011, 316). It is nevertheless beneficial to use generative music in combination with composed music, as "it is impossible to replace the human creativity a real composer brings to interpreting and augmenting a scene" (Farnell 2010, 326). It is also possible to use granular synthesis, where "an intelligent engine could use grains of sound to adapt algorithmically in real time to what is occurring in the gameplay" (Collins 2008, 151). This adaptation can be accomplished by using scalable parameters for musical elements like volume, pitch, texture, and tempo and attach these to gameplay proceedings, so that, for example, an increased level of combat could result in a faster tempo of the music. Although these techniques for variation are available, their use is of yet limited in MMOs.

7.4 Interactive Music in MMOs like *WoW*

In *World of Warcraft*, the music interacts with players when they move their character into "trigger spots." Trigger spots are areas of varying size, which cause specific music to play. There are two main types of music that launch from trigger spots. The type most often used is *zone music*, or "zonetracks," where each zone has a specific soundtrack of approximately five songs that play in random order. The zones have subzones, such as cities, buildings, and bosses that can each have their own music. *Cataclysm*, for example, included two new races and starting zones, the "worgen" (friendly werewolves controlled by players) and "goblin," where the composers had to make the music of these places and races distinct. Besides a few exceptions, like the car radio in the Goblin starting zone "Kezan", which the players can turn on using a radio "spell," zonetracks are the backbone of this music too. These are often general mood-makers, made to work with different gaming situations. For instance, one of the harpsichord songs in the worgen starting zone "Gilneas" uses a "climbing" melody, and it triggers when the stairs of a tower are ascended. In an old house with a ghostly looking, elderly lady, the same song adds a Victorian mystique to her character.

The other main type of music in *WoW* is *event music*, where music that is not specific to the zone triggers at certain events in the game. One example is the event of "Prince Liam Greymane" leading his people in an attack on the city Gilneas. The citizens of Gilneas have had to flee because they were attacked by werewolves. Now they have gathered up in the outskirts of the city with various weapons and the aid of worgen, in order to try to retake it. Before the attack, there is solemn, melancholic music, tailored to Greymane's last-stand speech so that they have a strengthening effect on each other.

As he launches the attack at the end of the speech, battle music plays, empowering the player. The event becomes almost cinematic in terms of describing and adding to the emotions on screen.

While the songs in *WoW* seem to change little after they start playing, apart from cross-fading with other songs, some events seem to have more interactive music than others. For instance, in the airship hovering above "Keel harbor," a brass drone plays when players land on it, as to tell them that they are now invading hostile territory and that fighting is about to start. As players run below deck to plant explosives, new heroic action music starts. When the ship is blown up, the Gilneas zone music recommences, to inform players that the quest has been fulfilled. Such musical cues are an effective way of making the music more interactive and flexible to the gameplay.

7.5 Benefits and Limitations of the Music Currently Used in *WoW*

Although event music is a refreshing addition, the zonetracks can sometimes be better suited to the game. There are quests and gaming situations that cannot be dramatized with music, and times when players need a break from dramatic music and events. Music previously unheard in the game can often attract attention to itself, making it more noticeable and potentially reducing immersion.

Zonetracks that were pleasing in the beginning of gameplay can, also, turn out to be tedious or perhaps even annoying after many repetitions. This repetitiveness may be one of the reasons that certain songs were removed or changed in the *Cataclysm* update, while others were kept or expanded. Another reason for the change in music could be due to the content change and the lack of suitability of the original music to the new content. Since enemies and landscapes were both changed, the original music might sound out of place in this new environment. For instance, music that was originally composed for a desert may no longer be suitable if that desert was flooded and turned into a lake. The original music may also clash with the new music. A commentator on the forums remarked: "Honestly, the big difference in the music is... that some was recording [*sic*] in 2010, and some in 2004. Quality difference is massive, and it [is] quite jarring" (From the forum thread "I truly miss the old music," November 29, 2010). Another wrote: "I notice that in many zones, there are two soundtracks battling for supremacy: the old music and the new music" (from the forum thread "New music conflict," December 6, 2010). For example, the woodwind tunes and string arrangements in the low-level zone "Loch Modan" used to evoke a calming feeling. After *Cataclysm* and the music update, the zonetrack became more dramatic and orchestral. The original music has nevertheless been kept around certain towns and buildings, leading to a patchwork approach to the music. It is possible that the disjointed approach to the music could influence player immersion (see e.g., Wharton and Collins 2011).

One of the reasons for the musical "mismatch" could be that the team of composers changes. For instance, lead composer Jason Hayes and composer Tracy W. Bush of the original *WoW* soundtrack were not part of the teams that made the music for the expansions *The Burning Crusade, Wrath of the Lich King*, and *Cataclysm*, and senior composer and sound designer Matt Uelmen worked only on *The Burning Crusade*. Also, the live orchestras and real instruments that were used for the expansion packs might sound different than the virtual or synthesized instruments that were used for the original game (Chiappini 2007).

Some of the music in *WoW* has many layers and rich instrumentation and this complexity can extend the music's appeal by reducing the listener fatigue sometimes associated with repetitive music. This listener fatigue effect, however, is still dependent on the ways in which the music is used and its context. The symphonic music of the login screen has been changed with each expansion of the game, approximately every second year, although it is a collage of many tunes within the game with many layers and rich orchestration. In contrast, the "simple" tavern tunes with three or four instruments are still the same as they were at the start of the game, as these are mainly heard in inns where players frequent less often.

WoW music uses the concept of leitmotifs, "a musical phrase, either as complex as a melody or as simple as a few notes, which, through repetition, becomes identified with a character, situation, or idea" (Kalinak 1992, 63). Leitmotifs are useful for building stories and tying narrative together through many hours, and are therefore well-suited to games like *World of Warcraft*. For example, the first theme from the login screen music is also associated with the dragon "Deathwing" (whose image is seen at the login screen of *Cataclysm*). The theme is heard in various versions in zones and dungeons until players finally fight Deathwing in the raid "Dragon Soul." Through many hours of play, this repetition of leitmotifs can provide the players with a feeling that the dragon is a menace and a threat, and when they are powerful enough to fight him the theme can make them feel that they have come full circle since they first heard it. The repetition of themes through years of play also helps to create emotional and mental flashbacks to the many previous hours spent in the game. As a result of this, the music of *WoW* can gain nostalgic value for players. As a player commented before the release of *Cataclysm*, "every time I go to old Azeroth and hear the old music tracks it really gives me that warm, nostalgic feeling, and it would be a shame to lose that" (from the forum thread "Cataclysm Music," December 9, 2010).

Online multiplayer games, with their extensive virtual worlds, can feel like a new world in the beginning, and the first zones of the game bring a newness and excitement that may be later associated with the music. In a sense, music in these zones can have some of the same effect as the music one cherished growing up, creating a soothing and nostalgic feeling when heard again later (DeNora 2000, 41–2). Although the music in *WoW* changes as one gains higher levels and reaches new zones, some of the music is repeated later in the game. For example, the event "Brewfest," which takes place in the low-level zone "Dun Morogh," repeats annually. Of course, one issue with keeping the same music in a zone year after year is that, just as it can recall positive experiences, it can also bring back negative memories for players.

The fact that contemporary videogame consoles have titles from the 1980s and 1990s available for purchasing or downloading implies that nostalgia is important for players. This nostalgia can also be seen on the Internet where "abandonware like games from the early 1990s is living a zombie life" (Parikka 2012, 3).[2] Nostalgia in MMOs can be compared to the nostalgia for such games; the players want to relive or be reminded of the excitement that they felt the first time they played them. Nicolas Esposito suggests that it takes about twenty years for players to create nostalgia for games (2005). In my experience, for MMOs, this nostalgia seems to be created faster, perhaps within a couple of years, due to the length of time spent with the game. Nostalgia can be related to specific gameplay periods like low-level zones or original game content before expansions (Simon, Boudreau, and Silverman 2009). The feeling is comparable to the way players are nostalgic about the first games in a genre because it is like "being very close to the genre spirit... it is like coming back to roots" (Esposito 2005). Perhaps most importantly, and unlike their offline counterparts, since updates and expansions of MMOs cannot be avoided, nostalgia for these games can be greater, since the changed game content will not be available again in its original form.

The updates and changes in *WoW*'s music help to give the game world a sense of history. Revisiting or playing a new character in a zone previously played at a particular point in the past reminds the player of these times, while helping to build a sense of tradition and reinforcing the idea that this is, indeed, another world. For instance, the music in the zone "Nagrand" was part of the first expansion called *The Burning Crusade*, released in 2007. As Uelmen describes it, "Nagrand, because it is so pastoral, was an opportunity to do a lot more of the kind of sweet orchestral sound that we had in the original release" (Blizzard 2007). Due to changes to the game, it is no longer necessary to visit Nagrand, so the beautiful and sad music associated with Nagrand is now seldom heard. In a sense, it has become more like a museum artifact for people who played there, a zone for audiovisual sightseeing. It could be suggested that the game developers picked up on this sense of history, since *Cataclysm* introduced an excuse for such sightseeing with the new in-game profession "archeology," which requires players to do "excavations" in zones that can otherwise be too low level for advanced players.

The music can become a part of the players' identities, and can be compared to other music genres' influence on identities, formation of subcultures, and so on. What makes the music in a virtual world different is that it is associated with a space constructed by game designers, as opposed to the negotiated imagined space of a genre that is extended or limited by "the competing definitions and understandings... promoted by fans, business interests, critics and others" (Walser 1993, 29). Duke describes how the game music can also become associated with life and memories outside the game: "The music in *WoW*... reaches beyond its original meaning in the context of the land, or the zone, or the story,...because that music permeates you so much while you are playing, it takes on a meaning outside of the game" (Blizzard 2010). In September 2008, Blizzard Entertainment made their music available for purchase on iTunes, suggesting that the music may become a part of the players' lives (and subsequently their identities) decontextualized from the game. As Simon Frith (1996, 110) describes, identities are thought to be more like processes than things, constantly in flux in negotiation with their environment, which could suggest that as the players' identities evolve, the music may also need to evolve and adapt to changing interests of the players.

Composer Jason Hayes said the following about the challenges of composing music for a virtual world like *WoW*: "If someone is hanging around in a location of the world for hours, it's very difficult to even conceive how you would approach that aesthetically from a musical standpoint" (in Chiappini 2007). As opposed to film, where composers know what they are scoring, the composers of music in an MMO do not know how long the players will be in a specific area or what they will be doing at a specific time. Making the music interactive is therefore helpful as it gives the composers some control of what will play when, as well as tools for variation. There are, however, potential problems associated with interactive music and I will address some of these below.

7.6 Potential Problems of Using Interactive Music in MMOs Like *WoW*

As discussed in the introduction, interactivity can lead to repetition. When the player moves over the same trigger spots, the same music will play. Brower addresses repetition for the *WoW* expansion *Wrath of the Lich King*: "I think the first rule of music in a game is to enhance the mood, the setting and the story, but not to annoy people and not to have it run on and on incessantly" (Blizzard 2008). This was a reason why they created a system for music breaks. In areas where the players do not spend too much time this system can work, but in others where players return daily it might not be as efficient. City themes, for example, play when entering or logging on in a city. They then fade out if players enter the auction house or another shop in the city, but sometimes start over within seconds after exiting the shop. As the music is bombastic, its use can be rather misplaced after gameplay "chores" like repairing and trading. To reduce over-repetition, there could be a limit to how often songs will play. This could be achieved by mechanisms similar to that in *Halo: Combat Evolved*, which composer Marty O'Donnell refers to as the "bored now" switch (in Battino and Richards 2005, 195), where the music fades out if players are still in the same area five minutes after they should have reached the next area or musical piece.

Sometimes, no music can be a powerful effect in dramatic situations. For example, in *Resident Evil 4* where the music normally warns the player about nearby enemies, combat situations not preceded by music can be an effective way of scaring the player. Rob Bridgett (2008, 130) writes about the importance of having a dynamic range of sound in games as well as silence. Making the music more interactive should therefore include music breaks. These are needed to give the players room to "breathe," heightening drama when music is played. Before *Cataclysm*, the songs in *WoW* were seldom more than one minute long and the music breaks could often be five minutes (Lomeland 2009, 48). After *Cataclysm*, the songs have become longer, often over two or three minutes long, while the breaks have become shorter, seldom longer than a couple of minutes. The pauses should nevertheless avoid becoming too long. As Lagim (2002) describes the music breaks in *Anarchy Online*, "too long pauses would make the music ineffective in maintaining the feel."

The shorter music breaks in *WoW* could suggest that there are currently too many trigger spots in the game. Outside the small town "Stormglen" in the zone Gilneas, for example, there were three different songs cross-fading over a walking distance of five meters, before the music stopped altogether, all happening within ten seconds. Although this is an "extreme" example, trigger-spot interactivity problems could probably be avoided by using a 50 meters' gap of silence between trigger spots, like Larson suggests (2007). The trigger spots may need even larger gaps when players are moving fast by air. In Dun Morogh, for instance, one song is interrupted by another after flying for three or four seconds. In the zone "Badlands," the music suddenly shifts from calm to menacing because players fly over the subzone "Camp Kosh." Since players are not in danger when they fly above zones, such cross-fades are unnecessary. The zonetracks could instead be stable while flying, as it will allow the players to hear more of the songs, taking focus off the (potentially uninteresting) traveling. "Airborne" trigger spots can warn players where the subzones are relevant, for instance where players risk being shot down by enemies.

Interactivity can risk becoming problematic when it "Mickey Mouses" each step the player takes. Mickey Mousing is a term originally used in cartoon animation, where the music is so synchronized with what is happening on screen that it becomes comical (Collins 2008, 148). By always increasing tempo or intensity during combat, this could risk becoming annoying if it is too obvious. Interactive music should therefore not necessarily follow the pattern: more action equals more musical intensity. Another problem with synchronizing the music to the gameplay is the risk of the music becoming ephemeral and uninteresting.

Finally, interactivity can interrupt the original composer's work if interactive systems allow for players to substitute in their own music. *WoW* players have created modifications that allow them to swap out the game music with their own selections. Wharton and Collins have found that this substitution has various implications for the game experience, as it may change the gameplay pacing and alter the level of player anxiety (2011). Although user-generated content creates more variety, such music customizations will not necessarily improve the gaming experience as there may be a disconnect between the emotional intent of the game's designers and the affective experience of the music listening.

7.7 Ways of Improving a Virtual World like *WoW* with Interactive Music

Brower describes of *The Burning Crusade*: "Sound is a big broad brush with as much potential as the visuals, as the story, and as the gameplay to affect your experience as a player" (Blizzard 2007). It is therefore important to find ways to realize this sonic potential, for instance by examining where and when interactivity can improve the music. As it is now, a zonetrack with five tunes can start wearing on the player after an hour or less. Common situations when repetition can lead to listener fatigue are during questing, socializing, trading, character tuning, play style practice, or consecutive visits to the same battleground or raid boss fights. The music in cities and battlegrounds is particularly vulnerable since players revisit them until they reach the highest level and beyond. Having parts of the music change every ten levels by using layering could be one solution, at least for battlegrounds which are already separate for players above level 10, 20, 30, and so on.

The music in battlegrounds where players fight one another could also change in accordance with which side is leading. When bases need to be captured, each base could be represented by specific layers of instruments for both sides, so that the music would vary in relation to how many bases each side occupied. In battlegrounds with flags to be captured the music could shift depending on which side held it. There could also be music layers related to which side was killing the most players, like tribal percussion for horde, and marching drums for alliance. There are currently different sound effects that let the players know which side captures a base or a flag, and this type of gameplay feedback could be extended to the music to make it more interactive.

When a player has spent a certain amount of time in the same area, the music could either fade out and stay muted until the player moves on, like in *Halo: Combat Evolved* above, or switch to generative variations of the zonetracks. In dangerous situations where the player is attacked by several enemies of the same level or above (which in *WoW* would mean enemies marked with yellow or red color), the music could rise in intensity, for example by increasing tempo or adding more layers that could fade up. If the player is in a zone where the enemies are at a slightly lower, or much lower level (enemies marked with green or grey color in *WoW*), the music could relate more passively to the enemies. A music system similar to this was used in *Anarchy Online*, where the music would change in intensity depending on whether the enemy was large, medium or small (Lagim 2002). In *Asheron's Call 2: Fallen Kings* the intensity of the music was modified by both the number of players and monsters in an area, subtracting intensity for each player and adding intensity for each monster (Jason Booth 2004, 478).

Using several shorter songs in a zone can be a better solution than having a few long songs, as short songs can be better suited for immersing the players. Short songs are also more flexible to gameplay changes. An example is the short songs in the Goblin starting zone that vary with different instrumentation like accordion and marimba. However, if the order of many short songs does not vary, like that of the car radio in the Goblin starting zone, this can instead become fatiguing for the player. An alternative can be to use branching of long songs so that the content and order will vary, like in *Anarchy Online* above.

Raids require a lot of attention from the players, especially in the beginning. Sometimes the music provides clues, for example at the first boss in the raid "The Siege of Wyrmrest Temple," where battle music plays each time the players need to run and hide. Often, though, bosses in raids either have "passive" background music that fades in during the fight or no music at all. Brower provides a reason why music is often in the background: "It is definitely a balancing act to make sure that we do not create music that is so foreground or so demanding of attention that we take away . . . your ability to communicate with [other players]" (Blizzard 2008). A good way to avoid this in raids could be to wait before introducing music or parts of it until the players are experienced and no longer need to communicate as much. This could be done by using layering to adjust the music to the achievements log of the players, creating refreshing variety for returning players while not taking too much attention from new players. Since boss fights in raids often last several minutes and have different phases relating to time or the boss' health bar, they are possible to orchestrate interactively with different musical phases tied together by transitions.

Songs with few instruments, like those in inns and ships, could be a good place to start experimenting with techniques for variation. It would make the visit to an inn more exciting and lifelike if "the band" sometimes changed their "set." Algorithmic random variations within the right scales and play styles, for example an Irish-sounding air or jig, could represent the Irish tradition of jamming in pubs. Three-dimensional positioned music played by NPCs could also be used to make the game more lifelike. As the volume of the NPC music would be higher near the stage than in the bedrooms of the inns, this could be both more calming and immersive. Three-dimensional positioned music is already used with drummers in troll villages.

Small towns and villages often use the same music as the larger cities, for instance "Lor'danel," which uses the same music as the city "Darnassus." This can make the cities less unique. Generative music could be used to create variations of city themes for small towns, making both the cities and the small towns unique. Changing the instrumentation and applying digital signal processing (DSP) is another way to create musical nuances, which can also be used to vary different quests and gaming situations. One example is the alternative version of the "Dragonblight" zonetrack that has been altered with DSP in the dungeon "End Time" to give a feeling of "time travel."

7.8 Conclusions

Virtual worlds like *World of Warcraft* offer many challenges to composers. The game requires the players to spend a long time in the same areas listening to more or less the same music, and this can contribute to listener fatigue. Interactive music can remedy this by keeping track of how long the players have been listening to the same soundtrack and provide variations or music breaks when necessary. Variety can be achieved with different composition techniques. The music can be made more adjustable to the players by using layering techniques so that it is less obtrusive for inexperienced players and more dynamic for experienced players. By adjusting to the development of boss fights or battlegrounds, the music can avoid becoming predictable. Generative and branching music can provide effective ways for creating better-suited music for specific gameplay situations, whether it is fighting dangerous enemies or gathering resources. An alternative can be to have parameters for musical elements like tempo and pitch adjust according to different gameplay situations. Such parameters could still risk becoming predictable, so an element of randomness would be beneficial. When combined with pre-recorded music, techniques for interactive music and variation can be an effective way to reduce listener fatigue, while keeping the nostalgia and "hum-along" ability of hearing familiar songs from time to time.

Notes

1. This song can be heard at http://www.youtube.com/watch?v=uvW-QTiZLQo.
2. Abandonware is a product that is no longer available for purchase and whose copyright ownership may be unclear.

References

Battino, David, and Kelli Richards. 2005. *The Art of Digital Music: 56 Visionary Artists and Insiders Reveal their Creative Secrets*. San Francisco: Backbeat.

Blizzard Entertainment. 2007. *World of Warcraft: The Burning Crusade: Behind the Scenes DVD*. Blizzard Entertainment.

——. 2008. *World of Warcraft: Wrath of the Lich King: Behind the Scenes DVD*. Blizzard Entertainment.

——. 2010. *World of Warcraft: Cataclysm: Behind the Scenes DVD*. Blizzard Entertainment.

Booth, Jason. 2004. "A DirectMusic Case Study for *Asheron's Call 2: The Fallen Kings*." In *DirectX 9 Audio Exposed: Interactive Audio Development*, edited by Todd M. Fay, Scott Selfon, and Todor J. Fay. Plano, Texas: Wordware Publishing.

Bridgett, Rob. 2008. Dynamic Range: Subtlety and Silence in Video Game Sound. In *From Pac-Man to Pop Music: Interactive Audio in Games and New Media*, ed. Karen Collins, 127–133. Aldershot, UK: Ashgate.

——. 2010. *From the Shadows of Film Sound: Cinematic Production and Creative Process in Video Game Audio: Collected Publications 2000–2010*. Self-published: Blurb.

Chiappini, Dan. 2007. Q&A: World of Warcraft Composer Jason Hayes. June 8, 2007. http://www.gamespot.com/news/qanda-world-of-warcraft-composer-jason-hayes-6172231.

Cifaldi, Frank. 2011. *World of Warcraft* Loses Another 800K Subs in Three Months. November 8. http://www.gamasutra.com/view/news/38460/World_of_Warcraft_Loses_Another_80K_Subs_In_Three_Months.php.

Collins, Karen. 2008. *Game Sound: An Introduction to the History, Theory and Practice of Videogame Music and Sound Design*. Cambridge, MA: MIT Press.

DeNora, Tia. 2000. *Music in Everyday Life*. Cambridge: Cambridge University Press.

Esposito, Nicolas. 2005. *How Video Game History Shows Us why Video Game Nostalgia Is so Important now*. University of Technology of Compiègne. http://www.utc.fr/~nesposit/publications/esposito2005history.pdf.

Farnell, Andy. 2010. *Designing Sound*. Cambridge, MA: MIT Press.

——. 2011. Behaviour, Structure and Causality in Procedural Audio. In *Game Sound Technology and Player Interaction: Concepts and Developments*, ed. Mark Grimshaw, 313–339. Hershey, PA: Information Science Reference.

Frith, Simon. 1996. Music and Identity. In *Questions of Cultural Identity*, ed. Stuart Hall and Paul du Gay, 108–127. London: Sage.

Geelen, Tim van. 2008. Realizing Groundbreaking Adaptive Music. In *From Pac-Man to Pop Music: Interactive Audio in Games and New Media*, ed. Karen Collins, 93–102. Aldershot, UK: Ashgate.

Kalinak, Kathryn Marie. 1992. *Settling the Score: Music and the Classical Hollywood Film*. Madison: University of Wisconsin Press.

Kaluszka, Aaron. 2007. Koji Kondo's GDC 2007 Presentation. *Nintendo World Report*, March 13. http://www.nintendoworldreport.com/feature/13118.

Lagim, Bjørn Arve. 2002. The Music of Anarchy Online: Creating Music for MMOGs. *Gamasutra*, September 16. http://www.gamasutra.com/view/feature/131361/the_music_of_anarchy_online_.php.

Larson, Kurt, Charles Robinson, Stephen Kaye, Nicholas Duveau, Guy Whitmore, Jennifer Lewis, Simon Ashby, Tom White, Jocelyn Daoust, Karen Collins, Barry Threw, Scott Snyder, and Aaron Higgins. 2007. Group Report: Overcoming Roadblocks in the Quest for Interactive Audio, Appendix B: Case Study for Music and SFX Interactivity in a Massively-Multiplayer game. From *The Twelfth Annual Interactive Music Conference Project Bar-B-Q 2007*. http://www.projectbarbq.com/bbq07/bbq07r6.htm.

Lomeland, Jon Inge. 2009. Musikk i *World of Warcraft*: Kjensler, Narrativ, Rasar og Lydlandskap. Master's thesis, University of Bergen.

Parikka, Jussi. 2012. *What Is Media Archaeology?* Cambridge: Polity.

Simon, Bart, Kelly Boudreau, and Mark Silverman. 2009. Two Players: Biography and "Played Sociality" in EverQuest. *Game Studies* 9 (1). http://gamestudies.org/0901/articles/simon_boudreau_silverman.

Tolkien, J. R. R. 1954–5. *The Lord of the Rings*. London: George Allen and Unwin.

Walser, Robert. 1993. *Running with the Devil: Power, Gender, and Madness in Heavy Metal Music*. Middletown, CT: Wesleyan University Press.

Wharton, Alexander, and Karen Collins. 2011. Subjective Measures of the Influence of Music Customization on the Video Game Play Experience: A Pilot Study. *Game Studies* 11 (2). http://gamestudies.org/1102/articles/wharton_collins.

World of Warcraft. English forums. http://us.battle.net/wow/en/forum/.

CHAPTER 8

SOUND AND THE VIDEOLUDIC EXPERIENCE

GUILLAUME ROUX-GIRARD

WHILE it took some time before scholars in the field of game studies paid any attention to the sonic aspect of videogames, the last few years have provided interesting perspectives on the subject. However, most of these viewpoints adopt a practical approach to sound by either presenting the technologies and techniques employed for the games' sound design (see Farnell 2011; for a historical account, see Collins 2008), by offering an insider's look of the game industry (Childs 2007; Marks 2009), or by attempting to model the structure and composition of game audio (see Folmann 2004; van Tol and Huiberts 2008). These approaches are certainly useful to study the *videoludic* object itself (or other interactive sound practices) but are not fully adequate to portray the relationship that is taking place, through gameplay, between the games and the gamers.[1] To fully circumscribe the questions inherent to interactive sound, a general study about the sonic dimension of videogames has to incorporate a reflection that foregrounds the notion of experience. But how can we define this videoludic experience, and why should we pay any attention to it?

In "The Filmic Experience: An Introduction," Casetti (2007, 1) defines the term "experience": It "indicates on one hand the possibility of perceiving reality as if for the first time and in the first person ('to experience'), and, on the other hand, the acquisition of knowledge and competence which allow an individual to face reality and create meaning from it ('to have experience')." The author explains that "by analogy, we can define the filmic experience as that particular modality through which the cinematographic institution allows the spectator to perceive a film and to [convert] the perception into knowledge and competence" (1–2). This double definition allows the creators—in a top-down/bottom-up fashion—to create "reflexive and projective relationships between the spectators and themselves and between the spectators and the world... leading them to a 'knowing how' and a 'knowing that' they are seeing the film both as a film and as a reality represented" (2). The same thought can be applied to videogames. Indeed, like films, videogames provide the gamers with a particular perceptual experience and allow them to translate their perceptual experience into knowledge and competences. However, because videogames are interactive objects, neither this experience (as an aesthetic perception) nor this gain of experience (as knowledge), can be attained or achieved without the continuous physical and cognitive involvement of the gamers. This involvement substantially alters the modalities of perception as well as the nature of the acquired knowledge. Accordingly, we can assume that the gamers' perception of sound within a videoludic context and the way it is understood are equally modified.

Yet, following an intermedial logic (Bolter 2005, 14), the specificity of the videoludic medium, as the vehicle of artistic and cultural practices, also results from the connection it maintains with other media and artistic forms (architecture, cinema, television, music, etc.) as well as other cultural practices (computer science, gaming, etc.). Consequently, while it is interesting to study the videoludic experience to comprehend how videogames become "the site of an experience which has reshaped the meaning of experience," (Casetti 2007, 2) the one we make of the world, it is also relevant to evaluate how the gamers' personal experience of the world preconditions their videoludic experience. Any reflection about the sonic experience of video games should be approached in a similar fashion. But how can we define this relationship of interdependency between "experiencing" and "having experience" in a videoludic auditory context? On the one hand, we need to assess how, through auditory perception, it is possible to collect information, which, once understood, helps to shape the gamers' experiences. Then, conversely, it becomes necessary to determine how the listening experience of the gamers is built according to a specific horizon of expectation (technological, economic, cultural, social, historical, generic, narrative, etc.) that was forged within a frame that goes beyond the mere videoludic medium.

8.1 Listening to Videogames

To answer the question "what is the sonic experience of a videogame?" we must necessarily turn our attention toward the notion of listening, and most of all, its modalities. We believe that every listening situation is dual in nature.[2] On the one hand, listening is oriented by the recognition of learned sonic formulas, some assimilated from the given reality of our everyday life, others based on the rhetorical formulas of media languages. In such cases, the game's sonic experience is partly founded on a principle of imitation.

On the other hand, listening focuses specifically on the sounds' materiality and treatment as well as the relationships between the sounds themselves and the other dimensions of the game (image, interactivity). As such, listening is focused on the movements, energy, colors, and other qualities that animate sounds. Gamers are also attentive to the arrangements of the sounds as well as the sound's propagation in the virtual environment. Finally, special attention is paid to the way sound participates in creating this simulated environment and the events that populate it.

While it is possible to polarize these listening patterns for comprehension purposes, they remain interdependent and are homogenized in a more general listening. It also appears that this dual scheme of listening finds echo in the two significations of the term "experience" we extracted from Casetti's research. The perception of sound bases itself on the gamers' experience to create meaning (imitation), and the gamers' knowledge is constantly enriched through new perceptual experiences (assimilation). But how can we further explain the relationship between a videoludic soundscape and a gamer who possesses specific knowledge? This is once again achieved in a dual fashion.

First, while playing a game, the gamers make use of a spectrum of specialized listening skills (everyday, formal, filmic, ecological, computer related, musical, videoludic, etc.) that contribute in the creation of their gaming experience. The refinement of these skills varies inevitably from one individual to another according to each person's familiarity with the formal and rhetorical structure of the aforementioned cultural soundscapes. Indeed, during our life, we develop our listening skills differently, consciously or not, depending on different sonic contexts. For example, the development of everyday listening, which begins while we are still in the womb (see Céleste, Delalande, and Dumarier 1982), does not work exactly according to the same imperatives as filmic listening, which may have been influenced by the soundscape of the Hollywood action film. Consequently, the more knowledge an individual possesses about a cultural practice, the more his or her specialized listening skills will benefit from this experience.

Although we speak of a specialized listening, we must specify that the gamers do not necessarily realize that they are making use of one type of listening more than another. In fact, an individual is rarely aware of their act of listening. Indeed, conscious awareness may prevent them from adequately performing the actions requested by the game, thereby limiting the quality of their experience. As Daniel Deshays (2010, 50, freely translated) recalls, "a shortcut is formed between perception and action, we avoid the detour through the consciousness of the action, which would inevitably slow us down if we questioned any action engaged." Nevertheless, gamers constantly use cognitive schemes—based on their general and specific knowledge—that articulate their judgment according to their expectations and the perceived sensory information. As a result, videoludic listening partly develops in relation to countless cultural practices of listening, some being more prominent than others—and homogenized in a general listening—that interact to allow the gamers to reach the "aesthetic experience of the game" (Arsenault 2011).

However, to fully grasp the notion of videoludic listening, we must also establish what distinguishes it from other listening practices. Following our previous hypothesis, listening is also organized according to the soundscape of the experienced object, and, thus, is consistent with the specificities of the videoludic language. The constitution of a videoludic listening, and the nature of some sonic effects—mostly the values associated with these effects—are internal to the game, and by extension, to the medium itself. In other words, if some of our understanding of sound comes from our expectation toward certain types of soundscapes and by the recognition of certain sonic patterns (intermedial or videogame specific), we are also sensitive to the specific soundtrack of the game we play. This statement is later supported by Thérien's (1992) affirmation when he says we must pay attention to the *lisibilité* of the games.

Synchronization points are a good example of the shift in meaning an audiovisual effect can be subjected to from one media context to another. Indeed, in addition to marking a privileged moment of encounter between sound and image, a videoludic synchronized sound is often tied to the gamers' actions, creating a link between the act of pressing a button on the controller and, for example, a gunshot fired on the game's diegetic axis. For that reason, there is a significant difference between the value of a synchronization point in a videogame and that present in a film. In a movie, synchronization points always have a strong value, making the amalgam of sound and image an aesthetic event while supporting the narrative and emotional dimensions of the film. Such an effect occurs partly because the viewer is fully subjected to the suddenness as well as the autonomous and independent "life" of the synchronization point. Thus, in a filmic context, the synchronization point of a gunshot is all the more salient because it is beyond the spectators' control. Accordingly, the material shock is also a perceptive and emotional one. In a videogame, synchronization points can have a similar value if used in concordance with an event that is beyond the control of the gamers. However, synchronization points that coincide with the gamers' actions turn out to be less aesthetic and more pragmatic as they become the product of the gamers' will in action. These synchronization points become concrete evidence of the gamers' influence on the digital world of the game, as they participate significantly in the creation of the gamers' feeling of agency. In turn, this feeling likely contributes to an effect of presence within the diegetic world of the game.

In videogames, these highlights of synchronism play a dual role and encourage a refinement of listening, which results in an adjustment of the decisions the gamers make in a given context. Overall, these two listening schemes allow the gamers to get in contact with a game's soundtrack and make sense of it. An approach that takes both modalities into account is therefore necessary to analyze the sonic experience of videogames.

8.2 A Methodology for Analyzing the Sonic Experience of Videogames

Because videogames as a medium propose so many types of experiences, and because there are so many listeners, we believe that building a framework to represent the gamers' sonic experience is a fruitless exercise. We would rather opt for a more flexible methodology. We believe that such an approach allows the analyst to avoid the pitfalls of a large general theory that attempts to embed a heterogeneous phenomenon into a mold.

At the same time, this flexibility does not confine the methodological tools to specific case studies—although we believe such tools might also facilitate these kinds of analysis.

The methodology that we propose combines four interconnected approaches that we feel are necessary to describe the nature of the videoludic experience as well as the modalities of listening we presented earlier: (1) a historical contextualization of the analyzed objects to reposition them within a broader media and cultural context; (2) an analysis of the games' reception to evaluate the social and cultural consensus that surrounds them; (3) a formal and gameplay analysis of those games to determine how the images and sounds functions in connection to the interactive nature of the games; and (4) an analysis of the sonic experience of videogames.

The purpose of the historical contextualization is to evaluate the power relationships active between the technological, economical, cultural, generic, serial, and intermedial aspects of the games. From a gameplay point of view, it also allows the analyst to properly assess the sonic dimension of the games in relation to a certain horizon of expectations. Indeed, the horizon of expectations that gamers maintain with a videoludic object remains the prime factor determining the production of meaning toward sound, as it initiates its construction even before the first session of play. As Hans Robert Jauss (1982, 22) states, "The analysis of the literary experience of the reader [or the videoludic experience of the gamer] avoids the threatening pitfalls of psychology if it describes the reception and the influence of a work within the objectifiable system of expectations that arise for each work in the historical moment of its appearance, from a pre-understanding of the genre, from the form and themes of already familiar works, and from the opposition between poetics and practical language." Analyzing the sonic experience derived from a game designed for the 8-bit Nintendo Entertainment System and another one conceived for the contemporary Sony PlayStation 3 cannot be performed on the same basis, as the audiovisual and gameplay styles of the games are tied to a different set of constraints. The technological resources available at the time the games are conceived certainly represent one of the defining factors. The eternal battle between technique and creativity, in which the limits of technology have often seemed to have the upper hand, had its share of consequences.

Moreover, many aesthetic features are intrinsically tied with the amount of time and money that was invested in the games' design and marketing (Collins 2008). The relationship that the objects maintain within the pool of cultural practices that is active at the time of their emergence is also a determining factor in the creation of the gamers' horizon of expectations (as these expectations often go beyond the videoludic media). Finally, each type of game has its particular gameplay needs, and sound must therefore fulfill several roles depending on the rules of the simulation, the objectives of the game, and the representation of the diegetic world (if the game makes use of a diegesis). Each of these factors impacts the gamers' expectations regarding a game and, by extension, influences their listening. Their experience of the game is therefore affected by these factors.

To recreate the aesthetic of reception tied to the historical context of the games, it becomes necessary to pay attention to a certain amount of data collected from the paratextual material surrounding the games. This step is useful to determine if some of the sonic aspects of videogames have, in different historical periods, captured the attention of the gaming community. Reliable sources such as magazines, reviews as well as specialized websites and blogs, represent the main sources to study the reception of the games.

However, because reviewers sometimes fail to state the limits of—or properly "reconstruct"—this horizon of expectations, consensus on what defines a specific game is hardly reached. Criticism surrounding a title does not always privilege the same approach, and the use of different analytic frameworks consequently multiplies the presumptions and perceptions gamers have of the games. Likewise, those appraisals are sometimes biased by a bond that experienced critics or gamers may feel for specific genres or game developers. Speaking of cinema, Christian Metz explains that a movie might not be perceived exactly as what it is (the real object), but idealized and often confused with the imaginary object, what Metz (1984, 19, freely translated) calls "the movie as it pleased us." The same perceptual effect happens with videogames. To truly understand how the games are experienced, a return to the original "experience" is required. This logic follows a statement by Gilles Thérien (1992, 107, freely translated); according to him, is it imperative to theorize the disconnect between the reception of movies and their *lisibilité*,[3] which means, in this case, that every game must first be read and experienced "as a singular and complex object which cannot be reduced to its abstract, but has to be considered with respect to the particular functioning of its imagery" and sonic dimension. Consequently, as for videogames, one has to play them and study their formal structure and their gameplay. Through its formalist approach, this methodology aspires to bridge the gap between the general reception of the games and the *lisibilité* of the analyzed objects (see also Roux-Girard 2009).

Even though our reflection is incorporated within a much wider media and videoludic frame, the ultimate goal of this methodology is to trace the portrait of the sonic experience of videogames. A huge part of the analysis must focus on identifying and describing the sonic effects of the games and putting them in relation to the types of listening we described earlier.

To reach its objectives, our methodology also needs to make use of a plurality of conceptual tools borrowed from different fields of study. The first one, called "cultural series," has been developed by André Gaudreault (film studies) and will help us to contextualize the objects that we wish to study in their cultural and media contexts. Then, alongside the formal analysis, we will lean on the concepts (our second and third tools) of figures of interactivity and actional modalities (game studies) that were created conjointly by the members of research team Ludiciné at Université de Montréal to describe the type of tasks gamers are asked to perform within specific games. The fourth tool is Jean-François Augoyard and Henry Torgue's sonic effects (interdisciplinary in nature). The sonic effects were developed to analyze the listening activity and sonic experience of everyday sounds (mostly in urban spaces), but seems fully adaptable to a videoludic context.[4]

8.3 Cultural Series

To fully circumscribe a cultural object, it is crucial to place it within the historical, cultural and media context in which it appeared. According to Jauss (1990, 63, freely translated), a literary work needs to be put back "within the 'literary series' in which it belongs, so that we can determine its historical situation as well as its role and its importance in the general context of the literary experience." The term "series" is of importance here, especially in an intermedial context. As Rick Altman (2008, 38, freely translated) explains about cinema, "in its strongest sense, intermediality should point to... an historical step, a transitory state in which a form that is about to become a full-fledged medium is still shared among several existing media, to a point where its own identity remains in abeyance." Correspondingly, for André Gaudreault (2008, 112, freely translated), when studying the early film—kinematography attraction, as he calls it—"it is preferable to begin [the historical analysis] from the other media and other cultural spaces that greeted the new apparatus within their practice, and to develop an approach founded on the principle of intermediality, hoping our object of study permits us, in turn, to question the very notion of intermediality in its historical depth." To illustrate this intermedial phenomenon, Gaudreault proposes the notion of "cultural paradigm," to which are subordinated "several units of meaning (literature, painting, art and popular tradition, etc)... themselves being subsystems of the first" (2008, 114, freely translated). These units of meaning the author calls "cultural series."

For Gaudreault, before cinema became an autonomous institution—a full cultural paradigm—it was first absorbed and put into relation with other cultural series. For example, George Méliès was not making cinema per se, but was instead using the Lumière's cinematograph within the frame of another cultural paradigm—the "stage show"—and more precisely in connection with the cultural series *les féeries*. From this point of view, what Méliès did was in fact, not so much movies, but *féeries* on film. This observation is also applicable to videogames as, for example, the video arcade game that can be inscribed in both the history of videogames and in the history of the arcade itself (as a place for entertainment). As such, the videogame arcade can be understood as "being part" as well as "being an extension of" the penny arcade cultural series.

However, the relationship between a medium and other cultural series is not only active when the medium appears and is absorbed by other series, but also works in the opposite direction as it becomes institutionalized. As Gaudreault (2008, 123, freely translated) recalls, cinema as an institution "had developed in concordance with or against a certain number of other institutional forms (genres, cultural series, etc.) that it either absorbed, destroyed, marginalized or rejected... Institutional forms that, let us recall, tried at first to absorb it and could have as well marginalized or destroyed it." Once again, this can be applied to videogames, as it is particularly by evaluating how, for instance, the language of cinema, in its institutional and generic forms, was absorbed, destroyed, marginalized, or rejected by videogames that it is possible to determine which connections with cinema are promising venues or pitfalls. For example, full-motion video (a videogame imagery technique that makes use of prerecorded video files to represent the action in the game) was eventually rejected by the videogame cultural series, "destroying" at the same time the interactive movie cultural series (or movie-game if we follow Perron's [2003] reasoning), because the full-motion images were fundamentally incompatible with gameplay. However, the sonic treatment and dramatic fixed-camera shots of horror movies were integrated into *Alone in the Dark*, and then by the survival horror genre in general. By studying how different types of games came into contact with other cultural series, we hope to better understand the sonic dimension of games developed over the years.

As the last statements suggest, the videoludic medium becomes a vast and complex cultural practice, and to evaluate the "intermedial meshing" (de Kuyper 1997) between videogames and other media, a more focused approach might be necessary. The notion of a videoludic genre, understood as a "discursive phenomenon" (Arsenault 2011, 23), might be of help. Videoludic genres are constituted by many intermingled characteristics—types of gameplay, themes, viewing perspective, and the like, all of them being a testimony of the heterogeneous nature of videogames. But, according to Gaudreault, genres themselves can be considered as institutions. As he explains, "genre (...as a 'cultural series') would be an institution in the sense that it is, following the expression suggested by Jean-Marie Schaeffer, a 'regulating convention'" (Gaudreault 2008, 125). This idea means that up to the moment a genre becomes, as Arsenault (2011, 23) would say, "the temporary crystallization of a common cultural consensus," it is subjected to a process that is similar to the one every new medium goes through. Every genre is thus historically connected to an ensemble of other artistic practices (e.g., architecture, cinema, animation, film, television, music) as well as other cultural practices (e.g., computer science, gaming, trekking, speleology, professional sports). Accordingly, while some videoludic genres have partly absorbed some representational aspects or effects associated with these series, others have marginalized or rejected them. The concept of cultural series is therefore perfect to describe the "intermedialization" of videogames from a historical point of view. At the same time, it allows us to better explain how this intermedial dimension of the medium impacts on the gamers' listening and their sonic experience while playing a game.

8.4 Figures of Interactivity and Actional Modalities

If videogames are historically connected to a plurality of cultural series, they also possess their own language as defined by the specificities of their media. Games are indeed interactive objects and, for this reason, gameplay, the relationship that establishes itself between a rule-based system and the gamers, needs to be accounted for. As Jesper Juul explains in *Half-real*, "gameplay is not a mirror of the rules of a game, but a consequence of the game rules and the dispositions of the game players" (2005, 88). From an experiential point of view, this idea translates into the actions the gamers must perform to achieve the tasks proposed by the games, as well as the conditions of performance required by those tasks. Those were respectively named *figures of interactivity* and *actional modalities* (Perron et al. 2010).

The figures of interactivity specify the actions performed by the gamer as envisioned in the imaginary axis of the player-character's actions (more broadly, they represent the effective transposition(s) of the gamer's intervention). There are four categories of figures of interactivity: (1) spatial progression, in which the gamers perform various actions allowing the player-character or units to move in the game space; (2) confrontation, which forces the gamers to perform various manipulations, allowing the player-character units to confront enemies, hide, or flee from a threat; (3) item manipulation, representing the actions performed by the gamers that allow the player-character or its diegetic representation to interact with objects (sometimes contained in an inventory) or the environment; and (4) social interaction, through which the gamer's actions allow the player-character or its diegetic representation to enter into communication or connection with nonplayer characters. It should be noted that various figures of interactivity can be nested. The manipulation of an item to solve a puzzle, thus allowing spatial progression, would be a good example of this.

The actional modalities (automation, trivial implementation, execution, resolution, strategy) are defined from the conditions of performance, progression, and exploration experienced by the gamer; more specifically according to three components: (1) the type of skills the work requires; (2) the sequence of actions planned by the gamer at the precise moment of his experience, determined by the action's length (the range); and (3) the frame of actions as envisioned by the gamer at the time of planning a sequence, determined by the prescriptive (unique solution) or emergent (range of performance) nature of the rules system.

For example, execution relies mainly on sensorimotor skills. To time a jump, the gamers must execute a short-term sequence of actions quickly assimilated from a prescriptive frame of action. To solve a puzzle, the gamers must reconstruct a short- or middle-term sequence of actions from a prescriptive frame of action. Strategy is mainly based on cognitive skills, as the gamers must plan a mid- or long-term sequence of actions from an emergent frame of action.[5]

It must be understood that these modalities are not derived from the actual structure of the work, but are inferred from the game experience. Accordingly, figures of interactivity and actional modalities are perfect to describe the game's differences from a gameplay point of view and to better assess the type of expectations gamers might have toward sound.

8.5 Sonic Effects

Our final tool "should not be understood as a full 'concept' in its strict sense" (Augoyard and Torgue 2005, 8), but rather envisioned as a paradigm. Named "sonic effect," this qualitative tool was developed by Jean-François Augoyard and

Henry Torgue in collaboration with their colleagues at the Centre for Research on Sonic Space and the Urban in Environment, in response to the limitations of extant descriptions of everyday sound perceptions and actions: the concepts of "sound object" (Schaeffer 1966) and "soundscape" (Schafer 1977). For Augoyard and Torgue: "the concept of the soundscape seems too broad and blurred, while the sound object seems too elementary (in terms of levels of organization) to allow us to work comfortably... To use a linguistic analogy, the soundscape corresponds to the whole structure of a text, while the sound object corresponds to the first level of composition: words and syntagmas" (2005, 7). To circumvent these limitations, "the concept of the sonic effect seem[s] to describe this interaction between the physical sound environment, the sound milieu of a sociocultural community, and the 'internal soundscape' of every individual" (9). As the authors explain, the tool is oriented toward sound as an event and toward the activity of listening: "There is an effect to any sonic operation. The physical signal is under a perceptive distortion, a selection of information and an attribution of significance that depends on the abilities, psychology, culture, and social background of the listener" (8). In addition, it "produces a common sense because it gathers together into unified and harmonious listening what other disciplinary knowledge divides" (11).

This idea is precisely why we find it a perfect paradigm for our analysis of the sonic experience. The prevalent criterion for videoludic sound analysis is relative, not so much to the identification of a certain type of sounds—for example "auditory icons," "earcons" (see Grimshaw 2008), or "nonarbitrary auditory icons" (see Jørgensen 2009)—but it resides in "the effectiveness of the feeling caused in the listener" (Augoyard and Torgue 2005, 10), which is exactly what sonic effects describe. Being interdisciplinary in nature, such a tool also facilitates the study of sound in an intermedial context. Furthermore, the sonic effect provides the same degree of flexibility, adaptability, and rigorousness that we wish to apply to our own methodology. Even if we will not be able to address every sonic effect Augoyard and Torgue have listed and defined, one of the objectives of this chapter is to demonstrate how it is possible to identify, analyze, and adapt some of them to the imperatives of the videoludic experience. To do so, we will now test our methodology with an analysis of the game *Uncharted 2: Among Thieves*.

8.6 *Uncharted 2: Among Thieves*: The Cinematic Experience

Part of the action-adventure genre, *Uncharted 2: Among Thieves*, a AAA title developed by Naughty Dog and published in 2009 by Sony Computer Entertainment, combines elements from both the adventure genre—mostly puzzle resolution—and game mechanics from various action game genres such as third-person shooters and combat games (all of them being videoludic cultural series). *Uncharted 2* is also an extension of the "3D platformer" series, adding a very visceral dimension to the game's spatial progression. While, by definition, action-adventure games are not heir to the adventure film genre, the *Uncharted* series, like many other action-adventure games, has been identified as borrowing heavily from the formal language of films. Harold Goldberg, in his book *All Your Base Are Belong to Us*, describes the second game as such: "*Uncharted 2: Among Thieves*, with its nineteenth-century penny dreadful influence [a literary cultural series] on a story surrounding Marco Polo's lost treasure, let you feel as though you were in a melodramatic movie with all the spills and thrills of an *Indiana Jones* adventure" (Goldberg 2011, 304). The desire to generate a cinematic feel through the development of photorealistic displays—"that is . . . to make their digital characters and settings look more and more like live-action film" (see Bolter 2005, 26)—has been present since the birth of the commercial videogames. However, the absorption of a cinematic language would not have been made possible without the technological pivot that allowed a tridimensional representation of space. The adoption of a third-person perspective, a choice of representation that is consistent with the type of gameplay that characterizes the action-adventure and (recent) platformer genres, was also favorable to a connection with the action film cultural series.

Uncharted 2's spatial progression and confrontation figures mainly rely on an execution modality. For example, in the "falling train sequence," a gameplay segment that needs to be played twice over the course of the game, the gamers, through the actions their player-character Nathan Drake, must reach the top of a wagon that is barely hanging on the edge of a cliff. To get Nathan out of danger, the gamers must execute multiple jumps, putting their sensorimotor skills to the test. To amplify the emotions associated with Drake's perilous ascension, the game makes use of a virtual camera that smoothly follows the player-character's actions by simulating pans and tracking shots. The game shifts between camera angles as well, deconstructing the action and space of the train through a replication of "continuity editing," a discursive layer that is mostly associated with movies. The *Uncharted* series also makes use of numerous fluid transitions between gameplay and cinematic sequences. While some gamers might feel frustrated by the frequent temporary loss of control cinematic sequences impose, the latter are essential to generate the spectacular effect of the game, and help to generate the desired visceral emotions tied to the execution modality. In addition, the cinematic camera helps the gamers to meet the expectations created by what might be considered an *Indiana Jones*-inspired game.

One of the roles of sound is therefore to meet these expectations. Corollary to the first category of listening we described in the chapter, sound *in Uncharted 2* (it could be completely different for another game depending on the cultural series in which the game has roots) mostly relies on the imitation of the sonic language of film and previous action-adventure games. As Collins (2008, 134) describes, "In many ways the realism aspired to in games is not a naturalistic realism in the sense of being a simulation of reality, but a cinematic realism that relies on established motion-picture convention. The 'cine-real' is a sense of immersion and believability, or verisimilitude, within a fantasy world." Augoyard and Torgue describe the imitation effect as "a semiotic effect referring to a sound emission that is consciously produced according to a style of reference" (2005, 59). By using the imitation effect, "sound designers use aural memory to authenticate sounds that they have been asked to reproduce or create for a film, a radio program,…a television show, [or a videogame]" (60). The imitation effect sometimes exploits sonic stereotypes, but is generally achieved through the sound's qualities, by creating sound images.

The imitation effect is thus related to some of the elementary and compositional effects that are active within the game. It is the case with resonance, reverberation, and filtration effects. The gunshots, explosions, punches, and other form of impacts are designed to create a cinematic feel that is usually associated with the Hollywood action film. The treatment applied to the sounds must create an effect of *rendering*. As Michel Chion explains, "the sound heard in films…hardly translates the real sound…but instead the physical, psychological, even metaphysical impact of the act" (2003, 214). The sound must instead seem "real, efficient and adapted" to "recreate the sensation…associated with the cause or with the circumstance evoked in the [game]" (Chion 1990, 94, freely translated). This is why Drake's punches sound so loud, contributing, at the same time, to the gamers' feeling of agency.

The imitation effect can also lead to an *anamnesis* effect, which is described by Augoyard and Torgue as "an effect of reminiscence in which a past situation or atmosphere is brought back to the listener's consciousness, provoked by a particular signal or sonic context" (2005, 21). As the researchers explain, the effect can span a short period of time (when a sound previously heard in the game is heard again) or longer (an entire life). Accordingly, this process also implies that the effect can be internal as well as external to the game. For example, the use of a musical leitmotif as well as the repetition (thus creating a repetition effect) of action sounds during combat sequences can bring back the emotion generated by the previous battle. The anamnesis effect can, however, transcend the limits of the game. In chapter 4 of the game, just before jumping off a cliff, Sully, Drake's friend says, "Hold on there, Sundance. You gotta be outta your mind," which is a direct reference to the movie *Butch Cassidy and the Sundance Kid* (1969). For a fan of the movie, this sentence will bring back not only the memory of the movie but the emotion that it generated by the intertextual reference. Thus, to be perceived, "the imitation [and by extension the anamnesis] effect implies a previous…culture [of connecting cultural series and objects]. Sometimes only the initiated will have access to this effect and be able to understand the allusion…In all perceived cases, there is nonetheless an immediate change in sound climate, a modification in the quality of listening" (Augoyard and Torgue 2005, 61).

Uncharted 2's sonic experience is also tied to the internal soundscape and videoludic nature of the game. For instance, the *filtration effect* mentioned earlier also needs to simulate the delimitations of the game space. This effect is particularly perceptible during chapter 19 when Drake is caught in a firefight at the heart of a Tibetan village. During the *Siege* chapter, when Drake penetrates a house or hides behind a wall, a dynamic filtering effect is applied to the sound of the gunshots fired by the enemies. When transitioning from the exterior to the interior, this effect is also accompanied by a cut-out effect that drops the external ambiance. Damian Kastbauer (2012), sound designer for such games as *Star Wars: The Old Republic* and *Uncharted 3: Drake's Deception*, explains that these filtering effects are applied as part of a sound propagation and are commonly referred to as obstructions or occlusions. While he was not able to confirm how audio implementation was achieved for *Uncharted* 2 (he did not work on the game), he explains that is it "common enough practice to use 'ray tracing' to determine obstructions/occlusions between the emitting point of a sound (i.e. gunshot) and the listener (in this case the [gamer]) in order to calculate an appropriate filter percentage." Then, environments "are usually authored using a 3D volume in a game editor/level editor." As Kastbauer clarifies, "in a situation where there are multiple environments, for instance an exterior and interior, there would be two volumes, each with their own defined ambiences. The sound designer would then author 'portals,' or additional volumes, that define locations where the sound can propagate between environments."[6] In *Uncharted* 2, the weakening of the higher frequencies, coupled with the change in ambiance informs the gamers that they are safe from the shooters. If the sound regains the frequency spectrum, this means that an enemy has entered the house. In *Uncharted* 2, these effects therefore can lead to a refinement of the gamers' listening skills that is essential to the player-character's survival. This refinement then contributes to the specialization of the gamers' videoludic listening skills that, in turn, will participate in the way they experience further games that use similar occlusion techniques.

We have shown here just some of the richness and analytical depth that we believe can be achieved by making use of the proposed methodology. Although succinct, this analysis has nevertheless demonstrated how, in a precise generic context, the experience and emotional response of the gamers are conditioned by both their familiarity with pre-existing sonic patterns (linked with a previous experience of the cultural series intersecting with the game), and by the effect that are internal to the game's soundtrack (effects that help the gamers to perform the tasks required by the game and to refine their videoludic listening skills).

8.7 Conclusions

The sonic experience of videogames is complex. Constructed by the gamers' listening activity, the sonic experience of a game is conditioned by an individual's acquired knowledge (from a historical, intermedial, and videoludic perspective) as well as his or her perception, and involvement with the audiovisual components of the games. For this reason, the sonic experience of videogames must be approached with flexible and adapted tools that can shed light on the relationship that is achieved through gameplay between the gamers and the videoludic sounds.

The methodology we have developed aims to analyze this experience. However, exhaustive analyses of other games will be required to evaluate its full potential. The videoludic phenomenon needs to be studied more broadly. A departure from the scope of a film-game connection, and an expansion to videoludic genres that are historically related to a variety of other cultural series, would constitute a first step in the right direction. For example, strategy games are an extension of the "strategy board games" cultural series, which, prior to its transmedialization onto computers, did not include a mediated audiovisual component. In *Sid Meier's Civilization VI*, the audiovisual dimension was grafted on to the game in order for it to be playable on a computer screen. In *Civilization IV*, every figure of interactivity (spatial progression, confrontation, item management, and social interactions) is respecified by a strategy modality that does not command the same audiovisual imperatives as the *Uncharted* series. According to its gameplay, *Civilization IV* employs effects that generate a sonic experience that is completely different from the one *Uncharted 2* provides.

In addition, the range of the analyzed effects demands to be expanded to the audiovisual and, ultimately the videoludic dimensions of the game. While the sonic effects represent a good starting point for the analysis of the gamers' experience, they are not sufficient to portray the relationship between the sounds and the images and between the sounds and gameplay. The example we provided about the synchronization point is a good example of the additions that this methodology would benefit from.

Notes

1. I chose to use the term "gamer" over the term "player" following a distinction made by Perron (2003). According to Perron (2003, 240–42), "gamer" and "player" are defined according to an attitude that is itself characterized by the type of gameplay a game proposes. I use videoludic as an adjective meaning "related to videogames."
2. I wish to thank Pierre-Olivier Forest who submitted this idea during a meeting of our research team on sonic creation at Université de Montréal.
3. For Thérien, a movie's *lisibilité* is the particular functioning of its imagery (and sounds). For the purpose of our study, *lisibilité* also includes the imperatives tied to gameplay.
4. Mark Grimshaw and Tom Garner also employ these sonic effects in their chapter "Embodied Virtual Acoustic Ecologies of Computer Games" to explain how "auditory processing is an embodied event, dependent upon the relationship between physical environment, memory, and physiology." For more details on the embodied cognition theory of computer games, please refer to Chapter 11.
5. For a more detailed list of actional modalities, please consult our terminological dictionary at www.ludicine.ca.
6. I would like to thank Damian Kastbauer for his enlightening insights.

References

Altman, Rick. 1999. De l'intermédialité au multimedia: cinema, medias, avènement du son. *Cinémas* 10 (1): 37–53.

Arsenault, Dominic. 2011. Des typologies mécaniques à l'expérience esthétique: Fonctions et mutations du genre dans le jeu vidéo. PhD diss., Montréal: Université de Montréal.

Augoyard, Jean-François, and Henry Torgue. 2005. *Sonic Experience: A Guide to Everyday Sounds*. Montreal and Kingston: McGill-Queen's University Press.

Bolter, Jay David. 2005. Transference and Transparency: Digital Technology and the Remediation of Cinema. *Intermédialités* 6: 13–26.

Casetti, Francesco. 2007. The Filmic Experience: An Introduction. http://www.francescocasetti.net/ENGresearch.htm.

Céleste, Bernardette, François Delalande, and Elisabeth Dumaurier. 1982. *L'enfant du sonore au musical*. Paris: Buchet/Chastel–INA

Childs, G. W. 2007. *Creating Music and Sound for Games*. Boston: Thomson Course Technology.

Chion, Michel. 2003. *Un art sonore, le cinéma: histoire, esthétique, poétique*. Paris: Cahiers du Cinéma.

——. 1990. *L'Audio-vision*. Paris: Nathan.

——. 1983. *Guide des objets sonores: Pierre Schaeffer et la recherche musicale*. Paris: Buchet/Chastel–INA.

Collins, Karen. 2008. *Game Sound: An Introduction to the History, Theory, and Practice of Video Game Music and Sound Design*. Cambridge: MIT Press.

de Kuyper, Éric. 1997. Le theatre comme mauvais objet. *Cinémathèque* 11: 63–75.

Deshays, Daniel. 2010. *Entendre le cinéma*. Paris: Klincksieck.

Farnell, Andy. 2011. Behaviour, Structure and Causality in Procedural Audio. In *Game Sound Technology and Player Interaction: Concepts and Developments*, ed. Mark Grimshaw, 313–339. Hershey, PA: Information Science Reference.

Folmann, Troels. 2004. Dimensions of Game Audio. http://www.itu.dk/people/folmann/2004/11/dimensions-of-gameaudio.html

Gaudreault, André. 2008. *Cinéma et attraction: pour une nouvelle histoire du cinématographe*. Paris: CNRS Éditions.

Goldberg, Harold. 2011. *All Your Base Are Belong to Us*. New York: Three Rivers.

Grimshaw, Mark. 2008. *The Acoustic Ecology of the First Person Shooter: The Player Experience of Sound in the First-person Shooter Computer Game*. Saarbrücken: VDM Verlag Dr. Muller.

Jauss, Hans Robert. 1982. *Towards an Aesthetic of Reception*. Minneapolis: University of Minnesota Press.

——. 1990. *Pour une esthétique de la réception*. Paris: Gallimard.

Jørgensen, Kristine. 2009. *A Comprehensive Study of Sound in Computer Games: How Audio Affects Player Action*. Lewiston, NY: Edwin Mellen.

Juul, Jesper. 2005. *Half-real: Video Games between Real Rules and Fictional Worlds*. Cambridge, MA: MIT Press.

Kastbauer, Damian. 2012. Personal correspondence, June 13.

Marks, Aaron. 2009. *The Complete Guide to Game Audio for Composers, Musicians, Sound Designers, and Game Developers, 2nd edn*. Burlington, MA: Focal Press.

Metz, Christian. 1984. *Le signifiant imaginaire: Psychanalyse et cinéma*. Paris: C. Bourgois.

Perron, Bernard. 2003. From Gamers to Player to Game Players. In *The Videogame Theory Reader*, ed. Bernard Perron and Mark J. P. Wolf, 237–258. New York: Routledge.

Perron, Bernard, et al. 2010. Ludiciné's Dictionary of Terms for the Ludography of Horror Video Games. http://ludicine.ca/sites/ludicine.ca/files/ludicine_terms_horror_en_0.pdf.

Roux-Girard, Guillaume. 2009. Plunged Alone into Darkness: Evolution in the Staging of Fear in the Alone in the Dark Series. In *Horror Videogames*, ed. Bernard Perron, 145–167. Jefferson, NC: McFarland.

Schaeffer, Pierre. 1966. *Traité des objets musicaux: Essai interdisciplines*. Paris: Édition du Seuil.

Schafer, R. Murray. 1977. *The Tuning of the World*. Toronto: McClelland and Stewart.

Thérien, Gilles.1992. La lisibilité au cinema. *Cinémas, cinema etréception* 2(2–3): 107–122.

van Tol, Richard, and Sander Huiberts. 2008. IEZA: A Framework for Game Audio. *Gamasutra*. http://www.gamasutra.com/view/feature/3509/ieza_a_framework_for_game_audio.php?page=3.

CHAPTER 9

DESIGNING A GAME FOR MUSIC

Integrated Design Approaches for Ludic Music and Interactivity

RICHARD STEVENS
AND DAVE RAYBOULD

THE question of how interactive music should function in games is perhaps a misleading one, as there are many different types of games and many different types of players. One of the most compelling explanations for the huge popularity of videogames is that they meet people's intrinsic psychological needs quickly, with consistency, and with great frequency (Rigby and Ryan 2011). The apparent drivers of the development of games and their marketing—such as the fidelity of graphics and audio, or as the popular press would have us imagine, the degree of violence—are far less significant factors than the drive to increase our sense of well-being through meeting the basic needs of competence (or mastery), autonomy (or volition) and relatedness (social connection) (Przblinkski et al. 2010) or the desire to become immersed in narrative worlds (Cairns 2006). Since it is clear that player satisfaction is a product of "needs met" over "needs," it is important that we recognize that music should operate in different ways in different circumstances.

Players will choose a genre of game that best matches their intrinsic needs (Madigan 2012) and they will also adopt different gameplay strategies according to their personality type (Bartle 1996). A player's desire for relatedness or fellowship (Hunicke, LeBlanc, and Zubek 2004) might be met through music that rewards cooperative play (Kristian and Girard 2011) or that allows them the ability to perform music with others (Collins 2007), but is also likely to be met by hearing music of their preferred genre. Given the importance of music to a sense of social identity and group membership and the links between personality type and musical preference (North and Hargreaves 2007), it is perhaps not surprising that there appears to be a strong correlation between game genre and musical style (Summers 2011). So the next time we complain about the marketing department conducting its research on Facebook to identify the bands to use on the soundtrack to the latest racing game (Baysted 2012), perhaps we are missing the point. A comprehensive assessment of the psychological needs of the player and how these can best be met by music in games is beyond the scope of this chapter, but we raise this in our opening remarks to highlight that, although the remainder of the chapter will be focusing on "interactive" music, we appreciate that music should function according to the needs of the game and of the player, and that some of these needs may be perfectly well met by traditionally linear music.

Of the player needs mentioned above, the "innate desire to grow our abilities and gain mastery of new situations and challenges" (Rigby and Ryan 2011) is seen by many to be the most important determinant of enjoyment in games (Vorderer and Bryant 2006). Termed "hard fun" by Lazzaro (2008), the success of this "voluntary effort to overcome unnecessary obstacles" (Suits 2005) is thought to produce a release of chemicals in the brain (Bateman and Nacke 2010), strongly associated with reinforcement and motivation (Salimpoor et al. 2011). Finding oneself at the optimal point between being suitably challenged and having the skills to master those challenges is referred to as being within the highly desirable and immersive state of "flow" (Csíkszentmihályi 1992). The emotional state of "fiero" (or triumph over adversity; Ekman, 2004), brought about by overcoming obstacles, contributes to maintaining a state of flow by providing the positive reinforcement the player needs to continue to meet the increasing challenge, and is recognized as an important source of pleasure or "fun" (Koster 2005).

In contrast to meeting players' social needs (where the focus is on musical genre) or the narratologically immersive needs (met through the evocation of time, place, and mood), music that contributes to flow by helping players to achieve competence (by providing information, or by motivating and rewarding us) or music that guides and supports players by making them feel like they are acting of their own volition and that their actions are meaningful (fulfilling the need for autonomy) must be synchronized tightly to game events. The requirements to ensure that feedback is immediate (Bond and Beale 2009) and that music is congruent with the game action (Wharton and Collins 2011) represent the inherent conflict between interactivity and musical form. The compromise between "contextual responsiveness and musical integrity" (Bajakian 2010) continues to challenge composers and implementers trying to avoid awkward or clumsy musical results (Munday 2007). Such game-specific, ludic, or metonymic (Whalen 2004) music and the issues that arise out of music synchronization within an interactive medium will be the focus of this chapter.

9.1 Musical Structures vs. Interactivity

There are many ways in which music can evoke or induce emotions, but there is clear evidence that strong or "peak" emotions in response to music (such as chills, lump in the throat, etc.) are associated with the creation of, and confirmation or violation of, expectancy (Sloboda 1991). Given that musical training unsurprisingly leads to a heightened sensitivity (Dellacherie et al. 2011), it may be that many commentators with a background in music (such as ourselves) are prone to exaggerate the problems that arise when such patterns of expectancy are interrupted by the need to respond to game events, but there is strong evidence that no formal training is required to make automatic predictions of chord functions (Koelsch 2011), to be acutely aware of phrase boundaries (Nan, Knösche, and Friederici 2009) and expectations of metrical or pitch patterns (Huron 2006), and that breaking these patterns of expectation can cause disorientation (Margulis 2007) and negative responses (Steinbeis, Koelsch, and Sloboda 2006).

It is of course possible to evoke a variety of emotions through musical styles that are not heavily expectation-based, and that rather than relying upon schematic expectations (derived through familiarity with the musical syntax of style), expectations may be the product of familiarity with the specific piece or dynamically generated from the piece itself (Huron 2006). Indeed in some genres (such as platformers), it can be seen that learned schematic expectations have allowed musical forms that are much more flexible, responsive, and cartoon-like. In the horror genre, where the lack of a tonal center or metrical pulse is often used to destabilize the audience or player (Summers 2011) or to parallel the characters psychological crisis (Whalen 2004), the cross-fading between atonal, arhythmic music of different intensities can induce the appropriate emotional effects without breaking any musical expectations, since the musical form itself (or lack of it) does not imply any. Likewise static tonality or drone-based music can make it much easier to transition between different segments without upsetting the implicit expectations of chordal progressions (Stuart 2010).

While there are exceptions, such as those outlined above, it must be recognized that the player's significant exposure to the paradigms of film and television music (Nielsen 2011) and the wish to activate the strongly associated cultural codes (Gorbman 1987) mean that many games based within fictional narratives bring with them the expectations of a Hollywood-style soundtrack (Jackson 2011), a strongly tonal and expectation-based form almost uniquely unsuited to the temporal uncertainty of games.

A fundamental form of musical expectancy that can be easily "broken" through the need to represent, or at least remain congruent with, game events is that of pulse. Using parallel forms (sometimes referred to as vertical re-orchestration; Collins 2009), where layers or "stems" are composed such that they work in vertical combination, can be very effective in maintaining musical continuity while allowing for significant changes in texture and instrumentation (see Figure 9.2). In *Splinter Cell: Chaos Theory*, the layers act as a warning to indicate the proximity of enemies, and in *Fallout: New Vegas*, concentric circles of triggers attached to musical stems help the player to navigate the Wasteland (Lawlor 2012). Layers can tell the player whether special modes are active, notify them of the alertness state or current health of nonplayer characters (NPCs), or represent overall progress through a puzzle (*Portal* 2) or battle (*Tom Clancy's EndWar*). The attenuation of different layers of music to represent different game states or continuous variables can be highly effective in providing the player with information to support success (enhancing their skill within the flow state) and can increase layers of tension (to heighten the impression of challenge). However given that *Splinter Cell's* musical form is predetermined (composed to be essentially static and allowing the game to generate its dynamics; IGN 2006) it is less suited to providing reward (enhancing fiero), since it lacks the ability to respond to game events with specific timed musical gestures.

Feedback on actions or game events can be transmitted via music using ornamental (Figure 9.1), or transitional forms (Figure 9.3). It is frequently the case that we want to acknowledge events in the game but they are not significant enough to warrant a whole scale change in music. In this case, games typically use an ornamental flourish or stinger that might reward a successful jump (*Uncharted 3*), a successful attack (*The Legend of Zelda: Skyward* Sword), or shot (*The Operative: No One Lives Forever*). Typically these are not aligned to musical pulse but happen immediately over the top of the currently playing musical bed (e.g., *CryEngine3*).

The function of musical feedback could be viewed from a human–computer-interaction perspective (indicating confirmation or rejection of action; Jørgensen 2010), but it also carries an implicit emotional message. The ludic or metonymic is not separable from the metaphoric (that which relates to the game as a story or world; Whalen 2004). A piece of music may confirm that an action has been successful (defeat of the enemy) and thus provide the positive reinforcement important to flow, but at the same time the music is also providing an insight into character, as it does in film (Hoeckner et al. 2011). Since the player is the character, this music is informing them of their place in the fictional world, their heroism, and their role in shaping the events of the world around them, supporting the player's sense of autonomy by making their choices appear meaningful. Given the audiovisual expectations formed from a lifetime of narrative media mentioned above, we expect these musical responses to be both synchronized and dramatic.

The simple transitional cross-fade can work if music is composed in such a way as to avoid or at least lessen musical expectations, or musical transitions can be masked with sound effects (Porter 2010), but the most effective way to maintain musical expectations within transitional forms is to restrict the changes to musically appropriate times. By carefully constructing matrices of possible transitions between sections of music that take account of potential entry or exit points and the types of transition permitted (immediate, next measure, next beat, etc.; Selfon 2003), it is possible to construct highly "musical" scores (that maintain musical expectations). However the by-product of this musicality is that there is a "lag" between game events and the music's response (Collins 2007). Again we are attempting to "adhere to the sound of film music while losing sight of its *raison d'etre*; the heightened emotional impact provided by the close synchronisation of musical and visual events" (Munday 2007).

It is acknowledged by many in the game music industry that "interactivity = modularity" (Ashby 2008) and a focus on temporally aware cells of music (Figure 9.4) or "micro scores" (Folmann 2006) can allow music to more quickly respond to events while maintaining musical flow. However, the production of such cellular forms remains problematic, as when transitioning from one cell to another the musical parts need to retain their natural decay portions or "tails" in order to sound natural (Selfon 2009). Certain styles

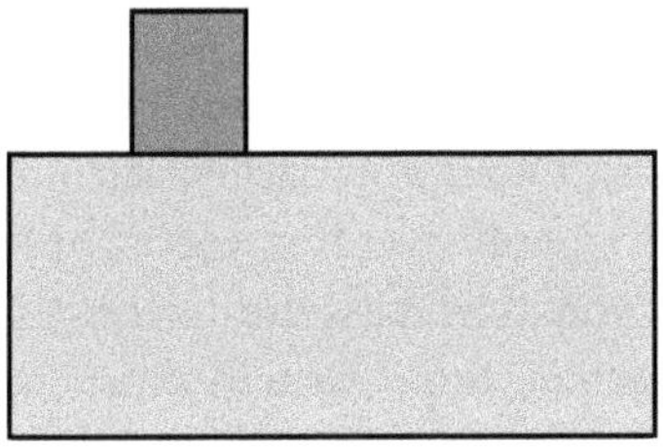

FIGURE 9.1 Ornamental forms.

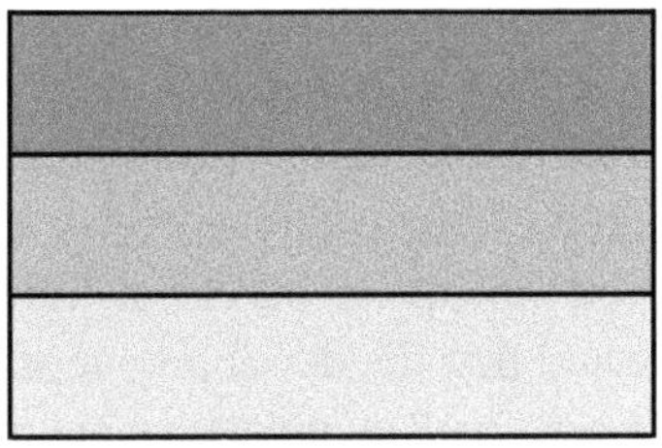

FIGURE 9.2 Parallel forms.

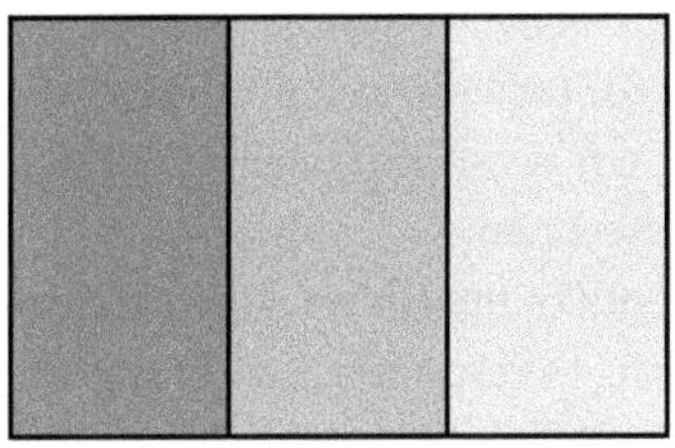

FIGURE 9.3 Transitional forms.

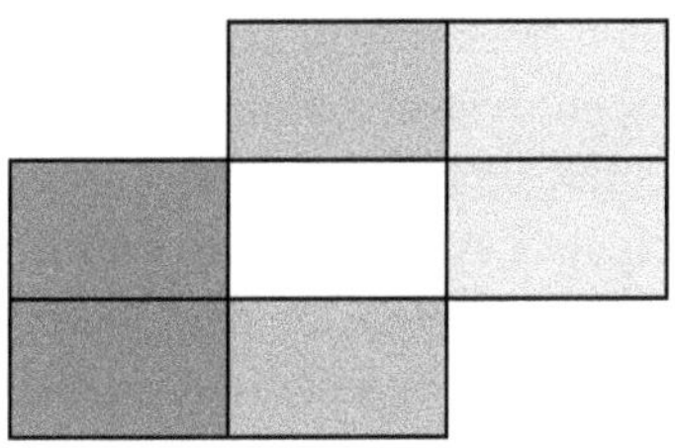

FIGURE 9.4 Cellular forms.

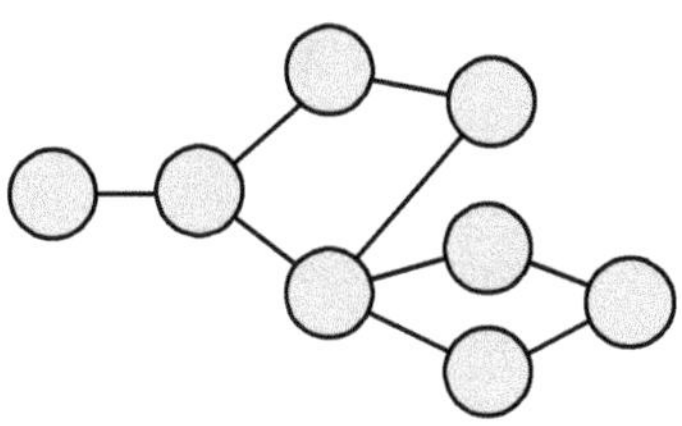

FIGURE 9.5 Algorithmic forms.

of music that have rigid time-based structures and short percussive elements (e.g., some "pop" music) can move effectively between segments or cells using short cross-fades (Durity and Macanulty 2010). Other approaches, such as Whitmore's dovetail technique,[1] or applying reverbs to smooth over transitions (by artificially creating decay tails in real time), can also work well, but these are rarely satisfactory for acoustic instrumental forms, as getting musicians to perform in short chunks (so you can capture the authentic decay within the correct acoustic space) is both time consuming and unnatural. The highly modular, or "granular," note-level approach of MIDI and sample-based systems resolves the decay problem (since the tail exists authentically within each sampled note) and also provides for the kind of parametric control ideally suited to interactivity (Collins 2009), but it has fallen spectacularly out of fashion within many genres as a victim of the quest for a Hollywood sound (Collins 2008). Senior figures within the game audio industry agree that the return of note-level or MIDI control in some form is the inevitable response to addressing questions of musical interactivity (Page and Kelly 2007), and others have suggested that the development of cloud-based processing and streaming might mitigate the perceived quality issues (in terms of addressing RAM for high-quality samples and processing for mastering) (Drescher 2010). There is an innate reluctance to replace activities seen as innately human, such as music composition, with processes or algorithms (Cope 2000) (Figure 9.5), but the potential for musical models (McAlpine 2009), stochastic (or generative) approaches (Weir 2011), and parameterized control (Livingstone and Brown 2005) adds weight to the need to move beyond the stereo wave file or the pre-rendered stem.

Although the return of granular, note-level control within games would undoubtedly improve the ability of the music to respond to, and support, game events more elegantly, it still remains a theoretical impossibility to align expectation-based musical structures with unpredictable events. If we imagine the music system as a black box containing a highly talented silent movie piano player, we can appreciate that he could quickly adapt the music to the action on the screen, using his highly evolved knowledge of musical harmony and form to neatly segue, via an appropriate passing chord or note, into a new "piece" or state. But it would not be immediate and irrespective of his skill: he could never build toward an anticipated event and synchronize precisely with the climatic point. In other words the synchronization of game fiero and musical peaks, paralleling the highly rewarding climax of many a Hollywood chase sequence, cannot happen, unless we reconsider the nature of the relationship between game design and music.

9.2 Interactivity?

Although there is general agreement that the umbrella term "dynamic" music somehow differs from the linear music of film (Collins 2007), the remaining terminology with regards to music in videogames is varied and confusing. The term interactive when applied to this field has a long history of ambiguity (Ross 2001), and although there is an inclination to use the term "adaptive" where the music may respond to game events without any direct input from the player (Fay 2004) (or at least when there is a degree of abstraction or a layer of interpretation between the player actions and the output; Farnell 2007), the usage of these terms is often interchangeable or contradictory. The shifting, or at least poorly defined, meaning of the term "interactive" is not unique to videogames (Aarseth 2003), and although there is little to gain from trying to impose a meaning here, it is worth pursuing briefly, as a number of definitions call for a reappraisal of what we might currently call interactive.

Although some commentators might consider any engagement with media to be interactive in some sense (Manovich 2002), our current common usage of the term within game audio to encompass all audio events that respond to user input (Selfon 2004) can detract from the idea of interactivity as a continuum, within which there are there are differing degrees. At one end of this scale is the notion, as yet unconsidered in many games, that interactivity is a cyclical process (Crawford 2003), where the agents within a system act upon each other (inter + act; Harper 2012), and that the receiver can also act as a transmitter (Gianetti 2007). McQuail (2005, 497) defines interactivity as "The capacity for reciprocal, two-way communication attributable to a communication medium or relationship. Interactivity allows for mutual adjustment, co-orientation, finer control and greater efficiency in most communication relationships and processes," and states that we might describe the degree of interactivity as being "indicated by the ratio of response or initiative on the part of the user to the "offer" of the source/sender" (2005, 144). If we consider the music, player, and game as components of a system, we can see that most current practice within music for games could be considered as simply "reactive," acting in response to events from the player, mediated by the game engine (shown as the dotted line in Figure 9.6), or in direct response to the game engine itself, "adaptive" (the dashed line in Figure 9.6).[2]

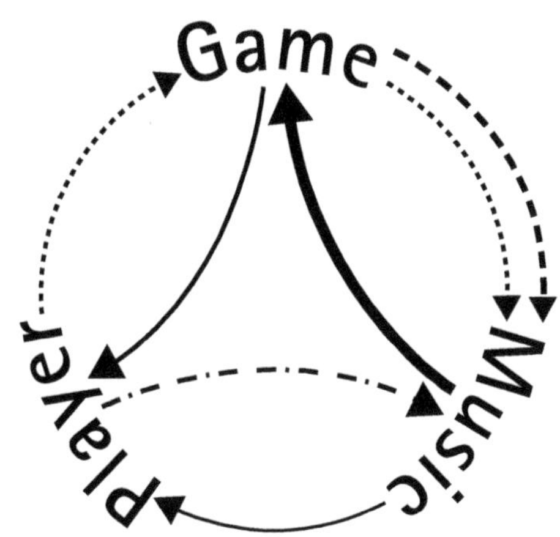

FIGURE 9.6 Game music systems.

By reserving the use of the term "interactive" for systems that are truly bidirectional, where the game's decision-making processes also take input from the music system as to its current state (indicated by the thick arrow in Figure 9.6), we raise the possibility of approaching the seemingly intractable interactivity vs. musical structure problem in a new way.

9.3 Thresholds, Windows, and Notifications

The game designer Clint Hocking (2012) refers to the design challenge of the "threshold problem" as being "any problem that arises as a result of a discrete state change that occurs at an arbitrary, designer-defined threshold in an analogue range," and points out that in order to avoid frustration these need to be clearly communicated to the player, or made "sticky," so that if they get near enough to the value they are automatically snapped to it.

In order to facilitate greater interactivity between the music and game state (so that moments of fiero can be heightened by synchronization with pleasurable structural points in music) we'd like to suggest that these arbitrary thresholds might instead be considered as windows of opportunity. When the game state is looking to take an action (the window is open) it might look at the condition of the music (which would be inputting its current state) to inform when that action might actually occur. This process would require a more integrated approach to music and game design that we will illustrate below with a few examples.

9.3.1 Example 1: Helicopter Gunship

You are in a helicopter attacking a fuel depot at the entrance to an enemy compound. The game system is set up so that it takes 100 direct hits with your weapon to destroy the depot (Figure 9.7).

Within a normal "reactive" system, when the direct hit variable equals 100, the depot explode animation is triggered. The currently playing music is cut off immediately and the triumphant brass music cue is played. "Interactively," when the direct hit variable equals 100 the game engine checks the music state. It sees that the music is currently at the fourth beat of the bar and, given that it knows the ideal (most pleasurable) musical transition point would be on beat one, it continues taking additional direct hits, until a musically appropriate musical time. Then the triumphant brass cue is played, and the depot explode animation is triggered simultaneously. The moment of fiero produced by the triumph coincides with the musical expectation implied by the 4/4 time signature, and therefore the pleasure

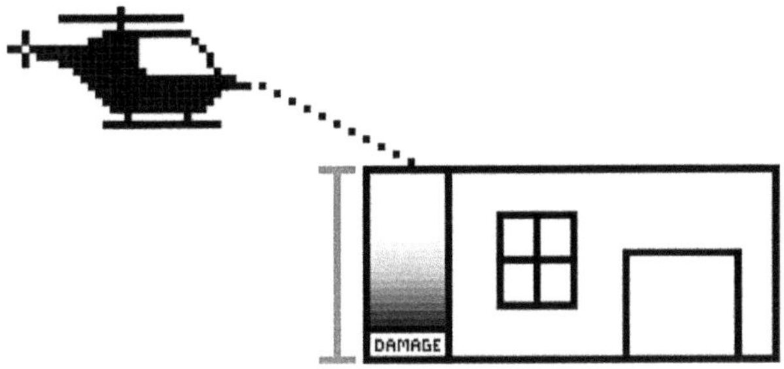

FIGURE 9.7 Helicopter gunship.

is heightened. To take this one step further, it might be appropriate to consider that a window may open up around the threshold (direct hits = 100), meaning that, if musically appropriate, the event may actually take place slightly earlier than the threshold point (e.g., direct hits = 97).

9.3.2 Example 2: Find the Enemy

Having gained entry to the enemy compound you need to find and detain the chief bad guy. On approaching the hut where he's hiding out, the game will jump to an in-game cut scene that shows your final steps up to the door, you kicking in the door, and gracefully leaping through, to the bad guy's surprise and horror (Figure 9.8).

In a reactive system, when the player passes the threshold (illustrated by the circle trigger around the hut) the in-game cut-scene is triggered. The currently playing music is cut off immediately and the cut-scene music is played. Interactively, we consider a window around the threshold point (indicated by the gray line) where the game state starts to look at the music state. Whenever the music state reaches the next appropriate musical juncture (for example approaching beat one of the bar again) the cut-scene is triggered to coincide with the musical change it also instigates at this moment.

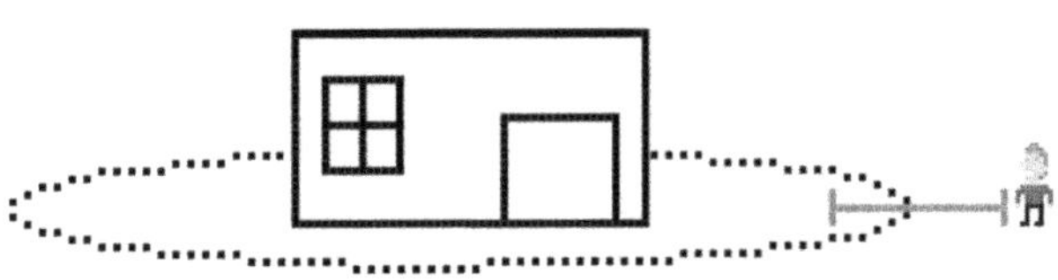

FIGURE 9.8 Cut-scene.

FIGURE 9.9 Death.

9.4 Timing and Animation

9.4.1 Example 3: NPC Death and Collapse

Unsurprisingly, the bad guy's henchman rushes to his aid. A thick-set man with an aggressive nature, he has a threshold of twenty blows before he will collapse and die (Figure 9.9).

Interactively, this could do the same as Figure 9.7 above, and actually trigger the event (death) to happen at nineteen or twenty blows, when close to a musical juncture. However the player may be attuned to the strength of the enemy and feel that this somehow does not feel right. Instead it may be possible to adapt the collapse animation, speeding up or slowing it down by interpolating differently between key frames, looking to the music system for timing, so that the impact onto the ground is timed to coincide with the appropriate transition point within the musical structure.

9.4.2 Example 4: Jump

In pursuit of the chief bad guy, who has now left the compound on a motor bike, you speed downhill toward a gaping chasm (Figure 9.10). We want to accompany your leap off, and landing, with an appropriately dramatic music cue, but you are weaving through a number of trees on your way down so we can make only a rough guess at your arrival time. Interactively, we could calculate the exact time required to hit the leap at an appropriate musical point. We then manipulate (constantly update) the speed of the vehicle to compensate for the player's turns so that they hit the jump in synchrony with the music, then also adjust their air speed and trajectory so that they land with a satisfying, musical, bump.

With the examples above we hope we have communicated some simple ways in which a more interactive and integrated approach to game design could exploit

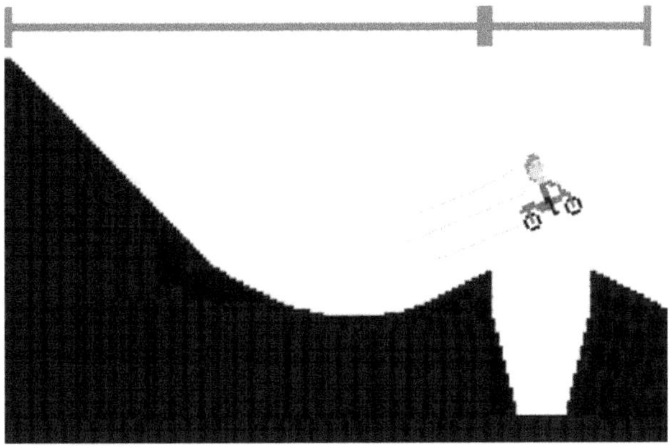

FIGURE 9.10 Jump.

the pleasurable benefits of aligning game events and musical structure. However, they also probably raise concerns as to the effect on the player's sense of autonomy or agency, raising the risk of this becoming another type of frustration-inducing "Quick Time Event" (Miller 2010), an attempt to add some limited interaction into what would otherwise be a passive cut-scene, typically through the sudden appearance of an onscreen icon prompting the player to "press X now...". The danger is that the satisfaction produced from the musical synchronization of game events will not be powerful enough to outweigh any frustrations that this wresting of control may induce. Anecdotal evidence from people already innovating in the area of integrated game and music design suggests that as long as players feel that their actions have been acknowledged, through some form of audio or visual feedback, they are happy to accept a momentary pause before the action (Kastbauer 2011). This feedback could be as simple as the rumble of a depot about to explode or the groan of an enemy about to die. This could also be accomplished with music through the introduction of a short stinger (star) and the fading in of a percussive part (ramp) that leads into the event measure (as illustrated in Figure 9.11).

The manipulation of animation and event timings and the use of opportunity windows rather than discrete thresholds are simple concepts to support two-way interactivity between game and music systems. In order to generate and support more innovation around this idea it is vital that attitudes, production processes, and tools are re-examined and developed.

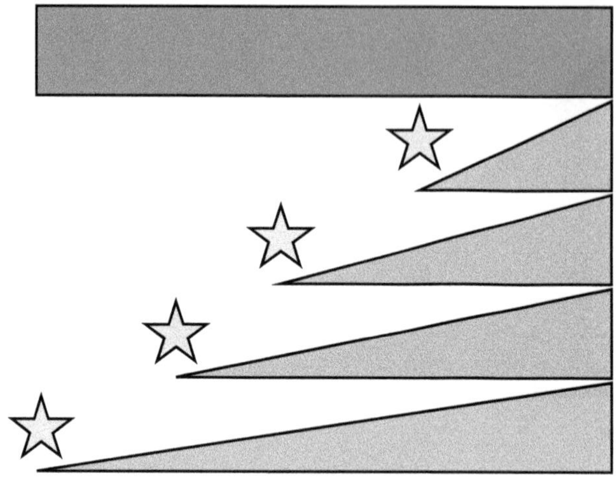

FIGURE 9.11 Feedback.

9.5 Requirements for Change

9.5.1 Attitudes and the Production Process

Excluded in part by music's cultural status as the mysterious preserve of specialists (Margulis 2007), in part by the sound isolation and acoustic treatment required for music production (Bridgett 2012), and poorly served by the game design literature, it is perhaps unfair to expect producers and game designers to be experts in understanding the contribution that music can make to games.

In the film-making process, the opportunity for the composer to play a role in offering their insight and suggestions is provided through the spotting process (Larsen 2007), and the ability of the director to try out different approaches in a hands-on way themselves is enabled through the common use of temp tracks throughout the editing process (Sadoff 2006). However, in games, the frequent outsourcing of music—often to composers from a linear film background—exacerbates the lack of integration between game design and music.

We have outlined above why we think there could be benefits to the game experience by aligning moments of fiero with structurally significant musical points in order to induce a heightened sense of pleasure in the player. The implementation of this concept requires a shift in both attitudes and production processes. To some, it is self-evident that the challenge of interactive music for games lies with the composer and that the implementation design should inform composition (Bajakian 2010), that one must spend hands-on time with a game in order to recognize its intrinsic rhythms (Kondo 2007), and that "The ability to understand game technologies and mechanics is becoming increasingly important for the composer" (Folmann quoted in Latta

2006). However, there appear to be a large number of composers who have little knowledge of games (Steighner 2011) and who do not consider it part of their remit to learn or understand the implementation tools (Graves 2011). Even if there were not inherent incentives in triple-A game development to go for the safest possible choice, using music in the tried and (un)tested way it has been used in previous titles within the genre, it is perhaps not surprising, given the common practice for such composers to be working remotely from images (Inglis 2012) or a few lines of instruction on a spreadsheet (Pham 2008), that more integrated design approaches are rare.

Although there are some companies that appreciate the importance of the in-house composer in creating a more integrated design approach (Broomhall 2011), there is much evidence that the practice of composers working in "the linear style that comes naturally" (Bajakian et al. 2000) remains problematic. Although in-house integrators may be (and often are), highly talented musicians themselves, it remains evident that it would be preferable for the composer to be more closely involved in the process of understanding how game variables might be translated into musical meaning. Furthermore they should not consider themselves to be above such "minutiae" (Mayer and Leary 2008) if music is to be composed with the medium in mind, rather than relying on the manipulation of pre-made assets. The claim that they "don't want interactivity to have a detrimental effect on the creativity of the composer" (Garry Schyman, quoted in Pham 2008) appears to parallel similar historical arguments from composers and theorists about the injurious effects on musical structure arising from having to compose to film events (Cooke 2008). Like the concert-hall composers before them who moved into film, the film composers who are now moving into games must also reappraise the role of music within the medium and become more involved in an integrated approach to finding solutions. As composer Guy Whitmore points out:

> If a composer simply supplies long, linear musical pieces for a game, that composer is not "scoring" the game; they are just providing music that is in the correct genre and style. Imagine if a film composer did the same thing—created music that had nothing to do with the images and action on screen. That composer would be fired! Scoring to picture is half the art in film composing, and the same applies to game scores. (Whitmore 2003)

Although we can be critical of the willful ignorance of film composers hired for marketing considerations, or a producer's personal preference (Broomhall 2012), it has long been recognized that judging music in isolation from the medium for which it was intended can be misleading (Gorbman 1987). The inclination to think that music should somehow be able to "stand alone" (Dabl 2010), together with the commercial incentives to promote the game "soundtrack" as a product (Kärjä 2008), further exacerbates the problems of considering music properly within its game context, which are already extant, given the lack of integration between the content-creation tools and implementation tools (Taylor 2012).

9.5.2 Tools

Although there have been significant advances in audio middleware tools in recent years, game development remains a fundamentally iterative process and it is desirable that the time necessary to test and iterate be as short as possible (Fullerton 2008).

The concept of affordances and constraints explores how the design of objects conveys messages about their possible uses and influences our abilities to carry out tasks with them (Norman 1988). The practice of contemporary composition is almost without exception carried out within what is commonly referred to as a digital audio workstation (DAW). It is rarely, as the name might suggest, a piece of hardware but in fact a personal computer and combined software sequencer and audio-editing package. By examining the spectrum of affordance (what is easy, and thus more likely, to what is difficult, and therefore less likely) (Mooney 2010) of a DAW it can be seen to be highly unlikely to produce music suited to interactivity, and that the production of interactive music is in spite of the tools, not because of them (see also Chapters 23 and 24 in this volume). It is worth noting that the unique and iconic style that is generally referred to when speaking of "game music"—that of the 8-bit chiptune era—is very much a product of the affordances and constraints of the sound chips on early games consoles (Collins 2008). The DAW has the granular note- and parameter-level controls ideally suited to interactivity, but lacks the stochastic capabilities and game engine integration of the middleware, while the wave file-based middleware lacks the granular control. This means that the iteration process involves, at the very least, the time-consuming rendering of all assets to wave files, importing of the wave files into middleware, the construction of interactive systems within the middleware, and the setting up of, and receipt of, the appropriate hooks (game variables) from the game itself. Any changes to the music after evaluation will then require a return to the DAW to modify and re-export the musical assets. It is worth reiterating that this is a best-case scenario: more typically this process is further worsened by the composer working remotely, by the involvement of live recording of musicians rather than the rendering from DAW, and by the evaluation process being undertaken without the composer's participation (Graves 2012). The original system within which the music is composed contains all of the control that is desirable for the iteration process (and for use in the final game) and yet the existing tools and processes involve rendering out material to the inflexible mediums such as Wave, MP3, or Ogg format files (Marks 2009).

To enable faster iteration and deeper integration of music in the game design process, there is a clear need to allow game engine variables to plug directly into DAWs, and for those DAWs to develop the compositional mechanisms and export formats to translate music into flexible formats for use in games. The aims of the interactive XMF (iXMF) working group (IASIG 2012) to establish a universal transfer format appears to have stalled, but perhaps there are initiatives to come from the new IASIG DAW working group, from the younger DAW pretenders (Kirn 2009), or indeed from the more unexpected direction of web audio (Rogers 2012).

9.6 Conclusions

Although we may question and debate the directness of the mapping of game information or actions to music from an aesthetic point of view, there are times at which the ludic function of music in providing information and motivational reward to the player, or the narrative function of enhancing the player's actions so they are seen to have a "spectacular influence" on the game (Nielsen 2011), emphasizes the need for it to be congruent with game events.

Through parallel forms we can provide information to the player within musical structures, and through ornamental gestures we can provide micro rewards to motivate and enhance the pleasurable flow state, but enhancing the peak emotion of triumph (fiero) when overcoming the frustration or stress invoked by major obstacles in the game (Hazlett 2006) requires the more powerful emotional responses associated with musical form. No matter what level of granularity or complexity of algorithm is involved, it is, and always will be, theoretically impossible to reconcile the indeterminate actions of the player with the kinds of expectation-based musical structures that induce such peak moments of pleasure.

We appreciate that a huge range of fascinating and brilliant games such as platformers, explicitly music-based games, or games that have audiovisual synaesthesia ideas as a core mechanic, already treat music as a highly integrated design element. However, within more narrative-situated games there are certain moments that deserve to deliver the powerful emotions associated with their Hollywood archetypes. Without the right tools, better integration of music into the iterative game design process is difficult, and without the right personnel and attitudes, the kind of *Gesamtkunstwerk* anticipated from the medium (Bridgett 2005) seems elusive, but by invoking a more nuanced interpretation of interactivity, that encompasses a range of possible exchanges, rather than accepting music in a purely reactive role, it is possible that new, as yet unexplored, possibilities will arise. It is our hope that the first game to fully use this interactivity to emotionally engage the player will provoke a paradigm shift in thinking about games and music.

Notes

1. In this technique the music cells start and end at performance boundaries that encapsulate a "pre" and "post" section, rather than simply containing the musical section itself. This means that the cells overlap when transitioning, allowing the decay of the current phrase to finish naturally (Whitmore 2003).
2. There is an additional form described in Figure 9.6, where the player acts directly on the musical form, such as in rhythm action games, termed here "performative"" (dotted and dashed lines).

References

Aarseth, Espen. 2003. We All Want to Change the World. In *Digital Media Revisited*, ed. Gunnar Liestøl, Andrew Morrison, and Terje Rasmussen, 415–439. Cambridge, MA: MIT Press.

Ashby, Simon. 2008. Interactive Audio for Video Games. Paper presented at Concordia Electroacoustic Studies Student Association, March 20, 2008, Concordia University, Montreal, Canada. http://cessa.music.concordia.ca/wiki/pmwiki.php?n=Presentations.080320SimonAshby.

Bajakian, Clint. 2010. Adaptive Music: The Secret Lies within Music Itself. Paper presented at the Game Developers Conference, San Francisco, California, March 9–13, 2010.

Bajakian, Clint, Peter Drescher, Duane Ford, Chris Grigg, Jennifer Hruska, Mike Kent, Ron Kuper, Mike Overlin, and Rob Rampley. 2000. Group Report: General Interactive Audio. *Project Bar-B-Q, 2000, Report, Section 7*. http://www.projectbarbq.com/bbq00/bbq0007.htm.

Bartle, Richard. 1996. Hearts, Clubs, Diamonds: Players who Suit MUDs. http://www.mud.co.uk/richard/hcds.htm.

Bateman, Chris, and Lennart E. Nacke. 2010. The Neurobiology of Play. In *Proceedings of the International Academic Conference on the Future of Game Design and Technology*, 1–8. New York: ACM.

Baysted, Stephen. 2012. Palimpsest, Pragmatism and the Aesthetics of Genre Transformation: Composing the Hybrid Score to Electronic Arts. Paper presented at Ludomusicology: Game Music Research [Royal Musical Association Study Day], April 16, 2012, St Catherine's College, Oxford, UK.

Bond, Matthew, and Russell Beale. 2009. What Makes a Good Game? Using Reviews to Inform Design. In *Proceedings of the 23rd British HCI Group Annual Conference on People and Computers: Celebrating People and Technology*, 418–422. Swinton, UK: British Computer Society.

Bridgett, Rob. 2005. Hollywood Sound: Part One. *Gamasutra*. http://www.gamasutra.com/view/feature/130817/hollywood_sound_part_one.php?page=3.

——. 2012. A Revolution in Sound: Break Down the Walls! *Gamasutra*. http://www.gamasutra.com/view/feature/170404/a_revolution_in_sound_break_down_.php.

Broomhall, John. 2011. Heard About: Batman: Arkham City. *Develop Magazine*, November 24, 2011 122, 44.

——. 2012. Heard About: Composition in Games. *Develop Magazine*, May, 127, 63.

Cairns, Paul, Anna Cox, Nadia Berthouze, Samira Dhoparee, and Charlene Jennett. 2006. Quantifying the Experience of Immersion in Games. In *Proceedings of Cognitive Science of Games and Gameplay Workshop at Cognitive Science*, Vancouver, Canada, July 26–9, 2006.

Collins, Karen. 2007. An Introduction to the Participatory and Non-Linear Aspects of Video Games Audio. In *Essays on Sound and Vision*, ed. Stan Hawkins and John Richardson, 263–298. Helsinki: Helsinki University Press.

——. 2008. *Game Sound: An Introduction to the History, Theory, and Practice of Video Game Music and Sound Design*. Cambridge, MA: MIT Press.

——. 2009. An Introduction to Procedural Music in Video Games. *Contemporary Music Review* 28 (1): 5–15.

Cooke, Mervyn. 2008. *A History of Film Music*. Cambridge, MA: Cambridge University Press.

Cope, David. 2000. *The Algorithmic Composer*. Madison, WI: A-R Editions.

Crawford, Chris. 2003. *Chris Crawford on Game Design*. Berkeley, CA: New Riders.

Csíkszentmihályi, Mihalyi, and Isabella Selega Csíkszentmihályi. 1992. *Optimal Experience: Psychological Studies of Flow in Consciousness*. Cambridge, MA: Cambridge University Press.

Dabl, Gideon. 2010. Editorial: Context Is Everything. *Original Sound Version*, August 10, 2010. http://www.originalsoundversion.com/editorial-context-is-everything/.

Dellacherie, Delphine, Mathieu Roy, Laurent Hugueville, Isabelle Peretz, and Séverine Samson. 2011. The Effect of Musical Experience on Emotional Self-reports and Psychophysiological Responses to Dissonance. *Psychophysiology* 48 (3): 337–349.

Drescher, Peter. 2010. Game Audio in the Cloud. Game. *O'Reilly Broadcast*, March 26, 2010. http://broadcast.oreilly.com/2010/03/game-audio-in-the-cloud.html.

Durity, Gordon, and Iain Macanulty. 2010. Contextually Driven Dynamic Music System for Games. Paper presented at the Vancouver Computer Music Meetings, Centre for Digital Media, Vancouver, Canada, October 6, 2010. http://www.metacreation.net/vcmm/#past.

Ekman, Paul. 2004. *Emotions Revealed: Recognizing Faces and Feelings to Improve Communication and Emotional Life*. New York: Holt.

Farnell, Andy. 2007. An Introduction to Procedural Audio and its Application in Computer Games. http://obiwannabe.co.uk/html/papers/proc-audio/proc-audio.html.

Fay, Todd. 2004. *DirectX 9 Audio Exposed: Interactive Audio Development*. Plano, TX: Wordware.

Folmann, Troels. 2006. Tomb Raider Legend: Scoring a Next-Generation Soundtrack. Paper presented at the Game Developers Conference, San Jose, California, March 20–24, 2006.

Fullerton, Tracy. 2008. *Game Design Workshop: A Playcentric Approach to Creating Innovative Games*. 2nd edn. San Francisco, CA: Morgan Kaufmann.

Giannetti, Claudia. 2007. Digital Aesthetics: Introduction. *MedienKunstNetz*, February 15. http://www.medienkunstnetz.de/themes/aesthetics_of_the_digital/editorial/.

Gorbman, Claudia. 1987. *Unheard Melodies: Narrative Film Music*. Bloomington: Indiana University Press.

Graves, Jason. 2011. Dead Space 2: Musical. Postmortem presented at the Game Developers Conference, San Francisco, California, February 28–March 4, 2011.

——. 2012. Audio Boot Camp. Paper presented at the Game Developers Conference, San Francisco, California, March 5–9, 2012.

Harper, Douglas. 2012. Online Etymology Dictionary. http://www.etymonline.com/.

Hazlett, Richard, L. 2006. Measuring Emotional Valence during Interactive Experiences: Boys at Video Game Play. In *Proceedings of the SIGCHI Conference on Human Factors in Computing Systems*, 1023–1026.

Hocking, Clint. 2012. In the Click of It: Living on the Edge. *Edge Magazine* 241: 152.

Hoeckner, Berthold, Emma W. Wyatt, Jean Decety, and Howard Nusbaum. 2011. Film Music Influences How Viewers Relate to Movie Characters. *Psychology of Aesthetics, Creativity, and the Arts* 5 (2): 146–153.

Hunicke, Robin, Marc LeBlanc, and Robert Zubek. 2004. MDA: A Formal Approach to Game Design and Game Research. In *Proceedings of the AAAI-04 Workshop on Challenges in Game AI*, July 25–29, 2004, 01–05. http://www.cs.northwestern.edu/~hunicke/MDA.pdf.

Huron, David. 2006. *Sweet Anticipation: Music and the Psychology of Expectation*. Cambridge, MA: MIT Press.

IASIG. 2012. Interactive XMF Working Group. http://www.iasig.org/wg/ixwg/index.shtml.

IGN. 2006. Michael McCann Interview. *IGN*. http://uk.music.ign.com/articles/741/741211p3.html.

Inglis, Sam. 2012. Music and Sound Effects for Videogame Mass Effect 3: Interview, Rob Blake (Bioware). *Sound on Sound*, June. http://www.soundonsound.com/sos/jun12/articles/mass-effect.htm.

Jackson, Leah. 2011. Nobuo Uematsu: Interview with a Legendary Video Game Composer. *G4tv*, September 9, 2011. http://www.g4tv.com/thefeed/blog/post/716221/nobuo-uematsu-interview-with-a-legendary-video-game-composer/.

Jensen, J.F. 1998. Interactivity: Tracking a New Concept in Media and Communication Studies. *Nordicom Review* 12 (1): 185–204.

Jørgensen, Kristine. 2010. Time for New Terminology? Diegetic and Non-Diegetic Sounds in Computer Games Revisited. In *Game Sound Technology and Player Interaction: Concepts and Developments*, ed. Mark Grimshaw, 78–97. Munich: Information Science Reference.

Kärjä, Antti-Ville. 2008. Marketing Music through Computer Games: The Case of Poets of the Fall and *Max Payne 2*. In *From Pac-Man to Pop Music: Interactive Audio in Games and New Media*, ed. Karen Collins, 27–46. Aldershot, UK: Ashgate.

Kastbauer, Damian. 2011. Audio Implementation Greats #10: Made for the Metronome. *Designing Sound*, January 3, 2011. http://designingsound.org/2011/01/audio-implementation-greats-10-made-for-the-metronome/.

Kirn, Peter. 2009. Inside the Rock Band Network, as Harmonix Gives Interactive Music its Game-Changer. *CreateDigitalMusic*, August 27, 2009. http://createdigitalmusic.com/2009/08/inside-the-rock-band-network-as-harmonix-gives-interactive-music-its-game-changer/.

Koelsch, Stefan. 2011. Response to Target Article "Language, Music, and the Brain: A Resource-sharing Framework." In *Language and Music as Cognitive Systems*, ed. Patrick Rebuschat, Martin Rohrmeier, John A. Hawkins, and Ian Cross, 224–234. Oxford: Oxford University Press.

Kondo, Koji. 2007. Painting an Interactive Musical Landscape. Paper presented at the Game Developers Conference, San Francisco, California, September 5–7, 2007.

Koster, Ralph. 2005. *Theory of Fun for Game Design*. Phoenix, AZ: Paraglyph.

Kristian, David, and Olivier Girard. 2011. Between 4 Ears : Splinter Cell : Conviction Co-op Sound Strategies. Paper presented at the Game Developers Conference, San Francisco, California, February 28–March 4, 2011.

Larsen, Peter. 2007. *Film Music*. London: Reaktion.

Latta, Westlee. 2006. CDM Interview: Tomb Raider: Legend Composer Troels Brun Folmann on Adaptive Micro-Scoring. *Create Digital Music*, October 11, 2006. http://createdigitalmusic.com/2006/10/cdm-interview-tomb-raider-legend-composer-troels-brun-folmann-on-adaptive-micro-scoring/.

Lawlor, Scott. 2012. The Music of the Wasteland: Interactive Music in an Open World. Paper presented at the Game Developers Conference, San Francisco, California, March 5–9.

Lazzaro, Nicole. 2008. The Four Fun Keys. In *Game Usability: Advancing the Player Experience*, ed. Katherine Isbister and Noah Schaffer, 315–342. San Francisco: Morgan Kaufmann.

Livingstone, Steven R., and Andrew Brown R. 2005. Dynamic Response: Real-time Adaptation for Music Emotion. In *Proceedings of the Second Australasian Conference on Interactive Entertainment*, 105–111. Sydney, Australia: Creativity & Cognition Studios.

Madigan, Jamie. 2012. The Psychology of Genres. *Edge Magazine* 241(June) 96–103.

Manovich, Lev. 2002. *The Language of New Media*. Cambridge, MA: MIT Press.

Margulis, Elizabeth Hellmuth. 2007. Surprise and Listening Ahead: Analytic Engagements with Musical Tendencies. *Music Theory Spectrum* 29 (2): 197–217.

Marks, Aaron. 2009. *The Complete Guide to Game Audio: For Composers, Musicians, Sound Designers, and Game Developers*. 2nd edn. Burlington, MA: Focal Press.

Mayer, Jonathan, and Keith Leary. 2008. Interactive Music Systems: Planning, Producing and Executing. Paper presented at the Game Developers Conference, San Francisco, California, February 18–22.

McAlpine, Kenneth B., Matthew Bett, and James Scanlan. 2009. Approaches to Creating Real-time Adaptive Music in Interactive Entertainment: A Musical Perspective. In *Proceedings of the 35th AES International Conference on Audio for Games*. New York: Audio Engineering Society.

McQuail, Denis. 2005. *McQuail's Mass Communication Theory*. Thousand Oaks, CA: Sage.

Miller, Ben. 2010. Immersive Game Design: Indigo Prophecy. In *Well Played 2.0: Video Games, Value and Meaning*, ed. Drew Davidson, 189–200. Pittsburgh, PA: ETC.

Mooney, James. 2010. Frameworks and Affordances: Understanding the Tools of Music-making. *Journal of Music, Technology and Education* 3 (2): 141–54.

Munday, Rod. 2007. Music in Video Games. In *Music, Sound and Multimedia: From the Live to the Virtual*, ed. Jamie Sexton, 51–67. Edinburgh: Edinburgh University Press.

Nan, Yun, Thomas Knösche A., and Angela D Friederici. 2009. Non-musicians' Perception of Phrase Boundaries in Music: A Cross-cultural ERP Study. *Biological Psychology* 82: 70–81.

Nielsen. 2011. State of the Media: Consumer Usage Report 2011. http://www.nielsen.com/content/dam/corporate/us/en/reports-downloads/2011-Reports/StateofMediaConsumerUsageReport.pdf

Norman, Donald. 1988. *The Design of Everyday Things*. Cambridge, MA: MIT Press.

North, Adrian C., and David J. Hargreaves. 2007. Lifestyle Correlates of Musical Preference: 1. Relationships, Living Arrangements, Beliefs, and Crime. *Psychology of Music* 35 (1): 58–87.

Page, Jason, and Michael Kelly. 2007. PS3 Audio: More Than Extra Channels. Paper presented at the Game Developers Conference, San Francisco, California, September 5–7, 2007.

Pham, Alex. 2008. Their Scores Can Be Huge. *Los Angeles Times*, December 8. http://articles.latimes.com/2008/dec/08/business/fi-composer8.

Porter, Tony. 2010. Goldeneye DS Dynamic Music. *Game Audio Forum* http://www.gameaudioforum.com/phpBB3/viewtopic.php?f=11&t=2457.

Przybylski, Andrew K., C. Scott Rigby, and Richard Ryan M. 2010. A Motivational Model of Video Game Engagement. *Review of General Psychology* 14 (2): 154–166.

Rigby, Scott, and Richard Ryan. 2011. *Glued to Games: How Video Games Draw Us in and Hold Us Spellbound*. Santa Barbara, CA: Praeger.

Rogers, Chris. 2012. Web Audio API: W3C Editor's Draft. https://dvcs.w3.org/hg/audio/raw-file/tip/webaudio/specification.html.

Ross, Rob. 2001. Interactive Music... er, Audio. *Gamasutra*. May 15, 2001. http://www.gamasutra.com/resource_guide/20010515/ross_01.htm.

Sadoff, Ronald H. 2006. The Role of the Music Editor and the Temp Track as Blueprint for the Score, Source Music, and Scource Music of Films. *Popular Music* 25 (02): 165–183.

Salimpoor, Valorie N., Mitchel Benovoy, Kevin Larcher, Alain Dagher, and Robert J. Zatorre. 2011. Anatomically Distinct Dopamine Release during Anticipation and Experience of Peak Emotion to Music. *Nature Neuroscience* 14 (2): 257–262.

Selfon, Scott. 2003. Linear Playback. In *DirectX 9Audio Exposed: Interactive Audio Development*, ed. Todd M. Fay, 17–40. Plano, TX: Wordware.

——. 2004. DirectMusic Concepts. In *DirectX 9Audio Exposed: Interactive Audio Development*, ed. Todd M. Fay, 3–16. Plano, TX: Wordware.

——. 2009. Interactive Music Techniques for Games. Paper presented at the 127th AES Convention, October 9–12, New York.

Sloboda, John, A. 1991. Music Structure and Emotional Response: Some Empirical Findings. *Psychology of Music* 19 (2): 110–120.

Steighner, Mark. 2011. Interview: Assassin's Creed: Revelations Composer Lorne Balfe. December 6, 2011. http://videogamewriters.com/interview-assassins-creed-revelations-composer-lorne-balfe-31008.

Steinbeis, Nikolaus, Stefan Koelsch, and John A. Sloboda. 2006. The Role of Harmonic Expectancy Violations in Musical Emotions: Evidence from Subjective, Physiological, and Neural Responses. *Journal of Cognitive Neuroscience* 18 (8): 1380–1393.

Stuart, Keith. 2010. Redemption Songs: The Making of the Red Dead Redemption Soundtrack. *The Guardian*, May 26. http://www.guardian.co.uk/technology/gamesblog/2010/may/26/red-dead-redemption-soundtrack.

Suits, Bernard. 2005. *The Grasshopper: Games, Life and Utopia*. Peterborough, ON: Broadview.

Summers, Tim. 2011. Playing the Tune: Video Game Music, Gamers, and Genre. *Act: Zeitschrift für Musik & Performance* 2, July. http://www.act.uni-bayreuth.de/en/archiv/2011-02/04_Summers_Playing_the_Tune/index.html.

Taylor, Michael. 2012. Interview with Michael Bross. *Designing Sound*, May 7. http://designingsound.org/2012/05/interview-with-michael-bross/.

Vorderer, Peter, and Jennings Bryant. 2006. *Playing Video Games: Motives, Responses, and Consequences*. London: Lawrence Erlbaum.

Weir, Paul. 2011. Stealing Sound: The Application of Generative Music. Paper presented at the Game Developers Conference, San Francisco, California, February 28–March 4, 2011.

Whalen, Zach. 2004. Play Along: An Approach to Videogame Music. *Game Studies* 4 (1). http://www.gamestudies.org/0401/whalen/.

Wharton, Alexander, and Karen Collins. 2011. Subjective Measures of the Influence of Music Customization on the Video Game Play Experience: A Pilot Study. *Game Studies* 11 (2). http://gamestudies.org/1102/articles/wharton_collins.

Whitmore, Guy. 2003. Design with Music in Mind: A Guide to Adaptive Audio for Game Designers. *Gamasutra*. http://www.gamasutra.com/view/feature/131261/design_with_music_in_mind_a_guide_.php?page=2.

CHAPTER 10

WORLDS OF MUSIC

Strategies for Creating Music-based Experiences in Videogames

MELANIE FRITSCH

SOUND and music in a videogame have to meet a range of requirements regarding technical, compositional, and functional demands. Although the music in many videogames is regarded by players as somewhat interesting, or even important for specific purposes as part of the gameplay (e.g., used in musical puzzles, as in *The Legend of Zelda* series), it is usually not the centerpiece. However, there are videogames in which music figures prominently. Besides famous examples of explicitly music-based games such as *Guitar Hero, Rock Band*, or *Dance Dance Revolution* franchises, some other types of videogames also involve music as a core feature. As Karen Collins noted, "for games like *Vib-Ribbon*, the music can literally *create* the structure of the gameplay" (2008, 131).

In the following, I examine strategies pursued in order to create music-based gameworlds. For that purpose, I discuss three examples originating from different genres, employing different musical styles, and three strategies of music-based "world-creation." The first example is a conceptual rhythm game, *Vib-Ribbon*. Even though it is no surprise that a rhythm game relies on music, the approach toward "world-creation" is notable, as we will see. The other examples are the humorous action-adventure game *Brütal Legend*, which features the heavy metal music genre, and the role-playing game *Eternal Sonata*, whose narrative centers on the life and music of the Polish composer Frédéric Chopin.

In the form of short case studies, I outline how these games engage players to interact with the music on different levels through the respective game and its music-based "gameworld," a term that will be given some attention. The main focus of the chapter is on the questions of how the "narratives" and "worlds" created rely on music, how this relationship can be addressed and analyzed, and how the music can influence the overall experience of the player. In order to find answers to these questions, it is necessary first to outline an approach for analysis and to define terminology.

10.1 Musical Game Worlds: Where Is the Entrance?

In her recent talk *The ALI Model: Towards a Theory of Game Musical Immersion*, Isabella van Elferen (2012) offers a framework to analyze game music in connection with the phenomenon of immersion. She suggests a "game musicology [as] an intermedial research methodology for audiovisual analysis that makes musical and game analysis compatible with one another... [G]ame musicological analyses are synesthetic in their design, identifying the convergence of musical, graphic, and interactive components in videogames as well as their cumulative effect" (van Elferen 2012). Van Elferen comes to the conclusion that "we can define optimal musical immersion as a form of augmented reality. As player involvement conflates the immersion in gaming with that in musical literacy, interaction and affect, the virtual reality of gaming is overlaid by a layer of specifically music-induced reality."

Following these reflections, music seems to be able to create a virtual layer or "world," enriching the overall experience of a game, but cannot be analyzed without taking the game itself into consideration. Given a videogame's interactive nature it is also necessary to keep the player in mind, because a game needs at least one player to be played, which again has implications for sound and music: "Although the goal of many game developers is to create an immersive experience, the body cannot be removed from the experience of videogame play... Unlike the consumption of many other forms of media, in which the audience is a more passive 'receiver' of a sound signal, game players play an active role in the triggering of sound events in the game" (Collins 2008, 3). Therefore, it is necessary to find an approach to game analysis that considers all of these aspects before turning toward the game examples and the question of how they create their "worlds" through music.

10.2 Pushing Open the Door: Videogames as Objects and Activities

In his book *half-real: Video Games between Real Rules and Fictional Worlds* (2005), Jesper Juul makes a clear distinction between the "real" rules and the fictional world of a videogame, both of which a game includes (see Juul 2005, 1). This structural distinction can help to circumscribe the area of focus when studying the videogame itself. Starting from that basic differentiation of real and fictional "game parts" the ensuing question of how videogames create game experiences through their rules *and* narrative often brings into play the highly controversial discussion about the concept of the "magic circle" (see Salen and Zimmerman 2004, 95; Zimmerman 2012). Briefly, this concept describes when starting to play a game, players enter a somehow separate space or world, where the game's rules are valid. In this space or world, the narrative unfolds, according to the rules pertaining there. Concepts like immersion, presence, and transportation are often discussed referring to this idea, in a hostile as well as in an approving manner.

Investigating critically this notion of the magic circle, Emmanoel Ferreira and Thiago Falcão have concluded that: "Thus constituted in the moment the *object* game becomes the *activity* game, the magic circle can be understood as a mediation structure... The second dimension [of this mediation] (ii) is related to the way the game shows itself in the moment of the gameplay—it concerns the game as activity; in the moment that the structure composed by rules and fiction... becomes available to potential players" (Ferreira and Falcão 2009, 2) This approach shifts the focus from thinking of the game as an object to analyzing it in the moment with respect to any kind of playful activity performed with it, within the borders of the rules, the given narrative and the possibilities provided by the hard- and software.[1] Even though the rules and the superordinate narrative stay the same, the activity of each play brings forth a unique structure, that is, the selection and sequence of (re)actions undertaken. But what is this structure, and how can it be addressed? Craig Lindley proposes an approach toward this issue by applying the term "gameplay gestalt." He focuses on playing a game as an action rather than trying to understand the game as an object, and develops an "alternative conception of gameplay as an interactive gestalt formation process" (Lindley 2002, 204).

Gestalt theory originally derives from an area of psychological theory introduced in the early twentieth century. It underwent changes in the different research areas by which it was adopted, like philosophy, biology, and systematic musicology. Essentially, gestalt theory stresses the idea of totality by taking into account not just the single parts and processes, but also their relationships. Mark Reybrouk describes this gestalt concept, focusing on how it was adapted by music theorists:

> So gestalt theory claims that in perception one can grasp *immediately* a configuration that is already organized... Music, in that sense, can be defined as a sound producing organism... and the musical experience should be the outcome of an interaction between the listener and the musical organism. Music, thus defined, is an organic structure and music analysis has to be broadened from a structural description to a description in terms of processes... An operational description of this idea is possible by substituting a system for the organism. (Reybrouk 1997, 58; emphasis in original)

Lindley applies this idea to the process of game playing by emphasizing the aspect of interaction: "The rules establish what as a player you can or cannot do, and what the behavioral consequences of actions may be within the world of the game... It is the central point of this paper to suggest that this is a matter of learning a *gameplay gestalt*, understood as a pattern of interaction. Playing the game is then a matter of performing the gestalt" (Lindley 2002, 207; emphasis in original).

This gameplay gestalt is nothing fixed or static. Instead it is processual, pointing to the act of playing a game as performative[2] activity based on the game as an object. On the part of the players, any information given by the game that is relevant for playing on the level of rules as well as on the level of narrative, need interpretation, before carrying out the appropriate bodily (re)action (see Lindley 2002). This interpretive ability and the learned set of adequate reactions ensue from the player's previous gaming experience, and also contribute to the gameplay gestalt: "It is a particular way of thinking about the game state, together with a pattern of perceptual, cognitive, and motor operations... the gestalt is more of an interaction pattern involving both the in-game and out-of game being of the player" (Lindley 2002, 207).

Lindley's findings resonate with those of Collins (2008), chiefly that the player's body cannot be removed from the gaming experience. Hence, the player's "out-of-game being" must always be taken into account in any analysis, because their aforementioned interpretation and bodily skills influence the process of playing itself and, consequently, the emerging gameplay gestalt and gaming experience. But can a gameworld, in which the player is immersed, be described as something detached when the player's body cannot be ignored? Is the gameworld limited to what can be seen on screen?

Based on these preliminary considerations, I describe the *performative space* in which all actions induced by the game take place, including those in front of the screen, as the gameworld. In this gameworld, the game's narrative, the specific and unique sequence of fictional and nonfictional events happening while playing the game, unfolds. In order to address the "world," which can be seen on screen, and set this apart from the gameworld, I will henceforth refer to this as the diegetic environment. I use the term diegesis here in the sense of Genette: "diegesis is not the story, but the universe in which it takes place" (1998, 201, my translation).

10.3 In the Thick of it: When Music Comes into Play

All three games explored here use music in different ways, letting players continuously interact with the music. But how do these games create gameworlds through music, and how is music implemented in order to bring forth a gameplay gestalt?

10.3.1 Vib-Ribbon

Vib-Ribbon is a rhythm videogame developed for Sony's PlayStation, released in 1999. In front of a black background the diegetic environment is figured as a white ribbon, which forms an obstacle course with loops, spiky waveforms, pitfalls, and blocks in various combinations. This course is generated in real time according to the beat of the accompanying music. The player can use the music delivered with the game, composed by the Japanese group Laugh and Peace. Similarly, it is possible to load any standard music CD into the PlayStation console, so that players can choose any music they want to play with. The objective is to guide the avatar, a rabbit named Vibri, drawn in white vector lines in the style of a stickman, through the game by pushing the correct buttons at the correct time to traverse different obstacles. Combinations of two obstacles require pressing a combination of two corresponding buttons.

In the case of *Vib-Ribbon*, the music, or more precisely the beat, is directly translated into the game's diegetic environment. According to the type of music chosen, the levels will be more or less challenging. The more difficult the diegetic environment, the more effort is required by the player to master the course, and the gameworld becomes more stressful. If the player has not yet acquired the skills matching this difficulty, further practice is required or the player may choose different, easier music. The player is given several options for selecting music for gameplay: preference of a certain musical genre, the desire for a more challenging diegetic environment, or for an entertaining rather than challenging overall experience.

A narrative is not provided in the game itself, so the player does not know what Vibri is meant to achieve. Depending on the music being selected for gameplay, the player might imagine a connection between lyrics and/or the music and the gameworld he or she player experience. It would be possible to even imagine an individual narrative around a certain musical piece and the game.

The player also learns to listen to music in a new way. Similar to other rhythm or rhythm-action games, a more active way of listening is required in order to split the music up into patterns, making it easier to foresee when a new obstacle will appear and when it is time to interact with the game and music by pushing the appropriate buttons on the controller. Kiri Miller states this very aspect regarding *Guitar Hero* players: "When asked how these games changed their listening experience, players explained that the combination of reading notation and the physical act of playing a particular part (guitar, bass, drums) made them hear songs differently, including songs they had never played in the games" (see Miller 2007, 410).

Therefore, the resulting music-based gameplay gestalt is created by a direct transformation of music into the game's diegetic environment. By reacting to the beat structure of the music, the player has direct bodily interaction with the music. This process can be understood as structural music-based gameplay gestalt.

10.3.2 Brütal Legend

Brütal Legend is an action-adventure game with real-time strategy elements released by Electronic Arts in 2009. The main character of the game is brawny roadie Eddie Riggs, who is a fan of 1970s heavy metal music. The character is named after artist Derek Riggs, creator of the *Iron Maiden* mascot "Eddie the head," and the voice-actor is actor Jack Black. The american comedian, actor and musician is well known for his strong liking of rock and heavy metal music, and is also the lead singer of the rock comedy band Tenacious D. The game starts with a live action-clip featuring Jack Black. He invites the player to follow him into a record store, where he looks for a special record. When he pulls out the record sleeve, it turns out to be the games menu. After the player chooses to start the game, another introductory cut-scene is replayed, an intermission of game-play in the form of a short film. During a cut-scene the player usually cannot or can only slightly influence the events on screen.

This cut-scene already presents game graphics, and serves as introduction to the narrative. After an accident on stage, which occurs during a concert, Eddie is transported by the ancient god Ormagöden into a diegetic environment inspired by 1970s and 1980s heavy metal record sleeves: the "Age of Metal." The landscape is cluttered with pieces of spiky scrap metal, huge rusty swords, bone piles, concert stages, monuments, or statues bound in leather or chains. It is populated with bullnecked metalheads, demonic beasts and other similarly fierce creatures. The humans living in this "Age of Metal" are threatened by the Tainted Coil. This group is headed by the evil Doviculus, assisted by his glam metal human minion, General Lionwhyte. Eddie is the "Chosen One," or hero of the game, although it is not clear at the outset whether this role makes him the savior or the destroyer of this world. In order to fulfill his destiny, he is "armed with the power of metal,"[3] namely a broad axe called The Separator and his Gibson Flying V guitar, Clementine. With Clementine, Eddie has the ability to cast magic spells by playing guitar riffs in a series of mini-games. In the case of *Brütal Legend* the mini-game is a short rhythm game akin to games like *Guitar Hero*, in which the player has to push a series of buttons at the correct time in order to perform the guitar riff, which creates the magic spell.

Some famous heavy metal musicians are characters in the game, including Ozzy Osbourne as The Guardian of Metal, Lemmy Kilmister as The Kill Master, Rob Halford as The Baron and also lending his voice to the evil General Lionwhyte, and Lita Ford as Zuma. While traveling through the diegetic environment with the Deuce (an armed Hotrod) players can choose from a huge range of 108 heavy metal songs to listen to, including tracks from seventy-five different bands, such as Tenacious D, Slayer, Testament, and Motörhead.

Every detail of the diegetic environment, the narrative, the background story, the characters, their behavior, appearance, and dialog is based on heavy metal music, the artwork developed around it, and its fan culture. For example, the in-game currency the player is rewarded with when having finished a mission is called "fire tributes." These are visualized by a row of silhouetted arms holding lighters and popping up at the bottom of the screen, a reference to the ritualized fan behavior, when ballads are being played. As another example, the appearance of the evil General Lionwhyte is modeled after David Bowie (but dubbed by Rob Halford) and his name is an allusion to the glam metal band White Lion.

Therefore, even if players turn off the game music, they interact with heavy metal all the time, because the entire game is constructed around visual and aesthetic cues derived from heavy metal music and culture. Knowledge of all these references is not necessary in order to play the game successfully, but it does enrich the overall playing experience. Heavy metal fans will recognize specific details, allusions, inside jokes, and hints throughout the game. The more the player interacts with these cues, the more he or she player understand the narrative. Of course, this process can also be interpreted the other way round: A player with little knowledge of heavy metal music and its cultural context can be introduced to it by playing the game. Deena Weinstein (2000, 4) underlines that heavy metal culture is a very complex structure, composed of several musical as well as social codes, and created by different agents (artists, audiences, and mediators) in the form of a *bricolage*, "a collection of cultural elements... Its parts exist for themselves as much as they do for the whole. They are held together not by physical or logical necessity but by interdependence, affinity, analogy, and aesthetic similarity" (Weinstein 2000, 4). She provides an overview on the diverse dimensions that contribute to what she calls "The Code of Heavy Metal."

In the case of *Brütal Legend*, sounds and images associated with heavy metal music and culture have been translated directly into a diegetic environment, narrative and gameplay, culminating finally in the creation of a distinct gameworld. Therefore, the focus in this game lies with the player's translation of the contextual or cultural pattern of a certain musical style rather than a direct transformation of a musical pattern. Everything the players see, hear, or undertake in the gameworld is contingent on heavy metal music and culture. In this regard, *Brütal Legend* is an example of what I'm calling a musical culture-based gameplay gestalt.

10.3.3 Eternal Sonata

Eternal Sonata is a Japanese role-playing game (RPG) released by Namco Bandai Games in 2007 with a turn-based fight-system, action game elements, and a considerable number of cut-scenes. Gameplay begins when the player dives into a dream that the Polish composer and main character of the game Frédéric Chopin experiences on his deathbed. In Chopin's dream world, subdivided into eight chapters, people with incurable illnesses like himself are imbued with magical powers. He meets a girl, Polka, who also suffers from an illness. Together with other party members[4] they set off to meet Count Waltz and ask him to lower down the shipment of a drug called "mineral powder" in favor of the more expensive, but traditional "floral powder."

Information regarding Chopin's life and music is provided in the cut-scenes, where a selection of his compositions played by pianist Stanislav Bunin is featured, though most of the in-game music was composed by Motoi Sakuraba. Players can also find thirty-two score pieces, or short musical phrases, scattered throughout the diegetic environment. In one mini-game, some nonplayer characters in the game offer to perform a musical composition building on these phrases. They require the player to match a score piece to a given phrase and the resulting composition is ranked. A good composition will be rewarded with a bonus item. All twelve playable characters are named after musical terms such as Polka, Beat, Allegretto, and so on. This idea holds equally true for places, for example, the town, Ritardando, the fort, Fermata, or the river, Medley.

In *Eternal Sonata* the player neither interacts with the contextual pattern translated directly into pictures on screen, nor are they able to play (with) Chopin's music, because the musical mini-games do not rely on pieces composed by Chopin. Nevertheless, the theme of the game is built on the life and music of this composer, and the player receives a great deal of information about him and has the opportunity to listen to his music.

In his recent talk, Tim Summers refers to this strategy as "texturing": "Music can make sonic semiotic reference to other media texts and cultural touchstones that are already well-established to bring particular referents to bear on the game in order to enhance the game experience. This effect may be termed "texturing," since it has the result of creating implied detail, textual depth, and rounded context to the surface level of gameplay activity" (Summers 2012). In the case of *Eternal Sonata*, this effect is achieved by referring to a real historical person, Chopin, and a selection from his musical repertoire.

The idea of a dream world as the underlying narrative context for gameplay is a romantic one, blurring the lines between fantasy and reality, in a way not dissimilar to other works in Romantic literature, music, and paintings. In this way, *Eternal Sonata* thematizes Romantic concepts such as the dissolution of boundaries, the romantic hero, escapism, and a blending of the mythical and the real worlds. Even though the player can also listen to the music of Chopin at some points, and find some references in the narrative, this game points to the idea that music is integrated into a broader sociocultural context rather than simply a localized discourse linked to the music within the game itself. A player who is not familiar with Chopin or his music will be introduced not just to the composer and his work, but rather to the entire ideational discourse bound to it by interacting with the game. This process is part of a broader learning experience, which is not focused on just gameplay itself, but further delivered through informative and educational cut-scenes.

While *Eternal Sonata* shares some similarities with *Brütal Legend*, here the music and its cultural context are not visualized in a diegetic environment, but rather thematized in a narrative context. Another key difference can be found with respect to how pivotal music is regarding the referent cultures. In the heavy metal subculture the music is the core. Therefore in *Brütal Legend* music is staged and referred to correspondingly. In the Romantic period instead music is only one possible manifestation of the romantic idea, and not its only expression. Literature or painting could have been chosen instead of music. But why does *Eternal Sonata* refer to a composer and his music instead of thematizing, for example, a romantic painter or poet? A reason why music has been privileged over other art forms can be found in the romantic discourse itself, of which the work of the romantic writer, music critic, and composer E. T. A. Hoffmann is a good example. According to Hoffmann, music, especially instrumental music, "is the most romantic of all the arts—one might almost say, the only genuinely romantic one—for its sole subject is the infinite. The lyre of Orpheus opened the portals of Orcus—music discloses to man an unknown realm, a world that has nothing in common with the external sensual world that surrounds him, a world in which he leaves behind all definite feelings to surrender himself to an inexpressible longing" (Hoffmann 1952, 35–6). For Hoffmann, music appears to be the perfect art form to transport Romantic ideals and ideas, which is also taken up in the game. Therefore, the role of music in *Eternal Sonata* can be understood as ideational music-based gameplay gestalt.

10.4 Conclusions

As I have demonstrated, music-based gameplay gestalt is understood as a concept of continual performative activity, which requires both the player's bodily and cognitive actions. This idea sees gaming not as an object or text, but as an *activity*, which takes place in a gameworld, as defined above. As this is an emergent process that involves or, more precisely, requires the player in front of the screen to be created, I would suggest that in these games, the "music-induced reality" Isabella van Elferen (2012) has mentioned can be paralleled with what I describe here as a gameworld.

Through music-based gameplay gestalt, a player becomes connected to the incidents and the diegetic environments shown on screen, and thereby immersed in the gameworld through the activity of playing. In the three examples presented, there are explicitly no simulations of reality, but they instead include explicit fantasy diegetic environments. *Vib-Ribbon* is depicted by simple vector graphics, *Brütal Legend* is presented in an exaggerated comic-book style, and the graphics in *Eternal Sonata* are those of Japanese anime. It is music with all its features and contexts that blurs the borders of fantasy and reality by being the "real thing" within these gameworlds.

Future studies using the theories described above might include analysis of other music-based videogames like *Patapon* or *Guitar Hero*, or particularly those games employing new technological approaches like *Child of Eden*[5]. Such games rely on gestural interfaces, and therefore use the player's body itself as the controller, while featuring an abstract diegetic environment in combination with music, which is in the case of *Child of Eden* even produced by virtual musicians. Similar to games like *Guitar Hero* or *Dance Dance Revolution*, the music games created for Kinect like *Child of Eden*, the dancing game *Dance Central*, or games like *Michael Jackson—The Experience* explicitly challenge players not just to perform the correct interactions using diverse interface devices in order to play the game, but transfer the visible results of the playing activity in front of the screen by the medium of the player's body. By doing so, these games bring concepts like immersion, simulation, or (virtual) reality into question again. Therefore, an approach toward analysis that explicitly comprises the player's body as presented here could be fruitful. Also, further research into videogaming's performative qualities, like processuality, unrepeatability, and so on, could potentially be significant, but would need more in-depth study. By bringing together theories from areas of performance and the theories described in this chapter, we may come to a better understanding of the use of music in games.

Notes

1. Players can decide to play the game as intended by the designers, but they can also choose to use it as a basis for their own playful agreements and to invent new games. For example, one might play *Quake III* in the usual way, or decide to compete in building towers with their avatars. For the sake of brevity I will omit this discussion, but it could be an interesting field for further research regarding game music, when people, for example, use games as a basis to create music.
2. In my line of reasoning, I adopt the use of the term performative in German Theaterwissenschaft including its reference to bodily acts. See Fischer-Lichte 2008: 26.
3. Quoted from the E3 2009 Cameo: http://www.ea.com/brutal-legend/videos/afc9dc98c5d91210VgnVCM100000ab65140aRCRD.
4. In role-playing games the term 'party' describes the group of avatars, of which the player directly controls one at a time. In the single player modus of a role-playing game, the player can usually switch between the party members to issue orders, which will be executed, e.g., how to behave in a fight or while exploring the diegetic environment. In coop-mode (two players play together using a split-screen view) or when playing online every party member can be controlled by players.
5. This chapter is based on a talk given at the IMS study group conference "Music and Media" in Berlin (2010), when *Child of Eden* was not released yet.

References

Arsenault, Dominic. *Dark Waters: Spotlight on Immersion*. Game On North America 2005 Conference Paper. Online. Available: http://umontreal.academia.edu/DominicArsenault/Papers/157453/Dark_Waters_Spotlight_on_Immersion. May 1, 2012.

Collins, Karen. 2008. *Game Sound. An Introduction to the History, Theory, and Practice of Video Game Music and Sound Design*. Cambridge, MA: MIT Press.

Ferreira, Emmanoel, and Thiago Falcão. 2009. Through the Looking Glass: Weavings between the Magic Circle and Immersive Processes in Video Games. Breaking New Ground: Innovation in Games, Play, Practice and Theory. *Proceedings of DiGRA 2009*. http://www.digra.org/dl/db/09287.45173.pdf.

Fischer-Lichte, Erika. 2008. *The Transformative Power of Performance: A New Aesthetics*. Trans. Saskya Irsi Jain. London: Routledge.

Genette, Gérard. 1998. *Die Erzählung*. Munich: Fink Verlag.

Hoffmann, E. T. A. 1952. Beethoven's Instrumental Music. In *Source Readings in Music History*, vol. 5: *The Romantic Era*, ed. Oliver Strunk, 35–41. London: Faber. (Originally published anonymously in *Zeitung für die elegante Welt*, 1813.)

Juul, Jesper. 2005. *half-real. Video Games between Real Rules and Fictional Worlds*. Cambridge, MA: MIT Press.

Lindley, Craig A. 2002. The Gameplay Gestalt, Narrative, and Interactive Storytelling. *Computer Games and Digital Cultures Conference Proceedings*, 203–215. Tampere, Finland : Tampere University Press. http://www.digra.org/wp-content/uploads/digital-library/05164.54179.pdf.

Miller, Kiri. 2012. Schizophonic Performance: *Guitar Hero, Rock Band*, and Virtual Virtuosity. *Journal of the Society for American Music* 3 (4): 395–429.

Reybrouk, Mark. 1997. Gestalt Concepts and Music: Limitations and Possibilities. In *Music, Gestalt, and Computing. Studies in Cognitive and Systematic Musicology*, ed. Marc Leman, 57–69 (Lecture Motes in Computer Science vol. 1317). Berlin: Springer.

Salen, Katie, and Eric Zimmerman. 2004. *Rules of Play: Game Design Fundamentals*. Cambridge, MA: MIT Press.

Summers, Tim. 2012. The Aesthetics of Video Game Music: Epic Texturing in the First-person Shooter. Paper presented at Ludomusicology: Game Music Research [Royal Musical Association Study Day], April 16, 2012, St Catherine's College, Oxford, UK. [By courtesy of the author.]

van Elferen, Isabella. 2011. ¡Un Forastero! Issues of Virtuality and Diegesis in Video Game Music. *Music and the Moving Image* 4 (2): 30–39.

——. 2012. The ALI Model: Towards a Theory of Game Musical Immersion. Paper presented at Ludomusicology: Game Music Research [Royal Musical Association Study Day], April 16, 2012, St Catherine's College, Oxford, UK. [By courtesy of the author.]

Weinstein, Deena. 2000. *Heavy Metal: The Music and its Culture. Rev. edn*. Boulder, CO: Da Capo.

Zimmerman, Eric. 2012. "Jerked around by the magic circle – Clearing the air ten years later." *Gamasutra—The Art & Business of Making Games*, February 7. http://www.gamasutra.com/view/feature/6696/jerked_around_by_the_magic_circle_.php.

SECTION 3

THE PSYCHOLOGY AND EMOTIONAL IMPACT OF INTERACTIVE AUDIO

CHAPTER 11

EMBODIED VIRTUAL ACOUSTIC ECOLOGIES OF COMPUTER GAMES

MARK GRIMSHAW
AND TOM GARNER

EVER since its humble beginnings, such as in Atari's *Pong* in 1972, game sound has used advancing technology to present increasingly dynamic, immersive, and both realistic and fantastical sonic landscapes. Our chapter introduces a new model, the embodied Virtual Acoustic Ecology (eVAE), for the understanding and design of sound in computer games, particularly First Person Shooter games. This model derives from a framework that incorporates a previous model of the first-person shooter as acoustic ecology and combines thinking on emotion and game sound with theories of embodied cognition. Such a model provides a way to think about the design of game acoustic ecologies in the context of new technologies for biofeedback that potentially allows for a closer and more real-time relationship between the player and the sound. The embodied Virtual Acoustic Ecology model we present has the potential to progress game sound design further and take artificial manipulation of the game's acoustic ecology beyond the ear itself.

There is an increasing acceptance of the embodied approach to sound design, and our embodied Virtual Acoustic Ecology model makes use of recent embodied cognition theories that are distinct from earlier models of cognition in several aspects. Prior to the embodied cognition approach, theories of cognition stressed the separation of mind and body and thus mind and environment. The body's motor and perceptual systems were distinct areas of enquiry from that which concerned itself with the central cognitive processing of the mind, and early models of computing, particularly those dealing with artificial intelligence, followed this trend in emphasizing the importance of the processing of abstract symbols. Embodied cognition theories, instead, place motor and perceptual systems within a model of cognition; indeed, some state that cognition *arises* from such systems' interactions with the environment. As Wilson states, "human cognition, rather than being centralized, abstract, and sharply distinct from peripheral input and output modules, may instead have deep roots in sensorimotor processing" (2002, 625).

The chapter comprises five sections. In the first, we briefly describe the relationship between game sound and player as an acoustic ecology and this leads to the second section, which looks at the potential for game sound to elicit emotions. The third section introduces aspects of embodied cognition theories where they are relevant to our thinking. The penultimate section introduces the embodied Virtual Acoustic Ecology model, a synthesis of our thinking on game acoustic ecologies, the engendering of player emotion through sound, and key points from embodied cognition theories. Finally, we discuss the use of our model for biofeedback and speculate on the theoretical and philosophical implications of such an approach.

11.1 The Acoustic Ecology of Computer Games

The acoustic ecology of any computer game may be summarized as the heard diegetic sounds of the game and, as an ecology rather than an environment, it presupposes that the player has a dynamic relationship to, and is able to participate in, that acoustic ecology and thus is a fundamental part of that ecology.[1] There are several terms and concepts here that we must explain further in order to arrive at a more complete understanding of the game's acoustic ecology, and proceed to our model. For this purpose, we take as our exemplar the first-person shooter (FPS) game, as it is this game genre, we argue, that most fully attempts an immersion of the player in the game world through its first-person perspective and first-person audition (e.g., Grimshaw 2012), and through its ability to manipulate emotion.

There is a range of sounds in the archetypal FPS (e.g., *Quake III Arena, Half-Life 2, Crysis*), each fulfilling a different function in the sound designer's mind and thus, by extension, providing the means by which the player can engage with the game world and all it entails. Such sounds are typically stored on the game media as audio samples; digital recordings either of real-world sounds (which may have been processed to a lesser or greater extent) or of artificially created (synthesized) sounds.

A number of authors have discussed the diegesis of game sound, deriving a variety of neologisms around the term to describe subtle variations in the function of sound in the game and the player's relationship to that sound. Grimshaw (2008a) uses terms such as *kinediegetic* (a sound the player can hear that is triggered directly by the player's actions), *exodiegetic* (a sound heard by the player but not triggered by the player), and *telediegetic* (a sound heard by one player whose subsequent response to that sound has consequence for another player who has not heard that sound). Jørgensen (2009, 7) describes a *transdiegetic* sound, particularly music, having no apparent source in the game world but still informing the player of events in that game world. Van Elferen (2011) refers to *supradiegetic* sound as sound, primarily music, that undergoes a diegetic shift from the gameworld to reality by way of a cross-fade from a sound suited to that game world's environment to one that is not.

All such definitions are rooted in the idea that game sound either derives from an apparent source in the game world ("apparent" means seen or experienced as being part of the game world despite the real sound source being the player's playback system) or they derive from elsewhere. Thus, a simple taxonomy of game-sound diegesis would be: *diegetic*, those sounds deriving from the internal logic of the game world, and *nondiegetic*, all other sounds that are part of the game (not the game world), such as menu-interface sounds and overlaid musical scores. For our purposes, disregarding Jørgensen's blurring of the distinction, we use the diegetic–nondiegetic definition, concentrating on diegetic game sounds.

In our definition of the game's acoustic ecology, it may seem curious to explicitly state that the acoustic ecology comprises *heard* sounds. After all, as far as any one player is concerned, it seems clear that *all* game sound is heard sound. However, in a multiplayer FPS game, for example, there are, as in the real world, many unheard sounds that are of relevance to that one player but yet may be heard by other players (and are thus part of the diegesis). This is the thinking behind the term *telediegesis*. In the multiple, physically disjunct, but virtually conjunct, acoustic ecologies of the multiplayer FPS game world, some of the sounds that only player B hears may lead player B to a course of action that has consequences for player A. While we do not pursue this particular line of reasoning further in this chapter, we do return to the concept of *unheard* sound in the concluding section, where we discuss psychophysiology and the possibility of directly stimulating the perception of sound.

Our definition further states that the player has a dynamic relationship to, and participates in, the game's acoustic ecology. This has been discussed in detail elsewhere (e.g., Grimshaw 2008a) so we will only deal with it briefly here. Diegetic game sounds primarily inform the player of game events and provide context. Event sounds may be, in the FPS game, footsteps, gunshots, or radio messages, for example, while context is not only provided for by such event sounds but also by what are typically known as ambient sounds. These latter sounds often refer to the visual spaces of the game world (reverberant dripping water in a cavernous space, for instance). Equally, though, they can bolster the historical setting or the more immediately temporal state of the world (from the authentic, or at least lifelike, sound of a Rolls-Royce Merlin engine in a World War II FPS game to the hoot of a nocturnal owl), while they also bring a sense of progression to the game. Sound requires linear time to be perceived; hear a sound, experience the passing of both game time and real-world time.

That the player has a relationship with all of these sounds is self-evident. The sound of footsteps or a scream from behind in an FPS game will invariably cause the player to turn (that is, to turn the character in the game world) to investigate the cause. Ambient sounds provide much-needed three-dimensionality and material life to the flat pixels displayed on the monitor, and this, with the ability to localize sound, helps position the player in the FPS game world with its first-person perspective and its first-person audition (a term used analogously with first-person perspective but referring to the perception of game sound; Grimshaw 2008a). But the player is also able to participate in the acoustic ecology through actions that contribute sounds, many of which are heard by other players in the vicinity (as their sounds are heard by this player). In the FPS game, such sounds include the firing of weapons, various vocalizations and footsteps, and, in some games (e.g., *Urban Terror*), variable breathing matching the in-game exertions of the character.

Grimshaw and Schott (2008), in describing the acoustic ecology of the FPS game, also suggest that there is a Virtual Acoustic Ecology. This definition accounts for a multiplicity of acoustic ecologies and the effects of telediegesis in a multiplayer FPS game. It is a model that integrates players and their sound environments, or *resonating spaces* (thus, the many acoustic ecologies of the game), with the game audio engine. What it lacks as a framework is a detailed modeling of the players' affect states so, while we will return to the concept of the VAE later, we first take a brief look at emotions in the context of game sound.

11.2 Emotion and the Game Acoustic Ecology

Our current work (e.g., Garner, Grimshaw, and Abdel Nabi 2010; Garner and Grimshaw 2011) examines and involves a closer integration and immediate two-way responsiveness between player and game sound through the use of biofeedback. That is, how can we measure the player's psychophysiology, using electroencephalography (EEG) and electromyography (EMG) for example, and then use that data, ideally representative of the player's affective state, in order to process and/or synthesize game sound and thus track, adjust, or change the player's emotions through sound. We are interested in emotions because games, such as horror survival games (a subgenre of the FPS game) operating on fear, manipulate emotions in order to engage the player and, with increasing efficacy, emotions may be inferred from psychophysiological data. This opens the door to increased personalization of games and to the real-time tracking of, and response to, player psychophysiology, as we discuss further below.

Interaction between humans relies heavily upon emotional communication and understanding, concepts that have been applied to human–computer interaction by Reeves and Nass (1996), who argue that natural and social factors are prevalent during interactions between man and machine. Emotional interactivity between software and user has influenced the consumer sales of computer technology (Norman 2004) and existing research has revealed significant positive correlation between user enjoyment and perceived suspense rating within a digital game context (e.g., Klimmt et al. 2009).

Emotionality is now an established concern of many game developers and a growing body of research supports its importance (Freeman 2003). Perron (2005) asserts that emotional experiences resulting from gameplay have great potential to improve the player experience and that the more intense the emotion, the greater the perceived experience. Perron also describes the experience of fear within a survival horror game as a pleasure and a significant incentive to play. In addition to a positive influence upon immersion, performance, and learning (Shilling and Zyda 2002), emotionality has the potential to grant players access to a wider spectrum of emotional states than can be easily achieved in reality (Svendsen 2008, 74) (see also Chapter 12 in this volume).

Sound is a critical component to consider when developing emotionality as it is directly associated with the user's experience of emotions (Shilling and Zyda 2002). Parker and Heerema (2008) suggest that sound carries more emotional content than any other part of a computer game and Shilling and Zyda (2002) quote industry professionals: "A game or a simulation without an enriched sound environment is emotionally dead and lifeless." Garner and Grimshaw (2011) present a framework of fear within a computer game context, supporting the capacity of affective game sound to significantly alter our physiological states and to determine the cognitive processing used to infer meaning from the sound data (primarily by way of determining the mode of listening).

To a certain degree, game engines already respond to player emotions and it has been argued elsewhere (Grimshaw 2008b) that game audio engines are sonification systems that sonify player actions. In an FPS game, the player moves or the player shoots a gun and there is an appropriate sound played; thus, nonaudio data (the player's actions) are sonified. It could be argued that this is already a sonic tracking of the player's affective state—a timid creeping around triggers the occasional furtive footstep sample, whereas the bold, excited player will leap into the fray all guns blazing. However, sound is not yet processed or systematically synthesized in order to manipulate emotions according to an analysis of the player's psychophysiology. If we were able to do such a thing, this would gain us several advantages that have significance for game sound design, the player's relationship to the acoustic ecology of the game, and the player's immersion in the game world. Some of these advantages are technical ones—for example, real-time synthesis would help to overcome the limitations of storage media as regards the provision of a wide variety of audio samples. Another advantage, a more interesting one to our minds, leads to our final point within this section: this relationship with, and this participation in, the acoustic ecology of the game leads to player immersion in the game world. There are other sonic factors that aid this, such as the advantage that game sound has over game image (for instance, the composite soundscape of the game is not limited to the small screen space of the monitor but surrounds the player), but it is our contention that this relationship and participation lead to immersion in the acoustic ecology (the player being an active element of that ecology) and that this is one of the main contributing factors to immersion in the game world itself.

Immersion, and the related concept of presence, is discussed in detail by a number of writers (e.g., Brown and Cairns 2004; Calleja 2007) and it is argued that sound has a role to play in that immersion (e.g., Grimshaw 2008a, 2012). What has not been comprehensively investigated is the role that emotion might play in immersion, particularly when that emotion is induced by a deliberate manipulation of sound for such a purpose. Our proposal is that a real-time synthesis or processing of sound according to the player's psychophysiology increases the opportunities for player immersion in the game world; in short, immersion may be enhanced by enabling the game audio engine to respond empathetically to the player by assessing the affective state of the player. Later in the chapter, in order to advance this proposal, we present a new model of the game's acoustic ecology that takes into account the player's psychophysiology. This model uses key concepts from embodied cognition theory and so the following section briefly introduces these, particularly where they are relevant to game sound.

11.3 Embodied Cognition and the Game Acoustic Ecology

A key title in the field of embodied cognition, Andy Clarke's *Being There: Putting Brain, Body, and World Together Again*, advocates the concept of integrated cognition, stating that, "the biological mind is, first and foremost, an organ for controlling the biological body" (Clarke 1997, 1). This bears some similarity to the notion of autopoiesis, especially in the autopoietic concept of a consensual domain; this domain is brought about by the structural coupling—the interplay—between mind, body, and environment (see Winograd and Flores 1986, 46–9). Von Uexküll's (1957) concept of *Umwelt* attempts to explain how the mind reduces incoming data to increase efficiency of processing with a perception filter that is determined by lifestyle, desires, and needs. The concept has been compared to Dawkins's *Extended Phenotype* (1982), the notion that our biological makeup can be fully understood only within the context of its interactions with the environment. In a similar vein, Rappaport (1968) coined the term *cognized environment*, referring to how an individual's cultural understanding may impact upon their perception of the natural environment.

Space precludes a full discussion of the many facets of embodied cognition (EC), so, here, we turn to a summary of the theories in order to tease out those of relevance to game sound. Margaret Wilson's (2002) documentation of the six views of EC theory offers a comprehensive outline and additionally provides a foundational framework that we can integrate with the notion of an FPS acoustic ecology. Wilson summarizes some of the principles of embodied cognition as comprising a cognition that: is situated (geographically and temporally—in the *here and now*); is time-pressured (cognition must be understood in real time); and is for the enabling and guidance of action. The environment is intrinsically connected to cognition by way of offloading (e.g., assigning markers on a map to plan and guide a journey). Although Wilson acknowledges that cognitive thought can theoretically occur even if the subject is detached from all sensory input (offline cognition), the activity of the mind remains "grounded in the mechanisms that evolved for interaction with the environment" (Wilson 2002, 626). This resonates with Gibson's (1979) concept of *affordances*, an intrinsic *demand*, characteristic of an object, that places our perception of entities within the environment inescapably within the concept of what they can and cannot do for us. What is, for us, the most crucial and encompassing view is that "the environment is part of the cognitive system" (Wilson 2002, 626), asserting that addressing the mind as a separate entity (*pace* Cartesian dualism) will not yield comprehensive results in attempting to understand it.

According to Wilson, the central notion of situated cognition is that all informational processing is susceptible to the continuous stream of incoming sensory data. Furthermore, any sensory information that is stored in long-term memory (alongside any relationship between the sensory input and associated objects, events, physiology, behavior, etc.) has the potential to influence future thoughts regardless of construal level or context. Wilson suggests that thought processing gradually builds a framework of automated subcortical routines. Regularities in comparable circumstances encourage an automated response generated by sensorimotor simulation; essentially a behavioral response, this precedes cognitive appraisal and is contextualized by conditioned representational links. This concept is supported by Garbarini and Adenzato (2004), who argue that cognitive representation relies on virtual activation of autonomic and somatic processes as opposed to a duplicate reality based in symbols. An embodied theory would not accept pure behavioral conditioning and would instead suggest that an object would first generate virtual sensory data, which characterize the stimulus, and then generate a threat interpretation. The entire process remains fundamentally cognitive, but only a fraction of the input data needs to be fully appraised as the simulated data is already directly linked to the autonomic nervous system through conditioning; this supports the concept of an efficiently responsive process achieved via reduced cognitive load.

The fundamental idea behind time-pressured cognition is that all human thought can be influenced by the concept of time as perceived by the individual and relating to objects or events. Liberman and Trope (1998) illustrate how an individual's perception toward a future event could change in response to different relative temporal distances. Personal evaluation has also been described as susceptible to psychological distance influence. As research by Freitas, Salovey, and Liberman (2001) has revealed, individuals are likely to employ a negative, diagnostic assessment when such an evaluation is expected in the more distant future but are more likely to prefer a positive, nondiagnostic assessment when it is perceived as imminent. For example, in preparation for a product unveiling, designers may employ a negative diagnostic assessment as there is time to address concerns. When the unveiling is imminent the designers may instead favor positive abstract assessments, as there is no time for corrections and confidence in the presentation is now a priority. Greater temporal distance encourages more generalized thought (one cannot see the trees for the forest) whereas immediacy evokes increased specificity (one cannot see the forest for the trees). Time, therefore, affects attention and becomes a significant factor in appraisal and decision making (Liberman and Trope 2008). Temporal distances are interrelated quantifiable values that, alongside hypotheticality, and spatial and social distance, establish psychological distance and influence higher-level cognitive processes such as evaluation and prediction (Liberman and Trope 2008).

Recollection of memories to deduce and arrange future plans is also embodied in sensory data. Existing research has argued that memory retrieval can cause a reexperiencing of the sensory-motor systems activated in the original experience, the physiological changes creating a partial reenactment (Gallese 2003). The notion of implicit memory, relating to perceptual fluency and procedural skill (Johnston, Dark, and Jacoby 1985), supports the developmental nature of embodied cognition. Wilson (2002) argues that implicit memory is automated action, acquired through practice whereby repetition instills conditioned movements and reduces the need for full cognition. She suggests that these processes of perception and action have the potential to become "co-opted and run 'off-line,' decoupled from the physical inputs and outputs that were their original purpose, to assist in thinking and knowing" (Wilson 2002, 633). A potential consequence of this theory is that any prior thought process that generated representations and relations between objects can impact upon any future thoughts regardless of construal level.

The information presented above strongly asserts that cognitive thought is heavily influenced by immediate sensory input to the degree that the environment must be integrated into any framework of function. Thus, the mind should be studied as part of an ecology. However, it could be asserted that the mind is capable of interpreting internally generated data (though the source of that data can ultimately be traced back to the environment). The existence of mirror neurons (outlined in Garbarini and Adenzato 2004) suggests that sensory observation "of another individual's action evokes a specular response in the neural system of the observer, which is activated as-if he himself were carrying out the action that he is observing" (102). This could be extended to assert that mirror neurons could respond to an imagining of another individual's action, facilitating an action simulation in response to an internal source. We suggest that the mind is able to reflect upon internalized scenarios and respond with virtual interactions.

Augoyard and Torgue (2005) describe a number of auditory phenomena as sonic illusions that support the notion of an embodied theory of cognition. These auditory effects include anamnesis (recollection of past memory in response to sound), narrowing (the sensation that the surrounding environment is shrinking), the Lombard effect (dictation of listener's vigilance level), phonomnesis (unintentional perception of an imagined sound as real), the Tartini effect (perception of a sound that has no physical existence), and remenance (the perceptual continuation of a sound that is no longer being propagated). If we are to acknowledge the existence of such effects, it is logical to consequently assume that auditory processing is an embodied event, dependent upon the relationship between physical environment, memory, and physiology.

Having briefly described the game's acoustic ecology, pointed to the relevance of player emotion for the perception of game sound, and summarized salient aspects of embodied cognition theory, we can now proceed to our model. The model is a synthesis of these three components and is intended to provide an embodied Virtual Acoustic Ecology framework for the design and understanding of game sound.

11.4 The Embodied Virtual Acoustic Ecology framework

Figure 11.1 visualizes the eVAE framework as a procedural chain to better elucidate the looping mechanisms and interrelating variables that impact upon our perception of game sound within an embodied framework. Critical elements of the VAE construct remain (such as soundscape, resonating space, sound functionality, and perceptual factors) but specific constructs within the player are now presented that suggest the

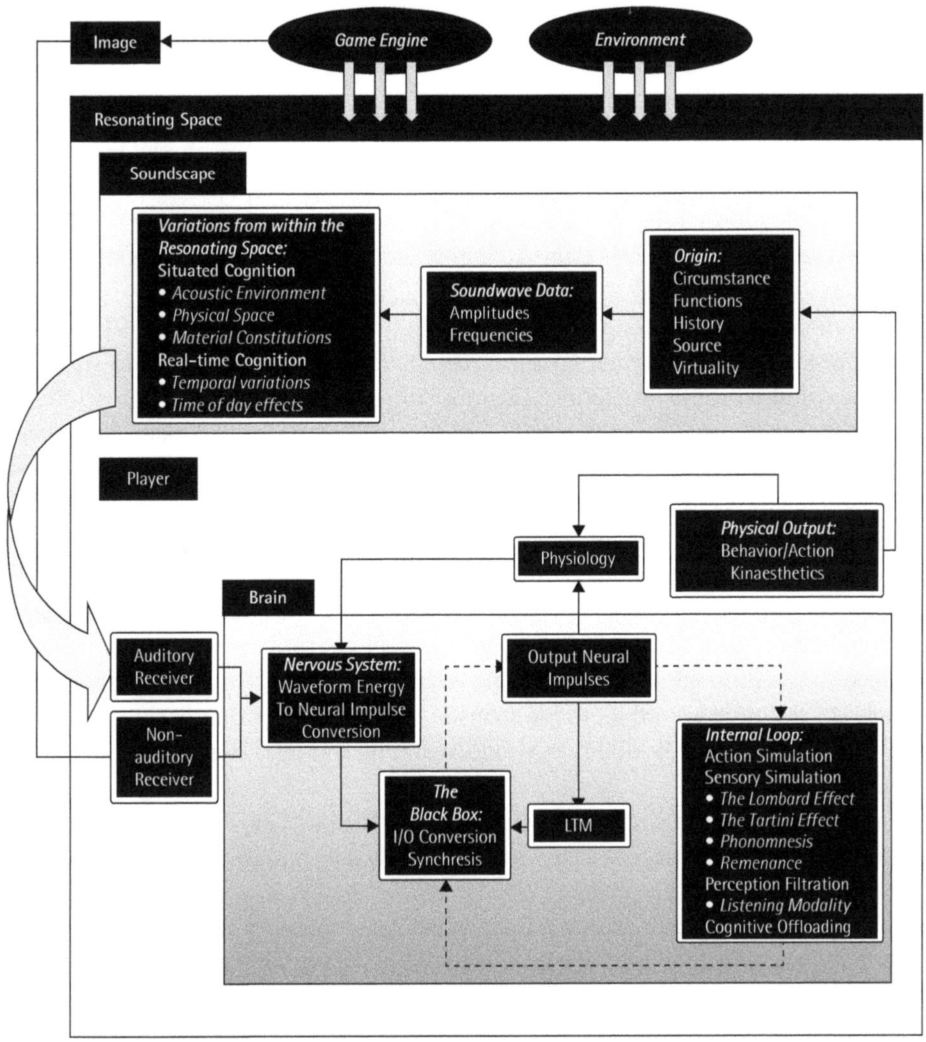

FIGURE 11.1 The embodied Virtual Acoustic Ecology model.

functionality of embodied cognition. At the origin stage, soundwaves are acknowledged to result from a complex matrix of historical and circumstantial factors (asserting that the sound is not only dependent upon the here and now, but also upon a highly complex chain of past events that have led, by way of causality, to the present). But, irrespective of this stage, the resultant wave can always be reduced to waveform amplitude and cycle frequency. Resonating spaces are asserted as key determinants of the *here and now* of embodied cognition theory, in that the physical makeup of the environment may (through only minor perturbations in signal processing) dramatically alter the perceptual data extracted during cognition. The dynamic nature of resonating spaces further accommodates the notion of real-time cognition, as changes within the physical environment (shifting temperatures, position or density of reflecting surfaces, new materials entering or leaving the resonant space, etc.) have significant potential for signal attenuation or amplification, meaning that no two sonic waveforms should have precisely the same acoustic data outside of a heavily controlled laboratory environment.

The internal system map displayed here acknowledges the embodiment theory that the brain is continuously affected by incoming sensory data, the physiology, and the long-term memory (LTM) of the listener. The term *Black Box* alludes to the limitations of this mapping in that the actual process of converting neural input signals into output impulses (that drive both external action and internal looping systems) remains unknown.

One immediate application of this visualization is the highlighting of key points within the listening process that a designer could focus upon in an attempt to artificially replicate a desired sonic perception. The most apparent node to replicate or synthesize would arguably be the *soundwave data* (the acoustical information that constitutes a complete sound) and this is certainly a common choice within game sound design. Such a task when approached with synthesis is a difficult undertaking, due to the dynamic complexities of sound. Recording and mechanical replaying of naturally occurring sound offers a partial solution, as (depending on the quality of the equipment) such recordings reflect a substantial portion of the original sound's acoustic characteristics. Limitations with this approach include realism concerns and the static nature of the recording. Although some game audio engines (e.g., *CryEngine*) can process audio samples in real time according to the acoustic spaces and materials of the gameworld, this approach lacks the flexibility of sound synthesis and thus cannot facilitate a truly dynamic soundscape.

The amalgamation of circumstances required to facilitate even a simple sound contains a large enough number of elements that, if artificially replicated, would enable even a low-fidelity sound be perceived as real due to the support structure of circumstantial information (much in the same way that we can anthropomorphize an animated lamp on a television screen because its observable behavior implies hopes and fears). Take the specific soundwave generated from a gunshot as an example. Even before we consider the environmental impact upon the wave as it travels from the source to the ear, we must acknowledge that such an event cannot simply happen without a complex set of requirements being met. There needs to be a gun, a bullet, a shooter, and a target. There must be a motive, driven by incentive and/or disincentive, which itself requires a complex arrangement of entities, associations, and processes. Early game developers lacked the technology to artificially replicate a believable gunshot soundwave, but they could replicate the circumstances leading to that sound, artificially replicating the shooter as a player avatar, the target or weapon as a sprite graphic, while the motive was established via plot or simply the player's awareness that, for example, "this is a game and it is my job to shoot things." These techniques present the player with an associative dataset that, when combined with the soundwave data, can manifest a perception of the sound as real and replete with contextual meaning.

Currently, most of these approaches to developing perceptual realism of virtual sound could be described as noninvasive, in that they replicate only a segment, or segments of the data process that occur external to the human body. Regarding this, the eVAE model asserts that to create a virtual sonic environment that is truly indistinguishable from reality, we may need to push deeper into the brain itself. If it were possible to replicate either input impulses (converted from sensory data) or the output neural impulses (converted from input signals via the Black Box), it could essentially short-circuit the framework, enabling the internal loop to function without actual sensory input. One important question to consider is, which neural impulse node (in or out) should be replicated? The answer to this question could be dependent upon the comparative difficulty of distinguishing I/O signals from electrical noise. Our final section explores this further and concludes by briefly noting some of the implications raised when we remove sonic *sensory* stimuli and directly stimulate *perception* of sound in the brain.

11.5 Biofeedback and the Perception of Unheard Sounds

Currently, the primary functions of research into brainwave data analysis projects are linked to specific theoretical or practical interests, such as emotion recognition (Murugappan et al. 2010) or brain–computer interfaces (Chih-hung, Chang, and Yi-Lin 2011), as opposed to attempts to replicate the human perception process. Quantitative psychophysiological research has recorded biometric (facial muscle, cardiac, and electrodermal) activity in response to various sounds and revealed significant variation with different sounds. This not only identifies an area of research in need of further development, but also nominates sound as a feasible approach to manipulating a player's emotional response. Bradley and Lang (2000) collected EMG and electrodermal activity (EDA) data in response to various auditory stimuli. Experimentation revealed increased corrugator activity and heart-rate deceleration in response to unpleasant sounds, and increased EDA in reaction to audio stimuli qualitatively classified as arousing. Electrodermal activity has been used to differentiate between motion cues, revealing increased response to approaching sounds (Bach et al. 2009), and event-related potentials (collected via EEG) reveal changes in brainwave activity in response to deviant sounds within a repeated standard pattern (Alho and Sinervo 1997).

If we are to accept that our entire collective of sensations across all five sensory modalities can be reduced to electrical information, the following questions are raised: Could it be possible to artificially replicate these electrical signals? And: Are these electrical impulses reducible to a single format of data and, if so, can we use these data to directly stimulate the perception of sound? Here, two subprocesses are identified as areas of research interest: the mechanism by which input data (light, pressure variation, etc.) are converted into electrical information in the brain; and the procedure that facilitates classification and the analysis required to convert the electrical information into acquired knowledge and thus perception of a sound with meaning and context. In alignment with traditional cognitive theory, the basic description of the latter process could be compared to that of EEG analysis, in which the raw data is cleaned (noise and artifacts identified and removed) and relevant features are extracted and classified to infer further knowledge by cross-referencing the new data with existing data. Although the mathematical detail behind EEG analysis is complex, we do not presume that it would be anything more than highly simplistic when compared to the processes that occur within the human brain. We do, however, assert that these similarities between processes could allude toward a macro process that, with greater understanding in the fine detail, could support the development of an artificial replication of the inference stages of human thought processing within which neural impulses are organized, classified, and translated into output thoughts (action potentials, etc.). In short, could EEG acquisition and analysis techniques be developed to facilitate a method of artificially replicating neural impulse signals—essentially, a reverse EEG? It might also be asked, if one of the purposes of game sound is to engage and immerse the player in the game world and that this can be achieved in part by manipulating emotion, can we directly stimulate the emotion that would be triggered in response to such sound rather than directly stimulating the perception of that sound calculated to induce a particular emotional response?

Reis and Cohen (2007) experimented with transcranial stimulation (an artificially created electromagnetic field designed to stimulate brain activity) and its effects upon cortical activity and learning while transcranial stimulation during the early stages of a deep sleep has been revealed to improve declarative memory retention (Marshall et al. 2006). While current research does not claim successful thought manipulation via replicated neuroelectrical activity, it does reveal that the human brain responds and responds safely to such stimulation. EEG studies have also provided correlations between brain activity and task efficiency (Chouinard et al. 2003), perceptual feature binding (Schadow et al. 2007), emotional valence (Crawford, Clarke, and Kitner-Triolo 1996), discrete emotional states (Takahashi 2004), and attention and meditation levels (Crowley et al. 2010), to name a few. This research supports the assertion that quantitative neuroelectrical data systems are becoming capable of interpreting neural impulses in a process that could potentially be greatly similar to that of the Black Box node within the embodied cognition model.

The classical acoustic definition of sound states that sounds are waves produced by vibrating bodies; sound is thus a compression wave moving through a medium and may be described by a number of factors, including frequency and amplitude. Indeed, this definition is explicit in our eVAE model given the requirements in that model to process or synthesize sound. There are other theories, though, that describe sound through features that are not accounted for in acoustic theory. O'Callaghan (2009, 27) describes the view of sound as a property of objects (sounds are properties of bodies and objects that vibrate at particular frequencies and amplitudes) and introduces a new definition of sound as event (sound is "the act of one thing moving another"). These definitions serve different epistemological purposes and, in at least one case, a very practical purpose (that is, the synthesis and electronic reproduction of sound) and the latter two depend to some extent upon the first for the sensation and perception of those properties and events. In this chapter, our theoretical musings lead us to another possible definition that, too, has epistemological and practical purposes: sound is perception that does not require sensation. Thus, sounds can be unheard yet perceived. This has implications for theories of embodied cognition: Does the mind really require the body and environment to cognize or is it capable of independently cognizing, only requiring body and environment during experiential, learning periods? At this point, however, we should remind ourselves that the evidence for such a definition of sound and the rewriting of theory remain elusive given our current understanding and state of the art as regards biofeedback and the semantics of sound. If we are to more fully engage and immerse the player in the game world through the directly emotive use of sound, though, we feel that the path we have mapped out in this chapter is one worth exploring.

Note

1. While it might be argued that nondiegetic interface sounds, such as menu sounds, are part of the player's acoustic ecology, we do no not count these as part of the game acoustic ecology as they are not heard during gameplay.

References

Alho, K., and N. Sinervo. 1997. Pre-attentive Processing of Complex Sounds in the Human Brain. *Neuroscience Letters* 233: 33–36.

Augoyard Jean-François, and Henri Torgue. 2005. *Sonic Experience: A Guide to Everyday Sounds*. Montreal: McGill-Queens University Press.

Bach D. R., J. G. Neuhoff, Q. Perrig, and E. Seifritz. 2009. Looming Sounds as Warning Signals: The Function of Motion Cues. *International Journal of Psychophysiology* 74 (1): 28–33.

Bradley, M. M., and P. J. Lang, 2000. Affective Reactions to Acoustic Stimuli. *Psychophysiology* 37: 204–215.

Brown, Emily, and Paul Cairns. 2004. A Grounded Investigation of Game Immersion. In *Human Factors in Computing Systems*, April 24–29, Vienna.

Calleja, Gordon. 2007. Revising Immersion: A Conceptual Model for the Analysis of Digital Game Involvement. In *Situated Play*, September 24–28, Tokyo: University of Tokyo.

Chih-hung, W., J. L. Chang, and T. Yi-Lin. 2011. *Brain Wave Analysis in Optimal Color Allocation for Children's Electronic Book Design*. Taichung: National Taichung University of Education.

Chouinard, S., M. Brière, C. Rainville, and R. Godbout. 2003. Correlation between Evening and Morning Waking EEG and Spatial Orientation. *Brain and Cognition* 53 (2): 162–165.

Clarke, Andy. 1997. *Being There: Putting Brain, Body, and World Together Again*. Cambridge MA: MIT Press.

Crawford, H. J., S. W. Clarke, and M. Kitner-Triolo. 1996. Self-generated Happy and Sad Emotions in Low and Highly Hypnotizable Persons during Waking and Hypnosis: Laterality and Regional EEG Activity Differences. *International Journal of Psychophysiology* 24: 239–266.

Crowley, K., A. Sliney, I. Pitt, and D. Murphy. 2010. Evaluating a Brain-computer Interface to Categorise Human Emotional Response. In *10th IEEE International Conference on Advanced Learning Technologies*, July 5–9, Sousse, Tunisia.

Dawkins, Richard. 1982. *The Extended Phenotype*. Oxford: W. H. Freeman.

Freeman, David. 2003. *Creating Emotion in Games*. Indianapolis: New Riders Games.

Freitas, A. L., P. Salovey, and N. Liberman. 2001. Abstract and Concrete Self-evaluative Goals. *Journal of Personality and Social Psychology* 80: 410–412.

Gallese, Vittorio. 2003. The Manifold Nature of Interpersonal Relations: The Quest for a Common Mechanism. *Philosophical Transactions of the Royal Society of London*. Series B, *Biological Sciences* 358: 517–528.

Garbarini, F., and M. Adenzato. 2004. At the Root of Embodied Cognition: Cognitive Science Meets Neurophysiology. *Brain and Cognition* 56: 100–106.

Garner, Tom, and Mark Grimshaw. 2011. A Climate of Fear: Considerations for Designing an Acoustic Ecology for Fear. In *Audio Mostly 2011*, September 7–9, Coimbra, Portugal.

Garner, Tom, Mark Grimshaw, and Debbie Abdel Nabi. 2010. A Preliminary Experiment to Assess the Fear Value of Preselected Sound Parameters in a Survival Horror Game. In *Audio Mostly 2010*, September 14–16, Piteå, Sweden.

Gibson, J. J. 1979. *The Ecological Approach to Visual Perception*. London: Lawrence Erlbaum.

Grimshaw, Mark. 2008a. *The Acoustic Ecology of the First-person Shooter: The Player Experience of Sound in the First-person Shooter Computer Game*. Saarbrücken: Verlag Dr. Mueller.

——. 2008b. Sound and Immersion in the First-person Shooter. *International Journal of Intelligent Games and Simulation* 5 (1): 119–124.

——. 2012. Sound and Player Immersion in Digital Games. In *Oxford Handbook of Sound Studies*, ed. Trevor Pinch and Karin Bijsterveld, 347–366. New York: Oxford University Press.

Grimshaw, Mark, and Gareth Schott. 2008. A Conceptual Framework for the Analysis of First-person Shooter Audio and its Potential Use for Game Engines. *International Journal of Computer Games Technology* 2008.

Johnston, W. A., V. J. Dark, and L. L. Jacoby. 1985. Perceptual Fluency and Recognition Judgements. *Journal of Experimental Psychology: Learning, Memory and Cognition* 11 (1): 3–11.

Jørgensen, Kristine. 2009. *A Comprehensive Study of Sound in Computer Games: How Audio Affects Player Action*. Queenston, ON: Edwin Mellen.

Klimmt, C., A. Rizzo, P. Vorderer, J. Koch, and T. Fischer. 2009. Experimental Evidence for Suspense as Determinant of Video Game Enjoyment. *CyberPsychology and Behavior* 12 (1): 29–31.

Liberman, Nira, and Yaacov Trope. 1998. The Role of Feasibility and Desirability Considerations in Near and Distant Future Decisions: A Test of Temporal Construal Theory. *Perspectives of Social Psychology* 75 (1): 5–18.

——. 2008. The Psychology of Transcending the Here and Now. *Science* 322: 1201–1205.

Marshall, L., H. Helgadottir, M. Molle, and J. Born. 2006. Boosting Slow Oscillations During Sleep Potentiates Memory. *Nature* 444: 610–613.

Murugappan, M., M. Rizon, R. Nagarajan, and S. Yaacob. 2010. Inferring of Human Emotional States Using Multichannel EEG. *European Journal of Scientific Research* 48 (2): 281–299.

Norman, Donald A. 2004. *Emotional Design: Why We Love (or Hate) Everyday Things*. New York: Basic Books.

O'Callaghan, Casey. 2009. Sounds and Events. In *Sounds and Perception*, ed. Matthew Nudds and Casey O'Callaghan, 26–49. Oxford: Oxford University Press.

Parker, Jim, and John Heerema. 2008. Audio Interaction in Computer Mediated Games. *International Journal of Computer Games Technology* 2008.

Perron, Bernard. 2005. Coming to Play at Frightening Yourself: Welcome to the World of Horror Video Games. In *Aesthetics of Play*, October 14–15, Bergen, Norway.

Rappaport, Roy A. 1968. *Pigs for the Ancestors*. New Haven: Yale University Press.

Reeves, Byron, and Clifford Nass. 1996. *The Media Equation*. Stanford: Center for the Study of Language and Information.

Reis J., and L. G. Cohen. 2007. Transcranial Slow Oscillatory Stimulation Drives Consolidation of Declarative Memory by Synchronization of the Neocortex. *Future Neurology* 2 (2): 173–177.

Schadow, J., D. Lenz, S. Thaerig, N. Busch, I. Frund, and C. Herrmann. 2007. Stimulus Intensity Affects Early Sensory Processing: Sound Intensity Modulates Auditory Evoked Gamma-band Activity in Human EEG. *International Journal of Psychophysiology* 65: 152–161.

Shilling, Russell, and Michael J. Zyda.2002. Introducing Emotion into Military Simulation and Videogame Design: America's Army: Operations and VIRTE. In *Game On*, London.

Svendsen, Lars. 2008. *A Philosophy of Fear*. London: Reaktion.

Takahashi, K. 2004. Remarks on Emotion Recognition from Multi-modal Bio-potential Signals. *IEEE International Conference on Industrial Technology* 3: 1138–1143.

van Elferen, Isabella. 2011. ¡Un Forastero! Issues of Virtuality and Diegesis in Video Game Music. *Music and the Moving Image* 4 (2): 30–39.

von Uexküll, J. 1957. A Stroll through the World of Animals and Men. In *Instinctive Behavior*, ed. Claire H. Schiller, 5–80. New York: International Universities Press.

Wilson Margaret. 2002. Six Views of Embodied Cognition. *Psychonomic Bulletin and Review* 9 (4): 625–36.

Winograd, Terry, and Fernando Flores. 1986. *Understanding Computers and Cognition: A New Foundation for Design*. Norwood, NJ: Ablex.

CHAPTER 12

A COGNITIVE APPROACH TO THE EMOTIONAL FUNCTION OF GAME SOUND

INGER EKMAN

THE archetypal horror of the sound of nails on a chalkboard, a moving passage of music, or the reassuring voice of someone familiar: these are all sounds with emotional impact. But what, precisely, is the emotional power of sound applied to games, and how is it that even sounds with seemingly much lesser capacity to excite, scare, or soothe may become so emotionally effective? This chapter discusses the ways in which game sound is used to stir, enhance, and alter players' emotional responses in a game and seeks to explain why these techniques have such an emotional impact.

Research on sound's ability to elicit emotion often seeks to establish the emotional reaction to the sounds in isolation, and focuses on sounds such as music (Juslin and Laukka 2004), or emotional speech (Scherer 2003). In games, however, sound is typically present in combination with other modalities (visual, haptic), as part of a narrative, and embedded in a functional framework of play. The literature on film sound (e.g., Altman 1992; Chion 1994; Weiss and Belton 1985; Whittington 2007) can be very useful for approaching game sound, but cannot adequately account for some of the experiences with interactive sound. Sonic interaction design (e.g., Rocchesso and Serafin 2009), on the other hand, covers interactive sonic experiences but rarely considers the affective reactions to sound, or sound as part of a larger functional-narrative environment. Finally, empirical investigations on affective game sound (Nacke, Grimshaw, and Lindley 2010; Van Reekum et al. 2004) are somewhat ambiguous about the impact of sound on the gaming experience, and cannot specify the role of sound in the overall emotional experience of playing.

The incomplete knowledge about the effects of sound is due to two common limitations when considering game sounds: overlooking the complex structure in which sound influences emotion and failing to consider the multimodal aspect of sound and focusing on purely sound-based effects. To fully understand game sound, it must be considered as part of a larger contextual arsenal for emotional influence. In other words, to understand the effects of game sound, it is important to consider that sound is one of many components in the overall attempt to elicit emotion. Emotional game sound is not only purely sound-based effects, but such sound-involving structures that result in emotional effect within the context of the game. As part of a general process of orchestrating emotions in a game, sound is involved in many simultaneous functions. There is no one single "game sound" that, at any moment can be assigned an emotional quality, but many simultaneous sound-involving processes that may serve the same, or different, emotional ends. Hence, even if different game sounds have identifiable emotional affordances, no single acoustic property, sound type, voice quality, or tonal progression can be associated with a specific emotional power independently of its function in the system as a whole. Instead, the fundamental emotional power of game sound resides in the contextual bindings of how sound is embedded and presented in a game. Game sound counts as emotional when it demonstrates a power to trigger emotion, and also when it shows capability to enhance or modify ongoing emotional experiences. Following these premises, the attempt to categorize the emotional roles of sound is not ultimately about classifying the actual sounds, but rather about classifying the functions where sound influences the perception and understanding of a given playing context. Knowing the functions, it is then possible to examine the emotional response to sound in a more informed manner.

This chapter examines the emotional impact of game sound in the context of play. It offers an explanation for the emotional reactions to game sound based in cognitive appraisal theory. The chapter describes the distinction between sounds used for narrative and functional purposes and describes the different approaches employed within these two categories in order to achieve emotional impact. Moreover, it also covers sound-related perceptual and interpretative processes that, when triggered, have emotional consequences beyond the narrative and functional roles of sound, and it explains how these affective responses influence the overall emotional experience.

12.1 The Dual Role of Game Sound

A typical game will contain a number of different sounds in various roles. In production, a common differentiation is made between music, voice acting, and other sounds (which are sometimes separated further into localized sound effects and nonlocalized ambient sound). Theoretical distinctions typically employ a dimension of diegesis, borrowed from film sound theory, depending on whether the sound is diegetic—emanating from the game's "story world"—or not (Ekman 2005; Collins 2007, 2009; Jørgensen 2007, 2008; Grimshaw 2007). The distinction between diegetic and nondiegetic concerns not only how a sound is presented, but also that sound's meaning—what it signifies to the player. As an example, a hooting owl in a forest is plausibly a diegetic sound. But if the howl is given a significance that is beyond the diegetic framework (for example, sounding the hoot every time you have fulfilled your mission), it becomes a carrier of nondiegetic information, and it will be perceived as external to the game world. Game sound, then, also fulfills a functional role in supporting gameplay, complementing its diegetic or narrative position. This double influence endows game audio with a unique mix of qualities.

The distinction between diegetic and nondiegetic, as well as between functional and narrative sound is, nevertheless, often complicated by the fact that sounds may alternate between being diegetic and nondiegetic. Jørgensen examines the breach of the diegetic–nondiegetic barrier at length (Jørgensen 2007, 2008) and concludes that it is "not possible to categorically identify a certain sound signal as related to one specific informative function" (Jørgensen 2008). In fact, most games employ some form of communication across the diegetic boundaries. Diegetic breaches are not breakdowns, but omnipresent. Nevertheless, those two roles—the narrative and the functional—serve as a basis for how sounds are used in games. As we shall see, this distinction also underlies the two cognitive frameworks by which sound becomes emotionally potent.

12.2 Emotion by Listening and Playing

Emotions are a distinctive type of mental state, involving cognitive and bodily changes, behavioral tendencies, and experiential components. The mental and physical activations that constitute an emotion are purposeful; emotions prompt us to perceive the causing events as salient, and prepare us mentally and physically to take survival-enhancing action. The affective response typically helps making events more memorable, which is beneficial for learning from experience. Emotions are separate from other affective states by always being directed toward, or about, something. Armed with this general understanding of emotion, the next question becomes: how does game sound elicit emotions? The simple answer is that emotions arise as our perceptions of sound undergo a series of conscious and unconscious evaluations. In the following section I will embrace the dual process model of emotion, presented by Clore and Ortony (2000), to provide a more specific description of the type of evaluations that are involved in this process. This treatment will also make evident why affective evaluations differ not only depending on a sound's auditory quality, but also based on how the sound is (functionally and narratively) integrated into the game.

12.3 The Dual-process Theory of Emotion

According to Clore and Ortony (2000), emotions are based on cognitive appraisal along two simultaneous processes: one proceeding bottom-up and the other top-down. Bottom-up processing performs new situational evaluations (not necessarily consciously), assessing the stimuli in relation to a set of values and goals. The top-down response system, on the other hand, works by reinstating prior emotional experiences based on association. The two systems complement each other: bottom-up reasoning adds adaptability and flexibility beyond simple reflexive behavior, whereas top-down responding increases chances of survival in time-critical situations.

Bottom-up situational evaluation compares events in relation to values, by: (1) how the situation influences personal goals; (2) how the actions compare to a certain set of standards (moral, social, and behavioral norms); and (3) how the encounters match personal attitudes or taste, and by complex combinations of these values (Clore and Ortony 2000, 27). For example, the way we feel about another person's performance depends on our prior feelings toward that person: a failed task may elicit gloating or pity, a success either envy or joy. When bottom-up evaluations invoke multiple evaluation frameworks, this produces complex emotional reactions.

Top-down reinstatements of prior experience occur when perceptions trigger a particular "deep structure of situational meaning" (Clore and Ortony 2000, 33). "Meaning" here refers to the particular activation patterns that represent our knowledge of past events. Emotions are formed when the perceived qualities of a certain situation trigger stored representations of earlier experience, reinstating the affective state. For example, fear is the reaction to an appraisal of threat. The perception of threatening stimuli (or stimuli that share enough salient features with something that is perceived as threatening) will therefore trigger the emotion of fear. Due to the organization of memory, the triggers can be quite unintuitive and unexpected, and may occur when the perceived content is outside focal awareness, which explains why we can become surprised by our own emotions (Clore and Ortony 2000, 36). The associations made at the mere recognition of emotional stimuli also account for so-called unconscious emotions, which have earlier been taken as a sign that emotion precedes cognition (e.g., Zajonc 1980).

12.4 Emotion and Misattribution

Top-down and bottom-up emotional evaluations occur in parallel whenever we respond to stimuli. In fact, it appears that emotional evaluations are a necessary subprocess to all reasoning: we need emotions to function properly even in situations that appear completely nonemotional. Damasio (2005) explains that emotional value judgments (good or bad) provide a rapid weighting system for choosing alternatives out of a vast set of options. According to Damasio, without this capacity to sort through cognitively complex alternatives, even seemingly insignificant decisions (e.g., what to wear today) would crowd cognitive processing. Moreover, people show a general tendency to use their overall affective state (not only emotional reactions) as a source of information. The effect was initially demonstrated by Schwartz and Clore (1983), who found that when asked about their future prospects, people tended to feel better about their lives if the question was asked on a sunny day as opposed to asking the question on a day with bad weather. Subsequent research has found that feelings are states that greatly inform reasoning and that people use a wide range of subjective experiences—not only moods, but emotions, metacognitive experiences such as fluency, and other bodily sensations—in evaluative judgment (Schwartz 2012).

Feelings are readily misattributed to concern the task at hand, but feeling-as-information can be discounted if people (correctly or incorrectly) attribute an external cause to how they feel (Schwartz 2012). The influence of affective information is most potent when people are not aware of its source, since these feelings are less likely to be actively discounted. This is, in part, why sound is such a potent medium. With our focus mostly turned to the visual modality, sound tends to slip into our experience relatively unnoticed. This makes us especially receptive to the emotional influences of sound.

12.5 Bottom-up Appraisal of Game Sound: Narrative Fit and Functional Fit

The emotional outcome of bottom-up appraisal depends on the rule sets engaged in the evaluation process. Perron (2005) and Lankoski (2012) propose two cognitive frameworks are particularly relevant for gaming: narrative and goal-oriented. They both involve sound.

In games with a story, narrative comprehension becomes a strong source for emotion. According to Tan (1994), fictive emotions require maintaining an apparent reality capable of tricking the brain into mistaking the events for real. The narrative fit of game sound reflects how helpful sound is to storytelling and helps bring out the emotions inherent in the story. Sound is one of the building blocks for creating a coherent, plausible environment, and engaging the player with the narrative setting. But apparent realism is not the end goal of narrative fit. Sound has high narrative fit when it facilitates narrative comprehension, even if it does so with unrealistic sound (Ekman 2008). Much of the sound used for narrative purposes is diegetic, but nondiegetic sounds such as music or a narrator voice can also impact story comprehension and serve narrative purposes.

The functional fit of game sound is not about story comprehension, but about how sound supports playing. The goal-oriented framework is directly related to play, and consists of evaluations in relation to how the player progresses in the game. According to Lankoski (2012), the goal structure and action affordances in a game can be used to predict (and design) how that game elicits emotions. Game sound is part of the feedback system that provides information on player action, constantly signaling which actions help the player to progress toward the goals of the game. A high functional fit signifies sound that facilitates, supports, or furthers goal-oriented action. Functional sound, for instance, provides feedback about the success or failure of actions, informs the player about available options, and helps time actions correctly. Functional sound is free to challenge narrative fit. It need not blend in with the diegesis (although it can do so). The emotional significance of functional sounds is the information pertaining to play; when sound performs well at this function, the player gains access to the game.

The methods for achieving narrative and functional fit are quite different. More specifically, narrative fit relies on creating compelling fiction, whereas functional fit is about information design. I have argued elsewhere that individual game sounds tend to cater to either one or the other of these frameworks but rarely serve both purposes simultaneously (Ekman 2008). When it comes to eliciting emotions, narrative comprehension and goal-oriented evaluations are usually driven by separate, even contradictory, motivations. However, the sounds assigned to the player character offer a natural intersection between the two frameworks and accurate character descriptions can also inform the player about their own action capabilities (Ekman and Lankoski 2009). For an example, the way Heather is portrayed in *Silent Hill III* signals physical vulnerability (narrative comprehension), which is also informing the player of the limited damage their character can handle without getting killed (goal-oriented action).

12.6 Designing for Narrative Fit

The techniques for enhancing narrative fit rely to a great extent on synchrony. The drive to make sense of the world multimodally is so strong that we are bound to perceive synchronous events as one (Burr and Alais 2006). Synchronized events are perceptually grouped together, pooling information from multiple sources: boosting redundant information and using information from different senses to fill in the blanks of others. Typically, the multimodal grouping is dominated by visual primacy, such as in the "ventriloquist effect", where a visual cue will trick the mind to believe the voice is coming from a puppet. Burr and Alais (2006) suggest that primacy is always allocated to the sensory channel that provides most accurate data. Indeed, temporal processes tend to give auditory stimuli primacy. For example, when presented with a series of sound signals and light flashes, most people will determine the number of flashes and their timing using sound information.

Synchronic sound uses visual primacy in a way that masks the constructed nature of the picture. Sound effects attest to the robust and physical quality of the events in the two-dimensional picture, tricking the brain into thinking of the game environment as real enough for emotional impact. But the perception of synchronicity is also influenced by music, which, at least in film, seems to guide the overall temporal synchrony of onscreen events (Cohen 2001). To a large extent, enhancing narrative fit is about constructing apparent realism, or filmic realism (Collins 2008; Ekman 2009). In general, sound adds a sense of believability to audiovisual representations, and is considered important to experiences such as immersion (Ermi and Mäyrä 2005) and presence

(Sanders and Scorgie 2002). Presence ratings demonstrably correlate with stronger emotional reactions to sound (Västfjäll 2003).

In addition to adding a sense of reality, synchronic sound supports storytelling by promoting the kind of information that is most helpful for understanding the events shown on screen. Sound's ambiguous nature allows for attaching additional meaning to seemingly neutral events. Chion (1994) coined the term *synchresis* (a combination of the two words "synthesis" and "synchrony") to emphasize that the audiovisual bond creates new meaning. Indeed, the extent of the multimodal effect should not be underestimated. For example, visual stimuli change not only affective judgments of sounds (such as annoyance ratings), but also the perception of loudness (see Cox 2008a).

Since synchronic sound also guides attention by boosting perception of visual events, sound helps the viewer to focus on the parts of the narrative that are most central to understanding the story. The attention-grabbing role of sound becomes important in games where players are relatively free to explore the world and choose where to look. To maximize this effect, less significant events may simply be left without sound. To ensure story comprehension, important narrative events are often furthered in cut scenes. The most pivotal sounds in film are typically refined to give the sound extra emphasis, loading them with narrative, connotative, and symbolic meaning, and enhancing their attention-grabbing effect. The same applies to game sound, but there are some limitations to how obviously this effect can be used for sounds that are expected to be heavily repeated.

Synchrony also works on a structurally broader level. Anderson suggests that we have a general tendency to double-check interpretations across modalities in a way that is not limited to temporally instantaneous events, but may span longer sequences, such as when the viewer uses the musical tone and emotion of a scene for "confirming or denying the viewer's response to what is seen" (Anderson 1998, 87). Indeed, music has power to drastically change viewers' interpretation of narrative content, demonstrated in several empirical studies on film (Vitouch 2001; Tan, Spackman, and Bezdek 2007) and also with scenes from computer games (Moffat and Kiegler 2006). In film, simply the fact that the sound is continuous helps bind together consecutive scenes, assuring us that fragmented visual glimpses belong to the same story. And "merely having a constant soundscape in a game can help the player to focus on the task at hand in a distracting environment" (Collins 2008, 132). Due to player action, achieving tight structural synchrony is harder in game sound than in film. Certain game forms (such as racing games) make it somewhat easier to predict the temporal duration of events, whereas others (puzzle games) make forecasting the duration of a level or scene a rough estimate, at best.

12.7 Designing for Functional Fit

To enhance functional fit, sounds should communicate action affordances (in the game) and provide (goal-related) cues. The capacity of listeners to use and make sense of various types of information has been thoroughly explored within user-interface design, suggesting sound can use various levels of symbolic mappings. For example, sonification harnesses the capacity to monitor multiple ongoing sound processes by presenting nonauditory data as audio information. Earcons convey information through musical symbols and audio icons employ a variety of mappings to express abstract information with environmental sounds. Finally, sonic interaction design builds on the notion of constructing realistic simulations, taking advantage of the full power of everyday listening. In terms of audio design, games can be found to employ all these types of information structures and sound-meaning mappings.

In contrast to narrative fit, functional fit does not aim at realism. Instead, the goal is to find mappings that are as intuitive and as fast to learn as possible. Functional sound is often transdiegetic, in that it operates simultaneously within the diegesis but also provides player feedback: reactively by affirming player input or proactively informing the player of an altered game state (Jørgensen 2007, 116). The driving factor for creating functional fit is to consider the utility of sounds for play. For example, with the exception of driving games, simulation-level fidelity is rarely used. Instead, sounds are overly simplified, grouped together, and tend to match game actions categorically with earcons or auditory icons. If a game offers two paces of movement (say, walking and running), it suffices to have two types of movement sound as well, even if both walking and running pace naturally varies slightly. The difference between functional and narrative fit is very evident in *Uncharted III: Drake's Deception*, which uses two distinctly different strategies to action sound. Context-sensitive fighting involves performing a series of timed button presses. These button presses are silent, but when properly executed, they propel fight scenes with highly narrative sound. In contrast, shooting relies on functional fit, and offers feedback directly mapped to player action.

When the sounds of actions are not distinctive enough, the player can be helped along with added sound cues that make the game more comprehensible. Useful information is not limited to feedback regarding individual actions. For example, game sound can communicate abstract structure by what Collins (2007, 131) refers to as the "boredom switch," a drop to silence that tells the player they have spent more time on a particular segment of the game than is intended. Indeed, the game *L.A. Noire* explicitly informs the player that "music will fade down to indicate that all clues at a location have been discovered."

Player action also introduces a haptic component to the process of perceiving synchrony. Where audiovisual synchrony lends information to congruent visual events, sounds added to player actions are perceptually grouped with the physical actions of playing. The consistent and responsive action–sound coupling also contributes to a sense of agency and control in a game. Hug (2011) argues for the affective quality, and the "joy of self-hearing" involved in agency. Specifically, he alludes to instances when sound effects can shape a satisfying differential of power between a physical action and a sound, for example, when a relatively small action (the press of a button) has a huge effect in the game (big explosion sound) (Hug 2011, 402). In general, the functional design of sound has the capacity to enhance agency, and to break it when functional fit is low: "Interfering with the sonic feedback of actions decouples action from effect, removing the sensation of control within the game and replacing it instead with an experience of fiddling with the controller" (Ekman and Lankoski 2009, 188).

12.8 Creating Intentional Ambiguity

Narrative and functional fit are ways to attach sound to existing frameworks of meaning in a game. In games with little storytelling, such as *Minesweeper* or *Tetris*, there is no need for narrative sound. In other cases, a poor narrative fit becomes destructive to emotion by challenging the believability of the fiction. Nevertheless, certain levels of intentional ambiguity can be used to build tension and create emotion through intentional contrast. The practice of obscuring the diegetic linkage in horror games is discussed by both Ekman (2008) and Kromand (2008). For example, navigating the game world in *Project Zero* is made particularly unsettling by giving the character noisy footsteps and having loud, banging doors: there is no way to traverse the threatening environment quietly. The unsettling quality is hard to shake even when the player realizes that these sounds have no direct impact on threat, and that ghosts do not navigate toward sound (unlike the monsters in the *Silent Hill* series). Another ambiguity effect is the proposed use of the uncanny to create a mismatch between human and nonhuman appearance (Grimshaw 2009; Ekman and Lankoski 2009), for example by having a human character with a clearly mechanistic voice.

Typically, even ambiguity must be diegetically plausible. In rare cases, however, clear breaks in diegesis can serve as a key element in creating emotion. Humor deliberately uses overt mismatches between source and sound for comic effect. This effect is used in several of Blizzard's games, where continuously prodding units will eventually have them give an outlandish reply. For example, the Viking unit in *Starcraft II* will exchange its male character voice for a computerized female voice imitating an automated call-center message: "Welcome to Viking. If you want rockets, press four; if you want weapons, press five; if you know the enemy you want to kill, press seven."

Whereas intentional mismatches in narrative fit are relatively rare, challenging functional fit is commonly used to create tension and shape gameplay. Sound-based gameplay mechanisms manipulate sound-action mappings to adjust the difficulty of gameplay. For example, the hard-to-hear audio cues used in the lock-tinkering minigame in *Elder scrolls: Oblivion* provide crucial information about how to time button presses, but successful tinkering calls for both intense listening and fast reflexes in order to succeed. When the sounds used for such cues are enhanced or obscured, the emotional response is related to the ease or difficulty of using the sound as a clue. The unsettling effect of masking treatments of sound cues signaling enemies in survival horror games, where players are continuously offered information in a hard-to-listen-to format, is another typical example, as is the perceived loss of control when, for example, sounds of footsteps are not consistent (Ekman and Lankoski 2009).

12.9 Top-down Appraisal of Game Audio: Reinstatement and Unconscious Process

The top-down process of appraisal occurs when perception directly triggers emotional experiences. These evaluations are not experienced in isolation from bottom-up emotions, but they reflect a different process whereby emotion can be triggered in a more direct fashion than by bottom-up calculation. Since the processes are both rapid and associative, they often go unnoticed. Often these emotions end up influencing our subjective experience through misattribution. The affective quality attached to events may also enter the bottom-up process, where it serves as raw material, for example, for the aforementioned processes of narrative and goal-oriented reasoning.

The repertoire of sound appraisals that can be invoked and reinstated through associative triggering process begins with brain-stem reflexes (in a sense, evolutionarily "learned" emotions), but grows with experience, reflecting each person's own private history of past appraisals. The reinstatement process is capable of triggering many types of experiences, but of the most interest for game sound are the reinstatement processes that might be common between players. Additionally, whatever prior events players carry with them into the gaming situation, the act of gaming also serves as an arena for forming new reinstatements. During play, the game links events with sounds. Thus, playing builds upon the repertoire that is already existent, and adds new affective triggers. Accordingly, sound–meaning links that might be arbitrary in the beginning will become associated with repeated encounters. This linking underscores the way these sounds will trigger emotions in the game and will become more consistent than what we are likely to find for the sound piece prior to gaming.

12.10 Acoustical Properties of Sound, Mere Exposure, and Perceptual Fluency

Certain acoustical properties seem to have innate affective properties, such as the startling effect of a sudden loud onset or the displeasing quality of dissonant chords. The extent to which sounds can be said to have an innate affective quality is not fully known, nor is how these evaluations link to subjective experience. What appears clear, however, is that sounds can produce a number of affective responses, rapid enough to precede conscious thought, and with remarkably consistent effects across listeners. And while these responses may not fully qualify as "emotions," they apparently are capable of shifting the overall affective state enough to influence the subjective experience.

What can explain this phenomenon? Juslin and Västfjäll (2008), who specifically mention brain-stem reflexes as one source of musical emotion, point out that a sound undergoes a number of analyses even before reaching the primary auditory cortex, many of which are capable of signaling simple value judgments such as pleasantness and unpleasantness. Apparently, some affective evaluations arise simply as the result of how easily sounds are processed in the brain. It has been suggested that the processing fluency, the extent to which stimuli confirm to the perceptual organization in the brain, is the underlying mechanism for perceptions of beauty (Reber, Schwartz, and Winkielman 2004). This theory suggests that low-level differences that compromise the perceptual clarity of audio (think signal-to-noise ratio) directly influence the emotional impact of sound. For example, Cox (2008b) covers a number of explanatory theories that all link irregular harmonic distributions to a perceptual disadvantage of (particularly speech) sounds and that could explain the perceived unpleasantness of dissonance. However, a phenomenon dubbed "mere-exposure effect" demonstrates that all kinds of stimuli increase attractiveness with repeated exposures—we like familiar stimuli (Zajonc 2011). Thus, fluency-based evaluations are not statically defined, but change to reflect prior experience. Indeed, research on speech perception suggests that the perceptual space might, over time, self-organize around the prototypical sounds of our native language. Such a structure could explain why speech sounds that closely resemble prototype centers are more readily perceived than sounds closer to the category borders (e.g., Salminen, Tiitinen, and May 2009). If similar perceptual groupings form for all sounds, prototypes would be expected to reflect the general sound qualities of the listener's physical environment, but also to eventually incorporate the conventions of auditory expression, for example, in popular culture.

Sound design involves a number of practices that would seem to find an explanation in fluency. For example, the construction of Foley (and replacing the actual sounds of events with something that conveys the action even better than the real thing) may reflect an effort to match the prototypical idea of a certain sound group. There have also been some attempts to document the low-level acoustic parameters that carry certain affective meaning. Kramer (1994) describes the following ways to add affective qualities to auditory displays: "Ugliness" is increased by moving from smooth to harsh by adding high non-harmonic partials. Decreases in "richness" are achieved by mutating from the full frequency spectrum to a sound with only highs and lows, and dissonance gives sound an "unsettling" quality. Audio interface design also employs a number of metaphorical associations that may, in some contexts, carry affective quality, typical examples are the use of louder, brighter, and faster sound to equal "more"; or higher pitch for "up," or "faster" (Kramer 1994, 214). Similar types of meaning-mapping has also been assigned to digital signal-processing effects, attributing meaning such as "larger" or "older" for reverb, and "futuristic" for delay (Collins 2009). Finally, an experiment by Kajastila and me explored whether fluency could be manipulated intentionally to shift affective judgments, demonstrating that merely the ease by which a sound source can be localized in a room is enough to influence affective quality (Ekman and Kajastila 2009).

12.11 Embodied Experience, Mirror Neurons, and Affective Mimicry

Another source for associated meaning resides in the embodied experience, which is triggered via the perception of sound as certain types of actions. Brain research has found that certain parts of the brain, so-called mirror neurons, are activated similarly when we perform actions and when we perceive others performing those same actions (e.g., Rizzolatti and Craighero 2004). The mirror system links hearing a sound to an abstract somatic representation of the physical actions involved with sound production. This direct link to the body invokes a powerful repertoire of experiential mappings, tying sound perception directly to our lived experiences (for a comprehensive treatment of the implications of embodied cognition for sound design, see Collins 2011). This type of embodied perception offers an explanation for certain strong aversive reactions to sounds. As tentatively suggested by Cox (2008b), the unpleasant sensation upon hearing nails on a chalkboard may be understood as originating in an audio-haptic activation. Interestingly, if sounds automatically trigger haptic knowledge, this extends the source for reinstated auditory experiences far beyond the auditory domain. Through mirroring, sound gains access not only to affective evaluations from prior sound experiences, but also to a history of haptic experiences.

Mirror neurons play an important role in how we perceive other people. Modern research has also confirmed people may "catch" the emotions of others, a phenomenon called affective mimicry. Demonstrably, vocal expressions are capable of conveying affect in such a way (Neumann and Strack 2000). In games, character sounds in particular offer ample material for catching emotions through affective mimicry. Moreover, Collins (2011) points to affective mimicry as the source for anthropomorphic effects, whereby inanimate objects are perceived to carry animate characteristics. Indeed, the auditory mirror system and affective mimicry have been used to explain many of the reactions humans have to different types of sound, for example, music (Molnar-Szakacs and Overy 2006). Juslin's "super-expressive voice theory" (2001) argues that the particular expressive quality of musical instruments with voice-like qualities (such as the violin) is that they remind us of the voice, but go far beyond what the human voice is capable of (in terms of speed, intensity, timbre). By exaggerating human emotional speech factors, these kinds of instruments create a kind of superpotent emotional speech.

12.12 Musical Emotions

Throughout this chapter, music has been grouped along with other types of sound. As we have seen, some of the above examples already mention musical emotions, such as when passages of music are interpreted as affective mimicry. Purely musical emotion has been linked to, among other ideas, expectancy (Huron 2007; Meyer 1961) and linguistic processing (Slevc 2011). Juslin and Västfjäll (2008, 563) detail a total of six psychological mechanisms involved in the musical induction of emotions: (1) brain-stem reflexes; (2) evaluative conditioning; (3) emotional contagion; (4) visual imagery; (5) episodic memory; and (6) musical expectancy. Of these, mechanisms 1–3 are primarily about top-down (associative) reinstatement, whereas 4 and 5 relate more strongly on bottom-up evaluation. Musical expectancy operates at both levels.

There is no denying that music is a potent source for emotion in games. But whereas Juslin and Västfjäll (2008) consider the experience of music in general, in games the framing of the music is bound to influence how it will be evaluated (that is, which of the mechanisms will be most prominent in determining the emotional outcome). Thus, in games, the presentation of sound in the role of a nondiegetic score is expected to invoke primarily unconscious evaluation of musical attributes. The affective qualities of unconscious processing may in turn inform the other processes, particularly narrative comprehension, by providing feeling-as-information (cf. Cohen 2001 on film sound). On the other hand, when used for earcons, music serves the purpose of signaling events or informing the player of altered game states. This functional role (providing feedback on action) will guide evaluations in different directions, and the emotional outcomes will reflect the overall utility of the sounds for playing. The processing of music where the player is producing or causing the music by their own playing (musical games, such as *Guitar Hero*) is probably dominated by evaluations pertaining to audio-tactile synchrony and agency, even emotional contagion (cf. Collins 2011). Finally, embedding music into games forges new symbolic linkages through player action and gives game music additional meaning by establishing the symbolization of events, for example through melodic phrasing (Whalen 2004). This symbolization allows the invocation of episodic memory and visual imagery and, in the long term, shapes musical expectancy through, for example, the constructing of genre-expectancy (Collins 2008).

12.13 Conclusions

One open question with immense implications for sound design is how predictable, and how reliably reproduced, is the emotional reaction to sound? Extant design knowledge would suggest certain sound solutions have predictable emotional consequences, but to what extent that holds true for all sound designs remains unclear. The theories presented above indicate some new research developments. Regarding bottom-up appraisal, we find that emotion is bound to the evaluation of a sound in the context of specific frames of reference: its narrative and functional fit. When predicting the emotional outcome of bottom-up evaluations, a simpler evaluation structure suggests more predictable sound behavior. Inversely, the capacity to excite complex emotions comes with increased representational and functional complexity. We can assume that part of this complexity arises from the increased number of possible evaluative frameworks imposed on the situation, adding more interpersonal variation. In particular, the complexity increases whenever sounds are evaluated simultaneously within two competing frameworks: the functional and the narrative. Another source of unpredictability is the direct result of the dual process of appraisal, whereby emotions may arise both by reinstatement and cognitive evaluation. Alongside the frameworks for bottom-up evaluation, sound automatically activates a number of prior emotional evaluations. This chapter covers several sources for such associations, tracing affective reactions to perceptual fluency, embodied cognition, and musical experience. If the combined effect of top-down and bottom-up processes determines the final outcome, how do we predict which response determines the end result? Interestingly, the feelings-as-information theory suggests that when there is a conflict between instinctual feelings and cognitive bottom-up evaluation, feeling-based information is generally considered more trustworthy (Clore and Ortony 2000, 39). If attention is low, there is a chance for misattributed feelings to influence bottom-up cognitive process and dictate which emotion "wins," so to speak. Apparently, however, in order for this process to occur, feelings must be perceived as salient to the evaluation. If people become aware of misattributions, they can discount their effect.

Examining the affective responses to sound types, however, Cox (2008b) finds that when identifiable, source dominates sound meaning: physical signal qualities take on significance only when the source event is not identifiable. Likewise, gauging the effect of different forms of reproduction, Västfjäll and his colleagues (2008) propose that when a sound carries symbolic affective quality, tweaking the affective content with low-level acoustic processes has relatively little effect on the emotional evaluation of the sound. However, in an experiment with musical emotions, Waterman (1996) asked participants to press a button whenever they felt moved, in any way, by the music. He found that, despite providing vastly different individual explanations for their reactions, participants nevertheless tended to indicate the same passages in music. This research aligns well with anecdotal evidence that film music really seems to function in quite predictable ways, in spite of the variations in viewers' personal musical preferences.

To summarize, simple games provide predictable frameworks, whereas more complex game structures will be harder to predict. But as structure-based prediction becomes harder, the likelihood of players relying on structure-based evaluation for their spontaneous emotional evaluations is also lower. And in perception, unattended stimuli also tend to be processed in simple ways. This analysis suggests three things: First, game sound is not a single phenomenon. The different sound roles within a game greatly inform the emotional evaluation process, and these functions must be taken into account when examining the affective quality of sounds. Second, game sound researchers should take deliberate caution not to compromise the natural pattern of player attention in games. Since emotional judgments depend, in part, on unconscious processes, inadvertently turning the player's attention to these processes might compromise the validity of the findings. Third, it is predicted that structurally simple sounds (that don't allow much variation in interpretation) and unconsciously processed sounds behave in the most predictable manner. This makes them a particularly good starting point for taking up the systematic research into game sound emotions.

References

Altman, Rick. 1992. *Sound Theory, Sound Practice.* New York: Psychology Press.

Anderson, Joseph D. 1998. *The Reality of Illusion: An Ecological Approach to Cognitive Film Theory.* Carbondale: Southern Illinois University Press.

Burr, David, and David Alais. 2006. Combining Visual and Auditory Information. Progress in Brain Research 155: 243–258.

Chion, Michel. 1994. *Audio-vision: Sound on Screen. Translated by Claudia Gorbman.* New York: Columbia University Press.

Clore, Gerald L., and Andrew Ortony. 2000. Cognition in Emotion: Always, Sometimes, or Never. In Cognitive Neuroscience of Emotion, eds. Richard D. Lane and Lynn Nadel, 24–61. New York: Oxford University Press.

Cohen, Annabel. 2001. Music as the Source of Emotion in Film. In Music and Emotion, ed. Patrick Juslin and John A. Sloboda, 249–272. New York: Oxford University Press.

Collins, Karen. 2007. An Introduction to the Participatory and Non-linear Aspects of Video Games Audio. In Essays on Sound and Vision, ed. Stan Hawkins and John Richardson, 263–298. Helsinki: Helsinki University Press.

——. 2008. *Game Sound: An Introduction to the History, Theory, and Practice of Video Game Music and Sound Design.* Cambridge, MA: MIT Press.

——. 2009. Generating Meaningful Sound: Quantifying the Affective Attributes of Sound Effects for Real-time Sound Synthesis in Audio-visual Media. Proceedings of the 35th AES International Conference on Audio for Games. New York: Audio Engineering Society.

——. 2011. Making Gamers Cry: Mirror Neurons and Embodied Interaction with Game Sound. *Proceedings of the AudioMostly Conference*, 39–46. Coimbra, Portugal.

Cox, Trevor. 2008a. The Effect of Visual Stimuli on the Horribleness of Awful Sounds. Applied Acoustics 69: 691–703.

——. 2008b. Scraping Sounds and Disgusting Noises. Applied Acoustics 69: 1195–1204.

Damasio, Antonio. 2005. *Descartes' Error: Emotion, Reason, and the Human Brain.* London: Penguin.

Ekman, Inger. 2005. Meaningful Noise: Understanding Sound Effects in Computer Games. Paper presented at Digital Arts and Cultures, Kopenhagen.

——. 2008. Psychologically Motivated Techniques for Emotional Sound in Computer Games. Proceedings of the AudioMostly Conference, 20–26. Piteå, Sweden.

——. 2009. Modelling the Emotional Listener: Making Psychological Processes Audible. Proceedings of the Audio Mostly Conference, 33–40, Glasgow, UK.

Ekman, Inger, and Raine Kajastila 2009. Localisation Cues Affect Emotional Judgements: Results from a User Study on Scary Sound. Proceedings of the 35th AES Conference on Audio for Games, February 2009, London. CD-ROM.

Ekman, Inger, and Petri Lankoski. 2009. Hair-raising Entertainment: Emotions, Sound, and Structure in Silent Hill 2 and Fatal Frame. In Horror Video Games: Essays on the Fusion of Fear and Play, ed. Bernard Perron, 181–99. Jefferson, NC: McFarland.

Ermi, Laura, and Frans Mäyrä. 2005. Fundamental Components of the Gameplay Experience: Analysing Immersion. In *Proceedings of Chancing Views – Worlds in Play*, ed. Suzanne de Castell and Jennifer Jenson, 15–27. Vancouver: DiGRA and Simon Fraser University.

Grimshaw, Mark. 2007. The acoustic ecology of the first person shooter. PhD diss., University of Waikato, New Zealand.

Grimshaw, Mark. 2009. The Audio Uncanny Valley: Sound, Fear and the Horror Game. Proceedings of the AudioMostly Conference, 21–26, 2009, Glasgow.

Hug, Daniel. 2011. New Wine in New Skins: Sketching the Future of Game Sound Design. In Game Sound Technology and Player Interaction, ed. Mark Grimshaw, 384–415. Hershey, PA: Information Science Reference.

Huron, David. 2007. *Sweet Anticipation: Music and the Psychology of Expectation*. Cambridge, MA: MIT press.

Jørgensen, Kristine. 2007. What are Those Grunts and Growls Over There? Computer Game Audio and Player Action. PhD diss., Copenhagen University, Denmark.

——. 2008. Audio and Gameplay: An Analysis of PvP Battlegrounds in World of Warcraft. Gamestudies 8 (2). http://gamestudies.org/0802/articles/jorgensen.

Juslin, Patrick N. 2001. Communicating emotion in music performance: A review and a theoretical framework. In *Music and Emotion*, ed. Patrik Juslin and John Sloboda, 309–337. New York: Oxford University Press.

Juslin, Patrick N., and Petri Laukka. 2004. Expression, Perception, and Induction of Musical Emotions: A Review and a Questionnaire Study of Everyday Listening. Journal of New Music Research 33 (3): 217–238.

Juslin, Patrick N., and Daniel Västfjäll. 2008. Emotional Responses to Music: The Need to Consider Underlying Mechanisms. Behavioral and Brain Sciences 31 (5): 559–575.

Kramer, Gregory. 1994. Some Organizing Principles for Representing Data with Sound. In *Auditory Display, Sonification, Audification and Auditory Interfaces*, ed. Gregory Kramer, 185–221. Reading, MA: Addison-Wesley.

Kromand, Daniel. 2008. Sound and the Diegesis in Survival-horror Games. Proceedings of the AudioMostly Conference, 16–19, Piteå, Sweden.

Lankoski, Petri. 2012. Computer Games and Emotions. In *The Philosophy of Computer Games*, ed. John Sageng, Hallvard Fossheim, and Tarjei M. Larsen, 39–55. London, New York: Springer.

Meyer, Leonard B. 1961. *Emotion and Meaning in Music*. Chicago: University Of Chicago Press.

Moffat, David, and Katarina Kiegler. 2006. Investigating the affects of music on emotions in games. Proceedings of the Audio Mostly Conference, 37–41. Piteå, Sweden.

Molnar-Szakacs, Istvan, and Katie Overy. 2006. Music and Mirror Neurons: From Motion to "emotion." Social Cognitive and Affective Neuroscience 1 (3): 235–241.

Nacke, Lennart E., Mark N. Grimshaw, and Craig A. Lindley. 2010. More Than a Feeling: Measurement of Sonic User Experience and Psychophysiology in a First-person Shooter Game. Interacting with Computers 22 (5): 336–343.

Neumann, Roland, and Fritz Strack. 2000. Mood Contagion: The Automatic Transfer of Mood between Persons. Journal of Personal and Social Psychology 79 (2): 211–223.

Perron, Bernard. 2005. A Cognitive Psychological Approach to Gameplay Emotions. Proceedings of the DiGRA 2005 Conference: Changing Views: Worlds in Play.

Reber, Rolf, Norbert Schwartz, and Piotr Winkielman. 2004. Processing Fluency and Aesthetic Pleasure: Is Beauty in the Perceiver's Processing Experience? Personality and Social Psychology Review 8(4): 364–382.

Rizzolatti, Giacomo, and Laila Craighero. 2004. The Mirror-neuron System. Annual Review of Neuroscience 27: 169–192.

Rocchesso, Davide, and Stefania Serafin. 2009. Sonic Interaction Design. International Journal of Human-Computer Studies 67 (11): 905–906.

Salminen, Nelli H., Hannu Tiitinen, and Patrick J. C. May. 2009. Modeling the Categorical Perception of Speech Sounds: A Step toward Biological Plausibility. Cognitive, Affective, and Behavioral Neuroscience 9 (3): 304–313.

Sanders, Richard D., and Mark A. Scorgie. 2002. *The Effect of Sound Delivery Methods on a User's Sense of Presence in a Virtual Environment.* MA thesis, Naval Postgraduate School, Monterey, CA. http://www.dtic.mil/dtic/tr/fulltext/u2/a403676.pdf.

Scherer, Klaus R. 2003. Vocal Communication of Emotion: A Review of Research Paradigms. Speech Communication 40 (1): 227–256.

Schwartz, Norbert. 2012. Feelings-as-information Theory. In Handbook of Theories of Social Psychology, vol. 1, ed. Paul A. Van Lange, Arie W. Kruglanski, and E. Tory Higgins, 289–308. Thousand Oaks, CA: Sage.

Schwartz, Norbert, and Gerald L. Clore. 1983. Mood, Misattribution, and Judgments of Well-being: Informative and Directive Functions of Affective States. Journal of Personality and Social Psychology 45 (3): 513–523.

Slevc, Robert. 2012 Language and Music: Sound, Structure, and Meaning. Wiley Interdisciplinary Reviews: Cognitive Science 3 (4): 483–492.

Tan, Ed. 1994. Film-induced Affect as a Witness Emotion. Poetics 23: 7–32.

Tan, Siu-Lan, Matthew P. Spackman and Matthew A. Bezdek. 2007 Viewers' Interpretations of Film Characters' Emotions. Music Perception 25: 135–152.

Västfjäll, Daniel. 2003. The Subjective Sense of Presence, Emotional Realism, and Experienced Emotions in Auditory Virtual Environments. CyberPsychology and Behavior 6: 181–8.

Västfjäll, Daniel, Erkin Asutay, Anders Genell, and Ana Tajadura. 2008. Form and Content in Emotional Reactions to Sounds. Journal of the Acoustical Society of America 123 (5): 3721.

Van Reekum, Carien, Tom Johnstone, Rainer Banse, Alexandre Etter, Thomas Wehrle, and Klaus Scherer. 2004. Psychophysiological Responses to Appraisal Dimensions in a Computer Game. Cognition and Emotion 18 (5): 663–688.

Vitouch, Oliver. 2001. When your Ear Sets the Stage: Musical Context Effects in Film Perception. Psychology of Music 29: 70–83.

Waterman, Mitch. 1996. Emotional Responses to Music: Implicit and Explicit Effects in Listeners and Performers. Psychology of Music 24: 53–67.

Weis, Elizabeth, and John Belton, eds. 1985. *Film Sound: Theory and Practice.* New York: Columbia University Press.

Whalen, Zach. 2004. Play Along: An Approach to Videogame Music. Game Studies 4 (1). http://www.gamestudies.org/0401/whalen/.

Whittington, William. 2007. *Sound Design and Science Fiction.* Austin: University of Texas Press.

Zajonc, Robert B. 1980. Feeling and Thinking: Preferences Need No Inferences. American Psychologist 35: 151–175.

——. 2011. Mere Exposure: A Gateway to the Subliminal. Current Directions in Psychological Science 10 (6): 224–228.

CHAPTER 13

THE SOUND OF BEING THERE

Presence and Interactive Audio in Immersive Virtual Reality

ROLF NORDAHL
AND NIELS C. NILSSON

In recent years the concept "presence"—often defined as the sensation of "being there"—has received increasing attention from scholars belonging to a variety of different disciplines. Lombard and Jones (2007), for instance, reveal that over 1800 journal articles, books, and other works on the topic have been published since 1930. Notably, more than 1400 of these texts were published within the last fifteen years (Bracken and Skalski 2010). Many of the authors are proponents of the view that works of literary fiction may give rise to mental representations of space similar or identical to the sensation of presence (e.g., Ryan 2001). However, Neuendorf and Lieberman (2010) present the argument that cinema was the original medium of presence, since it was able to photographically represent events unfolding in time and space. To this, Neuendorf and Lieberman (2010) add that since its origin, film has been a medium striving to elicit ever-stronger sensations of presence on behalf of its audiences—a view that they believe to be endorsed by film makers, scholars, critics, and audience. On a similar note, Tamborini and Bowman (2010) argue that the vividness and interactivity of computer games make these qualify as an ideal presence-inducing medium. Indeed, they argue that presence must be regarded as central if we are to understand how players use and experience videogames. More generally, Hartmann, Klimmt, and Vorderer (2010) present the argument that presence and entertainment may be connected, or at least coinciding. In order for a user to feel entertained or a sensation of presence, the user needs to believe in the mediated reality (Klimmt and Vorderer 2003; Green, Garst, and Brock 2004). However, the causal relationship between presence and entertainment is not considered an established fact. Hartmann, Klimmt, and Vorderer (2010) describe that presence may amplify the user's experience of entertainment; or, conversely, that the state of mind accompanying entertaining experiences may positively influence the sensation of presence. It should be stressed that it is far from all who believe that the sensation of presence necessarily entails an entertaining experience, or vice versa (e.g., Slater 2004).

Many scholars believe the concept of presence to be of relevance in relation to media entertainment, but it also has many applications outside this domain. Historically, presence has primarily been studied by computer scientists and scholars developing and evaluating immersive virtual reality (Bracken and Skalski 2010). In line with the recommendations of Frederick P. Brooks Jr. during his contribution to the IEEE Virtual Reality 2010 conference panel discussion on the nature of virtual reality (Jacobson et al. 2010), we distinguish between the concepts Virtual Reality (VR) and Immersive Virtual Reality (IVR). We use VR in a vein similar to Blascovich and Bailenson (2011) when referring to any form of mediated reality, including but not limited to oral and written storytelling, representational paintings, sculptures, theatre, photographs, and film. On the other hand, we reserve IVR to describe systems relying on high-fidelity tracking and displays in order to facilitate natural perception and interaction within a computer-generated environment. While IVR may be entertaining in its own right, it also has a range of more serious potential applications. These include psychological research (Loomis, Blascovich, and Beall 1999), treatment of phobias (Bouchard et al. 2006), rehabilitation (Rose, Brooks, and Rizzo 2005), and training and education of individuals who perform real-world tasks that are dangerous, costly, or difficult to practice due to other real-world constraints (Psotka 1995). Notably, it would appear that IVR is valuable largely due to its capacity for making individuals feel and act as if they are in the simulated environment. To exemplify, IVR may be a great tool for training individuals to perform potentially hazardous tasks for at least the following three reasons: (1) The user may be exposed to a potentially dangerous scenario without facing any actual danger; (2) since a user engaged in some virtual scenario is able to perform actions that are similar or identical to their real world counterparts, the acquired skills may be more or less directly transferred to the real world scenario; and (3) the reverse may be true since users are able to rely on their knowledge of physical reality and therefore do not need to acquire a new skill set, such as learning how to use the interface. Finally, IVR may also be used to simulate hazardous events that are impossible to recreate in reality due to their sheer scale. Such events include natural disasters and mass biological or chemical attacks on cities.

The study of presence in IVR in the past has been dominated by a focus on the influence of visual stimuli. The significance ascribed to this modality can presumably be ascribed to the fact that vision is regarded as dominant for spatial localization (Radeau 1994) and the popular belief that generally assumes that vision governs human experience (Schifferstein 2006). This focus notwithstanding, the importance of multisensory stimulation has long been acknowledged within the presence community (e.g., Steuer 1992). According to Larsson and colleagues (2010), the auditory modality possesses unique features that may make it a deciding factor in achieving a full sense of presence. Unlike its visual counterpart, auditory perception is always "turned on," since we cannot "shut our ears." Thus, this sensory channel always flows with information about the surrounding environment, regardless of whether we are attentive to this information or not (Gilkey and Weisenberger 1995). Visual perception may be superior in terms of spatial resolution, but it is inherently directional. Our limited field of view entails that we have to turn our heads or bodies in order to perceive the surrounding environment. Auditory perception on the other hand, is omnidirectional (Pope and Chalmers 1999). Moreover, Larsson and his colleagues (2010) highlight that auditory cues are inherently temporal in nature—a sounding event is by definition an unfolding event. In sum, it appears that auditory displays constitute a relatively inexpensive and valuable (if not necessary) component of VR and IVR systems intended to represent multimodal virtual spaces and elicit a sensation of presence in these spaces.

In this chapter we present a review of past and present theories of presence and describe how auditory stimuli may be used to elicit this perceptual illusion of "being there" in a virtual environment. The remainder of the chapter is organized in seven sections. Section 13.1 serves as an introduction to the comprehensive topic of presence and outlines Lombard and Ditton's seminal taxonomy of presence. Section 13.2 details what one might consider the most recent significant development within presence theory, namely Slater's conceptual framework for describing why individuals respond realistically to IVR. Taking Slater's conceptual framework as our point of departure, the following four sections illustrate how sound production and perception relate to the four concepts forming the basis for the framework. That is immersion, illusions of place, illusions of plausibility, and body ownership. Finally, the conclusion summarizes the discussions detailed throughout the chapter.

13.1 At the Heart of it All

The concept of presence has not exclusively been used to describe the sensation of "being there" in some fictional or real location. Based on a literature review of different conceptualizations of presence, Lombard and Ditton define presence as "the perceptual illusion of nonmediation" (Lombard and Ditton 1997). That is to say, presence is the illusion occurring when an individual erroneously takes something mediated as real and responds accordingly. Notably, the illusion is not the result of some mental defect. Despite giving in to the illusion, the individual is consciously aware that the mediated stimuli are not real. According to the two authors, this definition is broad enough to include the various existing conceptualizations of presence. Lombard and Ditton have summarized these conceptualizations in their now seminal taxonomy of presence. This taxonomy includes six different, albeit interrelated, conceptualizations of presence: presence as social richness, presence as realism, presence as transportation, presence as immersion, presence as social actor within a medium, and presence as medium as social actor.

13.1.1 Presence as Social Richness

According to Lombard and Ditton, presence as social richness is defined by the extent to which individuals engaged in some form of mutual interaction, find the medium facilitating the interaction sociable, warm, sensitive, and personal. Thus, presence as social richness relates to a medium's ability to produce a sense of intimacy and immediacy during acts of interpersonal communication.

13.1.2 Presence as Realism

The second conceptualization of presence identified by Lombard and Ditton is contingent upon the user perceiving the virtual environment and the characters inhabiting it as realistic. Lombard and Ditton distinguish between two forms of realism that may contribute to the experience of presence, when perceived in isolation or in concert, namely social and perceptual realism. Social realism refers to "the extent to which a media portrayal is plausible or 'true to life' as it reflects events that do or could occur in the nonmediated world," while perceptual realism refers to the extent to which mediated artifacts appear like their real world counterparts (Lombard and Ditton 1997).

13.1.3 Presence as Transportation

The conceptualization of presence as transportation relates to perceptual illusions involving spatial repositioning of real or virtual objects. Lombard and Ditton's taxonomy includes three different types of presence as transportation: (1) "You are there" involves the feeling of being transported to some other location and has also been referred to as telepresence (Minsky 1980), virtual presence (Sheridan 1992), or physical presence (Biocca 1997; IJsselsteijn 2000). (2) "It is here" involves transportation of virtual or real objects and environments to the user and is related to the notion of object presence (Stevens and Jerrams-Smith 2001). (3) "We are together" is used to describe how two or more users may experience the sensation of being transported to some shared location. The latter is sometimes referred to as copresence (Zhao 2003; Mühlbach, Bocker, and Prussog 1995).

13.1.4 Presence as Immersion

Lombard and Ditton explain that presence sometimes is regarded as a product of user immersion. Notably, it is possible to distinguish between two forms of immersion; that is, perceptual and psychological immersion. Perceptual immersion is achieved by substituting for the stimuli originating within the real world artificial stimuli through head-mounted displays, spatialized sound systems, haptic gloves and shoes, and similar technological innovations. Immersion may, as suggested, also be described as a psychological phenomenon (e.g., McMahan 2003; Witmer and Singer 1998). Lombard and Ditton say that immersive presence may be dependent upon some form of attentional surrender on the part of the user. Presence as psychological immersion is thus measurable presence based on the amount of attention allocated to the virtual environment as opposed to events in the real world (Van Baren and IJsselsteijn 2004; Nordahl and Korsgaard 2008).

13.1.5 Presence as Social Actor within a Medium

The concept of presence may also be related to an individual's responses to characters that are obviously mediated, such as news anchors or virtual pets. Despite the conspicuousness of the mediation, it is possible that a "users' perceptions and the resulting psychological processes lead them to illogically overlook the mediated or even artificial nature of an entity within a medium and attempt to interact with it" (Lombard and Ditton 1997).

13.1.6 Presence as Medium as Social Actor

Drawing on the work of Nass and others (e.g., Nass, Steuer, and Tauber 1994), Lombard and Ditton state that because computers use natural language, interact in real time, and fill traditionally social roles (e.g., bank teller and teacher), even experienced computer users tend to respond to them as social entities (Lombard and Ditton 1997). Thus users may respond to the medium itself almost as they would to another human being (e.g., a user who has been misdirected by a satellite navigation system, may respond by scolding it, despite knowing that this is pointless).

13.2 Presence in Immersive Virtual Reality

Of the six conceptualizations featured in Lombard and Ditton's taxonomy, presence as transportation is the one most frequently used to describe the sensation of "being there" in immersive virtual environments. In the introduction the value of IVR was described as largely stemming from its ability to make individuals feel and act as if they really were in the virtual environment. Indeed, Slater and colleagues have defined presence as the phenomena occurring when individuals respond to virtual stimuli in the same way as they would if they were exposed to equivalent real-world stimuli (Slater et al. 2009). More specifically, this response should be similar on every level "from unconscious physiological behaviors, through automatic reactions, conscious volitional behaviors, through to cognitive processing—including the sense of being there" (Sanchez-Vives and Slater 2005). While Slater and colleagues have not abandoned this view altogether, they have refined the theory of users' responses to IVR in a manner suggesting that presence is not the sole factor that determines whether an individual responds realistically to virtual stimuli (Slater 2009, 3550). Slater (2009) presents the hypothesis that this response-as-if-real (RAIR) can be ascribed to the simultaneous occurrence of not one, but two perceptual illusions; namely, the place illusion (PI: the illusion that you are really there) and the plausibility illusion (Psi: the illusion that the unfolding events are really happening). Combined with notions of immersion and a virtual body, PI and Psi make up a conceptual framework for explaining how IVR potentially can transform our experience of space and ourselves (Slater 2009).

13.2.1 System Immersion

Slater and others use the term immersion to describe the system delivering the stimuli. Thus, immersion is an objectively measurable quantity defined by the extent to which the system is able to track the actions of the users and provide appropriate feedback in as many modalities as possible (e.g., Slater 2009). A principal factor in determining the immersiveness of a system is the range of facilitated sensorimotor contingencies (SCs). Based on the work of O'Regan and Nöe (2001), Slater provides the following description of sensorimotor contingencies: "SCs refer to the actions that we know to carry out in order to perceive, for example, moving your head and eyes to change gaze direction, or bending down and shifting head and gaze direction in order to see underneath something" (Slater 2009). As Slater has done elsewhere (Slater 1999), the term System Immersion is used to make clear that we are not referring any of the many conceptualizations of immersion as a psychological phenomenon (see Section 13.4)

13.2.2 The Place Illusion

According to Slater (2009), place illusion (PI) is tantamount to the subjective sensation of presence, that is, the qualia[1] of "being there" despite knowing that one really is not. When clarifying how PI relates to system immersion, Slater (2003) eloquently uses the metaphor of the relationship between the wavelength distribution of light and the perception of color. Just as a color can be objectively described based on its wavelength distribution, so too immersion can be described based on objective properties such as frame rate, fidelity of tracking, or size of the field of view. Even though wavelength distribution and immersion are objectively describable, they both lead to subjective experiences, namely perceived color and PI. Thus PI may be described as the human response to immersion (Slater 2003). In terms of sensorimotor contingencies (SCs) this means that PI "occurs as a function of the range of normal SCs that are possible" (Slater 2009). It is entirely possible for PI to differ from one individual to another, even if the two are exposed to identical systems. To exemplify, one person might test the limits of the system, say, by inspecting parts of the environment more carefully than the other. If the resolution of the displays cannot cope with such close inspection, then PI might be broken for the curious individual, while remaining intact for the other.

13.2.3 The Plausibility Illusion

Unlike PI, the plausibility illusion (Psi) is not the direct result of an individual's ability to perceive the virtual environment. Instead, this perceptual illusion arises as a result of what the individual perceives within this environment. More specifically, Psi occurs when the unfolding events are experienced as really occurring, despite the sure knowledge that they are not (Slater 2009). Rovira et al. (2009) describe that Psi may be dependent on the IVR, meeting at least the following three conditions: (1) the actions performed by the user have to produce correlated reactions within the virtual environment (e.g., a virtual character might avoid eye contact and step aside if the user stares and exhibits aggressive body language; Rovira et al. 2009); (2) the environment should respond directly to the user, even when the user is not performing an instigating action (e.g., a virtual character might react to the presence of the user without the user initially approaching or addressing this character; Rovira et al. 2009, 3); and (3) the environment and the events occurring within it should be credible, that is, they should conform to the users' knowledge and expectations accrued through a lifetime of real-world interactions (Rovira et al. 2009). Notably, it would appear that the system has to meet the users' expectations to everything from the laws of physics to social norms and conventions. While not necessarily identical, Psi does have some commonalities with presence as realism (see below). In order for the illusion to occur, it is required that the "media portrayal is plausible or true to life in that it reflects events that do or could occur in the nonmediated world" (Lombard and Ditton 1997).

13.2.4 The Virtual Body

Slater (2009) describes the body as "a focal point where PI and Psi are fused." During our interaction with physical reality we are continuously provided with information about our bodies through sight, hearing, and other sensory modalities. Slater (2009) argues that this ability to perceive ourselves serves as a strong confirmation of PI. That is to say, if we are able to perceive our body, then we must be there. The ability to provide users of IVR with a credible virtual body is therefore central to eliciting PI in IVR. Indeed Slater (2009) suggests that a correlation between the proprioception of one's real body and the visual representation of the virtual body may lead to a compelling sensation of ownership over the latter. However, it is important to recall that the ability to perceive our body within IVR, unlike in real life, is anything but a matter of course as it requires high-fidelity tracking and multimodal stimulation.

In summary, Slater (2009) conceives that individuals exposed to IVR will exhibit a response-as-if-real (RAIR) if they feel that the depicted events are really happening to them, despite the sure knowledge that they are not. This experience will emerge as a consequence of two illusions, namely the individual feels that he or she is there in the environment (PI) and the occurring events are indeed really happening (Psi). The former is a direct response to the level of immersion, and both illusions influence the sensation of (virtual) body ownership on behalf of the individual. Taking Slater's conceptual framework as our point of departure, the remainder of the chapter will illustrate how sound production and perception relate to immersion, PI, Psi, and body ownership, and thus contribute to users responding-as-if-real.

13.3 Auditory Immersion

System immersion, outlined above, is dependent upon the extent and fidelity of the displays delivering sensory stimuli. Tracking of users' actions can be achieved by means of an array of different technologies, ranging from sophisticated and costly digital optical motion capture systems such as the Vicon MX to consumer-level systems like the Microsoft Kinect. However, tracking is not modality specific. That is, both the global position and orientation of the user and the local positions and orientations of individual body parts may be used to control stimuli delivered in any and all modalities. We therefore restrict the current discussion of system immersion to auditory displays (for more information about tracking, see Stanney 2002).

Larsson and his colleagues (2010, 147) note that the spatial properties of auditory environments have been assigned importance since the first stereo systems were constructed during the 1930s. The authors go on to point out that the aim of spatial sound rendering "is to create an impression of a sound environment surrounding a listener in the 3D space, thus simulating auditory reality" (Larsson et al. 2010). Thus it would seem that research on spatial sound rendering and IVR share the common goal of producing illusions of place and plausibility. According to Larsson et al. (2010, 146), it is possible to distinguish between two different types of delivery methods for spatial audio, namely sound field-related methods (Rumsey 2001) and head-related methods (Begault 1994).

13.3.1 Sound Field-related Delivery Methods

Field-related methods rely on multichannel loudspeaker audio reproduction systems to create a sound field within which the sound is spatialized in a natural manner. The number of loudspeakers determines the size of this area, which also is referred to as the "sweet spot" (Larsson et al. 2010). When describing how such systems facilitate sound spatialization, Shinn-Cunningham and Shilling (2002) explain that the total acoustic signal arriving at each ear at any given moment is defined simply by the sum of the signals originating from the individual sound sources in the environment. Thus, by varying the properties of the signal produced from each speaker in the array, it possible to influence spatial auditory cues. These include binaural cues such as interaural time differences and interaural intensity differences, and anechoic distance cues such as the spectrum of the sound (Shinn-Cunningham and Shilling 2002). However, this process is far from easy because field-based methods do not allow the signals arriving at each ear to be manipulated completely independently of one another. Speaker placement and room acoustics are therefore essential if one wishes to use this type of method (Shinn-Cunningham and Shilling 2002). Surround systems have become standard in both home and movie theaters. Larsson's group (2010) note that it is possible to spatialize sounds even more naturally as the number of channels increases and with the use of more sophisticated spatial rendering methods, such as ambisonics (Gerzon 1985), vector-based amplitude panning (Pulkki 1997), and wave field synthesis (Horbach et al. 2002). Application of such methods in relation to IVR includes the use of vector-based amplitude panning to produce the soundscape for a virtual version of the Prague Botanical Garden (Nordahl 2006), and the use of ambisonics to render the dynamic soundscape accompanying the experience of being on a wooden platform overlooking a canyon, a river, and a waterfall (Nordahl et al. 2011).

13.3.2 Head-related Audio Rendering Methods

Contrary to field-based methods, head-related audio rendering systems, or binaural systems, make it possible to completely control the sound arriving at each ear typically through the use of headphones that isolate the signal intended for each ear, thus limiting any crosstalk (Larsson et al. 2010). In addition to offering more precise control of binaural cues, unwanted sounds such as echoes and reverberation are kept from reaching the ears of the listener. However, this reduction of environmental cues comes at a price, since the headphones may be experienced as intrusive by the user (Shinn-Cunningham and Shilling 2002). Shinn-Cunningham and Shilling distinguish three types of headphone simulation, namely diotic displays, dichotic displays, and spatialized audio. The first simply refers to the display of identical signals in both channels. This may lead to so-called "inside the head localization" (Plenge 1974), since the listener gets the sensation that all sound sources are located inside the head (Shinn-Cunningham and Shilling 2002), a phenomenon referred to as "lateralization" (Plenge 1974). Second, Shinn-Cunningham and Shilling refer to stereo signals that contain only frequency-dependent interaural intensity or time differences as dichotic displays. They describe this type of display as very simple since the effect can be achieved by scaling and delaying the signal arriving at each ear. Just as with diotic displays, this display does not enable proper spatialization of the sound sources since listeners may feel that the sounds are moving inside the head from one ear to the other. Finally, spatialized sound makes it possible to render most of the spatial cues available in the real world. This is achieved through filtering of the sound signal and thereby transforming it so as to mimic an acoustic signal that has interacted with the torso, head, and outer ears of the listener (Shinn-Cunningham and Shilling 2002; Larsson et al. 2010). This transformation is achieved through so-called head-related transfer functions (HTRFs). Ideally an HTRF unique to the listener should be used, but since this situation is very impractical, generalized HTRFs are often used (Larsson et al. 2010).

13.4 Auditory Illusions of Place

So far, we have introduced some of the technology that may be used to immerse users aurally in IVR. Above, it was suggested that the place illusion, PI, by and large may be regarded as the human response to immersion. Throughout the next section we will present existing research pertaining to the influence of sound on PI. Please note that in the current section we use the terms PI and presence interchangeably in order to stay true to the works cited. Despite the scarcity of research on sounds and PI, Larsson and his colleagues (2010) describe work belonging to four categories of auditory factors believed to influence presence: spatial properties of the sound, the auditory background, consistency within and across modalities, and quality and contents.

13.4.1 Spatial Properties

Since the spatial acuity of the auditory modality is inferior to both vision and proprioception (Shinn-Cunningham and Shilling 2002), one might think that it is insignificant in regards to PI, which is an inherently spatial illusion. However, even though spatial hearing lacks the precision of vision and proprioception, it is far from insignificant for our perception of the surrounding environment (see also Chapter 26 in this volume). Indeed from an evolutionary perspective, one of the oldest and most basic functions of hearing was to alert the listening organism. The ability to hear and localize potential predators and prey before these enter the organism's field of view must be considered a competitive advantage (Hermann and Ritter 2004). In addition to providing information about the environment beyond our field of view, sound also influences perception of visible and tangible events and objects. The ventriloquism effect makes up one example of how stimuli in one modality may influence spatial percepts in another. So despite the limits to acuity, spatial hearing is crucial to how we perceive space. Larsson and colleagues (2010) present empirical evidence suggesting that the spatial properties of sound positively influence PI. Hendrix and Barfield (1996) describe two studies performed with the intention of investigating how spatialized sound influences the sensation of presence. In one study they compared silent virtual environments to environments including spatialized sound, and in the second study they compared environments including auditory cues that were either spatialized or not. The results indicate that participants deprived of auditory stimuli are less likely to experience a sensation of presence and that spatialized sounds are regarded as more realistic and are perceived as originating from sources within the environment. Moreover, both room acoustic cues and binaural simulation may positively influence the sensation of presence (Larsson, Västjäll, and Kleiner 2003, 2008). To be more exact, one study showed that in virtual environments devoid of visuals, the presentation of room acoustic cues was superior to anechoic representations (Larsson, Västjäll, and Kleiner 2008). The second study revealed that an audiovisual virtual environment including binaural simulation elicited significantly stronger sensations of presence compared to environments including stereo sound reproduction. Both environments included room acoustic cues (Larsson, Västjäll, and Kleiner 2003).

13.4.2 Auditory Background

In the introduction it was suggested that one of the reasons why the auditory modality may be crucial for the sensation of presence is that it is never "turned off." Interestingly, Larsson and colleagues (2010) describe a relatable property of the auditory environment that may influence the sensation of presence, namely the so-called auditory background. The auditory background may be understood as the continuous stream of auditory information reaching our ears, thus forming the auditory backdrop to the percepts we are presently attending to. This backdrop may include sounds such as the ticking of a clock in the far corner of the room, leaves rustling in the wind, or the sound of our own and other's footsteps (Larsson et al. 2010; Ramsdell 1978). The previously presented study suggesting that silence negatively influences presence (Hendrix and Barfield 1996) arguably lends some credence to the claim that the auditory background has a positive influence. Notably, Murray and others (2000) report findings of a number of experiments involving individuals deprived of auditory stimuli through the use of earphones. These individuals performed a series of familiar (and real) tasks. Their experience of the sensory deprivation was subsequently assessed through self-reports. The results indicated that the auditory background is important for environmentally anchored presence, that is, the sensation of being part of the environment (Murray, Arnold, and Thornton 2000). Here it is interesting to note that complete silence within classical filmmaking is by and large considered to be a critical problem (Figgis 2003).

13.4.3 Quality and Contents

The third category of auditory factors that may influence presence is quality and contents (Larsson et al. 2010). Ozawa and others (2003) exposed participants to binaural representations of ecological sounds with the intention of determining how self-reported presence ratings were affected by sound quality, information, and localization. The last two appeared to be the most influential. Moreover, Larsson and colleagues (2010) report findings from studies indicating that changes to the sound pressure level might influence presence. One study indicated that the addition of more bass content to a rally car video sequence accompanied by synchronized audio increased the sensation of presence (Freeman and Lessiter 2001). Similarly, Ozawa and Miyasaka (2004) demonstrated that a scenario where sound comparable to that heard when inside a car, yielded higher presence when sound pressure level was at its highest. This may be seen as an indication that the higher sound pressure levels produced the sensation that the virtual car was vibrating (Larsson et al. 2010). Without disputing the validity of these findings, one cautionary note should be added: Following the view of presence outlined above, a clear distinction between form and content should be made with respect to what factors influence presence. Presence is the product of media form rather than content. That does not imply that content as it has been conceptualized above is not influential. Instead, it implies that PI has nothing to do with whether the user finds the unfolding events interesting or emotionally evocative. To borrow an example from Slater (2003), imagine listening to a live recording of a piece of classical music through an immersive auditory display. You may get a compelling sensation of "being there" in the concert hall, even if you are not interested in classical music and find the particular piece to be devoid of any emotional appeal. With that being said, it does seem to be important whether the presented content matches the expectations generated by the visuals, that is, whether it is consistent with stimuli presented in other modalities (Chueng and Marsden 2002). The study described by Nordahl and colleagues (2012), lends itself as an interesting example since the authors compared whether the addition of audiohaptic simulation of foot-ground interaction influences perceived realism and presence. While no significant differences in presence were found, the addition of audiohaptic feedback did make the interaction seem more realistic to the participants (Nordahl et al. 2012).

13.4.4 Internal and Cross-modal Consistency

Human experience is inherently multimodal. We experience the world around us through several sensory channels and the concurrent presentation of congruent or incongruent information in two or more modalities may positively or negatively influence both perception and information processing (Kohlrausch and van de Par 1999). This multimodality cannot be ignored by anyone working with human-computer interaction, including the application within IVR (Pai 2005; Lederman and Klatzky 2001). Larsson and others (2010) note that consistency across the visual and auditory modality is a recurring theme within presence research. The factors believed to influence presence include: the consistency between the spatial qualities of the delivered stimuli (Larsson et al. 2007); the extent to which the audiovisual stimuli represent the same space (Ozawa et al. 2003); and the degree of congruence between visually induced expectations and presented sounds (Chueng and Marsden 2002). Storms and Zyda (2000) have performed a study suggesting that the quality of the stimuli in one modality might influence the perceived quality of the other. They compared visuals of varying quality displayed on screen with auditory feedback of varying quality played in headphones. The quality of the visual stimuli was varied by altering the pixel resolution, while the quality of the auditory stimuli was varied by altering the sampling frequency. Moreover Gaussian white noise levels were varied in the case of both stimuli types. The results confirm what may be regarded as recognized facts within both the entertainment industry and the VR community, namely, that the quality of an auditory display can influence the perceived quality of a visual display and vice versa (Storms and Zyda 2000). Finally, consistency within one modality may also influence experiences of IVR, including the sensation of presence (Larsson et al. 2010). Nordahl and others (2008) found that semantic consistency between auditory feedback and the auditory environment might influence recognition of the former. This concept was specifically discovered during the evaluation of their physics-based sound synthesis engine. The synthesized audio was the sound of footsteps on solid and aggregate surfaces, produced in real time, based on the ground reaction force exerted by the participants during the act of walking. The evaluation of the system indicated that the participants in some cases found it easier to recognize the simulated surface materials when this material was consistent with the presented auditory context (Nordahl and Korsgaard 2008). It has also been demonstrated that auditory feedback may be superior to haptic stimuli, in similar, albeit not identical, recognition tasks related to the footstep sounds (Nordahl and Korsgaard 2010). Finally, it appears that consistency within and across modalities also may influence auditory illusions of plausibility.

13.5 Auditory Illusions of Plausibility

While presence has been studied for decades, the conceptualization of Psi outlined above has been subjected to comparably less scrutiny. This development naturally also implies that little explicit effort has been made to investigate how sound might influence the illusion that unfolding events are really happening. However, sound is, unlike visual stimulus, inherently temporal: "while a visual scene real or virtual may be completely static, sound is by nature constantly ongoing and 'alive'; it tells us that something is happening" (Larsson et al. 2010). Thus it seems reasonable to assume that the auditory modality may play an important role in producing compelling illusions of plausibility. With that said, just because events appear to be unfolding, that does not mean that they are perceived as plausible. Consider the narratives of many films and computer games. While these narratives indeed are unfolding in time, the occurring events and actions performed by the fictional characters need not be perceived as plausible. IVR technology may also be used to simulate such implausible events and actions. For example, say that the immersed user has been cast as the protagonist of an action adventure. He may feel present within the fictional universe where the implausible events are occurring, but realistic responses on the part of the user are not certain and perhaps not even desirable. Recall that Psi is believed to be dependent upon the IVR fulfilling at least three criteria: The actions performed by the user have the entailed correlated reactions within the environment; the environments should respond directly to the user even if the user remains passive; and the environment and events should conform to the user's expectations. If the IVR includes other active agents (autonomous or controlled by another user) their actions and reactions will naturally need to be conveyed in a plausible manner. Thus it seems plausible that speech intelligibility and more subtle auditory cues like voice inflection might be of the utmost importance for making the interaction seem plausible. Moreover, since Psi is contingent upon the events of the VE conforming to the user's expectations, it seems likely that factors such as the degree of congruence between visually induced expectations and presented sounds (Chueng and Marsden 2002), might also be relevant in connection to Psi. Similarly, it seems probable that consistency within the auditory modality might influence Psi. The factors believed to influence presence include consistency between individual sounds and the general auditory context and correspondence between the spatialization and nature of a sound (Larsson et al. 2010) describe. Here it is worth referring to Ramsdell (1978) who introduces the concept psychological coupling. Psychological coupling refers to the phenomenon that occurs when an individual feels as if she is able to exert influence on the surrounding environment and thus take in the role of an active participant. Perception of the auditory environment is believed to influence this phenomenon (Ramsdell 1978).

13.6 Auditory-induced Body Ownership

Body ownership as described above may emerge as a consequence of a correlation between proprioception of the body proper and sight of the virtual body. While to the authors' knowledge there exists no research explicitly related to auditory cues and body ownership, the illusion need not result solely from integration of proprioceptive and visual stimuli. It has been demonstrated that the combination of visual and tactile stimulation may produce body ownership, and brain-computer interfaces have even been used to elicit weaker variations of the illusion (Slater et al. 2008). Notably, the previously mentioned study of the effects of wearing earplugs while performing everyday tasks (Murray, Arnold, and Thornton 2000) appears to provide some relevant insights. That is to say, the auditory deprivation experienced by the individuals participating in these studies simultaneously intensified self-awareness and caused detriment to the sensation of presence. Larsson and others (2010) imply that auditory self-representation may negatively influence the general sensation of PI. It should be noted that one should not view their findings as an indication that self-generated sounds always will be detrimental to PI, or for that matter, that body ownership and PI are somehow incompatible. Indeed, Nordahl (2005) reports the results of an experiment indicating that self-generated sounds resulting from interaction between the environment and the body may positively influence presence. The study in question compared the experiences of individuals exposed to an IVR including self-generated footstep sounds with one including no such auditory feedback. The results indicated that the condition including self-generated sounds facilitated significantly stronger sensations of presence. With that being said, it should be noted that it presumably also was a factor that one IVR was devoid of all auditory feedback as in the experiment described by Hendrix and Barfield (1996) (see below). Moreover, Nordahl (2006) describes a study that potentially may be of relevance. Previously it was considered a problem that individuals exposed to IVR do not exhibit much head or body movement. Nordahl proposed that this problem might be alleviated though the addition of auditory cues. In order to put this claim to the test, Nordahl performed an experiment investigating how different combinations of auditory feedback influenced the user movement and the sensation of presence. While no difference in presence was found, the results indicated that individuals will move more if exposed to IVR including a soundscape, spatialized moving sound sources, and an auditory self-representation, that is, footstep sounds (Nordahl 2006). Thus, it would seem that self-generated sounds—such as the sound of one's voice (Porschmann 2001)—may have a positive effect on natural behavior and PI, if these sounds are chosen and delivered in a manner corresponding to what the user would expect to encounter within the given virtual environment.

13.7 Conclusions

In this chapter we have introduced the concept of presence and the six different, yet interrelated, conceptualizations of presence proposed by Lombard and Ditton (1997): presence as social richness, presence as realism, presence as transportation, presence as immersion, presence as social actor within a medium, and presence as medium as social actor. Of these six conceptualizations, presence as transportation is most applicable when attempting to describe the sensation of "being there" accompanying exposure to IVR. The value of IVR largely comes from its ability to make individuals feel and behave as if they really were inside the virtual environment. While such responses generally have been viewed as a sign of presence, Slater (2009, 3554) has proposed that this response-as-if-real (RAIR) is the result of two perceptual illusions: "If you are there (PI) and what appears to be happening is really happening (Psi), then this is happening to you! Hence you are likely to respond as if it were real" (Slater 2009). Together with the concepts immersion and the virtual body, illusions PI and Psi make up a conceptual framework for describing how IVR may transform experiences of space and our selves.

Immersion is an objectively measurable property of a system. The afforded level of immersion depends on the degree to which the system is able to track the actions of the users and provide appropriate feedback in as many modalities as possible. Larsson and colleagues (2010) distinguish between two unique types of delivery methods for spatial audio, namely sound field-related (Rumsey 2001) and head-related methods (Begault 1994). The two, applied together with software used to generate the sound, define a unique way of immersing the user in sound.

PI is essentially the same as the subjective sensation of "being there" within an IVR. It was described that auditory stimuli may contribute to PI in several ways. According to Larsson and colleagues (2010), there exist at least four categories of auditory factors that may influence presence, namely the spatial properties of the sound, the auditory background, consistency within and across modalities, and the quality and contents. Considering that PI largely may be viewed as the human response to immersion, auditory PI relates to the maintaining of the sensory motor loop made up by human actions and perception on the one side and system tracking and displays on the other. The more that the auditory component of the sensorimotor loop is reminiscent of the one we would expect from our experiences with physical reality, the stronger the sensation of auditory PI.

Psi was described as the illusion arising when ongoing virtual events are experienced as really occurring, despite the sure knowledge that they are not. Considering that sound is inherently temporal—a sounding event is by definition a happening event—this modality may be of great importance to maintaining this illusion. During everyday life we are surrounded by a constant flow of auditory information—the auditory background—indicating that the environment is indeed "alive and breathing." However, as is the case with PI, it appears that the auditory stimuli need to conform to the knowledge and expectations of the user in order to elicit illusions of plausibility. Finally, it was noted that the compelling sensation of body ownership over the virtual body may arise if the user experiences a correlation between proprioception and one or more other modalities. While such illusions primarily have been elicited through visual stimuli, it seems plausible that the auditory and haptic modalities may produce similar illusions or at least intensify visually induced illusions. Indeed, it would seem that bodily interaction with IVR is inherently auditory and haptic, since it relies on different forms of physical contact, such as footsteps, which produce potentially audible vibrations.

In conclusion, it appears that auditory stimuli should be regarded as a necessary rather than simply a valuable component of IVR systems intended to make individuals respond-as-if-real through illusions of place and plausibility.

Note

1. Qualia can simply be understood as "the way things seem to us" (Dennett 1988).

References

Begault, Durand R. 1994. *3D-Sound for Virtual Reality and Multimedia*. Boston: AP Professional.

Biocca, Frank. 1997. The Cyborg's Dilemma: Embodiment in Virtual Environments. *Proceedings Second International Conference on Cognitive Technology Humanizing the Information Age*, 12–26. Washington, DC: IEEE Computer Society.

Blascovich, Jeremy and J.N. Bailenson. 2011. *Infinite Reality: Avatars, Eternal Life, new Worlds, and the Dawn of the Virtual Revolution*. New York: William Morrow.

Bouchard, Stéphane, Sophie Côté, Julie St-Jacques, Geneviève Robillard, and Patrice Renaud. 2006. Effectiveness of Virtual Reality Exposure in the Treatment of Arachnophobia Using 3D Games. *Technology and Health Care* 14 (1): 19–27.

Bracken, Cheryl Campanella, and Paul Skalski. 2010. *Immersed in Media: Telepresence in Everyday Life*. New York: Routledge.

Chueng, Priscilla, and Phil Marsden. 2002. Designing Auditory Spaces to Support Sense of Place: The Role of Expectation. In *Proceedings of the CSCW Workshop: The Role of Place in Shaping Virtual Community*, November 16, 2002, New Orleans, LA.

Dennett, Daniel C. 1988. Quining qualia. In *Consciousness in Contemporary Science*, ed. A. Marcel and E. Bisiach, 42–77. Oxford: Oxford University Press.

Figgis, Mike. 2003. Silence: The Absence of Sound. In *Soundscape: The School of Sound Lectures, 1998–2001*, ed. Larry Sider 1–14. New York: Columbia University Press.

Freeman, J., and J. Lessiter. 2001. Hear There and Everywhere: The Effects of Multi-channel Audio on Presence. In *Proceedings of the 2001 International Conference on Auditory Display*, July 29–August 1, 2001, Espoo, Finland, 231–234.

Gerzon, Michael A. 1985. Ambisonics in Multichannel Broadcasting and Video. *Journal of the Audio Engineering Society* 33 (11): 859–871.

Gilkey, Robert H., and Janet M. Weisenberger. 1995. The Sense of Presence for the Suddenly Deafened Adult: Implications for virtual environments. *Presence: Teleoperators and Virtual Environments* 4 (4): 357–363.

Green, Melanie C., Jennifer Garst, and Timothy C. Brock. 2004. The Power of Fiction: Determinants and Boundaries. In *The Psychology of Entertainment Media: Blurring the Lines Between Entertainment and Persuasion*, ed. L. J. Shrum, 161–176. Mahwah, NJ: Lawrence Erlbaum.

Hartmann, Tilo, Christoph Klimmt, and Peter Vorderer. 2010. Telepresence and Media Entertainment. In *Immersed in Media: Telepresence in Everyday Life*, ed. C Cheryl Campanella Bracken and Paul Skalski, 137–157. New York: Routledge.

Hendrix, Claudia Mary, and Woodrow Barfield. 1996. The Sense of Presence within Auditory Virtual Environments. *Presence: Teleoperators and Virtual Environments* 5 (3): 290–301.

Hermann, Thomas, and Helge Ritter. 2004. Sound and Meaning in Auditory Data Display. *Proceedings of the IEEE* 92 (4): 730–741.

Horbach, U., E. Corteel, R. Pellegrini, and E. Hulsebos. 2002. Real-time Rendering of Dynamic Scenes Using Wave Field Synthesis. In *Multimedia and Expo, 2002. ICME '02. Proceedings.* 1: 517–520.

IJsselsteijn, Wijnand A. 2000. Presence: concept, determinants, and measurement. *Proceedings of SPIE* 3959: 520–529.

Jacobson, Jeffrey, Chadwick A Wingrave, Doug Bowman, Frederick P. Brooks, Robert Jacob, Joseph J. LaViola, and Albert Rizzo. 2010. Reconceptualizing "Virtual Reality": What is VR? Statement of *Proceedings of the IEEE Virtual Reality 2010 Conference Panel.* https://sites.google.com/site/reconceptualizingvrprivate/public-discussion.

Klimmt, Christoph, and Peter Vorderer. 2003. Media Psychology "Is not yet There": Introducing Theories on Media Entertainment to the Presence Debate. *Presence: Teleoperators and Virtual Environments* 12 (4): 346–359.

Kohlrausch, A., and S. van de Par. 1999. Auditory-visual Interaction: From Fundamental Research in Cognitive Psychology to (Possible) Applications. In *Proceedings of SPIE*, 3644: 34.

Larsson, Pontus, Aleksander Väljamäe, Daniel Västfjäll, Ana Tajadura-Jiménez, and Mendel Kleiner. 2010. Auditory-induced Presence in Mixed Reality Environments and Related Technology. *Human-Computer Interaction* Series 1: 143–163.

Larsson, Pontus, Daniel Västjäll, and Mendel Kleiner. 2003. On the Quality of Experience: A Multi-modal Approach to Perceptual Ego-motion and Sensed Presence in Virtual Environments. *Proceedings of the First ITRW on Auditory Quality of Systems,* Akademie Mont-Cenis, Germany.

——. 2008. Effects of Auditory Information Consistency and Room Acoustic Cues on Presence in Virtual Environments. *Acoustical Science and Technology* 29 (2): 191–194.

Larsson, Pontus, Daniel Västfjäll, Pierre Olsson, and Mendel Kleiner. 2007. When what you see is what you hear: Auditory-visual integration and presence in virtual environments. In *Proceedings of the 10th Annual International Workshop on Presence,* October 25–27, 2007, Barcelona, Spain.

Lederman, S., and R. Klatzky. 2001. Designing Haptic and Multimodal Interfaces: A Cognitive Scientist's Perspective. In *Proceedings of the Collaborative Research Centre 453*, ed. G. Farber and J. Hoogen, 71–80. Munich: Technical University of Munich.

Lombard, Matthew, and Teresa Ditton. 1997. At the Heart of it All: The Concept of Presence. *Journal of Computer-mediated Communication* 3 (2): 20.

Lombard, Matthew, and Matthew T. Jones. 2007. Identifying the (Tele)presence Literature. *PsychNology Journal* 5 (2): 197–206.

Loomis, Jack, James Blascovich, and A. Beall. 1999. Immersive Virtual Environment Technology as a Basic Research Tool in Psychology. *Behavior Research Methods* 31 (4): 557–64.

McMahan, Alison. 2003. Immersion, Engagement and Presence. In *The Video Game Theory Reader*, ed. Mark J. P. Wolf and Bernard Perron, 67–86. New York: Routledge.

Minsky, Marvin. 1980. Telepresence. *Omni*, June 1980: 45–51 http://web.media.mit.edu/~minsky/papers/Telepresence.html.

Mühlbach, L., M. Bocker, and A. Prussog. 1995. Telepresence in Video Communications: A Study on Stereoscopy and Individual Eye Contact. *Human Factors* 37 (2): 290–305.

Murray, C. D., P. Arnold, and B. Thornton. 2000. Presence Accompanying Induced Hearing Loss: Implications for Immersive Virtual Environments. *Presence: Teleoperators and Virtual Environments* 9 (2): 137–148.

Nass, C., J. Steuer, and E. Tauber. 1994. Computers are Social Actors. In *Proceedings of the SIGCHI Conference on Human Factors in Computing Systems: Celebrating Interdependence*, 72–78: New York: ACM.

Neuendorf, Kimberly A., and Evan A. Lieberman. 2010. Film: The Original Immersive Medium. In *Immersed in Media: Telepresence in Everyday Life*, ed. Cheryl Campanella Bracken and Paul Skalski, 9–38: New York: Routledge.

Nordahl, Rolf. 2005. Self-induced Footsteps Sounds in Virtual Reality: Latency, Recognition, Quality and Presence. In *Proceedings of Presence 2005: The 8th Annual International Workshop on Presence*, ed. Mel Slater, 353–355. London: University College, London.

——. 2006. Increasing the Motion of Users in Photo-realistic Virtual Environments by Utilising Auditory Rendering of the Environment and Ego-motion. In *Proceedings of Presence 2006: The 9th Annual International Workshop on Presence*, ed. Cheryl Campanella and Matthew Lombard, 57–62.

Nordahl, Rolf, and Dannie Korsgaard. 2008. On the Use of Presence Measurements to Evaluate Computer Games. In *Proceedings of Presence 2008: The 11th Annual International Workshop on Presence*, ed. Anna Spagnolli and Luciano Gamberini, 174–177. Padua: Cooperativa Libraria Universitaria Padova.

——. 2010. Distraction as a Measure of Presence: Using Visual and Tactile Adjustable Distraction as a Measure to Determine Immersive Presence of Content in Mediated Environments. *Virtual Reality* 14 (1): 27–42.

Nordahl, Rolf, Stefania Serafin, Niels Nilsson, and Luca Turchet. 2012. Enhancing Realism in Virtual Environments by Simulating the Audio-haptic Sensation of Walking on Ground Surfaces. *Virtual Reality Short Papers and Posters*, 2012, IEEE, 73–74.

Nordahl, Rolf, Stefania Serafin, Luca Turchet, and Niels C. Nilsson. 2011. A Multimodal Architecture for Simulating Natural Interactive Walking in Virtual Environments. *PsychNology Journal* 9 (3): 245–268.

O'Regan, J. Kevin, and Alva Nöe. 2001. A Sensorimotor Account of Vision and Visual Consciousness. *Behavioral and Brain Sciences* 24 (5): 939–972.

Ozawa, Kenji, and Manabu Miyasaka. 2004. Effects of Reproduced Sound Pressure Levels on Auditory Presence. *Acoustical Science and Technology* 25 (3): 207–209.

Ozawa, Kenji, Yoshihiro Chujo, Yoiti Suzuki, and Toshio Sone. 2003. Psychological Factors Involved in Auditory Presence. *Acoustical Science and Technology* 24 (1): 42–44.

Pai, Dinesh K. 2005. Multisensory Interaction: Real and Virtual. *Robotics Research* 15: 489–498.

Plenge, G. 1974. On the Differences between Localization and Lateralization. *Journal of the Acoustical Society of America* 56: 944.

Pope, Jackson, and Alan Chalmers. 1999. Multi-sensory Rendering: Combining Graphics and Acoustics. *Proceedings of the 7th International Conference in Central Europe on Computer Graphics*, 233–242.

Porschmann, C. 2001. One's Own Voice in Auditory Virtual Environments. *Acustica* 87 (3): 378–388.

Psotka, Joe. 1995. Immersive Training Systems: Virtual Reality and Education and Training. *Instructional science* 23 (5): 405–431.

Pulkki, Ville. 1997. Virtual Sound Source Positioning Using Vector Base Amplitude Panning. *Journal of the Audio Engineering Society* 45 (6): 456–466.

Radeau, M. 1994. Auditory-visual Spatial Interaction and Modularity. *Current Psychology of Cognition*, 13(1), 3-51.

Ramsdell, Donald. A. 1978. The Psychology of the Hard-of-hearing and Deafened Adult. In *Hearing and Deafness*, ed. H. Davis and S. R. Silverman, 499–510. New York: Holt, Rinehart and Winston.

Rose, F., B. Brooks, and Albert Rizzo. 2005. Virtual Reality in Brain Damage Rehabilitation: Review. *CyberPsychology and Behavior* 8 (3): 241–262.

Rovira, Aitor, David Swapp, Bernhard Spanlang, and Mel Slater. 2009. The Use of Virtual Reality in the Study of People's Responses to Violent Incidents. *Frontiers in Behavioral Neuroscience* 3: 59.

Rumsey, Francis. 2001. *Spatial Audio*. Oxford: Focal Press.

Ryan, Marie-Laure. 2001. *Narrative as Virtual Reality: Immersion and Interactivity in Literature and Electronic Media*. Baltimore, MA: Johns Hopkins University Press.

Sanchez-Vives, Maria, and Mel Slater. 2005. From Presence to Consciousness through Virtual Reality. *Nature Reviews Neuroscience* 6 (4): 332–339.

Schifferstein, Hendrick N. J. 2006. The Perceived Importance of Sensory Modalities in Product Usage: A Study of Self-reports. *Acta psychologica* 121 (1): 41–64.

Sheridan, Thomas B. 1992. Musings on Telepresence and Virtual Presence. *Presence: Teleoperators and Virtual Environments* 1 (1): 120–126.

Shinn-Cunningham, Barbara, and Russell D. Shilling. 2002. Virtual Auditory Displays. In *Handbook of Virtual Environment Technology*, ed. K. Stanney, 65–92. Mahwah, NJ: Lawrence Erlbaum.

Slater, Mel. 2003. A note on presence terminology. In *Presence connect*, Volume 3.

——. 2004. Presence and Emotions. *CyberPsychology and Behavior* 7 (1): 121.

——. 2009. Place Illusion and Plausibility Can Lead to Realistic Behaviour in Immersive Virtual Environments. *Philosophical Transactions of the Royal Society*, Series B, *Biological Sciences* 364 (1535): 3549–3557.

Slater, Mel, Beau Lotto, Maria Marta Arnold, and Maria V. Sanchez-Vives. 2009. How We Experience Immersive Virtual Environments: The Concept of Presence and its Measurement. *Anuario de Psicología* (2): 193–210.

Slater, Mel, Daniel Pérez Marcos, Henrik Ehrsson, and Maria V. Sanchez-Vives. 2008. Towards a Digital Body: The Virtual Arm Illusion. *Frontiers in Human Neuroscience* 2: 6

——. 2009. Inducing Illusory Ownership of a Virtual Body. *Frontiers in Neuroscience* 3 (2): 214–220.

Steuer, Jonathan. 1992. Defining Virtual Reality: Dimensions Determining Telepresence. *Journal of Communication* 42 (4): 73–93.

Stevens, Brett, and Jennifer Jerrams-Smith. 2001. The Sense of Object-presence with Projection-augmented Models. *Haptic Human-Computer Interaction*, ed. Stephen Brewster and Roderick Murray-Smith, 194–198. Lecture Notes in Computer Science Volume 2058. Berlin: Springer.

Storms, Russell L., and Michael J. Zyda. 2000. Interactions in Perceived Quality of Auditory-visual Displays. *Presence: Teleoperators and Virtual Environments* 9 (6): 557–580.

Tamborini, Ron, and Nicholas Bowman. 2010. Presence in Video Games In *Immersed in Media: Telepresence in Everyday Life*, ed. Cheryl Campanella Bracken and Paul Skalski, 87–110. New York: Routledge.

Van Baren, J., and Wijnand IJsselsteijn. 2004. Measuring Presence: A Guide to Current Measurement Approaches. Deliverable of the OmniPres project IST-2001-39237.

Witmer, Bob G., and Michael J. Singer. 1998. Measuring Presence in Virtual Environments: A Presence Questionnaire. *Presence: Teleoperators and Virtual Environments* 7 (3): 225–240.

Zhao, Shanyang. 2003. Toward a Taxonomy of Copresence. *Presence: Teleoperators and Virtual Environments* 12 (5): 445–455.

CHAPTER 14

SONIC INTERACTIONS IN MULTIMODAL ENVIRONMENTS: AN OVERVIEW

STEFANIA SERAFIN

Most of our interactions with the physical world appear through a combination of different sensory modalities. When considering sonic interactions, obviously the sense of audition is involved. Moreover, often the sonic feedback is the consequence of an action produced by touch, and is presented in the form of a combination of auditory, tactile, and visual feedback. Let us consider for example the simple action of pressing a doorbell: the auditory feedback is given by the sound produced by the bell, the visual feedback is the motion of the bell, and the tactile feedback is the feeling of the displacement of the switch at the fingertip. It is important that these different sensory modalities are perceived in synchronization, in order to experience a coherent action.

In simulating realistic multimodal environments, several elements including synchronization need to be taken into consideration. However, technology gives some limitations, especially when the ultimate goal is to simulate systems that react in real time. Pai (2005) explains a tradeoff between accuracy and responsiveness, which represents a crucial difference between models for science and models for interaction. Specifically, computations about the physical world are always approximations. In general, it is possible to improve accuracy by constructing more detailed models and performing more precise measurements, but this comes at the cost of latency, that is, the elapsed time before an answer is obtained. For multisensory models it is also essential to ensure synchronization of time between different sensory modalities. Pai (2005) groups all of these temporal considerations, such as latency and synchronization, into a single category called "responsiveness." The question then becomes how to balance accuracy and responsiveness. The choice between accuracy and responsiveness depends also on the final goal of the multimodal system design. As an example, scientists are generally more concerned with accuracy, so responsiveness is only a soft constraint based on available resources. On the other hand, for interaction designers, responsiveness is an essential parameter that must be satisfied.

In this chapter, an overview is presented of how the knowledge in human perception and cognition can be helpful in the design of multimodal systems where interactive sonic feedback plays an important role. Sonic feedback can interact with visual or tactile feedback in different ways. While in this paper the focus is on the interaction between audition and other senses, the different interaction possibilities described below can happen between any combination of sensory modalities. As an example, *cross-modal mapping* represents the situation where one or more dimensions of a sound are mapped to a visual or tactile feedback (Norman 2002). An example of this situation is a beeping sound combined with a flashing light. *Intersensory biases* represent the situation where audition and another modality provide conflicting cues. When examining specific multimodal examples in the following section, several examples of intersensory biases will be provided. In most of these situations, the user tries to perceptually integrate the conflicting information. This conflict might lead to a bias towards a stronger modality. One classic example is the ventriloquist effect (Jack and Thurlow 1973), which illustrates the dominance of visual over auditory information. In this effect, spatially discrepant audio and visual cues are experienced as colocalized with the visual cue. This effect is commonly used in cinemas and home theaters where, although the sound physically originates at the speakers, it appears as coming from the moving image on screen, being for example a person speaking or walking. The ventriloquism effect occurs because the visual estimates of location are typically more accurate than the auditory estimates of location, and therefore the overall percept of location is largely determined by vision. This phenomenon is also known as visual capture (Welch and Warren 1980).

Cross-modal enhancement refers to the situation where stimuli from one sensory channel enhance or alter the perceptual interpretation of stimulation from another sensory channel. As an example, three studies presented in Storms and Zyda (2000) show how high-quality auditory displays coupled with high-quality visual displays increase the quality perception of the visual displays relative to the evaluation of the visual display alone. Moreover, low-quality auditory displays coupled with high-quality visual displays decrease the perception of quality of the auditory displays relative to the evaluation of the auditory display alone. These studies were performed by manipulating the pixel resolution of the visual display and Gaussian white-noise level, and by manipulating the sampling frequency of the auditory display and Gaussian white-noise level. Subjects were asked to rate the quality of the visual image of a radio with different pixel qualities, coupled to auditory feedback resembling sounds coming from a radio.

These findings strongly suggest that the quality of realism in an audiovisual display must be a function of both auditory and visual display fidelities inclusive of each other. Cross-modal enhancements can occur even when the extramodal input does not provide information directly meaningful for the task. A primary example was reported by Stein and others (1996). Subjects rated the intensity of a visual light higher when it was accompanied by a brief, broadband auditory stimulus than when it was presented alone.

The auditory stimulus produced more enhancement for lower visual intensities, and regardless of the relative location of the auditory cue source.

Cross-modal transfers or illusions are the situations where stimulation in one sensory channel leads to the illusion of stimulation in another sensory channel. An example of this is synesthesia, which in the audiovisual domain is expressed for example as the ability of seeing a color while hearing a sound. When considering intersensory discrepancies, Welch and Warren (1980) propose a modality-appropriateness hypothesis. Their model suggests that the various sensory modalities are differentially well suited to the perception of different events. Their model also shows that the dominance of a particular modality is relative to its appropriateness to the situation. Generally, it is supposed that vision is more appropriate for the perception of spatial location than is audition, with touch somewhere in between. Audition is most appropriate for the perception of temporally structured events. Touch is more appropriate than audition for the perception of texture, whereas vision and touch may be about equally appropriate for the perception of textures. The appropriateness is a consequence of the different temporal and spatial resolution of the auditory, tactile, and visual systems.

Apart from the way that the different senses can interact, the auditory channel also presents some advantages as compared to the other modalities. As an example, humans have a complete sphere of auditory receptivity around the head, while visual feedback has a limited spatial region in terms of field of view, or field of regard. Because auditory information is primarily temporal, the temporal resolution of the auditory system is more precise. We can discriminate between a single and a pair of clicks when the gap is only a few tens of microseconds (Krumbholz et al. 2003). Perception of temporal changes in the visual modality is much poorer, and the fastest visible flicker rate in normal conditions is about 40–50 Hz (Bruce, Green, and Georgeson 2003).

In contrast, the maximum spatial resolution (contrast sensitivity) of the human eye is approximately $\frac{1}{30}$ degree, a much finer resolution than that of the ear, which is approximately 1 degree. Humans are sensitive to sounds arriving from anywhere within the environment whereas the visual field is limited to the frontal hemisphere, and good resolution is limited to the foveal region. Therefore, while the spatial resolution of the auditory modality is cruder, it can serve as a cue to events occurring outside the visual field of view.

In this chapter we provide an overview of the interaction between audition and vision and audition and touch, together with guidelines on how such knowledge can be used in the design of interactive sonic systems. If we understand how we naturally interact in a world where several sensorial stimuli are provided, we can apply this understanding to the design of sonic interactive systems. Research on multisensory perception and cognition can provide us with important guidelines on how to design virtual environments where interactive sound plays an important role. Due to technical advancements such as mobile technologies and 3D interfaces, it has become possible to design systems that have natural multimodal properties similar to those in the physical world. These interfaces understand human multimodal communication and can actively anticipate and act in line with human capabilities and limitations. A large challenge for the near future is the development of such natural multimodal interfaces, and this requires the active participation of industry, technology, and the human sciences.

14.1 Audiovisual Interactions

Research into multimodal interaction between audition and other modalities has primarily focused on the interaction between audition and vision. This choice is naturally due to the fact that audition and vision are the most dominant modalities in the human perceptual system (Kohlrausch and van de Par 1999). A well-known multimodal phenomena is the McGurk effect (McGurk and MacDonald 1976). The McGurk effect is an example of how vision alters speech perception; for instance, the sound "ba" is perceived as "da" when viewed with the lip movements for "ga." Notice that in this case, the percept is different from both the visual and auditory stimuli, so this is an example of intersensory bias, as defined in the previous section.

The different experiments described until now show a dominance of vision versus audition, when conflicting cues are provided. However, this is not always the case. As an example, in Shams, Kamitami, and Shimojo (2000, 2002) a visual illusion induced by sound is described. When a single visual flash is accompanied by multiple auditory beeps, the single flash is perceived as multiple flashes. These results were obtained by flashing a uniform white disk for a variable number of times, 50 milliseconds apart, on a black background. Flashes were accompanied by a variable number of beeps, each spaced 57 milliseconds apart. Observers were asked to judge how many visual flashes were presented on each trial. The trials were randomized and each stimulus combination was run five times on eight naive observers. Surprisingly, observers consistently and incorrectly reported seeing multiple flashes whenever a single flash was accompanied by more than one beep (Shams, Kamitani, and Shimojo 2000). This experiment is known as sound-induced flash illusion. A follow-up experiment investigated whether the illusory flashes could be perceived independently at different spatial locations (Kamitani and Shimojo 2001). Two bars were displayed at two locations, creating an apparent motion. All subjects reported that an illusory bar was perceived with the second beep at a location between the real bars. This is analogous to the cutaneous rabbit perceptual illusion, where trains of successive cutaneous pulses delivered at a few widely separated locations produce sensations at many in-between points (Geldard and Sherrick 1972). As a matter of fact, perception of time, wherein auditory estimates are typically more accurate, is dominated by hearing. Another experiment in determining whether two objects bounce off each other or simply cross, is influenced by hearing a beep when the objects could be in contact. In this particular case, a desktop computer displayed two identical objects moving towards each other. The display was ambiguous to provide two different interpretations after the objects met: they could either bounce off each other or cross. Since collisions usually produce a characteristic impact sound, introducing such sound when objects met promoted the perception of bouncing over crossing. This experiment is usually known as motion-bounce illusion (Sekuler, Sekuer, and Lau 1997). In a subsequent study, Sekuler and Sekuler found that any transient sound temporally aligned with the would-be collision increased the likelihood of a bounce percept (Sekuler and Sekuler 1999). This includes a pause, a flash of light on the screen, or a sudden disappearance of the discs.

More recent investigations examined the role of ecological auditory feedback in affecting multimodal perception of visual content. As an example, in a study presented in Ecker and Heller (2005) the combined perceptual effect of visual and auditory information on the perception of a moving object's trajectory was investigated. Inspired by the experimental paradigm presented in Kersten et al. (1997), the visual stimuli consisted of a perspective rendering of a ball moving in a three-dimensional box. Each video was paired with one of three sound conditions: silence, the sound of a ball rolling, or the sound of a ball hitting the ground. It was found that the sound condition influenced whether observers were more likely to perceive the ball as rolling back in depth on the floor of the box or jumping in the frontal plane. Another interesting study related to the role of auditory cues in the perception of visual stimuli is the one presented in Thomas and Shiffrar (2010). Two psychophysical studies were conducted to test whether visual sensitivity to point-light depictions of human gait reflects the action specific cooccurrence of visual and auditory cues typically produced by walking people. To perform the experiment, visual walking patterns were captured using a motion capture system, and a between-subject experimental procedure was adopted. Specifically, subjects were randomly exposed to one of the three experimental conditions: no sound, footstep sounds, or a pure tone at 1000 Hz, which represented a control case. Visual sensitivity to coherent human gait was greatest in the presence of temporally coincident and action-consistent sounds, in this case the sound of footsteps. Visual sensitivity to human gait with coincident sounds that were not action-consistent, in this case the pure tone, was significantly lower and did not significantly differ from visual sensitivity to gaits presented without sound.

As an additional interaction between audition and vision, sound can help the user search for an object within a cluttered, continuously changing environment. It has been shown that a simple auditory pip drastically decreases search times for a synchronized visual object that is normally very difficult to find. This is known as the pip and pop effect (Van der Burg et al. 2008). Visual feedback can also affect several aspects of a musical performance, although in this chapter affective and emotional aspects of a musical performance are not considered. As an example, Schutz and Lipscomb report an audio-visual illusion in which an expert musician's gestures affect the perceived duration of a note without changing its acoustic length (Schutz and Lipscomb 2007). To demonstrate this, they recorded a world-renowned marimba player performing single notes on a marimba using long and short gestures. They paired both types of sounds with both types of gestures, resulting in a combination of natural (i.e., congruent gesture-note pairs) and hybrid (i.e., incongruent gesture-note pairs) stimuli. They informed participants that some auditory and visual components had been mismatched, and asked them to judge tone duration based on the auditory component alone. Despite these instructions, the participants' duration ratings were strongly influenced by visual gesture information. As a matter of fact, notes were rated as longer when paired with long gestures than when paired with short gestures. These results are somehow puzzling, since they contradict the view that judgments of tone duration are relatively immune from visual influence (Welch and Warren 1980), that is, in temporal tasks visual influence on audition is negligible. However, the results are not based on information quality, but rather on perceived causality, given that visual influence in this paradigm is dependent on the presence of an ecologically plausible audiovisual relationship.

Indeed, it is also possible to consider the characteristics of vision and audition to predict which modality will prevail when conflicting information is provided. In this direction, Kubovy and VanValkenburg (2001) introduced the notion of auditory and visual objects. They describe the different characteristics of audition and vision, claiming that a primary source of information for vision is a surface, while a secondary source of information is the location and color of sources. On the other hand, a primary source of information for audition is a source and a secondary source of information is a surface.

In Ernst and Bulthoff (2004) a theory is suggested on how our brain merges the different sources of information coming from the different modalities, specifically audition, vision, and touch. The first is what is called sensory combination, which means the maximization of information delivered from the different sensory modalities. The second strategy is called sensory integration, which means the reduction of variance in the sensory estimate to increase its reliability. Sensory combination describes interactions between sensory signals that are not redundant. By contrast, sensory integration describes interactions between redundant signals. Ernst and coworkers (Ernst and Bulthoff 2004) describe the integration of sensory information as a bottom-up process.

The "modality precision," also called "modality appropriateness," hypothesis, by Welch and Warren (1980), is often cited when trying to explain which modality dominates under what circumstances. This hypothesis states that discrepancies are always resolved in favor of the more precise or more appropriate modality. In spatial tasks, for example, the visual modality usually dominates, because it is the most precise at determining spatial information.

However, according to Ernst and Bulthoff (2004), this terminology is misleading because it is not the modality itself or the stimulus that dominates. Rather, the dominance is determined by the estimate and how reliably it can be derived within a specific modality from a given stimulus. The experiments described until now assume a passive observer, in the sense that a subject is exposed to a fixed sequence of audiovisual stimuli and is asked to report on the resulting perceptual experience. When a subject is interacting with the stimuli provided, a tight sensory motor coupling is enabled, that is an important characteristic of embodied perception. According to embodiment theory, a person and the environment form a pair in which the two parts are coupled and determine each other. The term "embodied" highlights two points: first, cognition depends upon the kinds of experience that are generated from specific sensorimotor capacities. Second, these individual sensorimotor capacities are themselves embedded in a biological, psychological, and cultural context (Dourish 2004).

The notion of embodied interaction is based on the view that meanings are present in the actions that people engage in while interacting with objects, with other people, and with the environment in general. Embodied interfaces try to exploit the phenomenological attitude of looking at the direct experience, and let the meanings and structures emerge as experienced phenomena. Embodiment is not a property of artifacts but rather a property of how actions are performed with or through the artifacts.

Audiotactile interactions, described in the following section, require a continuous action-feedback loop between a person and the environment, an important characteristic of embodied perception and sonic interaction design.

14.2 Audiotactile Interactions

Although the investigation of audiotactile interactions has not received as much attention as the audiovisual interactions, it is certainly an interesting field of research, especially considering the tight connections existing between the sense of touch and audition. As a matter of fact, both audition and touch are sensitive to the very same kind of physical property, that is, mechanical pressure in the form of oscillations. The tight correlation between the information content (oscillatory patterns) being conveyed in the two senses can potentially support interactions of an integrative nature at a variety of levels along the sensory pathways. Auditory cues are normally elicited when one touches everyday objects, and these sounds often convey useful informational regarding the nature of the objects (Ananthapadmanaban and Radhakrishnan 1982; Gaver 1993). The feeling of skin dryness or moistness that arises when we rub our hands against each other is subjectively referred to the friction forces at the epidermis. Yet, it has been demonstrated that acoustic information also participates in this bodily sensation, because altering the sound arising from the hand rubbing action changes our sensation of dryness or moistness at the skin. This phenomenon is known as the parchment-skin illusion (Jousmäki and Hari 1998).

The parchment-skin illusion is an example of how interactive auditory feedback can affect subjects' tactile sensation. Specifically, in the experiment demonstrating the rubber-skin illusion, subjects were asked to sit with a microphone close to their hands, and then to rub their hands against each other. The sound of hands' rubbing was captured by the microphone, manipulated in real time, and played back through headphones. The sound was modified by attenuating the overall amplitude and by amplifying the high frequencies. Subjects were asked to rate the tactile sensation in their palms as a function of the different auditory cues provided, in a scale ranging from very moist to very dry. Results show that the provided auditory feedback significantly affected the perception of the skin's dryness.

This study was extended in (Guest et al. 2002), by using a more rigorous psychophysical testing procedure. Results reported a similar increase in smooth–dry scale correlated to changes in auditory feedback, but not in the roughness judgments per se. However, both studies provide convincing empirical evidence demonstrating the modulatory effect of auditory cues on people's tactile perception of a variety of different surfaces. A similar experiment was performed combining auditory cues with tactile cues at the tongue. Specifically, subjects were asked to chew on potato chips, and the sound produced was again captured and manipulated in real time. Results show that the perception of potato chips' crispness was affected by the auditory feedback provided (Spence and Zampini 2006).

Lately, artificial cues are appearing in audiohaptic interfaces, allowing us to carefully control the variations to the provided feedback and the resulting perceived effects on exposed subjects (DiFilippo and Pai 2000; Nordahl et al. 2010; Van den Doel and Pai 1998). Artificial auditory cues have also been used in the context of sensory substitution, for artificial sensibility at the hands using hearing as a replacement for loss of sensation (Lundborg, Rosé, and Lindberg 1999). In this particular study, microphones placed at the fingertips captured and amplified the friction sound obtained when rubbing hard surfaces.

14.3 Multimodality and Sonic Interaction Design

As stated in the previous section, among scholars in perception and cognition there has been a shift in attention, from the human as a receiver of auditory and visual stimuli, to the perception-action loops that are mediated by multimodal signals. Some examples in this direction were presented in the previous section concerning auditory and tactile objects, although in that context the human was able to affect the quality of the produced signal by his actions (such as chewing a potato chip), but the loop did not close, in the sense that the human was not able to cyclically modify the feedback produced in an action-perception loop.

Several efforts in these directions were unified under the Sonic Interaction Design (SID) umbrella, thanks to a COST European cooperation action that started in 2006 and officially ended in 2011, which coined the term "sonic interaction design" that is now widely used to indicate interactive systems with a salient sonic behavior (Rocchesso and Serafin 2009; Rocchesso et al. 2008). This section presents different experiments in the field of sonic interaction design, arguing how a tight connection between users' gestures and sound is essential when designing interactive artifacts with a salient sonic behavior. SID is an interdisciplinary field recently emerging as a combined effort of researchers and practitioners working at the intersection of sound and music computing, interaction design, human–computer interaction, novel interfaces for musical expression, product design, music psychology and cognition, music composition, performance, and interactive arts. SID explores ways in which sound can be used to convey information, meaning, and aesthetic and emotional qualities in interactive contexts. One of the ultimate goals of SID is the ability to provide design and evaluation guidelines for interactive products with a salient sonic behavior. SID addresses the challenges of creating interactive, adaptive sonic interactions that continuously respond to the gestures of one or multiple users. At the same time, SID investigates how the designed gestures and sonic feedback are able to convey emotions and engage expressive and creative experiences.

SID also aims at identifying new roles that sound may play in the interaction between users and artifacts, services, or environments. By exploring topics such as multisensory experience with sounding artifacts, perceptual illusions, sound as a means of communication in an action-perception loop, and sensorimotor learning through sound, SID researchers are opening new domains of research and practice for sound designers and engineers, interaction and interfaces designers, media artists and product designers, among others. SID emerges from different established disciplines where sound has played an important role. Within human–computer studies, auditory display and sonification have been topics of interest for a couple of decades.

In sound and music computing, researchers have moved away from the mere engineering reproduction of existing musical instruments and everyday sounds in a passive context, toward investigating principles and methods to design and evaluate sonic interactive systems. This is considered by the Sound and Music Computing Research Roadmap to be one of the most promising areas for research and experimentation. Moreover, the design and implementation of novel interfaces to control such sounds, together with the ability to augment existing musical instruments and everyday objects with sensors and auditory feedback, are currently active areas of exploration in the New Interfaces for Musical Expression's community.

In the field of SID, continuous sonic feedback is an important element that mimics the way humans interact with the world (Rocchesso et al. 2008). In fact, most complex interactions in the world are essentially continuous, and multimodal interfaces need to be able to support such continuity. A particularly effective example of multimodal interaction where the auditory feedback plays an essential role is musical instruments. Let us consider, for example, the case of a person playing a violin: the player receives tactile feedback at the right hand given by the pressure of the bow on the strings, and on the left hand given by the fingers pressing on the fingerboard. Tactile feedback is also provided by the vibrations of the instrument's body in contact with the player. Auditory feedback is obviously the sound produced by the instrument, and visual feedback is the possibility of seeing the fingers and the bow moving.

When playing a musical instrument, there is clearly interaction with a physical object, and the sound is dependent on several interactions between player and instruments in complex ways. The player adjusts such sound by moving different parts of his body in an action-perception loop. This continuous physical interaction is one of the elements that makes playing musical instruments an engaging and challenging task. Moreover, cross modal enhancement is also an important element in musical instruments, in the sense that the different sensorial modalities are complementing and augmenting each other. When the information is not perceived as coherent among the different modalities, for example if some delay is perceived in one modality or if the different modalities are not perceived as synchronized, then the action-perception loop is broken. It is therefore extremely important that the overall interaction loop binds the channels together by the use of correlations between the channels. When a task is merely visual, the haptic and auditory channel can provide nondistractive informative feedback, as in the case of the pip and pop effect (Van der Burg et al. 2008).

When feedback provides information about data under analysis, or about the interaction itself that is useful to refine the activity, then we talk about interactive sonification (Hermann and Hunt 2005). A successful example of interactive sonification is the one proposed in Rath and Rocchesso (2005). Here, the task of balancing a marble ball on a wooden stick is improved by providing augmented auditory feedback given by rolling sounds.

Another interesting direction where knowledge on multimodal interaction can prove to be helpful is the design of auditory feedback for mobile devices. In Walker and Brewster (2000), the problem of visual clutter in mobile devices is addressed. The solution proposed uses spatial sound to provide information. Specifically, a progress bar was conveyed as a traditional graphic display, as well as a sonified spatialized display. User tests showed that participants performed background monitoring tasks better when auditory feedback was used.

Since mobile devices are already ubiquitous, and they all pose challenges in terms of limited size of the visual feedback, using high-quality auditory and haptic feedback presents interesting possibilities for sensory augmentation or even substitution. Furthermore, multimodal perception can be applied in the field of rendering of complex scenes in interactive virtual environments. Recent research in realism and efficiency in computer graphics and audio for virtual environments has embedded elements of human multimodal perception (see for example, Tsingos, Gallo, and Drettakis 2004). When complex scenes are rendered, it is not necessary to visually and auditorially reproduce every single detail. Moreover, if it is important to capture user's attention, for example if some element of a complex scene needs to be highlighted, it is possible to use results from multimodal attention, such as the pip and pop effect (Van der Burg 2008).

The understanding of how the senses interact is still mainly focused on simple stimuli such as beeps and flashes. Applications of this understanding to the design of immersive virtual environments and tangible interfaces where sound plays an important role is still open to several possibilities.

14.4 Conclusions

This chapter has provided an overview of several experiments whose goal was to achieve a better understanding of how the human auditory system is connected to the visual and haptic channel. A better understanding of multimodal perception can have several applications. As an example, systems based on sensory substitution help people lacking a certain sensorial modality to have it replaced by another sensorial modality. Moreover, cross-modal enhancement allows a reduced stimulus in one sensorial modality to be augmented by a stronger stimulation in another modality.

Nowadays advances in hardware and software technology allow us to experiment in several ways with technologies for multimodal interaction design, building, for example, tactile illusions with equipment available in a typical hardware store (Hayward 2008) or easily experimenting with sketching and rapid prototyping (Buxton 2009; Delle Monache, Polotti, and Rocchesso 2010). These advances in technology create several possibilities to discover novel cross-modal illusions and interactions between the senses, especially when collaboration between cognitive psychologists and interaction designers is facilitated. A research challenge is not only to understand how humans process information coming from different senses, but also how information in a multimodal system should be distributed to different modalities in order to obtain the best user experience.

As an example, in a multimodal system such as a system for controlling a tactile display, seeing a visual display and listening to interactive auditory display, it is important to determine which synchronicities are most important. At one extreme, a completely disjointed distribution of information over several modalities can offer the highest bandwidth, but the user may be confused in connecting the modalities and one modality might mask another and cause the user to attend to events that might not be important. At the other extreme a completely redundant distribution of information is known to increase the cognitive load and is not guaranteed to increase user performance.

Beyond the research on multimodal stimuli processing, studies are needed on the processing of multimodal stimuli that are connected via interaction. We would expect that the human brain and sensory system has been optimized to cope with a certain mixture of redundant information, and that information displays are better the more they follow this natural distribution. Overall, the more we achieve a better understanding of the ways humans interact with the everyday world, the more we can obtain inspiration for the design of effective natural multimodal interfaces.

References

Ananthapadmanaban, T., and V. Radhakrishnan. 1982. An Investigation of the Role of Surface Irregularities in the Noise Spectrum of Rolling and Sliding Contacts. *Wear* 83 (2): 399–409.

Bruce, Vicki, Patrick R. Green, and Mark A. Georgeson. 2003. *Visual Perception: Physiology, Psychology, and Ecology*. New York: Psychology Press.

Buxton, Bill. 2009. *Sketching User Experiences: Getting the Design Right and the Right Design*. San Francisco: Morgan Kaufmann.

Delle Monache, Stefano, Pietro Polotti, and Davide Rocchesso. 2010. A Toolkit for Explorations in Sonic Interaction Design. In *Proceedings of the 5th Audio Mostly Conference: A Conference on Interaction with Sound*. New York: ACM.

DiFilippo, Derek, and Dinesh K. Pai. 2000. The AHI: An Audio and Haptic Interface for Contact Interactions. In *Proceedings of the 13th annual ACM symposium on User interface software and technology*, 149–158. New York: ACM.

Dourish, Paul. 2004. *Where the Action Is: The Foundations of Embodied Interaction*. Cambridge, MA: MIT Press.

Ecker, A., and Heller, L. 2005. Auditory-visual Interactions in the Perception of a Ball's Path. *Perception* 34 (1): 59–75.

Ernst, Marc O., and Heinrich H. Bülthoff. 2004. Merging the Senses into a Robust Percept. *Trends in Cognitive Sciences* 8 (4): 162–169.

Gaver, William. 1993. What in the World Do we Hear? An Ecological Approach to Auditory Event Perception. *Ecological Psychology* 5 (1): 1–29.

Geldard, F., and C. Sherrick. 1972. The Cutaneous "Rabbit": A Perceptual Illusion. *Science* 178 (4057): 178–179.

Guest, S., C. Catmur, D. Lloyd, and C. Spence. 2002. Audiotactile Interactions in Roughness Perception. *Experimental Brain Research* 146 (2): 161–171.

Hayward, Vincent. 2008. A Brief Taxonomy of Tactile Illusions and Demonstrations that Can be Done in a Hardware Store. *Brain Research Bulletin* 75 (6): 742–752.

Hermann, Thomas, and Andy Hunt. 2005. Guest Editors' Introduction: An Introduction to Interactive Sonification. *Multimedia*, IEEE, 12 (2): 20–24.

Jack, Charles E., and Willard R. Thurlow. 1973. Effects of Degree of Visual Association and Angle of Displacement on the "Ventriloquism" Effect. *Perceptual and Motor Skills* 37 (3): 967–979.

Jousmäki, V., and R. Hari. 1998. Parchment-skin Illusion: Sound-biased Touch. *Current Biology* 8 (6): 190.

Kamitani, Y., and S. Shimojo. 2001. Sound-induced Visual Rabbit. *Journal of Vision* 1 (3): 478–478.

Kersten, D., P. Mamassian, D. Knill et al. 1997. Moving Cast Shadows Induce Apparent Motion in Depth. *Perception* 26: 171–192.

Kohlrausch, Armin, and Steven van de Par. 1999. Auditory-visual Interaction: From Fundamental Research in Cognitive Psychology to (Possible) Applications. In *Proceedings of SPIE*, volume 3644, 34.

Krumbholz, Katrin, Roy D. Patterson, Andrea Nobbe, and Hugo Fastl. 2003. Microsecond Temporal Resolution in Monaural Hearing without Spectral Cues? *Journal of the Acoustical Society of America*, 113: 2790.

Kubovy, Michael, and David Van Valkenburg. 2001. Auditory and Visual Objects. *Cognition* 80 (1–2): 97–126.

Lundborg, Göran, Birgitta Rosé, and Styrbjörn Lindberg. 1999. Hearing as Substitution for Sensation: A New Principle for Artificial Sensibility. *Journal of Hand Surgery* 24 (2): 219–224.

McGurk, Harry, and John MacDonald.1976. Hearing Lips and Seeing Voices. *Nature* 264, 746–748.

Norman, Donald. 2002. *The Design of Everyday Things*. Cambridge, MA: MIT Press.

Nordahl, Rolf, Amir Berrezag, Smilen Dimitrov, Luca Turchet, Vincent Hayward, and Stefania Serafin. 2010. Preliminary Experiment Combining Virtual Reality Haptic Shoes and Audio Synthesis. *Haptics: Generating and Perceiving Tangible Sensations* 123–129.

Pai, Dinesh K. 2005. Multisensory Interaction: Real and Virtual. *Robotics Research* 15: 489–498.

Rath, Matthias, and Davide Rocchesso. 2005. Continuous Sonic Feedback from a Rolling Ball. *Multimedia*, IEEE 12 (2): 60–69.

Rocchesso, Davide, and Stefania Serafin. 2009. Sonic Interaction Design. *International Journal of Human-Computer Studies* 67 (11): 905–906.

Rocchesso, Davide, Stefania Serafin, Frauke Behrendt, Nicola Bernardini, Roberto Bresin, Gerhard Eckel, Karmen Franinovic, Thomas Hermann, Sandra Pauletto, Patrick Susini and Yon Visell. 2008. Sonic Interaction Design: Sound, Information and Experience. In *CHI '08 Extended Abstracts on Human Factors in Computing Systems*, 3969–3972. New York: ACM.

Schutz, Michael, and Scott Lipscomb. 2007. Hearing Gestures, Seeing Music: Vision Influences Perceived Tone Duration. *Perception* 36 (6): 888–897

Sekuler, Allison B., and Robert Sekuler. 1999. Collisions between Moving Visual Targets: What Controls Alternative Ways of Seeing an Ambiguous Display? *Perception* 28 (4): 415–432.

Sekuler, Robert, Allison B. Sekuler, and Renee Lau. 1997. Sound Alters Visual Motion Perception. *Nature* 385: 6614.

Shams, Ladan, Yukiyasu Kamitani, and Shinsuke Shimojo. 2000. What You See is What You Hear. *Nature* 408: 788.

——. 2002. Visual Illusion Induced by Sound. *Cognitive Brain Research* 14 (1): 147–152.

Sound and Music Computing Research Roadmap. 2007. http://smcnetwork.org/roadmap.

Spence, Charles, and Massimiliano Zampini. 2006. Auditory Contributions to Multisensory Product Perception. *Acta Acustica united with Acustica* 92 (6): 1009–1025.

Stein, Barry E., Nancy London, Lee K. Wilkinson, and Donald D. Price. 1996. Enhancement of Perceived Visual Intensity by Auditory Stimuli: A Psychophysical Analysis. *Journal of Cognitive Neuroscience* 8 (6): 497–506.

Storms, Russell L., and Michael J. Zyda. 2000. Interactions in Perceived Quality of Auditory-visual Displays. *Presence: Teleoperators and Virtual Environments* 9 (6): 557–580.

Thomas, James Philip, and Maggie Shiffrar. 2010. I Can See you Better if I Can Hear you Coming: Action-consistent Sounds Facilitate the Visual Detection of Human Gait. *Journal of Vision* 10 (12): article 14

Tsingos, Nicolas, Emmanuel Gallo, and George Drettakis. 2004. Perceptual Audio Rendering of Complex Virtual Environments. *Proceedings of Siggraph 2004*.

Van den Doel, Kees, and Dinesh K. Pai. 1998. The Sounds of Physical Shapes. *Presence: Teleoperators and Virtual Environments* 7 (4): 382–395.

Van der Burg, Erik, Christian N. L. Olivers, Adelbert W. Bronkhorst, and T. Jan Heeuwes. 2008. Pip and Pop: Nonspatial Auditory Signals Improve Spatial Visual Search. *Journal of Experimental Psychology: Human Perception and Performance* 34 (5): 1053.

Walker, Ashley, and Stephen A. Brewster. 2000. Spatial Audio in Small Display Screen Devices. *Personal Technologies* 4 (2): 144–154.

Welch, R., and C. Warren. 1980. Immediate Perceptual Response to Intersensory Discrepancy. *Psychological bulletin* 88 (3): 638.

CHAPTER 15

MUSICAL INTERACTION FOR HEALTH IMPROVEMENT

ANDERS-PETTER ANDERSSON
AND BIRGITTA CAPPELEN

DURING the past decade, tangible sensor technologies have matured and become less expensive and easier to use, leading to an explosion of innovative musical designs within video games, smartphone applications, and interactive art installations. Interactive audio has become an important design quality in commercially successful games like *Guitar Hero*, and a range of mobile phone applications motivating people to interact, play, dance, and collaborate with music. Parallel to the game, phone, and art scenes, an area of music and health research has grown, showing the positive results of using music to promote health and wellbeing in everyday situations and for a broad range of people, from children and elderly to people with psychological and physiological disabilities. Both quantitative medical and ecological humanistic research show that interaction with music can improve health, through music's ability to evoke feelings, motivate people to interact, master, and cope with difficult situations, create social relations and experience shared meaning. Only recently, however, the music and health field has started to take interest in interactive audio, based on computer-mediated technologies' potential for health improvement. Here, we show the potential of using interactive audio in what we call *interactive musicking* in the computer-based interactive environment Wave. Interactive musicking is based on musicologist Christopher Small's (1998) concept "musicking", meaning any form of relation-building that occurs between people, and people and things, related to activities that include music. For instance, musicking includes dancing, listening, and playing with music (in professional contexts and in amateur, everyday contexts). We have adapted the concept of "musicking" to the design of computer-based musical devices.

The context for this chapter is the research project RHYME. RHYME is a multidisciplinary collaboration between the Centre for Music and Health at the Norwegian Academy of Music, the Oslo School of Architecture and Design (AHO), and Informatics at the University of Oslo. Our target group is families with children with severe disabilities. Our goal is to improve health and wellbeing in the families through everyday musicking activities in interactive environments. Our research approach is to use knowledge from music and health research, musical composition and improvisation, musical action research, musicology, music sociology, and soundscape studies, when designing the tangible interactive environments. Our focus here is interaction design and composition strategies, following research-by-design methodology, creating interactive musicking environments. We describe the research and design of the interactive musicking environment Wave, based on video documentation, during a sequence of actions with users.

Our findings suggest some interactive audio design strategies to improve health. We base the design strategies on musical actions performed while playing an instrument, such as impulsive or iterative hitting, or sustainable stroking of an instrument. Musical actions like these can also be used for musicking in everyday contexts, creating direct sound responses to evoke feelings that create expectations and confirm interactions. In opposition to a more control-oriented, instrument and interface perspective, we argue that musical variation and narrative models can be used to design interactive audio. The audio device is seen as an actor taking many different roles, as instrument, comusician, toy, and so on. In this way, the audio device, the interactive musicking environment, will change over time, answering with direct response, as well as nose-thumbing and dramatic response, motivating people to create music, play, and interact socially. Musical variation can also be used to design musical backgrounds and soundscapes that can be used for creating layers of ambience. These models create a safe environment and contribute to shared and meaningful experiences for people interacting. Altogether, our preliminary findings in the RHYME project are that the interactive musicking environment improves health, as it evokes feelings, motivates people to cope and master, and breaks isolation and passivity as people share actions and feelings with others.

In recent years, research and development in interactive music technologies has offered new forms of expression and new areas of use, within learning, play, and health. Often these areas have influenced each other. Musical interaction through variation and storytelling is used to motivate learning everything from mathematics to languages, for instance, the BBC's interactive online learning games *Bitesize* and the 2009 BAFTA award-winning *Mi Vida Loca: Spanish for beginners*. In these games, the pupils learn through interaction in a game environment, where assignments are given as interactive short stories with dialog and musical variation that create expectations and motivate interaction over time. Learning music composition and improvisation, adults and children are motivated by interactive technologies to teach themselves in fun, safe, and social ways, for example, in popular sound-editing application *Garage Band*, sound-synthesis performance instrument *Reactable* for iPad, and DJ application *Ableton Live*. In advanced music programming environments such as Pure Data and Max/MSP, playing and exploring the objects in the graphical interface and attaching tangible devices like keyboards, video cameras, and game controllers are essential parts of learning and creating music in the environment, as well as of sharing projects and programming code in a community. Often, music technologies integrate learning with play and gaming in a more direct way, where learning to operate the game controls or instrument is part of the playful gaming experience and the narrative of the game, for instance, the video games *Guitar Hero*, *Donkey Konga*, and *SingStar*. Games like these, with their tangible interfaces, build on our fascination for imitating musical actions such as physical guitar playing, drumming, singing, or dancing, to motivate gaming, play, and social relation-building. However, it is not solely the physical, tangible, and spatial music qualities of the activities that motivate the users to get immersed in the game. *Time-based* qualities of music are equally important. In particular, aesthetical rules of musical variation, repetition and montage techniques, previously developed in film music, with narratives developing over time, are used to motivate expectations of what will happen in the game, program, or learning platform over time. The music gives hints about new challenges and obstacles to arise further on, as well as confirming situations and supporting the role and action of the player of the game or the pupil engaged in learning. In *Sing Star*, the crowd is cheering or booing, depending on the player's skill. Therefore, sound's physical and time-based qualities increase the motivation of playing the game, creating and sharing gameplay and wellbeing. We argue that what makes this possible is the unique qualities of interactive music technologies—which combine music and narratives with computer-based interactivity with algorithms—that memorize, learn, respond, and challenge the user.

How, then, should interactive audio and music technologies be designed and composed to motivate interactions? Since 2000 the group MusicalFieldsForever (2012) has explored this question. The group consists of composer, programmer, and video artist Fredrik Olofsson and ourselves, industrial designer and interaction designer Birgitta Cappelen and musicologist and sound designer Anders-Petter Andersson. We have explored concepts and ideas from an artistic research and research-by-design methodology (Sevaldson 2010) and tried to come up with concrete suggestions in prototypes and exhibitions. We have chosen to take a multitude of perspectives and approaches from music, gaming, and interaction design, but also from the fields of sociology, cultural studies, philosophy, music and health, music therapy, and musicology. Why the need for many perspectives when the goal is to compose music and design interactions? A legitimate argument against many perspectives could be that musicians trained to improvise in groups already know how to collaborate in professional live situations. However, they are less trained to interact with amateur musicians outside of the traditional concert audience. Game designers were the first to apply interactive music technology for interacting amateurs, who most certainly have a different focus than the traditional music audience, as they let themselves get immersed in the gameplay. Composers can learn from game and interaction designers how to create motivating interactive systems. But even if game designers know about interactivity and the design of the physical things that the game uses, they do not always have the competence to understand how music and other time-based media motivate people to have expectations over time. Nor do they have knowledge of use outside of gaming contexts.

Why such concepts are difficult for composers and different from traditional music situations is because, in order to get motivated to interact, the audience (the amateur musician) has to get involved in *creating* the music. In one sense, he or she has to become a *co-creator* of the game and therefore the music. We have defined a co-creator as a person engaged in and shaping the music and the environment as part of an identity, and as a component of a relation-building activity with other people, things and music in the environment (Cappelen and Andersson 2003; Andersson and Cappelen 2008). The composer, being used to working with professional musicians, performing in front of a sitting audience, can't expect the same response if the person doing the action is an amateur in an everyday context such as meeting and playing with friends. Therefore, the composer has to alter their strategies and, somehow, rethink the musical structure, improvisation rules, and the design of music instruments, to suit the interactive. As we will describe in relation to assistive music technology, there is a risk in transferring the traditional control-oriented musical instrument without considering the change of context, abilities, and goals of interacting with amateurs.

Our experience is that interactive audio in game design and other related fields do not use the full potential of computer-based interactivity, either because the composer doesn't have experience in computer-based interaction or because the interaction and game designer doesn't know enough music and other time-based media. We believe that other perspectives can be helpful in solving what seems to be a locked position, between traditional music aesthetics based on the artwork and a perspective based on a technology. We argue that perspectives of musical actions, musicology, music and health, and music therapy can help us understand how to design for motivating interaction. And, as we have argued elsewhere (Cappelen and Andersson 2012b), to work artistically with interactive audio in a music and health field not only changes those perspectives, it questions our prejudices about music and music-making. It therefore empowers us in our roles as artists, designers, and researchers, making us rethink computer-based interaction, music, and our own roles as designers and composers.

Interactive music technologies such as musical games and programs for learning have also become popular to improve health and wellbeing for a broad range of people: giving rhythmical structure and motivating a person with limited physical abilities to move his or her body; relieving stress; motivating those with very low activity or even depression; stimulating memory in elderly people with dementia; and encouraging the use of voice for people with hearing difficulties. But music's use is more general and widespread, also outside of a professional, clinical, and therapeutic context, if we consider our own use of music to increase wellbeing and health (DeNora 2000). We also use music in the social arena (Stensæth and Ruud 2012), for expressing personal identity within a social relation in the family, at work, and among friends. The health effects of music for a number of illnesses have been thoroughly documented, in a biomedical tradition and in humanistic and ecological research. In the latter, games' motivating movements, such as popular *Guitar Hero*, have been used to empower individuals, to develop strategies for strengthening their own wellbeing. But for the most part, such interactive music technologies have improved health because they offer possibilities to engage in social activities and build relations.

The popularity of interactive music technologies and computer games for learning, play, and health, is due to the capabilities of the computer. The computer opens new interaction possibilities because it is not mechanistic, but dynamic and built on variation. There is not (as in acoustic instruments) a mechanical relationship between stimuli and response effect, input and what comes out, in sound, images, and so on. The computer can be programmed to learn, recognize, and answer, according to rules. These can be musical rules for communication that give rise to all sorts of musical and narrative variation. For instance, a single weak stimulus can result in many strong and repeated responses. Responses that change over time, dynamically and according to the musical-variation and the interaction strategies practiced by the person interacting.

In this chapter, we show how to design interactive music environments that improve health. The chapter is structured as follows: First we explore relevant relations between music and health and look at how music can promote health and wellbeing. Here, we draw on knowledge from an ecological, humanistic health approach used all over the world and extensively in Scandinavian countries (Stensæth and Ruud 2012; Rolvsjord 2010; Stige 2010; Bonde 2011). Further, we draw knowledge from musical actions applied in musicking, activities in a context and everyday environment that have potential for improving health. In the second part of the chapter, we suggest the health potential for interactive music and cross-media technology. Finally, we describe how we have created interactive musicking things and environments, and we discuss the results in the context of interactive audio for health.

15.1 Music and Health

15.1.1 What Do We Mean by "Health"?

The definitions of health and its research methods have developed in more or less two traditions: medical and humanist (Blaxter 2010). In medicine, health is pathologically defined as absence of illness. There are advantages with such a definition, in that it is easier to give a diagnosis in order to treat the illness. There are also, however, disadvantages. A risk is that too much focus on illness stops people from living high-quality lives, and instead makes them develop depression, and psychological and physiological illness. An alternative to pathologically defined health is an ecological and humanist definition. Here, humans interact with others and stand in relation to others in a biological, social environment, a physical and social context. In this type of ecology, relations between people in everyday work, play, family life, and so on affect wellbeing and health. A number of activities uphold and strengthen the relations. For instance playing, engaging in sports and cultural and musical activities, like dancing, creating music, listening, and so on. The health effect of these activities is first of all strengthening and preventive, and health is therefore something a person constructs in relation to other people and things within a culture. Health is, according to a humanist ecological approach, something that takes time to develop and uphold in the everyday. Or, as music therapist Kenneth Bruscia says, "Health is the process of becoming one's fullest potential for individual and ecological wholeness" (Bruscia 1998, 44).

15.1.2 How Music Improves Health

The health potential of music has been thoroughly and scientifically documented during the last fifteen years (e.g., Ruud 1998, 2010; Bruscia 1987, 1998). The use of music is known to promote health in many ways: listening, playing dancing, creating music for regulating emotions. One example is Stephen Clift's study of community singing in choirs as a public health resource, at the Sidney DeHaan Research Centre for Arts and Health, Canterbury Christ Church University, Folkestone, in the United Kingdom. Clift and his colleagues stress the main health-improving effect of choir singing to "Involve learning, keeping the mind active, help deep breathing to avoid anxiety, avoid passivity and isolation, offering the choir members social support and friendship on a routine basis" (Clift et al. 2007). People engaged in singing go from being patients with difficult illness or pain to choir members creating music and developing social relations in groups that regularly meet. The illness is still present, but the music and group activities enable the individual to cope.

In one sense, we are moving out of a therapeutic situation, with hierarchical power relations between the therapist who knows best and the client who knows less, to a situation where the client becomes active. This leads to a situation where the therapist is also empowered by the musicking activities, leading to a mutual relation between the therapist and the client. Rolvsjord (2010) describes this process as resource-oriented. When the therapist sees the client as a resource for his or her own musicking activities, therapist and client alike start to value their own creative work and grow as musicians. In our own work within the interactive art group MusicalFieldsForever we have had similar experiences when moving an interactive exhibition from an arts context at the Museum of Modern Art in Stockholm into a rehabilitation centre with multi-sensory environments in the same city (Cappelen and Andersson 2012b). Against our prejudice of what an audience could do, we were empowered in our roles as artists. We met with people with severe disabilities and experienced their artistic approaches and uses of our interactive environments, which we wouldn't have experienced if we had kept to the sheltered art scene a few kilometers away, where we felt at home.

According to RHYME project member and music therapist Even Ruud, music improves health through a process that evokes emotions and strengthens our ability to act by creating expectations and responding to actions. Further it creates an arena for developing social relations, and allows us to share meaning socially (Ruud 2010; Stensæth and Ruud 2012). One question we asked each other, initiating the RHYME project, was what happens between the music sessions, the weekly therapy sessions? What happens in an everyday family situation where the therapist is absent and the focus isn't necessarily on the client, where nobody has time to be at hand and the person with a disability therefore is left alone, bored, and isolated for long periods of time? Many of the participants in Clift's and Rolvsjord's studies speak about doing more and extending the music into their everyday. Is there a way that the music could extend into space and time and beyond therapy sessions?

15.1.3 Musical Actions and Musicking for Everyday Wellbeing

From a humanist and ecological health approach, as in resource-oriented music therapy, we have learned that music improve health and wellbeing motivating action and emotions. Music that evokes emotions and strengthens the ability to act is found in many musical actions in traditional music making and improvisation. Musicologist Rolf-Inge Godøy and later Alexander Jensenius described the motivational and emotional relation between physical-visual gestures and musical gestures and actions, activating overlapping regions in the brain (Godøy 2001; Godøy, Haga, and Jensenius 2006; Jensenius 2007). One sensory modality strengthens the other, for example in the effect of a drummer's visually and musically impulsive hit, the guitarist's iterative and repeated plucking of the strings, or the cellist playing a long note with his bow creating a sustained, visual-musical-physical gesture (Godøy, Haga, and Jensenius 2006). In case of a lack of ability in one modality, the others help the person compensate. Partly, the motivating effect in visual-musical-physical gestures comes from the fact that it creates cross-media expectation of what will happen. By "cross-media," we mean a sequential montage of visuals, sound, and actions in tangible media, creating variations and expectations over time. It is the principal reason why live music often is more engaging than listening to a recording without the physical and visual feedback.

15.2 Musicking: Roles, Relations, Contexts, and Things

The term "musicking" integrates relational thinking in resource-oriented and humanist health methods, with cross-media musical actions in the design. The term comes from musicologist Christopher Small who sees music as a relational activity, rather than a division between subject and object (Small 1998). He expands music from being a noun to a verb. Musicking, in Small's sense, is a meaning-making activity, including everyday listening, dancing, creating, and performing music. Therefore, it also expands musical actions from a narrow professional and controlled music context, into an open, and sometimes messy, everyday context.

In the context of interactive audio for games and interactive installations, it is particularly interesting how music and things take on different roles, depending on people's different meaning-making activities and understanding of a situation. Music sociologist Antoine Hennion (2011) calls these meaning-making processes *musical mediation*, where music, things, and humans all affect and change each other. Hennion's term comes from his colleague, the philosopher Bruno Latour's term *technical mediation* (1999), which describes the process where things, technologies, and humans create "hybrid" artifacts, while developing different roles in relation to each other over time. Our focus is on design and interaction possibilities that lie in the musical, physical, and tangible artifact. Artifacts include interactive, changing, and learning computers, software, hardware, sensors, and networks, as well as everyday cultural and musical things. Instead of viewing these "hybrid" things as static objects, Latour suggests the term *actor*, whose role shifts with the change of focus and activities. Based on previous observations made by music therapists in the installation ORFI (Stensæth and Ruud 2012), as well as our own observations over several years, we argue that participants' possibilities for changing roles and interacting over time is strengthened in interactive music environments. It is strengthened by the interactive music and physical things' ability to shift roles, from being musical instruments giving a direct response when being played, to becoming toys to play with as with a friend or acting as ambient environments in which to lay down and be. As we shall see below, in the design of interactive music the computer creates unique potentialities for strengthening these relational, musicking, and meaning-making activities.

15.2.1 The Health Potential of Interactive Music

Applications of advanced interactive music, dynamically changing over time and with interaction have, until recently, been rare within health-improving assistive technologies. Instead, assistive technologies, or augmented and alternative communication, have been mainly text- and image-based. Nevertheless, less advanced music technologies have been developed for rehabilitation and play for people with disabilities (Magee 2011), such as popular commercial products like the switch-based Paletto (http://www.Kikre.com), electronic instrument and ultrasound sensor Soundbeam (http://www.soundbeam.co.uk), and OptiMusic (http://www.optimusic.co.uk). These devices are being sold all over Europe and the United States, with considerable amounts of money being invested by health organizations, schools, and rehabilitation centers. However, they have limitations concerning musical variation and interface. They all offer direct response sounds only and, therefore, they have limited possibilities to create musical variations over time. They all build on a control- and interface-oriented design that limits the range of possible roles people using the music technologies can take: They are instruments, or tools, offering direct, unambiguous response to interactions. These limitations reduce every potential musical variation or surprise in the gameplay that could motivate people to continue to interact and take other roles. A control-oriented interface makes sense if a thing is used only as instrument or tool (e.g., piano, alarm, computer keyboard), but not in a play and gaming context. In play and in games, the goal is to create expectations, challenge, and surprise. If we see a person as somebody involved in co-creating the music with passion and vitality, and as a resource in relation to other people and things, the interactive music system should support changes over time, in order to motivate people to change roles and activity levels over time.

Wendy Magee and Karen Burland are two researchers with a focus on music technologies in music therapy. In a study of music therapists' use of MIDI-based music technologies like Soundbeam, they conclude by stressing the importance for a client to understand cause and effect, before engaging in complex interactions and music making (Magee and Burland 2008, 132–3). But the writers also point out problems with fatigue and decreasing motivation, caused by too strong a focus on trying and failing to master the interface sensor. Elsewhere, we have argued that instead of strengthening the relations and empowering the individual, as Magee and Burland most undoubtedly strive for, the Soundbeam connected to a MIDI synthesizer might have an opposite, disempowering effect on people with severe physical disabilities (Cappelen and Andersson 2012a). However, we believe that it is disempowering because the physically disabled client fails, gets tired and demotivated too many times before he or she gets it right. There is a conceptual design flaw in the mechanical response and lack of complex variation over time, and the interface's demands that the client must master it in *one way* only. The client is forced to take the role of a tool and instrument user, and the inability to do so leads to disempowerment. To empower the users, we have to create an arena for positive, mutually shared musicking experiences (Stensæth and Ruud 2012). With a more advanced interactive music system and open interface, the client can play and make music immediately and in many ways. In opposition to a more control-oriented, instrument and interface approach, we argue that traditional musical variation and narrative models can be used to design interactive audio. The audio and the interactive musicking environments will change over time, answering with direct response, motivating people to create music, play, and interact socially. With such a resource-oriented and musicking approach, the client could be a person that, on his or her own terms, becomes a positive resource to other people.

15.3 Interactive Musicking Improving Health

Based on a resource-oriented approach, interactive music and cross-media installations should offer a multitude of positive musicking experiences. We introduce the term *interactive musicking*, making use of the motivational positive effects of creating musical actions, in computer-based interactive environments, for health improvement. The interactive environments have to be open to many interpretations, interaction forms, and activity levels, where there are no wrong actions. They have to offer many possible roles (Latour 1999) and be simple and complex at the same time. The software should build on musical, narrative, and communicative principles, to motivate and develop musical competence and musicking experiences for all users over time. Interactive musicking is our suggested approach for understanding and designing health-improving music technology for people with complex needs, so that people with diverse abilities and motivations can experience vitality, mastering, empowerment, participation, and co-creation. To achieve these ambitions the interactive music and cross-media environments should:

1. Evoke interest and positive emotions relevant to diverse people's interpretation of the interactive environments and the situation.
2. Dynamically offer many roles to take, many musicking actions to make, and many ways of self-expression.
3. Offer aesthetically consistent response and build relevant cross-media expectations and challenges over time and space.
4. Offer many possible relations with people, things, experiences, events, and places.

Technically and musically, this means that the interactive musicking things and environments should be able to respond related to several types of events, to evoke interest and positive emotions. The environments ought to have rhetorical knowledge (programmed musical, narrative and communicative rules) and competence, remembering earlier user interactions, in order to respond aesthetically consistently over time and to create coherent expectations. They should, physically or wirelessly, be networked to other people and things, to exchange value and build relations over time. The interactive music and cross-media environments should have physically and musically attractive qualities related to material, shape, sensory modalities, character, genre and identity, social and cultural setting.

15.4 Creating Things for Interactive Musicking

In order to understand how we can create things and environments for interactive musicking, we like to describe the contexts, perspectives, and methods we have developed and work with. The group MusicalFieldsForever (2012) was formed in 2000 (the same group that formed the development team in the research project RHYME). The group was established in the research studio for Narrativity and Communication at the Interactive Institute and the School of Arts and Communication at Malmö University in Sweden. We have diverse backgrounds in music composition, musicology, generative music, hardware and sensor development, industrial design, and interaction design. We share the use of networking models and the computer as our major working tool and material. We also share a vision for the democratic potential of these technologies. It means that we try to understand technical, material, aesthetical, and social forms of mediation. We try to understand the influence of these mediations on power structures and relations among diverse users. Over time, we have built up knowledge based on practical design, development of hardware, software, smart textiles (e-textiles), and music in ten interactive installations with different versions. We have also collected experiences from user interaction in thirty exhibitions of the installations in art and design contexts, as well as user tests, observations, presentations, and publications in an academic context.

The context for this chapter is the RHYME project (http://www.RHYME.no), funded by the VERDIKT program and the Research Council of Norway. RHYME is a unique multidisciplinary collaboration between the Institute of Design at Oslo School of Architecture and Design, Centre for Music and Health at the Norwegian Academy of Music, and the Institute for Informatics at the University of Oslo. The project goal is to improve health and life quality for persons with severe disabilities, through the use of interactive musicking. In the project, we develop new generations of interactive music and cross-media environments every year, focusing on different user situations and user relations. The 1st and 2nd years' focus was cross-media; the 3rd year was on mobile platforms; the 4th and 5th years' focus is on social tangible media.

RHYME is based on a humanistic health approach: The goal is to reduce isolation and passivity through use of interactive musicking in cross-media interactive environments. Through multidisciplinary action-oriented empirical studies, multidisciplinary discussions, and reflections, we develop new generations of interactive music environments and related knowledge. Our design research methodology is user-centered and practice-based, where we develop knowledge through the design of new generations of interactive environments. The second empirical study in the RHYME project, for which we give examples below, was of Wave (see Figure 15.1). We observed five children between seven and fifteen years old with complex needs in their school's music room with a closely related person, not professional music therapists. We performed four different actions over a period of one month. From one action to the next, we made changes based on the previous action, weekly user surveys, observations, and multidisciplinary discussions. All sessions were recorded on video from several angles to capture as much as possible. A study of the health aspects of first year's prototype have been described and analyzed in a separate paper by researchers and music therapists Stensæth and Ruud (2012).

15.4.1 Designing for Interactive Musicking in Wave

Wave is the second year's interactive environment that we have designed, based on the requirements presented above. Wave is a seven-branched, wired, interactive, soft,

FIGURE 15.1 Family musicking in interactive environment Wave. Sister musicking by singing into the glowing microphone in Wave. Brother patting the "bubble field," with tones as a direct response, also affecting the movement sensor in the arm, playing back the sister's voice with raised pitch variations. Father relaxing in the vibrating Wave carpet. Photographer: Birgitta Cappelen.

dark carpet, with orange velvet tips that glow when the user interacts with one of the arms (see Figure 15.1). One central arm contains a microphone and two arms contain movement sensors, with accelerometers that change the recorded sound. The girl in Figure 15.1 is talking and singing into the microphone. Her brother affects the motion sensor in the shorter arm and plays back the sister's voice, with an added raised-pitch modulation. Two arms contain bend sensors and create the rhythmical background music. One arm contains a web camera with another microphone making sound effects, ring modulation, and filtering. Currently Wave contains five software programs offering different music and dynamic graphics to show with the small Pico projector embedded in one arm, or on the full wall projection. The carpet contains two robust speakers and a strong vibrator placed as a soft stomach in the middle of the carpet. The father relaxing in Figure 15.1 lies in direct contact with the vibrator, experiencing every musical gesture as part of a vibrating ambient background. We have also created a glowing soft velvet "bubble field" (see brother interacting in Figure 15.1) of infrared sensors in the dark carpet and RGB LEDs that represent a unique input device, with which the user can interact in many ways. The brother in Figure 15.1 pats the bubbles and gets direct responses in tones. The advanced musical variations over time depend on user interactions and musical rules created in the advanced real-time synthesis programming language SuperCollider. The programming was developed by MusicalFieldsForever member, composer, and video artist Fredrik Olofsson. With its size, shape, texture, advanced software, and input and output possibilities, Wave offers infinite ways to interact and co-create musicking experiences.

15.4.2 Observing Interactive Musicking in Wave

"Wendy" is fifteen years old and has Down syndrome. She likes to sing, but is shy in the company of others. The first time Wendy entered the room where the Wave carpet was placed, she spoke carefully into the microphone arm when her companion, "Nora," bent it toward her. Wendy said "Hi" and laughed when Wave played back her voice, one octave higher, as Nora shook an arm with the movement sensor. Still laughing, Wendy continued to go through the words she had been practicing with her speech therapist the previous hour: "O," "P, Q, R," "Europe." Wave answered back at a higher pitch. Wendy was happy with her achievement and thought it was fun to listen to the variation. Instead of the same repeated tonal response, the pitch shift effect created an aesthetically rich variation that was motivating to play with. Since it was Friday afternoon, Wendy continued to add some of her favorite foods that she expected to eat over the weekend: "Say Taco," "Say Pizza," "Can you say ice cream?" Wendy addressed Wave not merely as a tool or piece of technology, but as an actor she was friendly with, talked to, and with whom she had begun to develop a relationship, even saying "goodbye" when she left.

In the second action the following week, Wendy threw herself onto the Wave carpet, recognizing the soft vibrating and glowing creature-like carpet. She wasn't shy anymore, but felt at home, safe, and excited. She used all of her body to explore and interact with Wave. She took the initiative and developed her competence as she co-created with Nora in several ways. They gathered around the glowing bubble field as if it were a cozy "fireplace," shook their bodies to the beat and stroked the soft and glowing velvet microphone. They took turns filming each other and playing with the camera arm. They imitated and mirrored each other, taking turns, by each interacting with one arm with a bend sensor; starting slowly, by taking turns, first Nora, then Wendy, Nora again, and so on. In contrast to many existing systems, with Wave it was not necessary to first focus on the technology, understanding cause and effect, before being able to create music and play with others. Wendy didn't get tired or bored by too much repetition, nor was she demotivated due to high thresholds for response. Instead, the aesthetically rich cross-media interaction strengthened and motivated her and Nora's ability to act, at the same time as it motivated positive emotions, co-creation and development of competence over time, with varying musical, graphical, and tangible musicking.

15.5 Conclusions

In this chapter, we have discussed how computer-based music technology offers health improving opportunities, because it can remember, answer, and develop over time. We have shown how interactive audio can promote health for diverse users by motivating them to feel positive emotions, master, create mutual relations to others, and to develop competence over time. We have discussed our design solutions in relation to music technology for videogames, interaction design, and assistive technology. We have explored perspectives from Christopher Small's musicking and Bruno Latour's technical mediation, actors, and roles, and Antoine Hennion's musical mediation. These perspectives challenge traditional views of music as aesthetic object, instead viewing music as a relation-building social activity. We have also explored the resource-oriented perspectives from Randi Rolvsjord, Even Ruud, and Karette Stensæth in music therapy and music and health research, to understand what could empower humans interacting with each other and the music technology. We have applied research-by-design methods based on the musical mediation perspectives, analyzing the roles a person takes upon themselves in relation to others and the music technologies, when musicking in everyday school and family contexts. Musical mediation has led us to articulate design qualities for environments for interactive musicking improving health. We have designed for the possibility of taking different roles: from musicians playing in the environment as instrument, to interacting socially with other people and the interactive environment as friends or actors answering back with shifting response over time. Or just relaxing in an ambient landscape.

We have found interactive musicking to be an alternative to the limitations of traditional music in assistive technologies, which has too much focus on control of the interface. We have also found that many strategies from traditional music improvisation work also within interactive musicking. However, the interface has to change and become more open and flexible. Also the full potential of the computer has to be taken into consideration, in order to truly empower people and thereby improve health and wellbeing. This is not relevant only for people with disabilities, but for diverse groups of people who are musical amateurs.

We designed Wave to offer many potential musicking contexts, accounting for differences among users, from being a soundscape and ambient carpet for persons relaxing, an instrument for playing music, exploring one's own voice, and physical gestures, to a playground for playing together with other persons and the Wave system itself. We believe that moving from a definition of a person interacting as lacking abilities to a definition based on the same person being a potential resource to others is crucial to improve the individual's health and empowerment.

In summary, our suggestion is that interactive music environments should offer: Diverse possible roles to take, many musical actions to make and musical variations

over time, to improve health. These interactive music environments offer the user many forms of interaction to perform, such as stroking, patting, singing, hitting, moving, relaxing. Further they offer many cross-media expectations to experience and make over time and space.

References

Andersson, Anders-Petter. 2012. *Interaktiv musikkomposition*. PhD thesis, University of Gothenburg. http://hdl.handle.net/2077/30192.

Andersson, Anders-Petter, and Birgitta Cappelen. 2008. Same but Different, Composing for Interactivity. In *Proceedings of Audio Mostly 08*, 80–85. Piteå: Interactive Institute.

Blaxter, Mildred. 2010. *Health*. Cambridge, UK: Polity

Bonde, Lars Ole. 2011. Health Musicking: Music Therapy or Music and Health? A Model, Empirical Examples and Personal Reflections. *Music and Arts in Action* 3 (2): 120–140.

Bruscia, Kenneth E. 1987. *Improvisational Models of Music Therapy*. Springfield, IL: Charles C. Thomas.

——. 1998. *Defining Music Therapy*. Gilsum, NH: Barcelona Publishers.

Cappelen, Birgitta, and Anders-Petter Andersson. 2003. From Designing Objects to Designing Fields: From Control to Freedom. *Digital Creativity* 14 (2): 74–90.

——. 2012a. Musicking Tangibles for Empowerment. In *Computers Helping People with Special Needs (13th International Conference, ICCHP 2012)*, ed. Klaus Miesenberger, Arthur Karshmer, Petr Penaz, and Wolfgang Zagler, 254–261. Berlin and Heidelberg: Springer-Verlag.

——. 2012b. The Empowering Potential of Re-staging. *Leonardo Electronic Almanac* 18 (3): 132–141.

Clift, Stephen, Grenville Hancox, Ian Morrison, Bärbel Hess, Gunter Kreutz, and Don Stewart. 2007. Choral Singing and Psychological Wellbeing: Findings from English Choirs in a Cross-national Survey Using the WHOQOL-BREF. In *Proceedings of the International Symposium on Performance Science*, ed. Aaron Williamon and Daniela Coimbra, 201–207. Utrecht, Netherlands: AEC.

DeNora, Tia. 2000. *Music in Everyday Life*. Cambridge, UK: Cambridge University Press.

Godøy, Rolf Inge. 2001. Imagined Action, Excitation, and Resonance. In *Musical imagery*, eds. Rolf Inge Godøy and H. Jørgensen, 237–250. Lisse: Swets & Zeitlinger.

Godøy, Rolf Inge, Egil Haga, and Alexander Refsum Jensenius. 2006. Exploring Music-related Gestures by Sound-tracing: A Preliminary Study. In *Proceedings of the COST287-ConGAS 2nd International Symposium on Gesture Interfaces for Multimedia Systems (GIMS2006)*, ed. Kia Ng, 27–33. Leeds, UK.

Hennion, Antoine. 2011. Music and Mediation: Toward a New Sociology of Music. In *The Cultural Study of Music: A Critical Introduction*, ed. Martin Clayton, Trevor Herbert, and Richard Middleton, 80–91. New York and London: Routledge.

Jensenius, Alexander. 2007. *Action–Sound: Developing Methods and Tools to Study Music-related Body Movement*. PhD diss. University of Oslo.

Latour, Bruno. 1999. *Pandora's Hope : Essays on the Reality of Science Studies*. Boston: Harvard University Press.

Magee, Wendy L. 2011. Music Technology for Health and Well-being the Bridge Between the Arts and Science. *Music and Medicine* 3 (3): 131–133.

Magee, Wendy L., and Karen Burland. 2008. An Exploratory Study of the Use of Electronic Music Technologies in Clinical Music Therapy. *Nordic Journal of Music Therapy* 17 (2): 124–141.

MusicalFieldsForever. 2012. Musical Fields Forever. http://www.musicalfieldsforever.com.

Rolvsjord, Randi. 2006. Therapy as Empowerment: Clinical and Political Implications of Empowerment Philosophy in Mental Health Practices of Music Therapy. *Voices* 6 (3). https://voices.no/index.php/voices/article/view/283.

——. 2010. *Resource-Oriented Music Therapy in Mental Health Care*. Barcelona: Barcelona Publishers.

Ruud, Even. 1998. *Music Therapy: Improvisation, Communication, and Culture*. Barcelona: Barcelona Publishers.

——. 2010. *Music Therapy: A Perspective from the Humanities*. Barcelona: Barcelona Publishers.

Sevaldson, Birger. 2010. Discussions & Movements in Design Research. *FORMakademisk* 3 (1): 8–35.

Small, Christopher. 1998. *Musicking: The Meanings of Performing and Listening*. Middleton, CT: Wesleyan University Press.

Stensæth, Karette, and Even Ruud. 2012. Interaktiv Helseteknologi—Nye Muligheter for Musikkterapien? [Interactive Health Technology: New possibilities for Music Thearapy?] *Musikkterapi* 2: 6–19.

Stige, Brynjulf. 2010. *Where Music Helps: Community Music Therapy in Action and Reflection*. Aldershot, UK: Ashgate.

CHAPTER 16

ENGAGEMENT, IMMERSION AND PRESENCE

The Role of Audio Interactivity in Location-aware Sound Design

NATASA PATERSON
AND FIONNUALA CONWAY

"LOCATIVE" or "location-aware media" describes the concept of situating artwork in a real space, where the physical location and movement of the user affect the narrative and experience of the artwork. Movement within this space directly affects the digital content, creating an embodied experience that requires physical interaction. Therefore, locating an experience within a real space, whereby the content is interactive to user movement, adds to a sense of engagement with the space and to subsequent immersion and presence. Location-aware applications can be site-specific or developed in such a way that the same digital overlay can be deployed onto a number of generic locations.

The concept of locating a creative experience in a real space is not a new phenomenon. Many artists and multimedia developers have created experiences in a physical space, in a manner that reflects the surroundings and that focuses on accessibility for the general public. Artists and designers working in the area of location-aware media have, to date, explored ways to use technology that could be mobile, and there are projects that employ laptops and external GPS (global positioning system). For example, *ARQuake* (2000), created by the Wearable Computer Lab, is a location-aware version of the popular *Quake* game that uses a head-mounted display, mobile computer, head tracker, and GPS system to provide inputs to control the game. However, with the development of smartphones with integrated GPS tracking capabilities and high-bandwidth network access, location-aware media experiences are now less bulky and more readily accessible. The cell phone has progressed from its traditional social networking and communication purposes to becoming a creative tool employed in the art domain and entertainment industries.

In multimedia experiences, the aim has generally been to encourage immersion and interactivity for the user in order to fully experience the media narrative (Packer and

Jordan 2001). Multimedia experiences attempt to perceptually and psychologically immerse the participant in the experience to convey a meaning, concept, or feeling in order to create a sense of presence. "Presence" is defined as a psychological state or subjective perception in which an individual's experience is generated by and/or filtered through human technology (ISPR 2000). This can occur only when the participant is engaged or involved and immersed in the virtual space without being aware of the technology. Presence is a psychological state, which is induced not only by means of interacting with the technology but also by psychological determinants (e.g., meaningfulness of the situation, perceived realism). Becoming involved and immersed in a virtual space can be achieved by using multimodal interfaces that support the environment, concept, and narrative flow of the information being conveyed, in order to create a sense of presence. The aim of technologically mediated environments, as stated by Lombard and Ditton (1997), is for the "perceptual illusion of non-mediation," such that the participant is so immersed in the experience that they are consciously unaware of the technology. In gaining a sense of presence, the participant must engage with the medium and become immersed in the experience (Brown and Cairns 2004). Engagement, immersion, and presence are therefore interdependent, and together they are fundamental for the multimedia experience.

One can expect that a typical multimedia experience will use a variety of rich media, including visuals and audio, to encourage a multimodal experience. Interactive audio has a significant role in promoting a sense of engagement, immersion and presence in multimedia experiences. Location-aware media can use a multimodal interface with visual, haptic and auditory stimuli, all contributing to the overall immersive affect of the narrative (see also Chapter 2 in this volume). Audio, in this context, has been found to enhance the experience and contribute to a feeling of immersion (Behrendt 2012). According to Cohen (1999), music increases immersion and a sense of reality, and may assist the development of imaginative immersion and attention to media narrative (Lipscomb and Tolchinsky 2005). Due to the limitations of the current display size of most portable devices and cell phone screens, audio can play an important role in location-aware applications for engaging a user and creating a sense of presence in the "blended" game space. Location-aware audio experiences aim to construct immersive and engaging spaces, with the addition of the unique embodied interactivity that real-world location technology can provide. This embodied audio interactivity, in response to a physical space, presents a unique type of interaction as it creates a blend of a real and virtual world that responds to location cues. Even though other media such as the cinema or gaming also allow for the participant to be situated simultaneously in a real and technologically derived world, location awareness provides a different engagement, where the developer or artist can situate their narrative onto a real-world physical "canvas" that can act as a contextual cue, allowing for a physical interaction with the media content. This embodied interaction may increase engagement (both physical and psychological) with the medium and create a more immersive and unique relationship within a given location. Location-aware applications that use sound have influences from a number of disciplines, such as music, film, and gaming in terms of their content, style, and design. The unique mode of interaction of location-aware applications and importance of the physical space in triggering media content require a different approach for sound design that encourages a new type of engagement for the immersive experience.

Location-aware media can be either specific to a physical location or overlaid onto a generic space. Certain narratives rely on particular physical surroundings for the storytelling process and may include local architectural landmarks or historical events, therefore limiting where the experience can be undertaken (e.g., tourism guides). Generic narratives, such as some gaming applications, do not always need local information in their narrative and therefore can be experienced anywhere. Various methods of triggering audio content must be determined to reflect the unique embodied experience that incorporates physical movement, location, and interactivity for triggering and altering audio. There are a number of technical constraints that affect the implementation of interactive audio on the mobile platform. For example, memory availability and processor power can dictate the manner of interactive audio implementation. Therefore established game audio techniques that take into account constrained platforms and audio interactivity could be investigated, such as generative and adaptive audio (Guerraz and Lemordant 2008). These processes enable audio interactivity, file reusability, and real-time audio creation and could be used in location-aware applications. Furthermore, the manipulation of individual sound parameters such as pitch and timbre needs to be explored in conjunction with alternate methods of real-time music composition that respond to physical movement within a real space. And finally, in creating an engaging space through audio the impact of psychoacoustic effects in the sound design should be considered, as reverberation and spatialization have been linked with immersion (Cater, Hull, O'Hara et al 2007) and therefore would be desirable additions to location-aware applications.

The aim of this chapter is to examine techniques for sound composition and design as it is used in response to physical movement in real-world locations for creating meaningful experiences. By mentioning definitions of engagement, immersion, and presence, our intention is to draw attention to the role of interactive audio in creating meaningful experiences. We do not intend here to investigate the scope of work in this area and the impact of location-aware audio on an audience. This chapter therefore outlines the role of interactive audio in location-aware media and the issues involved, for an audience interested in creating immersive interactive audio and music for location-aware applications. In order to explain the context for the artistic intention of this type of work, the chapter will present an overview of the progression of sound art from gallery installation to situated urban locations. Looking more closely at music and sound design techniques, it will briefly discuss the influence of film and game sound on composition. Also, it will look at techniques for triggering content for location-aware scenarios, new and established methods of soundscape composition and audio interactivity. Finally, we will present and discuss techniques for emulating perception and experience of sound in a real-world location and say how they can impact on location-aware sound designs.

16.1 Locative Sound Art

Like most contemporary art, sound art draws on a variety of influences and incorporates areas such as the science of acoustics and contemporary music composition, with the distinction between sound art and experimental music often being unclear. Sound artists often aim to encourage their listeners to listen more "deeply" to their surroundings, aiming to create work that has an effect on listener engagement and immersion (see Oliveros, 2005). Sound artists tend toward encouraging movement away from the traditional concert-hall setting and into locations where the listener can be surrounded by naturally occurring sounds and where the environment is key to the experience of the work (LaBelle 2006). Alvin Lucier's *I Am Sitting in a Room* (1969), for example, focuses on demonstrating the effect that the room's acoustics or resonance has on a repeated phrase. The piece features the process of Lucier recording himself narrating a text and then playing the recording back in the room and rerecording it repeatedly. As each space has its own unique resonance, the original recorded sound changes as certain frequencies become more audible, such that the words become distorted and one begins to hear the "sound" of the space. Sound artists also focus on recreating a chosen space, as can be heard in Janet Cardiff's *Forty-part Motet* (2001). The piece aims to recreate the performance of Tallis's *Spem in alium* by the Salisbury Cathedral Choir by using a forty-channel stereophonic experience. The forty voices are separated and sent to individual speakers arranged in a circle in a space. By standing in the center of the speaker array, the audience is aurally and physically surrounded by the virtual choir. A physical engagement or interaction in the space is of importance for sound art and is also evident in Don Ritter's *Intersection* (1993), which requires that visitors interact with sensors controlling the sound of four or eight lanes of traffic rushing across a dark space. As the visitor moves through the space, the traffic sounds change with the end result of the soundscape depending on the physical interaction. In these examples, the sound artists aspired to draw attention to the space in which we hear sound or music, by devising new ways for listeners to be engaged by the work and be immersed in the soundworld.

In the same way that sound art wishes to investigate spaces and the physical experience of sound, so does locative media art, but with the physical extension into the real-world location (Galloway 2004). Locative media art uses technology to assist in relocating the experience away from the traditional gallery installation (Tanaka and Gemeinboeck 2008), blurring the lines between the exclusive gallery exhibition and our daily surroundings. Popular narrative-led examples include Blast Theory's *Can You See Me Now?* (2006) and *Uncle Roy all around You* (2003), works that place the narrative in an urban space. In *Uncle Roy all around You*, street players move throughout the streets of a city and collaborate with online players moving through a virtual model of the same town, to find Uncle Roy's office. These works explore user interactions with the social and spatial relations of a narrative in a given space and their cultural understandings. Similarly, sound artists have developed works that are intended to be experienced in real-world locations and through familiar interactions. Christina Kübisch's work in the 1970s first began exploring the sonification of interacting electromagnetic fields using small cubes with built-in speakers, which had to be held to listeners' ears as they approached wires within an empty installation space (Tittel 2009). This work later extended to the incorporation of wireless headphones in the experience of the work, which led to the creation of *Electrical Walks* (2003). In *Electrical Walks*, the participant moves throughout an urban space in which the effects of electrical currents of cell phones, elevators, light systems, and other devices are sonified, thereby augmenting the real-world location and adding another layer to the participant's experience of the space. In locative sound art such as this, the location becomes integral to the artwork; the boundaries between the real and the digital become blurred, the unheard becomes heard and the urban space is experienced in an altered way. Another example that explores the possibilities of interactive soundscapes in physical locations is *Sonic City* (2004), which uses a wearable system to create music that responds via sensors to changes in movement through an urban environment (Gaye 2003). By using an urban location as an interface, *Sonic City*'s soundscape is generated by physical movement, local activity, and urban ambient sounds.

Locative sound art makes an important contribution to location-aware experiences as it encourages the transition from the gallery space to a real-world location. It makes the experience accessible to a larger audience and removes the social barriers that contemporary sound art may sometimes have. Locative sound art usually requires a physical interaction with a space and aims to explore the interplay of space and sound. This is an integral aspect of location-aware sound creation, as audio interactivity within a physical space is paramount for this type of experience. The exploration of the effect of room acoustics on the sound or how the sound is experienced can also have an aesthetic influence on location-aware sound designs. Therefore locative sound art has provided a platform that artists and designers can build upon to include new technologically-mediated experiences that push the boundaries of established work. Having explored the contribution of locative sound art, in the next section we will briefly discuss the influence of film music composition and gaming audio interactivity techniques on locative audio soundscape creation.

16.2 Influences of Film and Game Audio Composition

Creative multimedia experiences that incorporate the physical world trace some of their stylistic and content influences from film and gaming. In researching sound design for location-aware applications, we have observed that these influences help to inform soundscape creation and interactivity. Film soundtracks are typically divided into background sound, sound effects, and dialog. Game audio can also reflect film sound in style and content but differs in that its soundscape is dependent upon the interaction of the avatar. These structuring methods and approaches are apparent in location-aware sound design. This section will briefly present certain elements of film and gaming that can also be found in location-aware media experiences. *SoundWalk* is an international sound collective based in New York City that produces various audio walks that mix fiction and reality, encouraging the listener to discover various city locations by being immersed in a dramatic cinematic soundscape. The artists attempt to recreate an experience that is related to the physical location and that uses the approach of film sound outlined above. The dialog gives the narrative or storyline, sound effects reflect the space, and background music provides the emotional mood. In game audio, the player's movement through a game space, decisions they make, including game choices, actions, and direction of movement, can be reflected in the sound. This interactivity is important for immersion and involvement with the gameplay and virtual space. In location-aware media, this same audio interactivity is a prerequisite for the experience in a real-world scenario. Every movement and real-world location change can modify the audio content in order for the user to feel connected to the digital media and physical space. Therefore, game audio techniques that employ interactivity and reflect avatar movement and game choices (Collins 2008b) are important methods used in location-aware applications. An example of game audio interactivity techniques in a location-aware setting is Blue Brain's *The Violet Crown* (2012). This musical composition is overlaid onto an area of Austin, Texas, with the soundscape changing seamlessly as the listener moves through the physical space. The interaction of the listener with the space is akin to how they would move through the virtual world of a game. As well as sound interaction methods and techniques, the developments made by gaming developers with regard to storage and processor saving methods (such as file reusability and variation; Collins 2008b), have real implications for location-aware sound designs, where cell phone storage capabilities may be limited. The technical methods used in gaming can therefore offer new ways for the locative media artist or developer to create interesting musical and interactive soundscapes, in a manner that is less taxing on the technology than single, long-playback files.

Location-aware creative media projects can therefore draw on the long history of multimedia art, film, and gaming, as well as the technical developments that have been made alongside the advancement of these art forms. The remainder of the chapter will look more closely at current location-aware projects and applications, identifying how these approaches have been carried out and offering potential new directions and innovations for future location-aware sound designs.

16.3 The Mobile Platform and Location Awareness

With the addition of location technology, digital media has now become a personalized experience that is revolutionizing the way people engage and experience their everyday environment (Bull 2000). Furthermore, with continuing advances in augmented-reality technology, new methods of implementing and designing audio need to be developed that push existing boundaries, both aesthetically and technologically. For example, the integration of a GPS receiver, accelerometer, and three-axis internal compass as standard in smartphones has seen cell phone gaming move from games such as Nokia's *Snake* (1998) on the Symbian operating system to graphically rich and interactive games such as *Epoch* (2012) for the iPhone, which is reminiscent of a traditional console presentation. To take developments that step further and embrace locative awareness, the challenge is to transfer the complexity of established console game audio onto a mobile platform and combine it with the unique embodied interactivity and immersion that location technology affords. In the broad spectrum of locative audio, from artworks to commercial multimedia applications, the aim therefore is to incorporate the rich possibilities that location technology offers and to develop formal processes of design to create soundscapes that are interactive and participatory, and that respond to real-world locations.

The remainder of this chapter will explore the technical and aesthetic concerns of location-aware media experiences and suggest some possibilities for extending and developing methods of interactive audio implementation. The sound design framework must be such that it is unencumbered, smooth in its transitions, and automatic, in the sense that the user is not consciously aware of the technology in order to encourage immersion. The technical challenges presented to the artist or sound designer while trying to implement audio on the smartphone platform will be looked at first. Compositional processes and input data that will control audio parameters must also be considered and will be looked at next. And finally, the technical constraints of processing digital audio in real time and its effects will be presented.

16.4 Triggering Audio Content

Physical interactivity is a requirement for location-aware applications, therefore an approach to sound design that creates responsive audio, adaptive to the real world location and the user's movement, is more effective and desirable. The first concern in location-aware applications is establishing the means, which are various, to trigger audio content in a real-world situation where the location of the participant needs to be established. In the case of locative art in an urban space, GPS technology is the most easily accessible and commonly used. However, problems with this technology persist, such as network unreliability, where the person's actual location is incorrect, so the incorrect audio file is triggered or, worse still, no audio is triggered. This error can at times be an eight-meter location discrepancy (the amount of location error that the GPS reading gives in regard to the true position in the physical space; Paterson et al. 2012) and hence disrupts the sonic experience and desired immersive quality. We have found that attention needs to be given to location selection and that locations should both support the narrative context and provide sufficient open space for best accuracy to enable the GPS system to register three points from three different satellites. Additionally, the internal compass of the smartphone should be able to accurately determine the direction the person is facing, thereby enabling the sound artist to explore the possibilities of anchoring audio to specific architectural landmarks.

Skyhook (a location-aware services company) has developed a Wi-Fi positioning system for determining geographical location that offers an alternate location technology. The system uses GPS technology and also incorporates Wi-Fi access points and cell tower information to present a multilayered approach to determining location. This allows for multiple ways to trigger content in areas where, for example, GPS might be weak, as is found in urban spaces with limited network coverage, such as underground car parks and some indoor spaces. There are alternative methods of triggering content that can be explored, such as *SonicNotify*, a media delivery platform that uses inaudible frequencies from televisions and radios to trigger content on the smartphone. With the continual progression of smartphone technology, we will likely see the development of object recognition, using the phone camera function, to trigger audio. Currently, applications exist that can identify faces (e.g., *Recognizr*, 2012, by The Astonishing Tribe company) or logos and books (e.g., Google Goggles, 2012), and these facilities could be harnessed to trigger interactive audio. Physiological measurements such as heart (*Instant Heart Rate*, 2012) and breathing rates could also be incorporated into the sound design, with the biological state of a user triggering audio that is related to specific real-world locations and narrative. The future presents endless and exciting possibilities.

16.5 Aesthetic Concerns in Location-aware Sound Design

Location-aware sound designs can be seen as an extension of film and gaming audio and music composition, and can therefore draw heavily on these approaches and stylistic elements. Of most relevance is the way music and audio are categorized in film and gaming, dividing their use as background sound, ambient audio, dialog, and music, to broadly summarize the categorization. Background sound was previously discussed as an important aesthetic tool in creating and supporting the mood of a narrative or atmosphere of a media space. Once a method of triggering audio in a physical location is established, a variety of methods can be used to create a background sound and mood. A common method of creating background sound is to loop audio files using the various built-in smartphone media players (e.g., Android's *MediaPlayer* and *SoundPool*). However, using a repeated sound file may lead to familiarity with the audio content and therefore possible boredom for the listener. As game audio interactive techniques are well established and documented (Collins 2008a, 2008b) the techniques offer great insight to location-aware sound designers and artists interested in creating location-aware audio. One such example is the "open form," where linear background sound can be broken into segments or wavelets (Collins 2008b, 160, 171) and the soundscape generated and controlled by the program code in real time. The location-aware game *Viking Ghost Hunt* (Paterson et al. 2012) uses this method, breaking a long musical phrase into segments of one- to two-second durations and allowing files to be selected randomly. These numerous random files are then layered onto one another and, when combined with each other and additional musical elements, create the perception of a continuous, ever-changing background. This process addresses the need to maintain some unpredictability in the audio by allowing for the simultaneous playback of multiple files with a random "sleep" time or pause between files, thereby preventing the establishment of a predictable pattern (Paterson et al. 2012) and encouraging continued engagement within a space.

Wavelets are useful in sound design as they can present high-quality realistic recordings or samples of actual sound. However this high quality comes with a price, namely that the files are often too large given the limited memory available on a smartphone platform, thereby often making them unsuitable for location-aware projects. MIDI files present an alternative that is more adaptable and can be used alongside and sometimes instead of Wave files. MIDI sounds were previously only "synthesized" sound and based on a wavetable synthesis process, and therefore for some artists not as aesthetically pleasing. However, MIDI used to control soundbanks (sample-based synthesis) has the advantage of using sampled real sounds, providing better sound quality but requiring less processing power. For location-aware applications, sample-based synthesis MIDI can offer more adaptability in generating background sound compared to Wave files alone, while also maintaining audio quality and allowing for multiple simultaneous file playback and the reusability of files, which also requires less memory.

As location-aware media experiences are situated in real-world spaces, sounds from the surrounding environment are audible (unless the listener is wearing circumaural, closed, headphones), thereby forming an additional layer in the audio narrative that can be incorporated into the background soundscape. As this external audio is continuous, it can smooth over gaps in the cell phone audio continuity, which may occur due to the technical constraints of triggering multiple audio files. *Net_Dérive* (Tanaka and Gemeinboeck 2008) is an example of the use of the mobile platform and external GPS unit to incorporate recorded and processed external sounds of the city into the soundscape, creating an abstracted interactive experience of the urban space. This process is also extensively used by developers RjDj, whose research includes the development of sonic experiences such as the *Inception* (2010) application, based on the recent film of the same name. Gaming middleware (services beyond those available from the operating system) such as FMod can be used to control the triggering and playback of audio or MIDI files, with the many functionalities of gaming, in response to a user's movements and location. Variability is very desirable for interactive sound, and middleware programs such as this offer more functionality and solutions to the sound designer, such as the ability to vary pitch, tempo, and processing of recorded sounds. These solutions will likely play an increasing role in the development of sequencing audio for smartphone platforms.

16.6 Computer Music Composition

Using a computational approach such as MIDI and wavelets in creating a soundscape for location-aware applications requires a programmed and instruction-based method of creating a real-time soundscape that can offer a possible solution for technically constrained platforms. However, procedural techniques are also to be found in traditional music settings that are used for creative as well as technological purposes, and are described as algorithmic compositions. Algorithmic composition, sometimes also referred to as automated composition, according to Alpern (1995), "is the process of using some formal process to make music with minimal human intervention," viewing music procedurally. The concept of formal processes for music composition has as long history and with the advent of the personal computer has paved the way for more complex and innovative compositions. This approach to composition can also be transferred to the smartphone platform, where the synthesized or sampled sounds are controlled by the programmer's code. *Kepler's Orrery* (2010) generative music application (for the iPhone) is an example of an algorithmic process for sound composition that uses gravity equations to compose and play ambient music. Each "piece" of music is defined by planetary gravitational equations where the user can visually build new worlds and change planet positions. These changes also affect the soundscape, which varies with every new simulation, as each planet system has a different set of melodies that play on different instruments. As well as melodic phrases and instrumentation, other musical parameters are controlled by the equations such as pitch, tempo, rhythm, and harmonic patterns. In addition to the soundscape being affected by changes in physical equations, sensors within the phone (such as an accelerometer) vary the soundscape when the phone is tilted. Using an external input to control musical parameters is important for location-aware applications as it provides a method of altering sound by physical movement. Other sensory input related to physical movement within a space can also be used to control procedurally determined soundscapes and audio interactivity. Data such as the GPS location, compass readings, the location, climate, and time of day, can be retrieved from the smartphone and "sonified," that is, used to control the production of sound within a predefined algorithmic process. As the listener moves through a real-world space, the digital soundscape can be informed by the sonification of this input data, either by triggering given sound files or by controlling various musical parameters of the sound file.

Another technique from computer music that can be considered for location-aware sound design is granular synthesis, which describes the splitting of audio wavelengths into smaller pieces of around 1 to 50 milliseconds called "grains." These grains can then be layered, each playing at different speeds, volumes, and frequencies which combine to form a soundscape. Many different sounds, and resulting soundscapes, can be created by varying sample waveform parameters, such as the waveform envelope, and the number and density of grains. Granular synthesis offers interesting possibilities (Paul 2008) for location-aware audio, as it not only provides a new creative tool for interactively changing the timbre (texture and quality) of a sound but may also be an alternative to using multiple stored audio files, samples, or MIDI wavetable synthesis. For example, a small number of stored audio samples could be granulated to create a multitude of varied sounds that could instantly change in response to movement. The *Curtis* (2009) iPhone application is an example of granular synthesis on a smartphone platform that allows a person to sample and then manipulate recorded sound, using granular techniques to create varying soundscapes in real time. Therefore granular synthesis can offer a method of creating real-time soundscapes from a small set of audio files, including sounds that are already programmed into the phone and audio that is being captured during the experience by the listener. While other approaches, such as using MIDI control, offer limited control of pitch and tempo data, granular synthesis is very powerful and offers a unique aesthetic outcome. Granular synthesis real-time processing of audio can alter audio in a more complex way, evident in changes in timbre and the significant alteration of sound. These elements can be altered instantly, based on various inputs such as speed of movement and location, and ultimately need less memory for storage.

While computer music techniques offer new ways to think about location-aware real-time composition, technical constraints on the smartphone platform are still an issue, as these techniques require a heavier use of the processing unit. However, this remains an exciting area for further exploration.

16.7 Perception of Sound and Space in Location-aware Design

Psychoacoustics is an area of science that deals with how we physiologically and psychologically respond to and experience sound, and the space within which we experience sound. This knowledge has been applied to many fields, for example digital signal processing, where it has influenced the development of audio compression formats such as MP3, and the entertainment industry, where it has impacted on the design of accurate reproduction of music in theaters and homes. Musicians and music producers apply this knowledge in the composition and production stages, mixing out unwanted frequencies and creating immersive soundworlds by positioning sound in different locations in a space. It follows that the body of knowledge would be of considerable relevance to location-aware sound design (Paterson et al. 2010a). This section will present our thoughts on the application of reverberation and audio spatialization in location-aware sound design.

Reverberation is a valuable cue in understanding the type of space where a sound is occurring (Rumsey 2001). It explains the persistence of sound in a space after the original sound is produced. For example, in a cathedral the sound of footsteps walking will remain for longer than in a carpeted small room. As reverberation provides an important cue for our understanding of a space, it is a considerable aid in the design of sound that is meant to be a realistic representation of that space, one that is attempting to immerse the listener in that space. Two types of reverberation can be used in sound designs: "artificial" and "convolution-based" reverberation (the latter requiring the measuring, calculating, or approximating the room impulse response). Convolution-based reverberation is the process of simulating the reverberation of a physical (or virtual) space by using the audio response of a real-world space and is based on the mathematical process of convolution (multiplying two signals to create a third) (Begault 1994). Artificial reverberation is an approximation of real reverberation and involves controlling various parameters such as time delay, room size, and the number of early and late reflections. The game *Thief: Deadly Shadows* (2004) uses multiple simultaneous reverberation settings (echoes, delays) and occlusion effects in game locations to help simulate real-world aural properties. This differs from older reverberation models, which allowed only a single environment to be reverberated at a time, resulting in all sounds having the same reverberation in the same room. Furthermore, in *Thief: Deadly Shadows*, reverberation is an integral part of interactive gameplay, where sound cues not only tell the player of other characters in the vicinity, but also indicate how much noise the protagonist makes when moving about an area. This concept can be transferred to location-aware scenarios where real-world sound can be recorded and reverberated in real time and in response to a listener's surroundings. An example of this is *Dimensions* (2012), which uses Pure Data to apply reverberation to recorded environmental sounds. However, the reverberation is not responsive to the real-world location. It would be desirable for location-aware sound designs to incorporate reverberation parameters that respond to location information and a listener's movements. For example, if a person were in a physical space with many reflective surfaces, a longer reverberation time could be applied to the sound design in real time, thereby reflecting the concepts of physical space and interactivity of *Thief: Deadly Shadows*. Convolution-based reverberation also offers interesting possibilities for exploration in this area. Impulse response libraries containing samples from specific physical locations or similar acoustical properties could be used in real-time to generate the appropriate reverberation of that same space. However basic convolution (in the time domain) is computationally expensive and typically cannot respond at the interactive rates necessary for location-aware experiences. Other techniques such as convolution in the frequency domain or using the Graphics Processing Unit (GPU) instead of the Central Processing Unit could be explored as a way to enhance computational speed.

The ability to position audio in different locations within the space (spatialized audio) is also very valuable in the creation of immersive soundscapes. In the real world, sound is presented in a three-dimensional manner to enable the auditory system to recognize and locate where a sound is emanating from (Ashmead, Hill, and Taylor 1989). For immersion, realism in sound propagation is important (McMahon 2003). The ability to control real-time interactive changes of spatial audio in response to a person's movement, and in a manner that reflects real-world scenarios, is also very desirable. Situating and positioning sound in different locations within a soundscape is not only important in creating a believable and realistic sound design but can also be used in creating soundscapes that represent artificial or fantasy spaces, for example, in the film *Avatar* (2009). Additionally, using spatial audio techniques presents a method of separating competing sounds of similar frequencies into different spatial fields. For example, dialog and certain sound effects may both be situated in the mid-frequency range and because of this compete for the listener's attention (Collins 2008b). Positioning these sound sources in different locations within the sound field allows for the listener to separate the sources and to hear their content more clearly.

Spatial audio is especially interesting for location-aware applications. The ability to anchor sound to real-world locations regardless of the direction the person is facing, would support the concept that the soundscape is reflective of that particular physical location, a valuable cue in the creation of realistic and immersive sound for location-aware experiences. For example, if a sound is to be representative of an outdoor building or is perceived to be emanating from that location, it would make sense that it remained perceptually where the building was situated and not move as the person turns their head. This type of spatial accuracy requires the audio engine to use Head Related Transfer Function (HRTF) binaural audio filters in real time, with headphones for playback. These filters take into account the effect of the ear structure, head, and torso on the sound input before it reaches the eardrum for sound localization (Gardner and Martin 1995). *Demor* (2004) is a location-aware 3D audio first-person shooter game that uses real-time processing of spatialized sound but that requires the use of a wearable computer and head tracker. In this application, audio is reactive to the player's location, head position, and physical movements, with audio files being adjusted accordingly in real time on a dedicated audio engine. The audio engine designed for *Demor* most likely uses a generic HRTF database for the 3D audio representation in combination with GPS technology. This processing is taxing on the smartphone and currently not available on the platform as it stands, hence the need to design a customized audio engine. We are not aware of any *unencumbered* location-aware application, at this time, that can position sounds accurately.

Approximations of accurate spatial audio can simulate realistic binaural sound on the smartphone, by using software that combines HRTF-based audio panning with a simulated model of the effects a room or space may have on a sound (such as wall reflections, reverberation, and the effect of movement, such as the Doppler effect). An example of this is *Papa Sangre* (2010), a smartphone game played entirely through sound, using a complex soundscape that includes a real-time binaural effect. At present, however, it is not responsive to the listener's location. Also, with binaural audio, headphones must be worn in order to present a clean signal to each ear in order to remove the problem of crosstalk, whereby a sound signal that is meant to be transmitted to one channel crosses over or interferes with another, distorting the original audio "image." However, in experiments listeners still find it difficult to distinguish sounds that are in front from those that are behind the head (Begault 1994), even in the case of 3D audio systems that use HRTF calculations. Currently, the most effective method of spatializing sound in location-aware smartphone applications, that is responsive to a listener's movement and direction, has been to pan sound and use time delays to audio files in order to simulate spatial audio that responds to physical cues (Paterson et al. 2012). This approach uses the notion of approximating aspects of HRTF filters, inherently creating the effect of spatialization. Ultimately it would be desirable to overcome the technological constraints of mobile technology to include real-time binaural audio for a more interactive location-driven sound design (Martin and Jin 2009).

16.8 Pure Data and the Smartphone: Current Trends for Audio

Due to the technological challenge of implementing interactive audio on a smartphone platform, other avenues are being explored to control and manipulate audio files and their musical parameters other than middleware programs. Pure Data (Pd) is already popular with artists in creating interactive computer music and multimedia works, but it is only recently that this program has become available for smartphone devices. Pd "patches," modular multiplatform, reusable units of code, controlling various aspects of audio, can now run on smartphones by using libpd (Pd library) and RjDj code. Audio functionalities of Pd include the analysis of incoming audio and the ability to create pitch changes, using accelerometer data from the smartphone to set the background tempo and using granular techniques to change the timbre of a soundscape. Additionally there are aspects of algorithmic processes and synthesis that can be controlled by Pd in response to movement within a physical location. An example of using Pd in this way is *Dimensions*, a smartphone application using Hans Zimmer's music from the movie *Inception* (2010). Aspects of the soundtrack respond to a listener's movements, with accelerometer data from the smartphone inputting to Pd and altering tempo and rhythmic patterns according to the speed at which the listener moves, to completely fading away when the listener stops. Additionally, audio samples recorded with the smartphone microphone are scrubbed repeatedly forwards and backwards (slowly moving across the sound file) using granular techniques, stretching them out in time (changing the sonic quality or timbre) and adding to the overall soundscape.

Pd is fast becoming a powerful tool for filtering and manipulating sounds and could be used to process samples for reverberation and spatial audio. This is relevant to location-aware projects as it provides a means of controlling many of the audio manipulation technologies discussed in the previous section. Therefore, Pd can provide audio manipulation beyond the capabilities of middleware programs and can afford location-aware applications with sound design physical interactivity from a variety of inputs.

Even though Pd offers an innovative approach to audio interactivity for location-aware experiences, there are still a number of fundamental technical issues regarding functionality on the smartphone that hinder the advancement of how audio is used and designed. The number of samples that can be buffered simultaneously is limited, especially as all smartphone processors require control from the CPU, which of course also controls various other tasks that require prioritization. Additionally, location services require significant power to function effectively, which is problematic for location-aware applications. However, recent developments have aimed to improve battery efficiency. In response to processor limitations for interactive audio in console gaming, GPUs only used for processing graphics have been employed for audio processing (Tsingos, Jiang, and Williams 2011). This is an approach that smartphone processor developers are also undertaking, with GPUs being used for other parallelizable computing tasks, such as speech recognition, image processing, and pattern matching. Hence increased performance is being achieved by dividing tasks between the CPU and GPU units. All of these innovations signal a positive and exciting move for location-aware sound designs.

16.9 Conclusions

Sound designs that focus on interactivity between music and sound composition, control of psychoacoustic cues, and audio mixing in response to a listener's physical movement and location can assist in how immersion and presence are felt. As mobile technologies continue to progress, increasing processing power is affording composers and sound designers more avenues in creating complex, interactive, and immersive soundscapes. Established game audio techniques of adaptive, interactive digital audio controlled by real-time processing programs, together with alternative compositional tools, can meet the current requirements of mobile technological constraints. Future innovative location technologies may be used in creatively triggering content in spaces previously restricted to GPS technology. For artists and designers, the continued advances in technological and compositional authoring tools signal a new and exciting time for location-aware immersive experiences.

References

Alpern, Adam. 1995. Techniques for Algorithmic Composition of Music. http://citeseerx.ist.psu.edu/viewdoc/download?doi=10.1.1.23.9364&rep=rep1&type=pdf.

Ashmead, D. H., E. W. Hill, and C.R. Taylor. 2009. Obstacle Perception by Congenitally Blind Children. *Perception and Psychophysics*, 46 (5): 425–433.

Begault, Durand. 1994. *3D Sound for Virtual Reality and Multimedia*. San Diego, CA: Academic Press.

Behrendt, Frauke. 2012. The Sound of Locative Media. *Convergence: The International Journal of Research into New Media Technologies* 18 (3): 283–295.

Brown, Emily, and Paul Cairns. 2004. A Grounded Investigation of Immersion in Games. *ACM Conference on Human Factors in Computing Systems, CHI 2004*, 1297–1300. New York: ACM.

Bull, Michael. 2000. *Sounding out the City: Personal Stereos and the Management of Everyday Life*. Oxford: Berg.

Cater, Kirsten, Richard Hull, Tom Melamed, and Robin Hutchings. 2007. An Investigation into the use of Spatialised Sound in Locative Games. Paper presented at the *CHI 2007 Conference*, San Jose, CA. April28—May 3.

Cater, Kirsten, Richard Hull, Kenton O'Hara, Tom Melamed, and Ben Clayton. 2007. The Potential of Spatialised Audio for Location Based Services on Mobile Devices: Mediascapes. In *Proceedings of the Spatialised Audio for Mobile Devices (SAMD) Workshop at Mobile HCI*, September 2007.

Cohen, Annabel J. 1999. Functions of Music in Multimedia: A Cognitive Approach. In *Music, mind and Science*, ed. S.W.Yi, 40–60. Seoul: Seoul University Press.

Collins, Karen, ed. 2008a. *From Pac-Man to Pop Music: Interactive Audio in Games and New Media*. Aldershot, UK: Ashgate.

Collins, Karen. 2008b. *Game Sound: An Introduction to the History, Theory and Practice of Video Game Music and Sound Design*. Cambridge, MA: MIT Press.

Demor, accessed September 2012. http://www.student-kmt.hku.nl/~g7/redirect.

Electrical Walks, Christina Kubisch. http://www.christinakubisch.de/en/works/electrical_walks, accessed October 29, 2013.

Forty Part Motet, Janet Cardiff. http://www.cardiffmiller.com/artworks/inst/motet.html, accessed October 29, 2013.

Galloway, Anne. 2004. Imitations of Everyday Life: Ubiquitous Computing in the City. *Cultural Studies* 18 (2/3): 384–408.

Gardner, William G., and Keith D. Martin. 1995. HRTF Measurements of a KEMAR. *Journal of the Acoustical Society of America* 97 (6): 3907–3908.

Gaye, Layla, Ramia Mazé, and Lars Erik Holmquist. 2003. Sonic City: The Urban Environment as a Musical Interface. In *Proceedings of the 2003 Conference on New Interfaces for Musical Expression NIME-03), Montreal, Canada*, 109–115. Montreal: McGill University, Faculty of Music.

Guerraz, Agnès, and Jacques Lemordant. 2008. Indeterminate Adaptive Digital Audio for Games on Mobiles. In *From Pac-Man to Pop Music: Interactive Audio in Games and New Media*, ed. Karen Collins, 55–73. Aldershot, UK: Ashgate.

International Society for Presence Research (ISPR). 2000. The Concept of Presence: Explication Statement. http://ispr.info/about-presence-2/about-presence/.

Intersection, Don Ritter. [n.d.] http://aesthetic-machinery.com/intersection.html.

LaBelle, Brandon. 2006. *Background Noise: Perspectives on Sound Art*. New York: Continuum.

Lipscomb, Scott, and David Tolchinsky. 2005. The Role of Music Communication in Cinema. In *Musical Communication*, ed. Dorothy Miehl, Raymond MacDonald, and David J. Hargreaves, 383–405. Oxford: Oxford University Press.

Lombard, Matthew, and Theresa Ditton. 1997. At the Heart of it All: The Concept of Presence. *Journal of Computer-mediated Communication* 3 (2).

Martin, Aengus, and Craig Jin. 2009. Psychoacoustic Evaluation of Systems for Delivering Spatialized Augmented-Reality Audio. *Audio Engineering Society* 57 (12): 1016–1027.

McMahon, Alison. 2003. Immersion, Engagement, and Presence: A Method for Analysing 3-D Video Games. In *The Video Game, Theory Reader*, ed. Bernard Perron and Mark J. P. Wolf, 67–86. New York: Routledge.

Oliveros, Pauline. 2005. *Deep Listening: A Composer's Sound Practice*. Lincoln, NE: Deep Listening.

Packer, Randall, and Ken Jordan, eds. 2001. *Multimedia: From Wagner to Virtual Reality*. New York: Norton.

Paterson, Natasa, Katsiaryna Naliuka, Soren Kristian Jensen, Tara Carrigy, Mads Haahr and Fionnuala Conway. 2010a. Design, Implementation and Evaluation of Audio for a Location Based Augmented Reality Game. *Proceedings of the 3rd International Confererence on Fun and Games*, 149–156. New York: ACM.

——. 2010b. Spatial Audio and Reverberation in an Augmented Reality Game Sound Design. *Proceedings of the 40th AES Conference: Spatial Audio, Tokyo, Japan*. New York: Audio Engineering Society.

Paterson, Natasa, Gavin Kearney, Katsiaryna Naliuka, Tara Carrigy, Mads Haahr, and Fionnuala Conway. 2012. Viking Ghost Hunt: Creating Engaging Sound Design for Location-aware Applications. *International Journal of Arts and Technology* 6 (1): 61–82.

Paul, Leonard. 2008. An Introduction to Granular Synthesis in Video Games. In *From Pac-Man to Pop Music: Interactive Audio in Games and New Media*, ed. Karen Collins, 135–150. Aldershot, UK: Ashgate.

Rumsey, Francis. 2001. *Spatial Audio*. Oxford: Focal Press.

SonicNotify. http://sonicnotify.com/.

Tanaka, Atau, and Petra Gemeinboeck. 2008. Net_Dérive: Conceiving and Producing a Locative Media Artwork. In *Mobile Technologies: From Telecommunications to Media*, ed. Gerard Goggin and Larissa Hjorth, 174–186. New York: Routledge.

Tittel, Claudia. 2009. Sound Art as Sonification, and the Artistic Treatment of Features in Our Surroundings. *Organised Sound* 14 (1): 57–64.

Tsingos, Nicolas, Wenyu Jiang, and Ian Williams. 2011. Using Programmable Graphics Hardware for Acoustics and Audio Rendering. *Journal of the Audio Engineering Society* 59 (9): 628–646.

SECTION 4

PERFORMANCE AND INTERACTIVE INSTRUMENTS

CHAPTER 17

MULTISENSORY MUSICALITY IN *DANCE CENTRAL*

KIRI MILLER

HAVE you ever had a song stuck in your head? The chorus cycles around, repeating indefinitely, and might fade out only when you substitute something with an even catchier hook. Now, consider what it might be like to have a song stuck in your body. As one *Dance Central* player put it, "Every time I hear the song—or I download the song for myself to listen to it because I like the track so much—then I can't help but think of the moves. When I'm listening to the track on my way to work, or if I'm at home: it runs through my head, and I can't help myself. It's become basically attached."[1] This player is describing a dancer's habitual aural/kinesthetic experience of music. He explained, "For me, music isn't about just listening to music. There's always been a movement attached to the music. I can't listen to great music and not want to dance." But what does it mean to forge that sound/body connection by playing a video game?

The *Dance Central* games teach players full-body choreography routines set to popular club music. The first game in the series, released in 2010, was among the launch titles for the Microsoft Xbox Kinect, a motion-sensing infrared camera device that creates a gestural interface for the Xbox 360 game console. The Kinect was designed to allow players to interact with games using a full range of body movements, rather than by pressing buttons on a traditional hand-held controller or moving a motion-sensitive controller in space (the previous gesture-based innovation associated with the Nintendo Wii). These new affordances encouraged game developers to explore the potential of gesture-based user interfaces. Unsurprisingly, game design for first-generation Kinect titles generally focused on movement-related features rather than innovative audio. However, the fact that *Dance Central*, the system's most successful launch title, revolves around popular dance music offers a reminder that digital gaming is always multisensory. Compelling games integrate audio, visual, and kinesthetic elements in the service of immersive experience (Grodal 2003; Salen and Zimmerman 2004; Collins 2008; Miller 2012).

The *Dance Central* series was created by Harmonix Music Systems, the same company that developed the *Guitar Hero* and *Rock Band* games. By 2013, the series included three games: *Dance Central* (2010), *Dance Central 2* (2011), and *Dance Central 3* (2012). Each game has its own musical and choreographic repertoire of about forty songs. The musical selections range over several decades of club hits, with an emphasis on hip-hop and electronic dance music. Additional tracks are released regularly and can be purchased as downloadable content (DLC). The three games feature increasingly sophisticated multiplayer options and narrative components, along with more subtle changes in graphic design and dance pedagogy. However, the core gameplay experience is consistent across the series. Players begin by choosing a song from a list. Each song has its own dance routine, which can be learned and performed at three different difficult levels: Easy, Medium, or Hard. The entire song list also proceeds from easier to more difficult dance routines, categorized as Warmup, Simple, Moderate, Tough, Legit, Hardcore, or Off the Hook.[2] The resulting spectrum of difficulty levels offers options that suit dancers of widely varying abilities. Gameplay videos posted on YouTube include performances by small children, gym-sculpted club-going types, heavily pregnant women, self-identified hardcore gamers, and professional dance teachers.

Once players have selected a song, they choose from a selection of avatars (or use the song's default avatar) and decide whether to proceed in performance or rehearsal mode. In either mode, they perform the dance routine by mirroring an avatar's movements, aided by a series of flash cards on the side of the screen that provide a name and icon for each upcoming move (see Figure 17.1). In the rehearsal mode, players work through the routine one move at a time, repeating difficult sections as needed, while getting instruction and encouragement from a voiceover dance teacher: "Left, together! Right, together!... You almost got it!... That was off the hook!... I see you, I see you!" In the performance mode, these exhortations are replaced by cheers from an admiring crowd, as well as quantitative evaluation provided by a numerical scoring system.

FIGURE 17.1 Screenshot from *Dance Central 2*. Courtesy of Harmonix Music Systems, Inc., via http://www.dancecentral.com/press. All rights reserved.

Many of these design features will be familiar to the millions of people who have played the *Guitar Hero* and *Rock Band* games. All of these games are built around a graded repertoire of popular music tracks. They employ a distinctive onscreen notation system to guide players through songs as they unfold, they offer separate "practice" and "performance" experiences, and they cultivate new embodied knowledge at the intersection of virtual and visceral experience (Miller 2012). However, *Dance Central* differs from its rock-performance-oriented predecessors in that gameplay does not affect musical playback. *Guitar Hero* and *Rock Band* make players feel responsible for their musical performances by providing separate audio tracks for each instrumental part, interrupting playback when players make technical errors, and offering customizable sound effects and opportunities for improvised fills. These games provide textbook examples of interactive audio; as Karen Collins writes, "While [players] are still, in a sense, the receiver of the end sound signal, they are also partly the transmitter of that signal, playing an active role in the triggering and timing of these audio events" (Collins 2008, 3). A *Guitar Hero* guitar solo dissolves into twangs and clanks when an inept player picks up the game controller, a design feature that creates an intimate relationship between physical input and audio output. *Dance Central* is different: the songs don't react to good or bad dancing. Nor are there variable outcomes in the avatar's dance performance: the avatar offers a model for the player's dancing, rather than a mirror that reflects the player's movements. The on-screen dancer is an instructor, not a puppet—that is, not a conventional game avatar at all. If you miss a particular arm motion, the screen dancer's arm will glow red to show you where you are making a mistake, but the screen body won't actually perform the mistake. Meanwhile, the song plays on, just as it would at a club.

Thus, it seems that *Dance Central* is not oriented around interactive audio, at least not as it has traditionally been conceived. Indeed, given that the player's movements don't guide those of the on-screen dancer, some gamers have questioned whether *Dance Central* is truly interactive at all. As two commenters responded to an online review of the game,

> SIKEOSOSHULL: without a controller can we still call ourselves gamers?
> TILIAN: So you're not controlling anything then? Just trying to mimic an avatar? Lame.
>
> (GameTrailers.com 2010)

Yet dancing to music is fundamentally an interactive, sound-oriented experience, one that brings musical listening, patterned physical action, and affective experience into intimate alignment (Garcia 2011). Moreover, *Dance Central*'s rehearsal mode, which relies on verbal dance instruction and evaluative feedback, adds another distinctive audio element to the gameplay experience—one that reproduces the multichannel oral/visual/kinesthetic transmission process typical of dance pedagogy (Hahn 2007).

The *Dance Central* games challenge us to develop models of interactive audio that move beyond considerations of dynamic soundtrack music, spatializing sound effects, or musical performance simulators to address the role of sound in multisensory interactivity. *Dance Central* draws attention to "the modularity of sensory technologies . . . and of the relations between senses, subjects and technologies" (Sterne and Akiyama 2012, 547). Choreographers translate popular songs into dance routines. Game designers create a motion-capture data archive of real people performing the routines; these performances become the game code that will animate dancing avatars. The Xbox translates that code back into sounding music, verbal prompts, and graphics, so that players can learn how to experience all this multisensory data through their own moving bodies and listening ears. Some players take this process a step further by recording their performances and posting videos online, as well as engaging in vigorous debates about the choreography for each song and how well it suits the music.

Dance Central thus relies on multiple technologies of transduction, which "turn sound into something accessible to other senses" (Pinch and Bijsterveld 2012, 4). As Sterne and Akiyama observe, "this extreme plasticity lays bare the degree to which the senses themselves are articulated into different cultural, technological, and epistemic formations" (2012, 545). I will focus on three aspects of trans-sensory transformation and multisensory musicality in *Dance Central*: how designers turn song into dance, how players listen like choreographers, and how the games represent and foster a dancerly sensibility—a way of sensing like a dancer.

17.1 Turning Song into Dance

In April 2012, I attended PAX East, the annual Penny Arcade Exposition game convention in Boston, Massachusetts. Thousands of gamers and game industry employees milled around the Boston Convention Center, trying out new games and attending panel talks by game designers. The *Dance Central* booth featured a large stage in the middle of the exhibition floor, where Harmonix choreographers, designers, player-relations staff, and convention attendees danced in front of a huge and varied audience. Matt Boch, the *Dance Central* project director, agreed to an hour-long recorded interview, and we spent some time discussing the relationship between music and choreography in the games.

KM: I was curious about how you think of *Dance Central* as being about interactive audio, or as being about music? As compared maybe to *Guitar Hero* and *Rock Band*?

MB: What's interesting about dance to me is that it has all of these different facets . . . The core of *Dance Central 1* is really the dance class experience. It's very indebted to the process that you go through learning a dance in a dance class, and it's about mastery of choreography. Then there are these breaks, the freestyle times, where you're encouraged to do whatever . . . [The game presents] these two oppositional states, or I guess I wouldn't call them polar opposites but pretty different facets of dance. Sort of like "do whatever you want that is you reacting to

> the music" versus "do this thing that is someone else reacting to the music in the same way that they did it." . . . The audio reactive parts to me are really about the ways in which the choreographers distill complex music down to the things which speak most to them rhythmically.[3]

In this off-the-cuff response, Boch drew attention to aspects of "interactive audio" that were built into dance experiences long before anyone dreamed of dynamic game sound or motion-sensitive camera peripherals. He identified at least three distinct modes of kinesthetic interaction with music: improvisational "freestyle" dance, which entails embodied interpretation of music as it plays; crafting choreography that is intended to match or represent a particular piece of music, which entails analytical listening and attention to rhythmic structure; and mastering *someone else's* choreography, which entails channeling that person's musical analysis and his or her embodied interpretation of that analysis through one's own body—thereby experiencing a "sensual orientation that reveals the constructs of our individual realities" (Hahn 2007, 171).

Boch went on to describe the parallels between the core audio design features of *Guitar Hero* and *Rock Band* and the music-dance relationship in *Dance Central.* As he noted,

> If you take a look at the choreography . . . there are these moves that are very, very linked to a particular sonic element. And it can do this strange thing that I think *Guitar Hero* and *Rock Band* were great at, which is—I have a sandwich metaphor for it. It's like if you're eating some highly complex sandwich like an Italian sandwich and you're eating this thing and it tastes good, but it's made of a whole bunch of parts. And in playing *Rock Band*, I think that the musical education part of it that's strongest to me is the way in which it shows you what a given instrument does to make a rock song. What a given instrument's role is, what it's playing, by showing you its absence and then its presence. And I think that *Dance Central* can do the same thing in a lot of cases for the complex musical production that underpins all these songs. When the choreographers listen to all this stuff, some of them are reacting very lyrically, and you'll see songs like "Drop It Like It's Hot," which have almost miming elements to them. Then you have songs like "Down" or "Like a G6," where people are latching on to rhythmic elements and you are, to an extent, beat-matching, but what your beat-match is, is actually a dance that is distinctly aimed at musical elements of the song. So you are reacting to audio, like you're *reacting to someone else's reaction to audio*, if that makes sense . . . I think that dance, in its expressiveness, takes a song generally more holistically. So you have those outlooks of particular parts where you're calling out a particular rhythmic pattern or a particular melodic pattern, but then you have maybe the majority of the dance moves that are taking the song holistically.

These observations point to the distinctive forms of musical listening that inform both choreographic work in *Dance Central* and players' subsequent experiences. Creating a notation track for a particular instrument in *Rock Band* involves analytical transcription that highlights the specific musical role of that instrument.

Creating choreography for *Dance Central* may mean responding to lyrics, distinctive rhythmic, melodic, and timbral features, and phrase structure. As Boch explained further, focusing on rhythm,

> If you watch choreographers build the dances for Dance Central, they're sitting there, they have their headphones on, they're trying out different things, they're pointing out different things, and they have a unique verbal language for the thing, where they're talking about "the booms and the cats." And what they're talking about is usually the kick and the snare, or the hand clap, or whatever is subbing in for the bass and the percussive hit. And they're feeling out those boom-cats, is what they would tell you, and building moves around those patterns in the ways in which they understand the music. And then the player has the experience of dancing to the song and feeling those moments in the same way that the choreographer did.

As *Dance Central* choreographers carry out this task of "feeling out" each song, they are working within particular aesthetic and practical parameters. The preexisting popular music featured in the *Dance Central* playlists is crucial to marketing the games. Choreographers need to create a unique routine for each song—something that will feel right to players who are already fans of the music, and might make converts of those who aren't. Choreographic variety and novelty are huge factors in selling additional DLC tracks and game editions, so the choreographers must also avoid recycling too many individual moves from other songs (although some repetition across songs will make the routines easier to learn). This means that *Dance Central* choreographers have a special incentive to identify and kinesthetically amplify the distinctive sonic features of each track. As Deniz Peters notes, music possesses "a hue of haptic experience," often discussed "in terms of texture, physiognomy, tactility, and breathing, either in bodily terms (as if it had a body), or in terms of visceral experience (as felt in the body)" (2012, 19). In effect, *Dance Central* choreographers are charged with assembling a palette of these "hues of haptic experience" for each song, and using it to paint that song's choreographic portrait.

But where does "interactivity" come into play in this process, for choreographers or for players? Up to this point in our conversation, Boch consistently used the word "reactive" instead of "interactive" when referring to audio design in *Dance Central*. When I asked him whether these were two different concepts for him, his response pushed the concept of "interactivity" in another direction:

> MB: I would say it's interaction. I'd say the process of dancing to a song is interacting with it. It is not changing what the song is, but it is changing your perception of what the song is. And I think that is as valid. If you think about *Rock Band* doing the same thing, you hear the whole song and then, here's someone who has very little understanding of how rock music is made. You hear the whole song and now you're going to play a bass part to it and you keep messing up and now you

hear the song without the bass part. All of a sudden, all these things peel away and you're interacting with the audio in this very different way as a result of gameplay decisions that you made. I think your proprioceptive interaction with the game is also proprioceptive interaction with the music. And in feeling out with your body a given rhythm, I think it pushes your audio system to find the same pattern and to figure out where that is....

KM: That's really interesting, because I've been trying to think through, what's the analog to missing a note and not hearing that note? Which is that tiny but huge design move for *Rock Band* and *Guitar Hero*, which makes such a difference in your interactivity, perceived interactivity. So you're saying, it's like you miss the beat and you feel that you missed the beat?

MB: Yeah, or you hit the beat and you feel that that is a pattern in the song. You notice that there is a bass synth that is doing that rhythm. You understand that rhythm better. You hear that particular part of the song because that's the part of the song that the choreographer is hearing when they're making the move for it. So that very tight linkage between the song and the choreography for it explicates a fair amount of musical information to the player... I mean, you can also point to—we do direct audio manipulation and filter sweeps with your hands during freestyle, which is much more direct audio manipulation. But I think the interaction really comes in what is revealed to you and what is highlighted for you through specific rhythmic motion that then unpacks the song a bit.

In digital game discourse, "interactivity" usually refers to situations in which "the user/player is able to change the visual appearance of a computer screen (and/or sounds from speakers) by some motor action via an interface" (Grodal 2003, 142). A similar working definition applies in the art worlds of electronic music and digital performance; "interactivity" typically implies that human and machine are in a collaborative relationship, one that can generate perceptible effects. For example, in interactive dance installations, the dancers' gestures might generate changes in music, lighting, or an accompanying video projection; the dancers might respond to this multisensory feedback with new kinds of gestures. Experimental systems like the Embodied Generative Music project "lead movers to reconsider their 'natural' ways of connecting a certain movement with a sound" (Parviainen 2012, 79) and create "the 'feeling' of cybernetic connection to the digital media they activate" (Dixon 2007, 147; see also Kozel 2012).

While experimental digital media artists and digital media theorists often celebrate human–machine collaborations as partnerships, commercial game audio developers seem more inclined to emphasize human agency. As audio producer Lani Minella explains, "When players have a direct effect on what they hear, it's like they're the developers in some small way. They control the environment and have an audible impact and effect on it" (cited in Marks and Novak 2009, 150). Game audio pedagogy and scholarship often focus on this special quality of "adaptive," "interactive," "dynamic," or "nonlinear" audio, analyzing what happens when "the player can become a causal agent in the audio's playback" (Collins 2008, 168). Many authors invoke this quality in order to differentiate game audio design from cinematic scoring, thereby making a case for the distinctive value of game sound (e.g., Collins 2008; Marks and Novak 2009; Grimshaw

2012). As Mark Grimshaw notes, "Where the intended soundscape of a film is fixed at the point of production, digital game soundscapes are created anew at the point of reproduction" (Grimshaw 2012, 350). He goes on to argue that dynamic game audio plays a key role in generating gameplay immersion, suggesting that "the active relationship between the player and sound may be likened to the acoustic ecologies found in nature" (362; cf. Whalen 2004; Collins 2008, 133; Salen 2008).

Matt Boch's notion of proprioceptive interaction with music offers a different approach to conceiving of an "active relationship" between player and sound. In our interview, he acknowledged that *Dance Central*'s freestyle sections offer brief interludes of "kinetic gestural interaction" with the music (Collins 2008, 127), but he did not regard this feature as the core "interactive" aspect of the game (in fact, Boch observed that many players disliked the freestyle sections; in later game editions, players can turn off this feature). Rather than casting about for evidence of players' agency—their perceived control over the game technology, verified by their influence on musical playback—he pointed to how *Dance Central* gameplay changes *the players*. Playing this game has dynamic effects in real time, but these effects transpire on the players' side of the screen and speakers: in the actual world, not the virtual world (Boellstorff 2008, 19; Miller 2012, 8). Thus interactive audio in *Dance Central* is true to Torben Grodal's perception-oriented gloss of interactivity: "the creation of experiences that appear to flow from one's own actions" (Grodal 2003, 143). Players are *really dancing*, and their musical experience flows from that proprioceptive interaction. As Boch put it, "It is not changing what the song is, but it is changing your perception of what the song is." This form of interactive audio still has perceptible effects, but they play through other sensory channels. Players learn to "feel out" music through their bodies, as choreographers do.

17.2 Listening Like a Choreographer

As a commercial product, the *Dance Central* franchise has a symbiotic relationship with the songs and artists featured on its playlists. Some people will buy a game edition or additional DLC tracks because they already know and love the music; others will buy songs for listening or seek out artists' other recordings after encountering music in the games. (Harmonix gained experience developing these mutually beneficial licensing agreements while building the song catalogs for *Guitar Hero* and *Rock Band*.) But while an initial purchase might be driven primarily by name recognition—the promise of dancing to a familiar track by Lady Gaga—experienced players bring other criteria to their assessment of new repertoire. When an upcoming DLC release is announced on the Harmonix-sponsored *Dance Central* community forum, players immediately begin considering the song's possible choreographic affordances. When a preview of the choreography is released, they discuss how the choreographer's choices line up with their listening expectations. Finally, once players have purchased and played through the track (or have watched gameplay videos posted to YouTube), they offer detailed evaluative reviews of the routine.

For example, in May 2012 the Harmonix forum manager started a new discussion thread entitled "DLC discussion–Low by Flo Rida." ("Low," originally released in 2007, was Flo Rida's multi-platinum-selling debut single.) She posted a link to a thirty-second preview video for the song, which included the dance steps for the song's chorus: "She hit the floor / Next thing you know / Shorty got low, low, low" (Harmonix Music Systems 2012). By featuring this portion of "Low," the preview not only reminded players of the song's most recognizable musical hook but gave them an opportunity to assess the dance routine's signature moves: the chorus subroutine will repeat at regular intervals and must be associated with distinctive musical material. In this case, since the lyrics of the chorus explicitly describe movements on the dancefloor, the choreographer could be expected to draw on them. Players could speculate about possible physical enactments of "hitting the floor"—perhaps striding onto the dancefloor, or literally striking it with a hand or foot? And what about the title move, "getting low"? Would it entail bent knees, dipped hips, a limbo backbend, or a gesture connoting "low" sexuality?

The Harmonix forum manager seeded the discussion of this new track with a direct invitation for feedback: "Check out the sample of the new routine and share your thoughts in this thread. Once the DLC drops tomorrow leave your reviews here!" (DanceCentral.com 2012). Players immediately jumped into the fray:

HEYORADIO: Going to be immediately honest and say I was really disappointed with the use of Step Pump for "low lowlowlowlowlowlowlow" as I was hoping we'd have a fun new move that went along with the lyrics. Oh well.:/Nearly everything in the preview is a move we've seen before, so you could say I was pretty let down with this. [. . .] Here's hoping things are better outside of this little preview? I gotta keep *some* of my optimism. haha

LAUSON1EX: Saw it coming, therefore I'm not deceived. Just face it, people: the song has been advertised as being in the Moderate category. Not Tough, not Legit. Moderate. I'm surprised that you guys expected anything more than what you actually got!

WHITEMO: Honestly though, D.A.N.C.E., Pon de Replay, Rude Boy, Right Thurr, Oops (Oh My) and I Like It, these are also moderate level songs and they have amazing and mildly challenging choreographies. [. . .] Thus, we have the right to have high expectations for LOWer level songs [smile icon] (see what I did there?)

For the "low lowlowlow" part, I imagined something like the Topple move in Down [a song by Jay Sean that features the lyrics "Down, down, down, down"]. [. . .] All we can do is to wait until tomorrow [smile icon]

I've almost never been disappointed with a Chanel routine... and the song is pretty cool without the dance anyway.

BOSSPLAYER: This is the type of DLC I buy off the bat because I enjoy the song, not for the difficulty or choreography.

LAUSON1EX: I was expecting a Muscle Swish [link to YouTube video of gameplay featuring this move] at the "low lowlowlow" part *at the very least*, but now that you have mentioned it, the Topple move would have worked SO much better.[4]

This discussion demonstrates the expectations that experienced players bring to new DLC tracks, informed by their acquired knowledge of the existing choreographic repertoire. In the course of the discussion, many players mentioned the choreographer, Chanel Thompson, by name; several echoed the declaration that "You can obviously tell Chanel choreoed this song. It is written all over it" (Appamn). Players also acknowledged the practical constraints that shape the work of Harmonix choreographers. As Lauson1ex noted, songs assigned to the "Moderate" difficulty level simply cannot have showstopper routines. Another player observed, "Putting the song at a 'Moderate' difficulty level (probably to make it accessible to all skill/fitness levels due to the song's popularity) probably limited Chanel's options a little bit.[...] It isn't her best work, but I am still a Chanel fan and look forward to future DLCs by her" (Seanyboy99). ZJ11197 chimed in, "yeah you guys have to give Chanel some slack. [...] We cannot be selfish [....] they had to tailor low to be a song everyone at any level could play." Still, reviews of "Low" were mostly lukewarm, and after playing through the track players supported their evaluations in detail:

WHITEMO: The renewed Victorious move is really great, but so tiring that I'm actually glad we don't have to do it for a second time in a row. As for new moves, there aren't many—what we mainly get is a bunch of old moves freshened up a bit, and it isn't a bad thing at all, for they fit the song and don't repeat themselves unnecessarily.

As I mentioned before, this dance is very tiring, as it involves quick leg lifts, bending knees and waist and wide arm movements, but that's what we expect from a Chanel routine. The finishing move is interesting, but it's similar to Gonna Make You Sweat's finishing move.

SEANYBOY99: The Barreto Clap + Whatever Move (Crab Walk here specifically) combo feels better for slower songs, rather than faster songs like this one. I also felt that the Coconut Crab move was a lot similar to the Bobblehead Step move. The Freq Whip (/Jump) move does fit with the song, but I personally think that the slot where it appears would have been an excellent opportunity to do a new move and/or one with more flavor (I could totally think of a move that borrows from "Scenario." It would be called "The Slipper Slap" Tee Hee.)

APPAMN: Whenever I heard low on the radio, mainly the first thing I would note for this song was its heavy bass. I'd turn up the radio in my car and

> just have fun listening. Second, I would notice its badassness that it has. However, DC. . .Sort of made it a feminent, girly song. Don't get me wrong, I have no problem doing girly moves in the game, but DC ruined the reputation that this song had.

This critical analysis of new repertoire illuminates another facet of interactive audio in *Dance Central*, one that complements and informs players' visceral experiences in the moment of gameplay. Here players are engaging in what Eric Zimmerman calls *cognitive interactivity* and *meta-interactivity*: "interpretive participation with a text" and "cultural participation with a text," respectively (2004, 158). But while Zimmerman's analytical categories are meant to account for player's "interactions" with a game narrative (following in the footsteps of reader-response theory), *Dance Central* players are reflecting on their multisensory embodied experience as dancers and listeners.

Importantly, *Dance Central* offers players a basic vocabulary with which to discuss and critique choreography. Game discourse grows from a lexicon of move names, a list of choreographer credits, and a common experience of a shared repertoire, allowing players to compare routines and identify specific choreographic styles. Forum discussions also give players space to hash out conventions for discussing how routines *feel* in practice: "tiring," full of "flavor," "girly," "badass," a good fit or poor fit with the music. The online format also makes it easy for players to include links to illustrative video examples when words fall short.

As Susan Foster observes, "Any standardized regimen of bodily training . . . embodies, in the very organization of its exercises, the metaphors used to instruct the body, and in the criteria specified for physical competence, a coherent (or not so coherent) set of principles that govern the action of that regimen. These principles, reticulated with aesthetic, political, and gendered connotations, cast the body who enacts them into larger arenas of meaning where it moves alongside bodies bearing related signage" (1995, 8). The *Dance Central* franchise has brought more than five million players into one such "arena of meaning," where their gameplay experience and reflective discourse enter into interactive feedback loops with other received ideas about music, dance, and embodied or performed identity.[5]

17.3 Conclusions

Shortly after the release of the first *Dance Central* game, a commenter posted this skeptical rejoinder to a positive game review:

> I played the original version of this game just now. A song came on the radio while I was getting something to eat and I was like "this is fun" and started dancing a bit. The graphics were much better than the 360 version and it had less loading times. It also cost £0. I'd recommend it instead of buying this, I think it's called "Dancing in Real Life." (Stegosaurus-Guy-II, comment posted November 4, 2010, on Smith 2010)

Such criticisms invite us to consider what *does* distinguish *Dance Central* gameplay from "dancing in real life." It's a tricky question, since the split between virtual and actual performance functions very differently here than in most digital games. Again, the comparison to *Guitar Hero* may be useful: where *Guitar Hero* players serve as middlemen for a prerecorded musical track, *Dance Central* players are actually dancing. They are not controlling an avatar's movements, nor do their gestures shape musical playback. The proprioceptive interactivity and multisensory musicality fostered by the games could also be developed by dancing to the radio. So what does the game really contribute, besides an attractive commercial package, some limited feedback on the technical accuracy of one's moves, and the allure of trying out the latest motion-sensing interface?

Dance Central conjoins a dancerly sensibility with a gaming sensibility—a "lusory attitude" (Salen and Zimmerman 2004, 574). These software products are not simply learning-oriented interactive simulators, but are specifically designed, marketed, and experienced as games. They adhere to what Jesper Juul calls the "classic game model": "a rule-based system with a variable and quantifiable outcome, where different outcomes are assigned different values, the player exerts effort in order to influence the outcome, the player feels emotionally attached to the outcome, and the consequences of the activity are negotiable" (Juul 2005, 36). In this case, the "rule-based system" involves mastering a complex and minutely codified choreographic repertoire, including moves that many players would not perform of their own accord, set to music that might not suit their usual listening tastes. Crucially, *Dance Central* also offers completely private dance lessons—so private that even the instructor isn't really present, although the player still receives corrective feedback in real time. Players can work through a carefully organized dance curriculum without ever submitting themselves to human evaluation. They can leave behind anxiety about their technical skills, their body type, or whether their identity traits seem to "match" the games' hip-hop inflected club moves, musical repertoire, or expressions of gender and sexuality (Miller 2014).

Approaching *Dance Central* as a game, players are free to claim, "I'm only dancing this way because the game is making me do it": that is, they are dancing for the sake of earning points, getting to the next level, or completing all the game challenges, rather than because a particular song or routine accurately represents their own tastes and identity. A complex scoring system awards points for accurate execution of specific moves, plus bonus points for extended sequences. Additional score multiplier algorithms "ensure that it's really hard to get the same score as someone else" (mattboch 2012), contributing to a sense of individual accomplishment. Scores can be posted to online leaderboards, where players vie for the highest achievements on particular songs. Whether players are competing on the leaderboards or not, carefully graded difficulty levels lead them through satisfying "cycles of expertise": "extended practice, tests of mastery of that practice, then a new challenge, and then new extended practice" (Gee 2006, 180). The *Dance Central* games are oriented around rehearsal, repetition, and performance of an extensive song-and-dance repertoire; they reward long-term commitment, frequent practice sessions, and substantive critical and analytical engagement in the affinity space offered by the online community forum (Gee 2004, 85).

As Dena Davida and numerous dance ethnographers have demonstrated, "Dance is not an oral or written tradition for the most part, although its transmission does involve speaking and writing"; rather, dance "might be thought of as a 'kinaesthetic tradition,' one that is principally carried from body to body" (Davida 2012, 13, cf. Hahn 2007; Samudra 2008). *Dance Central* accomplishes the feat of transmitting a dance repertoire from body to body without having both bodies in the room at the same time. The games offer a new channel for the transmission of embodied knowledge, and for indexing that knowledge through popular music—"feeling out" music with one's body, as Matt Boch put it, and imagining how it feels in someone else's body. As players gain expertise in this specific repertoire, their new knowledge transforms their experience of music and dance: even when they are listening or observing, they may do so with a dancerly sensibility (cf. Foster 2011; Goodridge 2012, 122).

What dancers know intuitively, neuroscientists have been studying using fMRI scans. Their findings indicate that "action observation in humans involves an internal motor simulation of the observed movement" (Calvo-Merino et al. 2005, 1246). Moreover, significant "expertise effects" come into play: that is, "the brain's response to seeing an action is influenced by the acquired motor skills of the observer" (1245). Thus, when groups of expert ballet dancers and capoeira practitioners watched videos of people performing in these styles, "the mirror areas of their brains responded quite differently according to whether they could do the actions or not" (1248). Musicians and dancers will likely file this study under "scientists find sky is blue"; *of course* there is something qualitatively distinctive about listening to a piece of music that one knows how to play, or watching choreography built from moves that one has performed.

Moreover, as *Dance Central* players often report (in online social media contexts and interviews with me), learning a choreographic routine for a song may transform one's subsequent listening experiences. Since these games use existing popular songs, players often encounter the musical repertoire in the course of everyday life. As the player quoted at the start of this chapter told me, "The experience of the game has become attached to the song: so when I listen to the song, I experience the game again."[6] This is the same enculturated and embodied response that inexorably summons hip-hop dancers to the floor when they hear canonical b-boy tracks. As Joseph Schloss writes, "From the moment this ability becomes a part of any given breaker's disposition, that individual carries a piece of hip-hop history in his or her physical being and recapitulates it every time he or she dances" (Schloss 2006, 421). Again, neuroscientists offer mounting quantitative evidence that complements these ethnographic findings; for instance, recent studies indicate that listening to music that has previously been associated with a particular motor activity leads to improved retention and future performance of that motor activity (Lahav et al. 2012). That is, once organized sound has been associated with organized movement, the association has enduring effects that can be accessed via multiple sensory channels. Returning to the player who complained that the *Dance*

Central choreography had made "Low" into a "feminent, girly song," we might consider the implications of his internally rehearsing that "girly" choreography every time "Low" plays on the radio.

By creating powerful links between music and choreography, *Dance Central* inculcates these sound/body connections for people without prior dance training, as well as inviting dancers of all experience levels to engage in movement styles that might not match their own sense of self. The games teach players how to sense like a dancer, and lead many to reflect on and develop that new embodied understanding by engaging with a community of practice (Hamera 2007). This is *Dance Central*'s most fundamental dance lesson, one with broad implications for interactive audio. Moving forward, as we build on this foundation to consider particular instances of multisensory interactivity, we should not lose sight of the complex articulations of sound and kinesthetic repertoire with other cultural formations, including identity categories that are experienced through the body (Sterne and Akiyama 2012, 545; Born 2012, 165). "Reacting to someone else's reaction to audio" isn't only about channeling that person's analysis of musical structure; it may mean feeling out the embodied experience of someone of a different gender, race, sexual orientation, or cultural background. In the *Dance Central* context, "interactive audio" involves music and dance that grew out of urban African-American, Caribbean, and Latino youth culture. The "teacher" voice that guides players through the rehearsal mode is marked by a black vernacular accent and vocabulary; the choreographers are mostly people of color; and the governing dance aesthetic might best be located at the intersection of contemporary hip-hop and gay club culture. As Matt Boch told me, "The space is so diverse, it can allow for all sorts of different peoples from various backgrounds to have an experience with another type of dance culture that they wouldn't have otherwise... My hope is that people would be interested in and enlivened by their interactions there to make deeper cultural connections with the things that speak to them." *Dance Central* reminds us that interacting with sound—especially musical sound—always means interacting with culture, and that the "effects" that define interactivity may play out beyond the confines of the console hardware and game code.

Notes

1. Riffraff [username], recorded Skype interview with the author, August 24, 2011.
2. A complete song list for the *Dance Central* franchise—sortable by difficulty level—appears at http://www.dancecentral.com/songs.
3. Recorded interview with the author, April 6, 2012, in Boston, Massachusetts. All subsequent Matt Boch quotations are from this interview.
4. The complete forum discussion is available at DanceCentral.com (2012).
5. Franchise sales figures are from VGChartz.com (2013). See Miller (2012) for more examples of amateur-to-amateur online discourse and Miller (2014) for a discussion of *Dance Central* and gender performance.
6. Riffraff [username], recorded Skype interview with the author, August 24, 2011.

References

Boellstorff, Tom. 2008. *Coming of Age in Second Life: An Anthropologist Explores the Virtually Human*. Princeton, NJ: Princeton University Press.

Born, Georgina. 2012. Digital Music, Relational Ontologies and Social Forms. In *Bodily Expression in Electronic Music: Perspectives on Reclaiming Performativity*, ed. Deniz Peters, Gerhard Eckel, and Andreas Dorschel, 163–180. New York: Routledge.

Calvo-Merino, Beatriz, et al. 2005. Action Observation and Acquired Motor Skills: An fMRI Study with Expert Dancers. *Cerebral Cortex* 15 (8): 1243–1249.

Collins, Karen. 2008. *Game Sound: An Introduction to the History, Theory, and Practice of Video Game Music and Sound Design*. Cambridge, MA: MIT Press.

DanceCentral.com. 2012. DLC Discussion: Low by Flo Rida. (May 28, 2012). http://www.dancecentral.com/forums/showthread.php?t=8354.

Davida, Dena. 2012. Anthropology at Home in the Art Worlds of Dance. In *Fields in Motion: Ethnography in the Worlds of Dance*, ed. Dena Davida, 1–16. Waterloo, ON: Wilfrid Laurier University Press.

Dixon, Steve. 2007. *Digital Performance: A History of New Media in Theater, Dance, Performance Art, and Installation*. Cambridge, MA: MIT Press.

Foster, Susan Leigh. 1995. An Introduction to Moving Bodies: Choreographing History.: In *Choreographing History*, ed. Susan Leigh Foster, 3–21. Bloomington: Indiana University Press.

——. 2011. *Choreographing Empathy: Kinesthesia in Performance*. New York: Routledge.

GameTrailers.com. 2010. Dance Central Video Game, Review. (November 4, 2010). http://www.gametrailers.com/video/review-dance-central/707175#comments.

Garcia, Luis-Manuel. 2011. *"Can You Feel It, Too?": Intimacy and Affect at Electronic Dance Music Events in Paris, Chicago, and Berlin*. Ph.D. dissertation, Department of Music, University of Chicago. Chicago, Illinois.

Gee, James Paul. 2004. *Situated Language and Learning*. New York: Routledge.

——. 2006. Learning by Design: Good Video Games as Learning Machines. In *Digital Media: Transformations in Human Communication*, ed. Paul Messaris and Lee Humphreys, 173–186. New York: Peter Lang.

Goodridge, Janet. 2012. The Body as a Living Archive of Dance/Movement: Autobiographical Reflections. In *Fields in Motion: Ethnography in the Worlds of Dance*, ed. Dena Davida, 119–144. Waterloo, ON: Wilfrid Laurier University Press.

Grimshaw, Mark. 2012. Sound and Player Immersion in Digital Games. In *The Oxford Handbook of Sound Studies*, ed. Trevor Pinch and Karin Bijsterveld, 347–366. New York: Oxford University Press.

Grodal, Torben. 2003. Stories for Eye, Ear, and Muscles: Video Games, Media, and Embodied Experience. In *The Video Game Theory Reader*, ed. Mark J. P. Wolf and Bernard Perron, 129–156. New York: Routledge.

Hahn, Tomie. 2007. *Sensational Knowledge: Embodying Culture through Japanese Dance*. Middletown, CT: Wesleyan University Press.

Hamera, Judith. 2007. *Dancing Communities: Performance, Difference, and Connection in the Global City*. New York: Palgrave Macmillan.

Harmonix Music Systems. 2012. Preview Video: "Low" by Flo Rida. (May 28, 2012). http://www.dancecentral.com/preview-low.

Juul, Jesper. 2005. *Half-Real: Video Games between Real Rules and Fictional Worlds*. Cambridge, MA: MIT Press.

Kozel, Susan. 2012. Embodying the Sonic Invisible: Sketching a Corporeal Ontology of Musical Interaction. In *Bodily Expression in Electronic Music: Perspectives on Reclaiming Performativity*, ed. Deniz Peters, Gerhard Eckel, and Andreas Dorschel, 61–70. New York: Routledge.

Lahav, Amir, T. Katz, R. Chess, and E Saltzman. 2012. Improved Motor Sequence Retention by Motionless Listening. *Psychological Research* 88 (3): 310–319.

Marks, Aaron, and Jeannie Novak. 2009. *Game Audio Development*. Clifton Park, NY: Delmar.

mattboch. 2012. DC2 Perform It Scoring Clarification. *Dance Central*, May 25, 2012. http://www.dancecentral.com/forums/showthread.php?t=7894andp=24693andviewfull=1#post24693.

Miller, Kiri. 2012. *Playing Along: Digital Games, YouTube, and Virtual Performance*. New York: Oxford University Press.

——. 2014. Gaming the System: Gender Performance in *Dance Central. New Media & Society*. OnlineFirst DOI: 10.1177/1461444813518878. http://nms.sagepub.com.

Parviainen, Jaana. 2012. Seeing Sound, Hearing Movement: Multimodal Expression and Haptic Illusions in the Virtual Sonic Environment. In *Bodily Expression in Electronic Music: Perspectives on Reclaiming Performativity*, ed. Deniz Peters, Gerhard Eckel, and Andreas Dorschel, 71–82. New York: Routledge.

Peters, Deniz. 2012. Touch: Real, Apparent, and Absent: On Bodily Expression in Electronic Music. In *Bodily Expression in Electronic Music: Perspectives on Reclaiming Performativity*, ed. Deniz Peters, Gerhard Eckel, and Andreas Dorschel, 17–34. New York: Routledge.

Pinch, Trevor and Karin Bijsterveld. 2012. New Keys to the World of Sound. In *The Oxford Handbook of Sound Studies*, ed. Trevor Pinch and Karin Bijsterveld, 3–36. New York: Oxford University Press.

Salen, Katie, ed. 2008. *The Ecology of Games: Connecting Youth, Games, and Learning*. John D. and Catherine T. MacArthur Foundation Series on Digital Media and Learning. Cambridge, MA: MIT Press.

Salen, Katie, and Eric Zimmerman. 2004. *Rules of Play: Game Design Fundamentals*. Cambridge, MA: MIT Press.

Samudra, Jaida Kim. 2008. Memory in Our Body: Thick Participation and the Translation of Kinesthetic Experience. *American Ethnologist* 35 (4): 665–681.

Schloss, Joseph G. 2006. "Like Old Folk Songs Handed Down from Generation to Generation": History, Canon, and Community in B-boy Culture. *Ethnomusicology* 50 (3): 411–432.

Smith, Jamin. 2010. Dance Central Review. *Videogamer*, November 4, 2010. http://www.videogamer.com/xbox360/dance_central/review.html.

Sterne, Jonathan, and Mitchell Akiyama. 2012. The Recording that Never Wanted to be Heard and Other Stories of Sonification. In *The Oxford Handbook of Sound Studies*, ed. Trevor Pinch and Karin Bijsterveld, 544–. New York: Oxford University Press.

VGChartz.com. 2013. Game Database: Dance Central. http://www.vgchartz.com/gamedb/?name=dance+central.

Whalen, Zach. 2004. Play Along: An Approach to Videogame Music. *Game Studies* 4 (1). http://www.gamestudies.org/0401/whalen/.

Zimmerman, Eric. 2004. Narrative, Interactivity, Play, and Games: Four Naughty Concepts in Need of Discipline. In *First Person: New Media as Story, Performance, and Game*, ed. Noah Wardrip-Fruin and Pat Harrigan,. Cambridge, MA: MIT Press.

CHAPTER 18

INTERACTIVITY AND LIVENESS IN ELECTROACOUSTIC CONCERT MUSIC

MIKE FRENGEL

In today's electronically mediated landscape, human–machine interaction has become routine, whether it be to withdraw money from an automated teller machine, to obtain information through a customer service automated phone system, or to simply use a computer to check email. Interactivity has also made its way into practically all forms of art and entertainment, including television and film, games, the visual arts, dance, and music. In most cases, the principal aim of interactivity is to offer users an opportunity to participate in the unfolding events, shifting the role of the end-user from that of a passive consumer to one who is actively engaged in the experience.

Interactivity in the performing arts is distinctive because there is a third party involved—the spectator. In concert music performances, the interaction typically occurs between a performer and a system, but it is done for an audience that remains, in most cases, outside the interactive discourse.[1] But the human–machine relationship is important to spectators. Placing a performer on stage in an interactive environment frames not only the performer's actions, but also the interaction itself, often making comprehension of it necessary for both a full appreciation of the work and for an evaluation of that particular live performance. Because visual cues play a vital role in conveying information about the human–machine relationship in interactivity, one might argue that live performance is of significant import to interactive musical works.

The focus of this chapter is on interactivity in electroacoustic concert music—a contemporary genre which has grown out of the Western art music tradition, embraces technology as a means of advancing musical practice, and remains committed to the primacy of the concert experience. Four common models of interactivity are identified and discussed, with a particular emphasis given to the opportunities and challenges that arise in relation to liveness in electroacoustic music performance.

18.1 The Spectacle of Live Performance

If liveness is to be viewed as a significant attribute of interactive music, then it is instructive to begin with an examination of why people enjoy live performance in the first place. Within the domain of contemporary concert music, five commonly cited positive attributes of live performance are: *virtuosity, fallibility, spontaneous variability, the influence of visual cues*, and *presence*.[2]

Virtuosity has traditionally held a central position in live performance. Those who can perform actions that are beyond our own capabilities simply dazzle us. In many ways, watching a musical performance is similar to watching humans perform any activity that requires great skill and practice to master. But musical performance somehow feels different. Becoming a virtuoso musician requires mastery of both motor skills and musical language. Great musicians not only execute actions with seeming perfection, but they choose the right actions for the moment and apply them in just the right proportions. Virtuosity is as much about musical sensibility as it is about physical dexterity, and for this reason it can be recognized even in the performance of relatively simple musical passages. On top of exceptional musicianship is the emotive power of music itself, and a sense that the person playing is contributing to that emotive energy through the performance decisions being made. This ternary combination of physical control, musical sensibility, and the emotive element of the music itself leads to a highly charged, and at times, overwhelming perceptual experience.

Hand in hand with virtuosity is the possibility of error—a recognition that humans are fallible and at any moment the entire performance endeavor could collapse. While this may, at times, keep spectators on the edge of their seats, seeing others achieve greatness also inspires us. The narrative of performer faced with challenges that demand great skill to overcome is reminiscent of the classic hero who triumphs over obstacles—a tale found in many of the world's oldest myths (Campbell 1949). In the case of musical performance, the mythos is played out on a micro scale in front of us.

Another quality attributed to live performance is that it brings spontaneous variability to the music with each performance, and each performer brings something individual to a work through their interpretation of it. Live music changes from one performance to the next, from performer to performer, and also over longer time-spans with the evolution of performance practice in general. Many feel that the interpretive element that scored music affords brings life to the notes on the page, in contrast to music that is fixed on a medium and invariable.[3]

Bell (2008) explains that live performance is concerned with both *process* and *product*. Performances clearly generate products—in the case of music, the sounding results of the performer's actions. But performances emerge through coordinated activities, and the execution of those actions can be considered a goal in itself. Witnessing the performance of a musical work is a distinctly different experience than listening to a recording of that same performance because we also observe the process of its creation, which adds value to the experience. Whenever a performer is placed on stage, we inevitably highlight both their actions and the products of those actions, which are given the status of "performance" through "framing"—an invitation to perceive them as extraordinary (Bell 2008).

Aside from the emergence of an extraordinary experience, Bell is ambiguous as to what the added values of performance are. Witnessed virtuosity, coupled with the possibility of error, are certainly important contributions. In addition, observing a live performance allows spectators to see the bodily movements and energy going into the creation of the music and relate those to the sounding result. These visual cues can clarify the intentions of performers and the structure of their actions. Moreover, Cox (2011) argues that musical experience is rooted in imagined bodily action. As we listen to music, whether live or recorded, we imagine what it would be like to make the sounds we hear. Such vicarious performances are usually unintentional, subconscious and covert, but they significantly persuade a listener's interpretation of the music. The visual stimuli present in live performance can certainly enhance vicarious motor imagery when the performer's actions support the energy profiles in the sounds heard. A final added value, relevant to works incorporating electronics, is what Robert Wechsler (2006) has referred to as the "how'd-they-do-it?" factor—a tendency for spectators to shift attention to the role of the technology itself and how it functions. Live settings add clarity to the human–machine interaction, whereas audio recordings are more likely to conceal those relationships and thus attenuate those aspects of the work.

A final attribute of live performance is *presence*—merely being there at the moment of the music's realization. Auslander provides what he considers to be the classic definition of liveness: "physical co-presence of performers and audience; temporal simultaneity of production and reception; experience in the moment" (2008, 61).[4] According to this account, live performance is not only tied to a particular space and time, but it is transitory, fading out of existence as quickly as it emerged. Phelan agrees, echoing the centrality of transience: "Performance honors the idea that a limited number of people in a specific time/space frame can have an experience of value which leaves no visible trace afterward" (1993, 149). The impermanence of live performance surely adds to a feeling that those extraordinary experiences are that much more special (see also Chapters 19 and 20 in this volume).

18.2 The Reintroduction of Liveness in Electroacoustic Music

Since the origins of electronic and electroacoustic music, practitioners have had to come to terms with issues of liveness in their art. Many electroacoustic works involve no performer on stage; they are fixed on a medium, such as tape, CD, or digital sound file, and presented at concerts through speakers. In the case of early computer music, digital systems were not powerful enough to allow for real-time control of sound, so composers wishing to work with such systems had to create music for fixed-media out of necessity. Others, such as Pierre Schaeffer in Paris, embraced the fixed media format because it allows listeners to focus their attention solely on the sound of the music, without visual distractions. A rich aesthetic framework has emerged around this "acousmatic" mode of presentation that is largely concerned with what can be gained in a musical experience when the sources of the sounds heard cannot be seen. Despite the relative success of acousmatic music, the absence of a performer on stage continues to puzzle many concertgoers unfamiliar with the aesthetic concerns of the genre. Acousmatic music often involves live sound diffusion, but because the composer is typically not on stage and not the center of attention, they do not acquire a "performer" role. Live performance has traditionally been a central component of the concert music experience and many electroacoustic composers have felt the need to reintroduce the "live" into their music.

The mixed-work format, which combines traditional instruments with electronically mediated sounds, clearly shifts concern back to live performance. Historically, the performer would be required to play along with an electronic part that was fixed on a medium, such as tape. Although still in use today, this single-index technical format places severe interpretative constraints on the performer due to the fact that the electronics are inflexible. The performer must stay strictly synchronized to the playback medium, and thus loses much of their expressive potential. This is a well-known problem for performers, who feel straitjacketed by the temporal rigidity of the tape. It is also an issue for composers, who must create the music with synchronization issues in mind, providing salient cues for performers to indicate upcoming tempo changes, downbeats, or other events that require coordination between live and nonlive forces.

At the very least, interactivity offers a means of regaining temporal fluidity in electroacoustic music, freeing the performer from the unwavering chronometer of fixed media electronics. Systems that "listen" for particular cues from instrumental performers before advancing or that simply allow the performer to move through sections of a work using a foot pedal return temporal control to the live player, offering an effective alternative to the single-index fixed-electronics format. Moving beyond the issue of temporal freedom, interactive systems can introduce variability to electronic components by generating or modifying their outputs in real time and in response to actions taken by a performer. But perhaps the single most distinctive feature of interactivity is the potential it offers for novel performance interfaces and new paradigms for the presentation of music in live contexts.

18.3 Models of Interactive Music

In the field of electroacoustic music the term "interactive" is applied to a variety of electronically mediated systems that exhibit a wide range of behavioral qualities. Some systems function much like traditional instruments; they are played by performers and afford a great deal of control over their sounding output. Others are configured such that control is shared between the performer and the system. Still others function as autonomous virtual improvisers, generating original sound materials in accordance with the context. Four common models of interactive music systems are identified and discussed below, based largely on metaphors and classifications proposed elsewhere (Chadabe 2007; Rowe 1993; Winkler [1998], 2001). They are:

- The instrumental model
- The conductor model
- The reflexive model
- The virtual musician model

While it may be possible to recognize progressive trends—for instance, the electronic component becomes increasingly independent as we move from the instrumental to the virtual musician model—it would be misleading to view these models on any sort of continuum, as each embraces its own set of aesthetic aims and musical concerns. The models are better viewed as distinct approaches to interactivity.

18.4 The Instrumental Model

In the instrumental model of interactivity a system is designed to function much like a traditional instrument, affording the performer complete control over the output. Miranda and Wanderley (2006) have used the term digital musical instrument (DMI) to describe such systems, distinguishing them from traditional instruments by the fact that inputs can be freely mapped to a wide variety of sound parameters. While such freedom offers exciting opportunities for the design of novel instruments, it potentially poses challenges to traditional notions of instrumentality for the performer, and in some cases, it can obscure the significance of the player's actions for the spectator. A cursory examination of human interaction with acoustic instruments provides an instructive framework against which digital musical instruments can be contrasted.

Physical and sounding gestures are intimately linked in traditional instrumental performance. The sound produced by an instrument is coupled directly with the performer's physical gestures. For Cadoz (2009), the notion of instrumentality necessitates physical interaction with an object, which establishes an energy continuum from the gesture to the sound. In his view, a performer's perception of making a sound is not confined to the auditory domain, but rather it is distributed throughout the body in the form of tactile-proprio-kinesthetic (TPK) feedback. Musicians know how performance actions feel and adjust according to both the sound and the physical response of the instrument. Indeed, much of what a virtuoso musician knows about performance on an instrument is stored in the form of enactive knowledge, learned through actions and constructed on motor skills. Traditional instruments, along with their respective performance practices rooted in physical interaction, afford the mastery of such sensory-motor skills.

The acquisition of virtuosity also demands predictable behavior; a particular interaction with the instrument should always produce a similar output. This consistency is necessary for performers to develop skill. Richard Moore recognizes that performers learn to modify sound in subtle ways for expressive purposes. The more an instrument allows such subtlety to be reflected in the sound, the more musically expressive that instrument will be. According to Moore, such control intimacy is "simultaneously what makes such devices good musical instruments, what makes them extremely difficult to play well, and what makes overcoming that difficulty well worthwhile to both the performer and the listener" (Moore 1988, 22).

18.5 New Instruments and New Instrumental Paradigms

Today we find a plethora of generic electronic controllers on the market that are increasingly adapted to musical performance. While some are designed for musical applications and resemble instrumental interfaces, others, such as Nintendo's Wiimote and Microsoft's Kinect game controllers, are being appropriated for musical purposes, offering new paradigms for interacting with sound. Software tools such as Max/MSP and Open Sound Control make the task of linking data from nearly any digital control device to sound parameters trivial.

While the abundance of software and hardware tools makes the development of new musical instruments more accessible than ever, there are significant differences between traditional instrumental interfaces and generic controllers worth considering. Controllers do not produce sound, but instead generate data streams that must be mapped to parameters of a sound-generating algorithm or device. There is, by default, a division of labor between the performance interface and the sound-producing unit. It is certainly possible to establish perceptual links between the two, but any connection can

just as easily be disregarded. The relationship between physical and sounding gesture can be further obfuscated by the fact that the sonic parameters under a performer's control in a DMI may bear little resemblance to those typically associated with conventional instruments, and the mappings themselves may be complex one-to-many or many-to-one configurations. With such enormous mapping flexibility, restraint may be the most sensible operative methodology when designing digital musical instruments if a perceived link between physical and sounding gesture is the goal.

Although links between physical and sounding gesture remain a concern for some, the proliferation of new, generic controllers has ushered in a new age of instrument design—one in which performance gesture is often perceptually unrelated to the sounding result. The decoupling of physical and sounding gesture is not only disruptive to the traditional performer–instrument relationship, but it also affects the potential for spectators to predict the sounding results of a performer's actions. If we can envision for a moment the image of a cellist performing a tremolo—the bow moving in short, rapid, alternating directions—the spectator is not only able to form an expectation of the type of sound that will be produced, due to familiarity with the instrument and its idiom, but the actions themselves carry strong connotations as to the quality of the sound, namely, one that contains an iterative energy profile and high rate of spectral flux. By contrast, a performer using a generic controller, such as a QWERTY keyboard, can produce a similar sound with a keystroke that initiates playback of a sampled cello tremolo. Although a temporal association between the action and its effect may be retained, there is nothing in the action or the device that indicates the nature of the sound because keystrokes are not differentiable (Jensen, Buur, and Djajadiningrat 2005). Moreover, in the case of laptop performances, which are increasingly common in electronic music, the actions of the performer are hidden behind a screen, unable to be seen by the audience. Paine (2009) has observed that the laptop musician, much like a DJ, often appears to be broadcasting precomposed materials, leading the spectator to question authenticity, and in the worst cases, a perception of what he describes as a "counterfeit" performance.

How can we explain the lack of concern for the relationship between physical and sounding gesture in so many interactive works today? Could it be a mere oversight, or is it possible that younger generations of so-called "digital natives," having grown up with electronic interfaces, videogames, and virtual environments, do not feel the same need to couple physical and sounding gesture? D'Escriván (2006) points out that many spectators today are perfectly comfortable with the record-spinning of a DJ or with a laptop music performance. Undeniably, much of the music today, even that which is seemingly instrumental, is produced on computers and involves little or no acoustic instrument performance. However, one might argue that the significance of the link between physical and sounding gesture is not rooted in prior experience with traditional musical praxis, but rather due to experience with the physical world in general, and we have not yet managed to escape that. Regardless of the reasoning, one cannot ignore current trends in the field, and the ubiquity of laptop performance and new control interfaces requires an acknowledgment of these practices, and perhaps a redefinition of the very notion of instrumentality.

18.6 The Conductor Model

In his book *Formalized Music*, Iannis Xenakis provides a colorful depiction of composition with a computer: "the composer becomes a sort of pilot: he presses the buttons, introduces coordinates, and supervises the controls of a cosmic vessel sailing in the space of sound" (1971, 144). The type of interactive system that Xenakis describes is one in which control over the output is shared between the performer and system in a manner somewhat analogous to the way a conductor directs an ensemble of musicians; the musicians provide the sounding materials while the conductor guides them through it, exerting influence over particular parameters. Similarly, in the conductor model of interactivity a performer engages with a system to either modify parameters of a generative process or to affect its sounding output, thereby influencing the shape of the machine-generated material. Schloss (2003) refers to this as *macroscopic* control. Interacting at the macroscopic level, the performer relinquishes control over event-level details to focus on the development of larger structures and trajectories in the music.

Joel Chadabe's *Solo* (1977) provides a clear example of an interactive work that embraces the conductor model. Chadabe developed software to generate transformations of a melody based on a free-jazz clarinet improvisation, which is then arranged in eight voices divided by instrument-like timbres: flutes, clarinets, and vibraphones. The performer (usually Chadabe himself) stands on stage between two single-antenna Theremin-like devices. The proximity of his left hand to one antenna controls instrumentation by determining which voices are heard. The proximity of his right hand to the other antenna controls the overall tempo of the melodic material being generated. Chadabe (2000) discusses his particular concern that the interactivity in *Solo* should be comprehensible to the audience. To that end, he chose to use antennae because proximity is an easy attribute for a spectator to measure.

For control to be shared, conductor model systems must have some built-in representation of the music or a priori conception of what it will sound like. Embedded representations might be stored in the form of a predetermined score that the system steps through. Alternatively, they could take the form of indeterminate algorithms that establish some predefined timbral or behavioral attributes but leave others to be shaped according to parameters controlled by a performer. Schnell and Battier (2002) have labeled such systems *composed instruments*, referencing the fact that predetermined decisions regarding aspects of the work are embedded in the system itself, which distinguishes them from conventional instruments or interactive systems that adhere to the instrumental model. Embedded representations of the music will naturally tend to make conductor model systems work-specific.

Conductor model systems typically incorporate new controllers and thereby incur many of the same challenges related to liveness and audience expectation that pertain to the instrumental model. In the conductor model, these matters can be magnified because the performer's actions are only loosely tied to sounding results. Some sounds may have no corresponding physical gesture, while others may be the delayed result of actions taken some time ago. In a cause-and-effect chain, if the time between action and effect is beyond short-term memory, spectators are unlikely to perceive the relationship (Emmerson 2007). Sharing control between a player and a system can easily obfuscate the effects of performance gestures for the spectator, making it much more difficult to relate actions to sounding results.

Thus far we have examined interactive systems that the performer engages with directly. Instrumental and conductor models function as devices that are "played," and their outputs encompass the entire contribution of their performers. On the contrary, the remaining two models describe systems that performers interact with while playing instruments of their own. The latter are most commonly encountered in mixed works that combine traditional instruments with electronics.[5] We now turn to an examination of reflexive and virtual-musician models of interactivity.

18.7 The Reflexive Model

The reflexive model describes interactive systems that produce predetermined electronics in response to a performer's actions—the same input always produces a similar output. Two technical strategies are prevalent in reflexive systems: *real-time processing* and *sound-file triggering*. Real-time processing refers to the use of digital signal processing techniques to transform the sound emanating from the instrument on stage. Examples include the use of reverberation, echoes, filters, and asynchronous granulators. Sound-file triggering involves the playback of prepared sound files at particular moments throughout a work, which can be instigated using a foot pedal, keystroke, or other device, either by the performer onstage or by the composer or sound technician offstage. More sophisticated score-following systems are capable of tracking a performer's position within a work and triggering the sound files automatically at the appropriate times.

Reflexive models are frequently encountered in mixed works that combine acoustic instruments with electronics. Because traditional instruments are involved, live performance retains a central position and with a character that is much in line with conventional notions of liveness. Composers can write virtuosic instrumental parts with confidence that there are performers able to play them, and audiences can be expected to recognize virtuosity due to prior knowledge of the instruments and a clear connection between the performer's actions and the sounding results. In addition, both reflexive strategies free the performer from the temporal rigidity associated with the "instrument and fixed electronics" mixed-work format.

The differences between real-time processing and sound-file triggering are noteworthy, the most obvious being the medium of interaction itself. In real-time processing, interaction is rooted in the sounding output of the instrument, the audio signal, whereas in the case of sound-file triggering it may involve bodily interaction with a device such as a foot pedal or keystroke. Compositionally, the strength of real-time processing is to be found in the potential that it affords for coherence between live and nonlive sound sources, as nuances specific to a particular performance can make their way into the electronic part. This is particularly useful in indeterminate contexts where it may not be possible to predict what the performer will be playing at a given moment. Indeed, in works that involve improvisation, real-time processing may be the only way to achieve coherence between the instrument on stage and the electronics.

A strong argument for preparing sound files in advance is that they tend to sound better. In a studio setting it is possible to engage in a level of critical listening and attention to production detail that is simply not possible in a live setting. Working out of real time also allows composers to be selective about the sounds that end up in the work. Large amounts of material can be generated and the composer can then sift through it, condensing it down to only the best moments. Even more, prepared electronics can take on far greater independence. Since they do not originate from the live performance they are timbrally and behaviorally detached from the instrumental sound and can function as a truly distinct voice in the musical texture.

From an audience perspective, it can be difficult to distinguish between real-time processing and sound-file triggering strategies, and for many composers, the distinction is unimportant. Composers will implement the strategy most appropriate to achieve their desired goal. Fortunately, these strategies are not mutually exclusive and both are frequently integrated in the same work.

18.8 The Virtual-musician Model

Creating digital systems that behave like virtual musicians is perhaps the most technically ambitious of the four interactive models discussed. Numerous composers have developed virtual musician systems, two notable cases being Robert Rowe's *Cypher* (1993) and George Lewis's *Voyager* (1993). More recently, Tim Blackwell and Michael Young at Goldsmiths College in London have been working on the development of "live algorithms," which they describe as autonomous idea generators that can collaborate and interact with human performers, making apt and creative contributions to the music (Blackwell and Young 2006). Design considerations of virtual musician systems have been detailed extensively elsewhere (Handelman 1995; Lewis 2000; Rowe 1993) and are beyond the scope of this chapter. Instead, the current discussion focuses on behavioral qualities desirable in a virtual musician.

Unlike the other models, virtual musician systems are intended to function as autonomous players, and typically in an improvisatory context where both inputs to and outputs from the system are unpredictable. As Blackwell and Young (2006) explain, free improvisation rejects a priori plans in favor of open, emergent patterns of behavior. Performers assume and cast roles and pursue shared goals as they progress, sometimes rapidly, through a dynamic web of musical relationships. Musical structures emerge as a consequence of these behaviors from the bottom up.

Blackwell and Young's account of free improvisation reflects traditional social theories of human–human interaction. George Herbert Mead (1934) emphasized the influence of shared cognitions on the formation of responses when engaged in a social interaction. He introduced the notion of the "generalized other" to refer to an individual's conception of the general attitudes and values of others within the environment. According to Mead, social interaction requires the individual to assume the role of others—to put him- or herself in their shoes—when considering how the individual's own actions might influence the group dynamic. The importance of the generalized other is that it functions as a constraining influence on behavior, because an individual will generate responses to a given situation based on the supposed opinions and attitudes attributed to the others.

Mead's concept of the "generalized other" also plays an important role in the interaction between musicians in free improvisation. Players do not merely respond, but instead they imagine where the music might go next and then take actions in an attempt to influence other players toward those goals. For a machine to engage in free improvisation, as it is described above, the system must be able to "listen to" and make sense of the sounds around it, as well as to both respond to and instigate meaningful discourse, all of which can be computed only against a framework of musical knowledge that is general and shared between all agents involved. Interactivity at this level falls under the purview of artificial intelligence (AI).

The terms "strong" and "weak" are commonly applied to systems in the field of artificial intelligence, and Blackwell and Young (2006) have aptly enlisted them to describe interactive systems. They align weak interactivity to the reflexive model, where an incoming signal is analyzed and a resultant action is taken according to some predetermined process. To the contrary, strong interactivity involves creativity—the ability to imagine possible scenarios and to respond in an unpredictable, yet meaningful way. Alan Turing, an early proponent of AI, argued that the only real test for strong AI is to see if anyone can tell the difference between the performance of a machine and that of a human being. If a machine is capable of performing as well as an intelligent human then, for all practical purposes, it is intelligent.

Clearly, a host of technical challenges must be addressed in order to develop an effective virtual musician system. The aim here is not to tackle those issues, but simply to formulate a wish list of behavioral qualities desirable in a digital performer. Taking into account theories of social interaction, free improvisation, and artificial intelligence, we might conclude that the ideal virtual musician should be able to:

- Analyze the material of other players, breaking it down into constituent components (pitches, rhythms, dynamics, densities, and so on);
- Interpret the analyzed material against an embedded knowledge of musical relationships that is, at least in some ways, similar to musical knowledge that human performers possess (i.e., rooted in music of the past);

- Make both short-term and long-term assumptions about the intentions of others based on what they are currently doing;
- Maintain some degree of continuity and/or directed motion within its own output;
- Respond to others in an appropriate and meaningful way;
- Initiate appropriate and meaningful discourse;
- Make assumptions about what others might do given the system's own actions.

Even when all of the above objectives are achieved, a system's performance will be limited to its embedded musical knowledge, the domain of which is constrained by the designer's own knowledge of music, the ability to describe that knowledge programmatically, and the design intentions, which may focus on a particular musical style or genre. Systems that are developed for a particular genre are bound to the relevant audio and syntactical attributes of that style and will be unable to interpret or respond meaningfully to diverse musical contexts. Similarly, the output of human performers is bound to their own embedded knowledge of music, but an experienced musician will likely have been exposed to a wide range of musical styles and will be able to call upon that knowledge when engaging in musical discourse. In theory, it may be possible to equip a virtual performer with knowledge of multiple musical styles, but effectively modeling just a single style is an enormous challenge.

Rather than embedding musical knowledge explicitly, virtual-musician designers might focus their efforts on modeling how musical knowledge is acquired in humans through exposure to music and performance experience. David Cope has made great strides in this direction with his Experiments in Musical Intelligence (EMI) software, which is capable of analyzing past works and then composing original pieces in those styles. Systems that acquire knowledge of music over time would be limited only by their prior experiences. Each time a learning-capable system engages in an improvisation, information about that performance is stored in the system's memory to be factored into future performances. Advances in machine learning and neural networks offer realistic opportunities in this area. Perhaps the greatest obstacle to the design of a learning-capable virtual musician is our incomplete understanding of the cognitive processes involved in musical language acquisition in humans.

It is also important to note that a virtual-musician system capable of performing as well as a human does not circumvent many of the issues of liveness that have been discussed. Such a system may produce musical material that is indistinguishable from that of a human player, but there is still no bodily engagement with an instrument.[6] The absence of physical interaction deprives spectators of many of the attributes that were associated with liveness earlier. How can we speak of machine virtuosity when we have come to expect machines to perform beyond our own capabilities and with perfection? Machines are certainly not flawless—there may be "bugs" in their logic—but those errors tend to be characteristically different than human performance errors. The significance of virtuosity, fallibility, visual cues, and presence all seem to be somewhat negated in machine performance.

18.9 The Spectator's Perception of Interaction

Interactive musical works performed in a concert setting often frame the human–machine relationship, and in many works an understanding of the interactivity becomes an important criterion for a spectator's apprehension of what they are witnessing.[7] Nowhere is this more evident than in the evaluation of the live performance itself. The ability to distinguish a good performance from a bad one, to recognize virtuosity or to identify errors, necessarily rests on a clear understanding of the performer's contribution to the music. Inspired by an analysis of human–human communication, Bellotti and others (2002) offer five relevant questions that might inform the design of interactive music systems, but the questions could just as easily be asked of a spectator:

- How does the spectator know that the performer is communicating with the system?
- How does the spectator know that the system is responding to the performer?
- How does the spectator think the user controls the system?
- How does the spectator know that the system is doing the right thing?
- How does the spectator know when the performer or the system has made a mistake?

Gurevich and Fyans (2011) agree that the spectator's perception of interactions must be considered, but point out that, in the case of works that employ digital musical instruments, perception may be inaccurate, different from that of the performer, or it may vary significantly between spectators. They conducted a study on the spectator's perception of performances on digital musical instruments and their findings are enlightening. While a detailed discussion of the study is beyond the scope of this chapter, their findings can be summarized as follows:

Spectators had difficulty understanding the interaction between a performer and a relatively simple DMI.

- When spectators were unable to understand the human–machine interaction their ability to assess skill and identify errors in the performance was compromised.
- When spectators were unable to understand the human–machine interaction they were more attuned to visual signals, such as body language and facial expressions, and an intellectual understanding of the technology involved.
- When the performer was perceived as controlling a process rather than immediate events (the conductor model), spectators often perceived the performer as being "immune from errors."
- When the human–machine interaction was clear, spectators tended to focus more on the performer's perceptual-motor skills.

Gurevich and Fyans conclude that one of the key features of digital musical instruments is that there is no universal experience. The study showed that spectators perceived different modes of interaction, sometimes simultaneously. Spectators may simultaneously engage with a performance in terms of technical action, bodily movement, facial expressions, soundscape, and environmental conditions. They suggest that successful digital musical instruments are more likely to be those that account for this diversity and capitalize on the flexibility that digital devices afford.

18.10 Conclusions

Auslander (2008) points out that our conception of liveness and what it means to perform changes with technological developments. Interactive systems and interfaces for controlling sound that challenge established notions of instrumentality, musicianship, and liveness in performance are exciting precisely because they force us to reconsider traditional musical praxis, and in some cases, redefine the boundaries of those practices. George Lewis (2009) has suggested that computers can guide us forward in music, reasoning that human–computer interaction based on coherent and intelligible logic that is specifically not modeled on traditional musical practice may ultimately become a part of our human musical language. In other words, as musicians increasingly engage with machines in musical performance we may begin to play *like them*.

While interactive systems do offer exciting opportunities in the field of contemporary music, they also run the risk of being perceived as novel demonstrations of technology in which the artistic merit of the interactivity may not match the spectacle of performance. To some degree, a predisposition toward issues of performance over those that are purely concerned with sounds and their relationships might be expected, since it is in the area of performance that interactivity distinguishes itself the most. However, composers should strive for balance in their consideration of content and presentation. In the best of cases, interactivity leads to music that is every bit as innovative as the system and mode of performance used to present it.

Notes

1. There are exceptions, such as Ben Englert's *Please turn on your cellphones* (2011), where audience members can influence the work directly through text messages. However, such cases are not the norm.
2. I am referring here to musical practices that have grown out of the classical music tradition. Live performances of contemporary music are, in significant ways, different than live performances of popular music. I refer the reader to Auslander (2008) for a wonderful examination of these differences.
3. There are counterarguments that can be levied to the claim that fixed music is invariable. Composers of fixed works often diffuse their music live through various speakers

distributed in the concert space, exploiting the unique qualities of the particular space and speaker arrangement available, both of which are bound to be significantly different from one concert to the next. Furthermore, each hearing of a work, fixed or not, brings with it a new perceptual experience in which the listener attends to different elements in the music. The philosophical divide between composers of scored and fixed music points to a deeper ontological debate over where one places the musical object itself—in the score, the act of performance, the resultant sound, or in perception.

4. Auslander not only acknowledges that this definition has been expanded with emergent technologies, but argues that it is no longer valid in some highly mediatized fields, including popular music.
5. The performer's "instrument" could just as well be a DMI, but the combination of traditional instruments with electronics is so common that it remains the focus of this discussion.
6. Virtual reality technologies may offer a means of establishing simulated bodily engagement. However, this discussion remains focused on live performance in a physical concert hall setting.
7. The spectator's comprehension of live interaction is not a concern shared by all composers. Some are content with an audience more interested in the music alone. However, in many interactive works the interactivity is more than a means to an end; it is an integral component of the work itself.

References

Auslander, Paul. 2008. *Liveness: Performance in a Mediatized Culture*. New York: Routledge.

Bell, Elizabeth. 2008. *Theories of Performance*. Thousand Oaks, CA: Sage.

Bellotti, Victoria, Maribeth Back, W. Keith Edwards, Rebecca E. Grinter, Austin Henderson, and Cristina Lopes. 2002. Making Sense of Sensing Systems: Five Questions for Designers and Researchers. *Proceedings of the SIGCHI Conference on Human Factors in Computing Systems*, 415–442. New York: ACM.

Blackwell, Tim, and Michael Young. 2006. Live Algorithms for Music Manifesto. http://www.timblackwell.com.

Cadoz, Claude. 2009. Supra-instrumental Interactions and Gestures. *Journal of New Music Research* 38 (3): 215–230.

Campbell, Joseph. 1949. *The Hero with a Thousand Faces*. Princeton, NJ: Princeton University Press.

Chadabe, Joel. 1997. *Electric Sound: The Past and Promise of Electronic Music*. Upper Saddle River, NJ: Prentice Hall.

——. 2000. Devices I Have Known and Loved. In *Trends in Gestural Control of Music*, ed. Marcelo M. Wanderley and Marc Battier. Paris: IRCAM.

——. 2002. The Limitations of Mapping as a Structural Descriptive in Electronic Instruments. *Proceedings of the 2002 Conference on New Instruments for Musical Expression*, 38–42. Dublin, Ireland.

——. 2005. The Meaning of Interaction. *Proceedings of the 2005 HCSNet Conference*. Macquarie University, Sydney, Australia.

——. 2007. A Brief Interaction with Joel Chadabe. *SEAMUS Newsletter* 2: 2–3.

Cox, Arnie. 2011. Embodying Music: Principles of the Mimetic Hypothesis. *Music Theory Online* 17 (2). http://www.mtosmt.org/issues/mto.11.17.2/mto.11.17.2.cox.html.

d'Escriván, Julio. 2006. To Sing the Body Electric: Instruments and Effort in the Performance of Electronic Music. *Contemporary Music Review* 25 (1/2): 183–191.

Djajadiningrat, Tom, Ben Matthews, and Marcelle Stienstra. 2007. Easy Doesn't Do It: Skill and Expression in Tangible Aesthetics. *Personal and Ubiquitous Computing* 11 (8): 657–676.

Emmerson, Simon. 2007. *Living Electronic Music*. Aldershot, UK: Ashgate.

Englert, Ben. 2011. *Please Turn on your Cellphones*. https://soundcloud.com/bengl3rt.

Gurevich, Michael, and A. Cavan Fyans. 2011. Digital Musical Interactions: Performer-system Relationships and their Perception by Spectators. *Organised Sound* 16 (2): 166–175.

Handelman, Eliot. 1995. Robert Rowe, Interactive Music Systems: Machine Listening and Composing [book review]. *Artificial Intelligence* 79: 349–359.

Jensen, Mads V., Jacob Buur, and Tom Djajadiningrat. 2005. Designing the User Actions in Tangible Interaction. *Proceedings of the 4th Decennial Conference on Critical Computing: Between Sense and Sensibility*, 9–18. New York: ACM.

Lewis, George E. 1993. *Voyager [CD]*. Japan: Avant.

——. 2000. Too Many Notes: Computers, Complexity and Culture in Voyager. *Leonardo Music Journal* 10: 33–39.

——. 2009. Interactivity and Improvisation. In *The Oxford Handbook of Computer Music*, ed. Roger T. Dean, 457–466. New York: Oxford University Press.

Mead, George Herbert. 1934. *Mind, Self, and Society*. Chicago: University of Chicago Press.

Miranda, Eduardo R., and Marcelo Wanderley. 2006. *New Digital Musical Instruments: Control and Interaction beyond the Keyboard*. Middletone, WI: A-R Editions.

Moore, F. Richard. 1988. The Dysfunctions of MIDI. *Computer Music Journal* 12 (1): 19–28.

Paine, Garth. 2009. Gesture and Morphology in Laptop Music Performance. In *The Oxford Handbook of Computer Music*, ed. Roger T. Dean, 214–232. New York: Oxford University Press.

Phelan, Peggy. 1993. *Unmarked: The Politics of Performance*. New York: Routledge.

Rowe, Robert. 1993. *Interactive Music Systems: Machine Listening and Composing*. Cambridge, MA: MIT Press.

Schloss, W. Andrew. 2003. Using Contemporary Technology in Live Performance: The Dilemma of the Performer. *Journal of New Music Research* 32 (3): 239–242.

Schloss, W. Andrew. and David A. Jaffe. 1993. Intelligent Musical Instruments: The Future of Musical Performance or the Demise of the Performer? *Journal of New Music Research* 22 (3): 183–193.

Schnell, Norbert, and Marc Battier. 2002. Introducing Composed Instruments, Technical and Musicological Implications. *Proceedings of the 2002 Conference on New Instruments for Musical Expression*, 156–160. Dublin, Ireland.

Wechsler, Robert. 2006. Artistic Considerations in the Use of Motion-tracking with Live Performers: A Practical Guide. In *Performance and Technology: Practices of Virtual Embodiment and Interactivity*, ed. Susan Broadhurst and Josephine Machon, 60–77. New York: Palgrave Macmillan.

Winkler, Todd. (1998) 2001. *Composing Interactive Music: Techniques and Ideas Using Max*. Cambridge, MA: MIT Press.

Xenakis, Iannis. 1971. *Formalized Music: Thought and Mathematics in Composition*. Bloomington: Indiana University Press.

CHAPTER 19

SKILL IN INTERACTIVE DIGITAL MUSIC SYSTEMS

MICHAEL GUREVICH

It has been said that one of the primary reasons for attending musical performances is to experience skill (Schloss 2003): to see and hear musicians performing in ways that the spectator cannot or would not, in doing so demonstrating the fruits of years of laborious training and practice. Of course, this is but one reason among many to go to a concert, but it raises questions of how performers develop instrumental skill, how skill is expressed between performers and spectators, and how spectators draw upon their knowledge and experience to make sense and meaning of skilled performances. This chapter deals with these issues as they pertain specifically to performances with interactive digital music systems.

Interactive digital music systems have the potential to foster different types of relationship, of which skill is one important facet, in the ecosystem that exists between performers, instruments, spectators, and society. The simple question, "How do we know if a performance was skillful?"—the answer to which may seem to be intuitive or self-evident in most acoustic music situations—becomes quite thorny when it comes to performances with interactive digital systems. It would be futile to attempt to produce a universal checklist of criteria that could be used to answer this question. Instead, this chapter develops a framework for understanding how performers and spectators may arrive at a shared sense of what constitutes skill in a given situation, from which all may form their own opinions. This in turn will offer insight into how we can design interactive performance situations that foster a greater ability to develop, recognize, discuss, and critique skill.

19.1 Toward a Definition of Skill

Skill as a general phenomenon appears to be nearly universally understood instinctively, especially in its extreme cases: a pole-vaulter launching himself six meters over a bar; a chess player defeating twenty-five opponents in simultaneous matches; a nonswimmer struggling to stay afloat in a pool. Yet, it is important to specify what the term "skill" entails, which I undertake by way of discussing the essential characteristics of skill that are generally agreed upon by researchers in psychology and human motor control (e.g., Magill 1993; Proctor and Dutta 1995).

Fundamentally, skill involves "goal-directed" behavior (Proctor and Dutta 1995). It is evident in sports or crafts that skill should lead to a desired outcome or artifact that can be measured in retrospect, such as an arrow shot through a small target or a structurally sound, symmetrically woven basket. Yet where, as in dance, skilled activity is manifested as a continuous process—where the outcome may be ephemeral and unquantifiable—it remains that the practitioner has a goal in mind, however difficult to verbally specify, and that increasing skill will lead to more desirable performance more frequently.

There is a subtle but important distinction between two senses of the word "skill," highlighted by Magill (1993, 7). In the first sense, *a* skill is a goal-oriented act or task to be performed—whistling, snapping your fingers, finding the roots of a quadratic equation, or baking a pie. In the other, which is more useful for the purposes of the present discussion, skill is an environmentally situated human trait that leads to qualitative differences in performance. Skill in this sense fosters variability within and between performances, dependent in part on proficiency, but also on a range of environmental factors. This situated, qualitative notion of skill also suggests a challenge in measuring or characterizing an individual's skill. Above I hinted at two indicators—the desirability of an outcome and the frequency of positive results, the latter of which Magill (1993, 8) refers to as "productivity." Regardless of the dilemma of assessing skill, it is generally agreed that a hallmark of any skilled activity is some degree of efficiency (Welford 1968), what Proctor and Dutta (1995, 18) call "economy of effort." Several people may be able to produce a sophisticated knot with indistinguishable results, but a more skilled rigger would be able to do so with less exertion and possibly in a shorter amount of time.

Implicit in this and all of the previous illustrations is that skill exists within some domain of practice. Certain domains are more clearly demarcated than others, and some may overlap—one may conceive of a continuum from "baseball player" to "left-handed knuckleball pitcher"—but at some point, skill within one domain does not necessarily equate to skill within another. Although all involve coordinated rhythmic activities, many musicians are famously poor dancers, and may be even less skilled table tennis players. This is in part because skill is acquired and develops over time. Although individuals may begin with different abilities and may progress at different rates, novices will improve through practice, which may be a complex, multifaceted activity beyond simple repetition. Several authors have proposed distinct stages or levels that characterize skill development over time. Fitts and Posner (1967) describe three such stages primarily in terms of perceptual-motor qualities that can change with practice. Dreyfus (2004) identifies five stages from novice to expert, taking a wider, phenomenological view that accounts for a range of emotional, cognitive, neurological, sensory, and motor developments.

Even seemingly commonplace human activities like running and talking represent acquired, organized, goal-directed behavior, and are thus included under the umbrella of skill. Drawing on Dreyfus and Dreyfus (1986), Ingold (2000, 316, 353), emphasizes that skill is actionable knowledge—"knowledge how" as opposed to "knowledge that"—and as such can be learned only through doing, not through the transmission of abstract ideas. He illustrates this with an example of a futile experiment in which participants were given verbal or static visual instructions for tying a knot. Only in retrospect, after successfully tying the knot themselves, could participants make meaning of the instructions (Ingold 2001). This concept of "know-how" (Dreyfus and Dreyfus 1986) can be traced to Polanyi's (1966) term "tacit knowledge," which encapsulates the notion that the body can carry out activities that cannot be otherwise symbolically expressed or verbally articulated.

19.2 Cognitive and Sensorimotor Skill in Music

Skill research tends to distinguish between cognitive and sensorimotor skills (Colley and Beech 1989). The former broadly involve "intellectual" activities in which desirable outcomes are symbolic, whereas the latter, which are at times further subdivided into perceptual and motor skill components (e.g., Welford 1968), result in physical action. Although many activities include aspects of both cognitive and sensorimotor skill, and there is evidence that they may have common mechanisms of acquisition (Rosenbaum, Carlson, and Gilmore 2001), researchers tend to limit their scope to one domain or the other in part "as a matter of heuristic convenience" (Newell 1991, 213). Music is thus precisely the type of behavior that confounds yet provides rich fodder for researchers, as both cognitive and sensorimotor skills are deeply involved (Palmer 1997). As Gabrielsson (1999, 502) states, "Excellence in music performance involves two major components: (a) a genuine understanding of what the music is about, its structure and meaning, and (b) a complete mastery of the instrumental technique."

Researchers in music performance (e.g., Clarke 1988) have historically broken down the process of performance along the lines of this dichotomy, into a preliminary stage of "planning," a largely cognitive process based on knowledge of the music that is to be performed, which informs the subsequent "execution" by the motor system. The enactivist view (e.g., Varela, Thompson, and Rosch 1991) argues that the separation between these stages is also largely a conceptual convenience. Knowing what to play (as well as when and how to play it) is not a matter of merely selecting a sequence of events informed by an abstract understanding of what the body is able to play; it is fundamentally conceived in terms of the embodied relationship between the performer and instrument. Indeed, Ingold (2000, 316) describes skill as "both practical knowledge and knowledgeable practice." In his own account of playing the cello, Ingold (2000, 413) argues that the conventionally "mental" concepts of intention and feeling do not exist a priori to physical execution; they are immanent in and not abstractable from the activity of playing.

Nonetheless, skill psychologists and enactivists can at very least agree that both cognitive and sensorimotor processes, however inseparable they may be, play significant roles in skilled music performance.

As Gabrielsson's (1999) formulation implies, with few exceptions skilled music performance involves substantial physical interaction with an instrument external to the performer's body. Several useful models have been proposed to distinguish between fundamental types, levels or degrees of skilled interaction with technology in general. Prominent among these are Heidegger's (1962) *Vorhandenheit* (presence-at-hand) and *Zuhandenheit* (readiness-to-hand) (see also Dourish 2001); Fitts's (1964) cognitive, associative, and autonomous stages of skill development; Anderson's (1982) model of progress from declarative to procedural knowledge in skill acquisition; Rasmussen's (1983) framework of knowledge-based, rule-based, and (sensorimotor) skill-based behavior; and Norman's (2004) troika of reflective, behavioral, and visceral mental processes. Although not identical in substance or application, the endpoints of these theories generally align with the poles of cognitive versus sensorimotor skill from psychology. In spite of the obvious role of cognition, skilled performance with a musical instrument is often held as a prime example of one of these extremes—a visceral, autonomous activity in which the instrument is present-at-hand; one in which the performer plays *through* their instrument rather than *with* it. The requisite cognitive, reflective, or intellectual skill required for expert music performance is invisible to the observer, overshadowed by potentially stunning physical feats and their ensuing sonic manifestations.

19.3 The Problem of Skill in Interactive Digital Music Systems

The burgeoning trend of music performance with interactive digital systems has prompted observers to question to what degree skilled performance with such systems is the same as with acoustic instruments. As in other cases where digital technologies become entwined with a venerated cultural realm, there appears to be an instinctive sense that a critical and uniquely human aspect of music making is in danger of being lost. Perhaps the most pervasive challenge in the literature surrounding the nascent field of "new interfaces for musical expression" (NIME) is in addressing the notion that interactive digital music systems ("new" seems to imply "digital"), by virtue of functionally separating human action from the sound-producing mechanism, limit the potential for skilled practice and human expression that are associated with conventional acoustic instruments. From the NIME field have emerged cries of "whither virtuosity?" (Dobrian and Koppelman 2006) and questions of how performances with interactive digital systems can be meaningful, perceptible, and effortful (Schloss 2003; Wessel and Wright 2002). From very early in their development, authors expressed misgivings about the tendency for interactive digital music systems to diminish or obfuscate both the apparent effort of the performer and the relationship between their actions and ensuing sounds (e.g., Ryan 1991). Several authors have adopted the position that designers of digital music systems should aim to facilitate the type of *intimacy* that exists between performers and acoustic instruments (Cook 2004; Moore 1988; Wessel and Wright 2002). Intimacy is itself a difficult quality to define, but it is revealing that a term normally reserved for the most personal and delicate human bonds has become the standard for instrumental relationships against which digital systems are measured. Regardless of the specific term we adopt, there clearly exists a concern that the relationship between a performer and an interactive digital music system is somehow impoverished, which negatively impacts the musical experience. The following sections will attempt to dissect this concern and frame the problem in terms of skill.

19.3.1 Multiple Actors, Multiple Perspectives

The phenomenon of skill with interactive digital music systems must be considered from the perspectives of different actors in the performance environment, including those of the performer and the spectator. I contend that many of the unresolved problems in the existing NIME literature stem from confusion between these two distinct perspectives and from presumptions surrounding the relationship of the two. This is not to say that performers and spectators can be treated in isolation: they of course ultimately coexist within the same ecosystem, but they do have somewhat different and at times conflicting perspectives and concerns.[1] Performers want to be able to develop skill, to feel improvement in their ability to achieve increasingly complex goals in their performance as they practice over time. Performers also want their skill to be observed and to be appreciated by an audience. Insofar as music listening can be seen as vicarious experience (Cone 1968; Trueman and Cook 2000), spectators, among other goals, desire in turn to recognize, identify with, and appreciate the skill of a performer. But merely possessing skill is no guarantee it will be effectively communicated across a performance ecosystem, nor that it will be effectively apprehended by any given spectator. Below I consider first the phenomenon of skill as it exists between the performer and the interactive digital music system, and subsequently how that relationship is expressed or communicated between performers and spectators. Finally, I discuss what spectators themselves carry with them to the performance that impacts their experience of skill.

19.3.2 Performers

Many of the concerns around skilled digital music performance have emanated from musicians who are accomplished performers with acoustic musical instruments but who find the experience with their digital counterparts to be somehow deficient (e.g.,

Wessel and Wright 2002). The lack of intimacy is especially prominent among these authors. If we attempt to unpack this notion of intimacy, it appears at least in part to be facilitated or characterized by sensorimotor skill. Moore (1988) describes an intimate relationship with an instrument in terms of a feedback-control system involving a performer's perceptual and motor faculties and the instrument's dynamic behavior. Fels (2004) elaborates to describe intimacy in terms of a relationship where the performer embodies the instrument, reflecting the Heideggerian state of *Vorhandenheit* and Fitts's (1964) autonomic phase of skilled practice. This is a phenomenon that is well documented (e.g., Ingold 2000; Ihde 1979), one in which the instrument feels as if it has become an integral part of the body and ceases to be perceived as an external entity. Other authors who aspire to attain a similar connection between performer and interactive digital music system prominently discuss *gesture* (Wanderley and Battier 2000), *tangibility* (Essl and O'Modhrain 2006), and *effort* (Bennett et al. 2007), all suggesting that skilled sensorimotor activity is seen as essential in music performance.

That so many see a similar challenge or deficiency with regard to sensorimotor skill in this context suggests that the nature and/or implementation of interactive digital music systems may truly be problematic. Many authors point to the fact that these systems, at least as they presently exist, rely too heavily on cognitive skill and thus do not afford the cultivation of sensorimotor skill. Nowhere is this critique more apparent than in relation to the phenomenon of laptop music performance, in which performers use only the native input capabilities of a laptop. Somewhat tongue in cheek, Zicarelli (2001) identifies "two characteristics of the computer music process: it is driven by intellectual ideas, and it involves office gestures." Magnusson (2009) argues that even the tangible interfaces that digital musical instruments present to the world are merely arbitrary adornments to a fundamentally symbolic computational system, thus demanding a different modality of engagement—a *hermeneutic* relationship between the human performer and the instrument. In other words, interactive digital music systems allow the performer to specify only symbolic goals, and thus facilitate cognitive but not sensorimotor skill. Green (2011) admits this is often the case, but refutes the disembodied relationship that Magnusson (2009) and many others ascribe as a *necessary* or *essential* condition of interactive digital music systems, suggesting the concepts of *agility* and *playfulness* as indicators or manifestations of musical skill that transcend the acoustic and the digital.

Cadoz (2009) offers a more nuanced spectrum of relationships between performers and interactive digital music systems than Magnusson's (2009) embodied–hermeneutic duality, but similarly contends that the nature of the technology prescribes fundamentally different kinds of interactions. But, like Green (2011), Cadoz disagrees that *instrumental* interactions are solely the province of acoustic systems. Rather, instrumental relationships are characterized by what he calls *ergotic* interactions (Cadoz and Wanderley 2000), ones in which physically consistent, realistic exchanges of energy occur between elements of the system. However, the energetic relationships need not be manifested in actual mechano-acoustic systems in order to facilitate instrumental interactions; they may include any combination of material or simulated objects situated in real or virtual environments with human or nonhuman actors (Cadoz 2009).

Although not framed explicitly in terms of skill, the implication is that sensorimotor skill can indeed exist outside of strictly physical, acoustic interactions with instruments. A recent investigation of the user experience of an interactive virtual music environment based on physical simulation in fact revealed three distinct modalities of interaction between performers and the system: instrumental, ornamental and conversational (Johnston, Candy, and Edmonds 2008). These can be thought of as representing a fluidly shifting balance of cognitive and sensorimotor skill.

As mentioned above, *effort* is regarded as a quality in skilled sensorimotor interactions that is missing in digital systems that afford primarily cognitive engagement. The blame is assigned to the very nature of digital systems but also to their designers. "Too often controllers are selected to minimize the physical, selected because they are effortless. Effortlessness is in fact one of the cardinal virtues in the mythology of the computer" (Ryan 1991, 6). The lament for the loss of sensorimotor skill with digital devices is echoed outside of the musical context as well. Djajadiningrat, Matthews, and Stienstra (2007, 660) attempt to "chart the increasing neglect of the body with respect to human–product interaction," a phenomenon they attribute in part to the preoccupation with "ease of use" in interactive product design. Devices that simplify user actions shift "the complexity from the motor actions to the decision process of what to do. It is exactly because button pushing is so simple from a motor point of view that learning is shifted almost completely to the cognitive domain" (Djajadiningrat, Matthews, and Stienstra 2007, 659). Jensen, Buur, and Djajadiningrat (2005) attribute this shift to the proliferation of what Norman (1998) calls "weak general" products: those in which a user's actions are neither distinct from one another, nor are they associated with unique outcomes. Quite unlike traditional acoustic instruments, such devices preclude the development of specific sensorimotor skills that are particular to the interaction or to an intended result.

The critique of interactive systems at times extends beyond the notion that they make the human body's job "too easy," to assert that they may in fact overtake or overshadow much of the work of the human performer. Magnusson (2009, 175) contends that "software has agency" and thus digital instruments reflect the culture, identity, and skill of their designers as much as, if not more than, those of performers. Indeed, digital systems may be imbued with so much "intelligence" as to limit the possibility for intervention by human performers to simply setting processes in motion or adjusting high-level parameters (Schloss 2003). The notion that in replacing an acoustic instrument, the interactive system itself (and by proxy, its designer) may supplant the role of the skilled performer is reflected in Ingold's (2000, 300–302) synthesis of views on the difference between tools and machines. Although they clearly lie on a continuum (the potter's wheel and the sewing machine being somewhere in the middle), the historical concern is that in progressing from tools, which are guided and powered by the physical and volitional impulses of a skilled craftsman, to machines, which are externally powered and pushed along predefined paths by operators, the richness and reward offered to the skilled human practitioner is lost. The injection of computers, their mechanistic baggage in tow ("machine learning," "human-machine interaction"), into such a refined human tool-using activity as music has historically led to a concomitant decline in the directness between a performer's actions and sonic outcomes (Cook 2004) that no doubt fuels some of the concerns over the diminishing role of skill.

19.3.3 Between Performers and Spectators

Though motivated in part by dissatisfaction with their own experiences, the performer-centered critiques of interactive digital music systems are also informed by performers' own experiences and expectations as spectators. If a disconnect exists between the performer and their digital instrument, another appears between the performer-instrument system and the spectator. In the broadest terms, the challenge between performers of digital instruments and their audiences is framed as one of *expression*. There is a growing body of literature on musical expression both within and outside of the digital context that is too large to summarize or explore in depth here (see e.g., Gabrielsson and Juslin 1996; Juslin and Sloboda 2010), but it is necessary to discuss expression in as much as it pertains to the present discussion of skill. Although there is some question as to whether this is a reasonable universal expectation in new music (Gurevich and Treviño 2007), the very appearance of the term as the "E" in NIME (Dobrian and Koppelman 2006), suggests that spectators largely desire interactive digital music systems to support expression by performers.

I contend that "expression" in this context is largely a proxy for "sensorimotor skill." In general, the range of potential physical realizations of a particular sensorimotor skill is far more restricted than for a cognitive skill. Playing a violin inherently imposes greater constrains on the performer's actions than does playing chess: one can play chess masterfully regardless of how one holds or moves the pieces, or even by instructing another person to move the pieces; the same cannot be said of playing the violin (Rosenbaum, Carlson, and Gilmore 2001). Consequently, the relatively more subtle variations in performance take on greater significance in activities where sensorimotor skills are prominent. These differences in performative action are seen as meaningful regardless of whether they are expressive of any idea or emotion in particular. Indeed, many authors highlight the affective, emotional, or communicative potential of the kind of intimate, embodied relationship with an instrument that sensorimotor skill engenders (Fels 2004; Moore 1988; Trueman and Cook 2000; Wessel and Wright 2002). Here again, Djajadiningrat, Matthews, and Stienstra (2007) take the wider view that any activity involving refined sensorimotor skill has potential expressive and aesthetic value. Others have illustrated that seemingly mundane skilled technical actions such as preparing coffee (Leach 1976) or pouring a beer (Gurevich, Marquez-Borbon, and Stapleton 2012) can communicate cultural or personal values between actors and spectators.

In terms of the characteristics of skill described at the outset of this chapter, it would seem that efficiency is a primary obstacle when it comes to the negotiation of skill between spectators and performers with interactive digital music systems. In order for an observer to appreciate the "economy of effort" that comes with skilled performance they must be able to apprehend the potential difficulty. Imbedded in the adage that a skilled musician makes their performance "look easy" is the notion that for a less skilled musician a similar performance would visibly require a great deal more effort; for most (i.e., the average spectator) it would be impossible. In the case of traditional acoustic instruments, this phenomenon hinges on the performer's direct sensorimotor involvement in the sound-production mechanism. Even where the precise details of a performer's actions are not visible, such as when a pianist's hands are obscured, the spectator is on some level aware that the precise temporal and acoustic characteristics of each sound event are under the performer's direct control. When the spectator experiences a desirable performance, they are consequently aware that it is a result of the performer's skillful execution.

But when an interactive digital music system does not demand significant sensorimotor skill, the distinction between a performance *looking easy* (exhibiting the economy of effort that is a hallmark of skill) and actually *being easy* (requiring minimal effort altogether) may not be evident to a spectator. Cognitive skills do not generally involve physical exertion, and their outcomes may not be temporally or spatially immediate. Therefore the skill, effort, and difficulty of a cognitively demanding performance, as in the case of live coding (Collins et al. 2003), may not be apprehended by a spectator who can only see the performer's actions and hear the resulting sounds. Cognitive skill's lack of specificity of action and immediacy of outcome can be compounded by the potential for agency on the part of the interactive system, giving rise to the possibility that the spectator may confuse the performer's and system's contributions.

19.3.4 Spectators

The role of the spectator in the interactive performance ecosystem is perhaps the least well studied or understood. Yet they are active participants; their very presence and attention provide the impetus for performers to play, and they bring a set of expectations, experiences, and skills (of which performers are on some level aware) that they draw upon to make meaning of the performance. Whereas the previous two sections of this chapter dealt respectively with the performer's skilled relationship to their instrument, and with the consequences of that relationship for the spectator, this section focuses on what the spectator brings to the interaction and how it may impact their experience of skill.

In spite of the apparent desire for greater displays of sensorimotor skill in interactive music performance, we know that spectators do willingly experience and enjoy performances of cognitive skills in other domains. Television quiz shows offer not just the suspense and vicarious thrill of prize money won and lost, but as in music performances, the appreciation of a display of skill—cognitive skill in this case—beyond what most spectators can attain. Although chess is already a well-worn example of cognitive skill, it is illustrative of an important extension of this point. Large audiences routinely attend chess matches between highly skilled players, yet we do not hear protestations about the players' lack of expression or their physical detachment from the chessboard. Spectators remain engaged in what is almost entirely an intellectual, cognitive enterprise, but this is surely only true in cases where they arrive equipped with a prior understanding of what constitutes skill in the domain of chess. "Knowing the game" would seem to be crucial in the spectator experience of cognitive skills. Even a chess match at the highest level would be meaningless for a spectator who does not at very least know how the pieces move or what constitutes victory; *Wheel of Fortune* would not be very rewarding for a spectator who neither speaks English nor reads Roman letters. This is a fundamental difference from some sensorimotor skilled activities, which do not strictly depend on the spectator possessing knowledge or experience external to the experience at hand. A child need not arrive at the circus with a procedural explanation of the mechanics of juggling, nor need they have ever attempted to juggle. The embodied nature of many sensorimotor skills means that spectators can appreciate them in terms of their own bodily knowledge, even without direct experience of the activity in question. A growing body of evidence from the field of action perception, including the discovery of mirror neurons (Rizzolatti and Craighero 2004), supports the idea that we experience the physical behavior of others quite literally in terms of our own bodies (for reviews see e.g., Blake and Shiffrar 2007; Decety and Grèzes 1999).

This is not to say that a spectator's own prior knowledge and bodily skill cannot enrich the experience of sensorimotor skilled performances. In fact, there is evidence to the contrary. Even with small amounts of musical training, music listeners exhibit brain activation in the same motor control areas that would be used to perform the music they are listening to (for a review, see Zatorre, Chen, and Penhune 2007). Moreover, as we have established, even acoustic music performance is not purely a sensorimotor skill. Indeed, a spectator's own cognitive skills play an important role in forming an assessment of a performer's skill; to some extent "knowing the game" is important in music as well. An understanding of music theory, knowledge of the body of musical repertoire surrounding the work, and awareness of the social and cultural context in which a piece of music was conceived can all drastically impact a spectator's overall experience of a performance. These are in turn mediated by a spectator's perceptual skill in listening to the music and watching the performer, and possibly their sensorimotor skill from prior performance experience.

Recent studies of spectators of electronic and acoustic music performances have shown that spectators do indeed draw upon their perceptions of sensorimotor skill but also upon knowledge of stylistic conventions and performance practice in forming assessments of skill (Fyans and Gurevich 2011; Gurevich and Fyans 2011). Significantly, even when spectators in these studies had some basis for assessing embodiment and sensorimotor skill, they were unable to confidently form judgments of overall skill without intimate knowledge of the musical context. Furthermore, this phenomenon persisted whether the instrument in question was acoustic or digital, familiar or not. Thus it would seem that spectators' judgments of skill are indeed informed by factors well beyond performers' displays of speed, control, timing, and dexterity. Spectators, like performers, participate in the sociotechnical systems from which musical performances emerge. Indeed, Lave and Wenger (1991) propose that participation in a community of practice helps give meaning to learning and skill development. In performative domains such as music, it is important to recognize that spectators, in learning to experience, assess, and form opinions of skilled practice, are ultimately participants in the same community as performers (see also Chapters 18 and 20 in this volume).

19.4 On Virtuosity

Especially in the musical domain, skill is frequently uttered in the same breath as *virtuosity*. It seems we all know instinctively that virtuosity requires skill, yet the two terms are not exactly interchangeable: accomplishments involving high degrees of skill are not necessarily virtuosic. For one thing, virtuosity tends to be confined to the arts; apart from usages for rhetorical effect, we don't often hear of virtuoso sprinters or airline pilots, although both can be highly skilled. This is true in part because virtuosity requires not only "high technical proficiency," or sensorimotor skill, but also "critical skill," which Howard (1997, 46) describes as imaginative "interpretive judgment" in the execution of technical skill. In the musical domain, this interpretive judgment may be synonymous with "musicianship." Imaginative interpretive judgment may of course be applied in a number of intellectual domains without virtuosity—history or philosophy, for example—thus, what confines virtuosity to the province of the arts is the employment of imaginative interpretive judgment in the execution of sensorimotor skill (Howard 1997).

According to Mark (1980), it is an artwork's quality of having of a subject—an artwork is *about* something (even if it is about nothing)—that enables it to be virtuosic. In this formulation, a work of virtuosity then must require and demonstrate technical skill, but must also make skill its subject. In other words, virtuosic performances are fundamentally *about* skill. Therefore, the apprehension and attribution of skill are central to a spectator's ascription of virtuosity. This suggests that, like skill, musical virtuosity is socially situated, depending not only upon the performer's skill and musicianship, but also the audience's ability to reflect upon these with respect to both a broader community of musical practice and the perceived limitations of skilled action.

As a more constrained and specific manifestation of skill, virtuosity therefore presents special challenges for interactive music systems. In order to facilitate virtuosity, such systems must of course afford the development of extreme sensorimotor skill but also allow enough room for imaginative interpretive judgment so that performers can exhibit musicianship. However, beyond these, virtuosity requires a musical culture that allows spectators to reflect on how great is the technical and musical accomplishment. This is a difficult proposition for interactive music systems that may be unfamiliar and unique, and that may blur the distinction between human and machine contributions. Reflecting upon the emerging notion of machine musicianship (e.g., Rowe 2001), Collins (2002) considers that plausible "machine virtuosity" would have to be rooted in human sensorimotor and psychoacoustic abilities. A virtuosic machine performance would have to appear to extend human abilities, to transform from human to inhuman, and to be susceptible to mistakes. Although it may be difficult for some spectators to attribute interpretive judgment to the machine performer itself, rather than its programmer, such a performance could certainly fulfill Mark's (1980) criterion of being about the skills that are on display. But it is less evident how virtuosity may emerge in a performance between a human and a machine, where the attribution of skill and interpretive judgment may be fluid or vague. By potentially divorcing a complex sonic outcome from the necessity for high technical skill, interactive systems may leave the performer to rely upon musicality or judgment, which are in themselves insufficient for virtuosity.

19.5 Breakdowns in the Social Construction of Skill

The prevalence of calls for greater and more refined development and expression of skill in performances with interactive digital music systems suggest a number of potential breakdowns in the performer–instrument–spectator ecology. In what follows, I frame these breakdowns in terms of the essential characteristics of skill laid out at the beginning of this chapter.

The most evident breakdown can occur between the performer and the instrument, most likely because the instrument is unable to support attainment of increasingly complex or desirable goals through sustained practice. This situation is an instance of the dilemma of ceilings, floors, and walls: How can we design systems with a low floor to support easy initial access, high ceilings to support sustained skill development, and wide walls to support an acceptably broad range of activities (Resnick et al. 2009; Wessel and Wright 2002)? Although normally framed as a challenge for the development of skilled practice in general, there is a tendency to conflate this breakdown with the aspiration for specifically sensorimotor skill. An incomplete list of properties, some of which I have previously mentioned, that authors suggest are crucial for sensorimotor skill development includes: mapping between gesture and sound (Fels, Gadd, and Mulder 2003), jitter and latency in the system's temporal response (Moore 1988; Wessel and Wright 2002), tangibility (Essl and O'Modhrain 2006), specialization and simultaneity of action (Djajadiningrat, Matthews, and Stienstra 2007), force feedback (O'Modhrain 2001), and effort (Bennett et al. 2007).

Yet the challenge of the floors–ceilings–walls problem can also be addressed through interactive systems that involve primarily cognitive skills. Live-coding laptop practice is a domain in which performers regularly display dazzling feats of cognitive skill in performance (Collins et al. 2003). The primary breakdown in the development and expression of skill may therefore not occur exclusively between the performer and the interactive digital music system, where most tend to locate it, but rather in spectators' perceptions of cognitive skills. Recalling that skill develops within domains of practice that are circumscribed by finite bounds, there may exist a mismatch between the spectator's embodied cultural knowledge and the domain of practice in which a skilled performer is operating. Just as being a skilled distance runner has minimal bearing on my ability to play football, being a skilled oboist may have a very tenuous connection to my skill as a turntablist or practitioner of live coding. "Music" is an excessively broad domain when it comes to skilled practice, and this applies to spectators as well as performers. That skill is a goal-oriented activity that exists within a domain of practice means that in order to apprehend skilled performance, spectators must be aware (or made aware) of how that domain is circumscribed, and be able differentiate between more and less desirable performances according to the performer's goals.

Accordingly, it has been argued that spectators lamenting the feeling of disconnection, disembodiment, or lack of sensorimotor skill between performers and interactive digital music systems are unrealistically transposing their expectations from one subdomain of music to another (Stuart 2003). Perhaps they are failing to understand what constitutes the primarily cognitive domain of skilled practice in which a performer is operating. In the context of laptop music, Stuart (2003) asserts that at least some digital music performances are fundamentally aural phenomena in which, unlike acoustic music, the performer's bodily relationship to sound is unimportant. The onus is thus placed on the listener to overcome their misplaced desire for sensorimotor skill.

A further mismatch may exist between performers' and spectators' notions of what constitutes a desirable outcome. This is always a potential concern in a performative domain, one that is especially salient in contemporary music. Stirring a listener's emotions or displaying physical dexterity may not be among the goals guiding a performer's activity; misapprehension of these goals may lead to another breakdown in the ecology of skill.

Finally, especially in cases where the interaction is largely cognitive, it may be difficult for the performer's skill to be separated from that of an instrument builder, designer, composer, or software programmer. Spectators of acoustic music performances generally understand the bounds between the contributions of instrument makers and performers; it still takes a highly skilled performer to make even a Stradivarius sound good. But insofar as the interactive digital system has greater potential for spontaneity, programmability, or agency, it can be difficult to attribute the outcomes of the system to the skill of the performer or to properties that were built into the system.

19.6 Authenticity

Auslander (2008, 98) contraposes Stuart's (2003) renunciation of the necessity of the visual with Schloss's (2003) emphasis on perceptible effort. He situates the "decorrelation" of visual evidence of music performance from the means of sound production within the larger frame of a supposed ontological distinction between "live" and "mediatized" forms of performance, one that he ultimately rejects (Auslander 2008, 5). In this view, calls for intimacy, transparency, and evidence of skill in the relationship between performers and interactive systems may be seen as a demand for *authenticity*, analogous to the function of live performance in validating the credibility of rock performers whose primary outputs are recordings (77). Although most music created with interactive digital systems lies outside of rock culture, the classical or "new music" culture from which it tends to derive has its own norms and expectations for authenticity on the part of performers, which include demonstrable skill in live performance. It is clear that for some spectators, a display of sensorimotor skill is a necessary constituent of an "authentic" performance with an interactive system.

It is interesting to note that Schloss and Jaffe's (1993) earlier article positing "the demise of the performer" emerged at exactly the same time as the crisis of authenticity in rock music that is Auslander's (2008) primary case study reached its apex. Auslander chronicles the Milli Vanilli lip-syncing scandal of 1990 and the role that *MTV Unplugged*—in particular Eric Clapton's performance and *Unplugged* album that earned six Grammy awards in 1993—played in restoring a semblance of authenticity to the rock music establishment. Although there is no evidence that this episode directly affected Schloss and Jaffe's writing, it foregrounded questions of musical authenticity within the wider societal consciousness, and, as Auslander traces, contributed to a subsequent cultural reassessment strengthening the need for apparent authenticity, even in nonrock music. It is worth considering to what extent the broader cultural discourse on authenticity and its relationship to "liveness" (Auslander 2008) forms the background for expectations of demonstrable skill in interactive music performances.

19.7 Conclusions

From this complex suite of relationships emerges a picture of skill not just as a property of a performer to be assessed by a spectator, but rather as a situated, multidimensional, socially constructed phenomenon that emerges within the performance ecosystem. It is a phenomenon for which society has largely been able to converge, if not upon universally agreed judgments, then upon at least a basis for informed critique within certain well-established traditions of music performance, but a basis that remains almost completely untamed in the jungle of interactive digital music systems. Although there is an undeniable tilt toward the relative importance of the cognitive versus the sensorimotor in digital music performance, this binary opposition is inadequate for fully characterizing and problematizing the phenomenon of skill as it applies to interactive music.

Overcoming the potential breakdowns in the ecology of skill cannot solely be a matter of imbuing interactive systems with greater potential for sensorimotor engagement, nor one of spectators needing to overcome an anachronistic desire for physical performativity and immediacy. Skill emerges from a performance ecosystem that includes a performer, instrument, and spectator, all as active participants that also exist within a society and draw upon cultural knowledge. Anything resembling a consistent conception of skill between a performer and spectator relies on some degree of shared understanding of the performer–instrument relationship, confluence between the performer's and spectator's goals and expectations, commonality of cultural experience, and participation in overlapping communities of practice. Of course, this framework represents just a single spectator. For informed discussion or shared experience of skill to emerge between different spectators, these relationships must extend outward to the larger social ecosystem of the audience.

Although I have painted a picture of an undeniably complex and fragile system, the intention is not to say that all hope is lost. In fact, quite the contrary: as a society we have already managed to negotiate this ecosystem rather effectively (and somewhat organically) in a large number of acoustic musical performance situations. There is no doubt that we can accomplish the same as we set out to incorporate new interactive technologies into skilled music practice, as long as we bear in mind the complexity and potential for disruption to the existing ecosystem. We must expect that new forms of technological relationships between performers and instruments require simultaneous reconsideration and recalibration of what skill means throughout the performance ecosystem and how design can facilitate its emergence.

Note

1. At this point it is worth highlighting that there are valid and accepted musical situations in which skill is unimportant or unnecessary from both the spectator and performer perspectives (e.g., certain experimental pieces by John Cage, Cornelius Cardew, and members of Fluxus), but this chapter is specifically concerned with circumstances in which skill is desirable.

References

Anderson, John R. 1982. Acquisition of Cognitive Skill. *Psychological Review* 89 (4): 369–406.

Auslander, Philip. 2008. *Liveness: Performance in a Mediatized Culture*. New York: Routledge.

Bennett, Peter, Nicholas Ward, Sile O'Modhrain, and Pedro Rebelo. 2007. DAMPER: a Platform for Effortful Interface Development. In *Proceedings of the 7th International Conference on New Interfaces for Musical Expression*, 273–276. New York: ACM.

Blake, Randolph, and Maggie Shiffrar. 2007. Perception of Human Motion. *Annual Review of Psychology* 58: 47–73.

Cadoz, Claude. 2009. Supra-Instrumental Interactions and Gestures. *Journal of New Music Research* 38 (3): 215–230.

Cadoz, Claude, and M. M. Wanderley. 2000. Gesture-music. In *Trends in Gestural Control of Music*, ed. M. M. Wanderley and M. Battier, 71–93. Paris: IRCAM–Centre Pompidou.
Clarke, Eric F. 1988. Generative Principles in Music Performance. In *Generative Processes in Music*, ed. John A. Sloboda, 1–26. Oxford: Clarendon Press.
Colley, Ann M., and John R. Beech. 1989. Acquiring and Performing Cognitive Skills. In *Acquisition and Performance of Cognitive Skills*, ed. Ann M. Colley and John R. Beech, 1–16. New York: John Wiley.
Collins, Nick. 2002. Relating Superhuman Virtuosity to Human Performance. In *Proceedings of MAXIS*, Sheffield Hallam University, Sheffield, UK.
Collins, Nick, A. McLean, J. Rohrhuber, and A. Ward. 2003. Live Coding in Laptop Performance. *Organised Sound* 8 (3): 321–330.
Cone, Edward T. 1968. *Musical Form and Musical Performance*. New York: W. W. Norton.
Cook, Perry R. 2004. Remutualizing the Musical Instrument: Co-design of Synthesis Algorithms and Controllers. *Journal of New Music Research* 33 (3): 315–320.
Djajadiningrat, Tom, Ben Matthews, and Marcelle Stienstra. 2007. Easy Doesn't Do It: Skill and Expression in Tangible Aesthetics. *Personal and Ubiquitous Computing* 11 (8): 657–676.
Decety, J., and J. Grèzes. 1999. Neural Mechanisms Subserving the Perception of Human Actions. *Trends in Cognitive Sciences* 3 (5): 172–178.
Dobrian, Christopher, and Daniel Koppelman. 2006. The "E" in NIME: Musical Expression with New Computer Interfaces. In *Proceedings of the 2006 Conference on New Interfaces for Musical Expression*, 277–282. Paris: IRCAM–Centre Pompidou.
Dourish, Paul. 2001. *Where the Action Is: The Foundations of Embodied Interaction*. Cambridge, MA: MIT Press.
Dreyfus, Hubert L., and Stuart E. Dreyfus. 1986. *Mind over Machine: The Power of Human Intuition and Expertise in the Era of the Computer*. New York: Simon and Schuster.
Dreyfus, Stuart E. 2004. The Five-Stage Model of Adult Skill Acquisition. *Bulletin of Science, Technology and Society* 24 (3): 177–181.
Essl, Georg, and Sile O'Modhrain. 2006. An Enactive Approach to the Design of New Tangible Musical Instruments. *Organised Sound* 11 (3): 285–296.
Fels, Sidney. 2004. Designing for Intimacy: Creating New Interfaces for Musical Expression. *Proceedings of the IEEE* 92 (4): 672–685.
Fels, Sidney, Ashley Gadd, and Axel Mulder. 2003. Mapping Transparency Through Metaphor: Towards More Expressive Musical Instruments. *Organised Sound* 7 (2): 109–126.
Fitts, Paul M. 1964. Perceptual-motor Skill Learning. In *Categories of Human Learning*, ed. A. W. Melton, 243–285. New York: Academic Press.
Fitts, Paul M., and Michael I. Posner. 1967. *Human Performance*. Belmont, CA: Brooks/Cole.
Fyans, A. Cavan, and Michael Gurevich. 2011. Perceptions of Skill in Performances with Acoustic and Electronic Instruments. In *Proceedings of the 2011 Conference on New Interfaces of Musical Expression*, 495–498. Oslo, Norway: University of Norway and Norwegian Academy of Music.
Gabrielsson, Alf. 1999. The Performance of Music. In *The Psychology of Music*, ed. Diana Deutsch, 501–602. San Diego: Academic Press.
Gabrielsson, Alf, and Patrik N. Juslin. 1996. Emotional Expression in Music Performance: Between the Performer's Intention and the Listener's Experience. *Psychology of Music* 24 (1): 68–91.
Green, Owen. 2011. Agility and Playfulness: Technology and Skill in the Performance Ecosystem. *Organised Sound* 16 (2): 134–144.

Gurevich, Michael, and A. Cavan Fyans. 2011. Digital Musical Interactions: Performer-System Relationships and Their Perception by Spectators. *Organised Sound* 16 (2): 166–175.

Gurevich, Michael, Adnan Marquez-Borbon, and Paul Stapleton. 2012. Playing with Constraints: Stylistic Variation with a Simple Electronic Instrument. *Computer Music Journal* 36 (1): 23–41.

Gurevich, Michael, and Jeffrey Treviño. 2007. Expression and Its Discontents: Toward an Ecology of Musical Creation. In *Proceedings of the 7th International Conference on New Interfaces for Musical Expression*, 106–111. New York: ACM.

Heidegger, Martin. 1962. *Being and Time*. Translated by John Macquarrie and Edward Robinson. New York: Harper.

Howard, Vernon A. 1997. Virtuosity as a Performance Concept: A Philosophical Analysis. *Philosophy of Music Education Review* 5 (1): 42–54.

Ihde, Don. 1979. *Technics and Praxis*. Dordrecht, Holland: D. Reidel.

Ingold, Tim. 2000. *The Perception of the Environment: Essays on Livelihood, Dwelling and Skill*. London: Routledge.

——. 2001. Beyond Art and Technology: The Anthropology of Skill. In *Anthropological Perspectives on Technology*, ed. Michael B. Schiffer, 17–31. Albuquerque: University of New Mexico Press.

Jensen, Mads V., Jacob Buur, and Tom Djajadiningrat. 2005. Designing the User Actions in Tangible Interaction. In *Proceedings of the 4th Decennial Conference on Critical Computing: Between Sense and Sensibility*, 9–18. New York: ACM.

Johnston, Andrew, Linda Candy, and Ernest Edmonds. 2008. Designing and Evaluating Virtual Musical Instruments: Facilitating Conversational User Interaction. *Design Studies* 29 (6): 556–571.

Juslin, Patrik N., and John A. Sloboda, eds. 2010. *Handbook of Music and Emotion: Theory, Research, Applications*. Oxford: Oxford University Press.

Lave, Jean, and Etienne Wenger. 1991. *Situated Learning: Legitimate Peripheral Participation*. New York: Cambridge University Press.

Leach, Edmund R. 1976. *Culture and Communication: The Logic by Which Symbols Are Connected: An Introduction to the Use of Structuralist Analysis in Social Anthropology*. Cambridge, UK: Cambridge University Press.

Magill, Richard A. 1993. *Motor Learning: Concepts and Applications*. 4th ed. Madison, WI: Brown and Benchmark.

Magnusson, Thor. 2009. Of Epistemic Tools: Musical Instruments as Cognitive Extensions. *Organised Sound* 14 (2): 168–176.

Mark, Thomas C. 1980. On Works of Virtuosity. *Journal of Philosophy* 77 (1): 28–45.

Moore, F. Richard. 1988. The Dysfunctions of MIDI. *Computer Music Journal* 12 (1): 19–28.

Newell, K. M. 1991. Motor Skill Acquisition. *Annual Review of Psychology* 42 (1): 213–237.

Norman, Donald A. 1998. *The Invisible Computer: Why Good Products Can Fail, the Personal Computer is so Complex, and Information Appliances are the Solution*. Cambridge, MA: MIT Press.

——. 2004. *Emotional Design: Why We Love (or Hate) Everyday Things*. New York: Basic Books.

O'Modhrain, Maura Sile. 2001. Playing by Feel: Incorporating Haptic Feedback into Computer-based Musical Instruments. PhD diss., Stanford University.

Palmer, Caroline. 1997. Music Performance. *Annual Review of Psychology* 48 (1): 115–138.

Polanyi, Michael. 1966. *The Tacit Dimension*. Garden City, NY: Doubleday.

Proctor, Robert W., and Addie Dutta. 1995. *Skill Acquisition and Human Performance*. Thousand Oaks, CA: Sage.

Rasmussen, Jens. 1983. Skills, Rules, and Knowledge; Signals, Signs, and Symbols, and Other Distinctions in Human Performance Models. *IEEE Transactions on Systems, Man, and Cybernetics* 13 (3): 257–266.

Resnick, Mitchel, John Maloney, Andrés Monroy-Hernández, Natalie Rusk, Evelyn Eastmond, Karen Brennan, Amon Millner, et al. 2009. Scratch: Programming for All. *Communications of the ACM* 52 (11): 60–67.

Rizzolatti, Giacomo, and Laila Craighero. 2004. The Mirror-neuron System. *Annual Review of Neuroscience* 27: 169–192.

Rosenbaum, David A., Richard A. Carlson, and Rick O. Gilmore. 2001. Acquisition of Intellectual and Perceptual-motor Skills. *Annual Review of Psychology* 52 (1): 453–470.

Rowe, Robert. 2001. *Machine Musicianship*. Cambridge, MA: MIT Press.

Ryan, Joel. 1991. Some Remarks on Musical Instrument Design at STEIM. *Contemporary Music Review* 6 (1): 3–17.

Schloss, W. Andrew. 2003. Using Contemporary Technology in Live Performance: The Dilemma of the Performer. *Journal of New Music Research* 32 (3): 239–242.

Schloss, W. Andrew, and David A. Jaffe. 1993. "Intelligent Musical Instruments: The Future of Musical Performance or the Demise of the Performer?" *Interface* 22 (3): 183–193.

Stuart, Caleb. 2003. The Object of Performance: Aural Performativity in Contemporary Laptop Music. *Contemporary Music Review* 22 (4): 59–65.

Trueman, Dan, and Perry R. Cook. 2000. BoSSA: The Deconstructed Violin Reconstructed. *Journal of New Music Research* 29 (2): 121–130.

Varela, Francisco J., Evan Thompson, and Eleanor Rosch. 1991. *The Embodied Mind : Cognitive Science and Human Experience*. Cambridge, MA: MIT Press.

Wanderley, Marcelo M., and Marc Battier, eds. 2000. *Trends in Gestural Control of Music*. Paris: IRCAM–Centre Pompidou.

Wanderley, Marcelo M., and Philippe Depalle. 2004. Gestural Control of Sound Synthesis. *Proceedings of the IEEE* 92 (4): 632–644.

Welford, Alan Traviss. 1968. *Fundamentals of Skill*. London: Methuen.

Wessel, David, and Matthew Wright. 2002. Problems and Prospects for Intimate Musical Control of Computers. *Computer Music Journal* 26 (3): 11–22.

Zatorre, Robert J., Joyce L. Chen, and Virginia B. Penhune. 2007. When the Brain Plays Music: Auditory–motor Interactions in Music Perception and Production. *Nature Reviews Neuroscience* 8 (7): 547–558.

Zicarelli, David. 2001. Keynote speech presented at the International Computer Music Conference, Havana, Cuba, September 15, 2001. http://finearts.uvic.ca/icmc2001/after/keynote.php3.

CHAPTER 20

GESTURE IN THE DESIGN OF INTERACTIVE SOUND MODELS

MARC AINGER
AND BENJAMIN SCHROEDER

GESTURE is fundamental to music. Gesture initiates and forms sound. If sound is "the carrier of music" (composer Morton Subotnick, in Schrader 1982), then gesture can be thought of as the animating force of music—the force that brings life to music, the force that enables us to impart meaning to sound. We can think of a violinist's bowing, a drummer's strokes, or a pianist's fingers moving across the keys. As an extreme example, the significance of gesture in music can be seen today in the popularity of "air guitar," where performance consists solely of gesture (Godøy 2006).

Musicians have developed an astonishingly wide range of gestural control over their instruments, from the over-the-top "windmill" pyrotechnics of Pete Townshend to the subtle finger movements of a violinist such as Itzhak Perlman. Each of these gestures enables a specific type of sound and expression, from the most aggressive and overt to the most delicate and subtle. Performers work all of their lives to develop this wide range of expressive control. Similarly, traditional instrument designers work all of their lives to build this wide range of expressive potential into their instruments, producing an art that has a corresponding range of expressive depth.

Given this central role of gesture to music and music making (Gritten and King 2006), anyone who designs computer-based instruments is faced with a fundamental problem. How can we design systems that can transform physical gesture into sound in a way that is as intuitive and rich as earlier mechanical designs (such as the piano or violin), while also taking advantage of the unique properties of the digital medium that the gesture is driving? This chapter will present a survey of some of the issues involved with the transformation of gesture into sound in digital systems. We will see that it is not only a question of designing an intelligent interface (a nasty problem by itself), but, really, it is a problem of designing an entire system—a holistic *interactive sound model*—of which the intelligent interface is just one inseparable, organic part.

First we will discuss these three interrelated topics: Gesture as a multimodal phenomenon; the role of gesture in the performer–instrument–listener relationship; fidelity to reality vs. flexibility in the design of reality in interactive sound models. After introducing these initial topics, we will then talk about the design of two traditional mechanical instruments (the piano and harpsichord) and the design of an early iconic electronic (nonmechanical) instrument (the Theremin). The two traditional instruments will stand in contrast to the Theremin (which, as an early electronic instrument, may be understood here as an early instance of an interactive sound model). By contrasting these instruments, we will begin to understand how our initial three topics come into play in the design of interactive sound models.

Next, we will discuss the way in which some early designers of interactive sound models decoupled gesture and sound production. This was done both for practical reasons (in order to reduce computational loads and to conceptually simplify instrument design), and, in some instances, for artistic purposes. For instance, because you *can* have a keyboard that sounds like a saxophone, this is something that some people are interested in exploring. Finally, we will introduce the use of a physical modeling technique (*direct simulation*) as one specific technique that is well suited to the creation of rich and intuitive interactive sound models. Of course, there are many techniques that are available for interactive sound model design, and many techniques may be equally well suited. Furthermore, any technique is only as good as its implementation. However, we wish to present a particular instance of a technique, and physical modeling has several positive attributes that conform to our foregoing discussion: Physically based models work naturally in multimodal environments; physically based models provide performers with instruments that are physically rooted and thus interactive in intuitive ways; physical models allow us to describe the physical world in more or less precise ways, but they also provide us with conceptually direct ways of transforming the physical world in ways that are possible only in the virtual world.

20.1 Gesture as a Multimodal Phenomenon

Marc Leman writes "The multimodal aspect of musical interaction draws on the idea that sensory systems—auditory, visual haptic, and tactile, as well as movement perception—form a fully integrated part of the way the human subject is involved with music during interactive musical communication... Corporeal articulation should thus be seen as a unified principle that links mental processing with multiple forms of physical energy" (2008, 141). Leman goes on to state that, therefore, "It is straightforward to assume that any technology which mediates between mental processing and multiple physical energies should be based on multimedia... These tools can function as an extension of the human body, the natural mediator between musical energy and mental representation."

The Pete Townshend windmill sweep of the guitar produces a vivid sound; but it also produces a vivid visual, and, for the performer, a vivid feel and vivid haptic feedback. Perlman's violin likewise produces a vivid sound (although certainly a sound of a different nature), but he will also be highly attuned to the way the instrument feels and the haptic feedback that he receives from his instrument. This multimodal nature of sound is well known to performers. Performers work endlessly on not just the sound that they produce, but also the way their performance looks and feels. All of these aspects of gesture work together to create an integrated multimodal experience that we call "sound," but that is in reality a metaevent that includes sound.

20.2 The Role of Gesture in the Performer–Instrument–Listener–Audience Relationship

There is a complex network of relationships among the performer, the instrument, and the listener (i.e., the audience). Let us take the audience out of the relationship for a moment and consider performers practicing or playing for their own enjoyment. As performers initiate a sound, they listen, feel, and watch (even if the performer's eyes are closed, they should have a mental image of the visual shape of their movements), all the while making adjustments (often microadjustments) to their gestures in response to the multimodal stimuli that they are receiving from their instrument (see also Chapters 18 and 19 in this volume).

If we place the audience back into the relationship, then we have listeners who possess varying degrees of ability to "decode" the information that the performer "encodes" into the sound (Leman 2008). Audience members who have dabbled with an instrument will decode the sound in one way. Audience members who are familiar with the instrument but not with the music played by the performer will decode the sound differently than audience members who are familiar with both the music and the instrument. There are many possible audience members and many different levels of experience among them. However, the performer hopes to reach them all, which is to say that the performer hopes that each audience member will be able to decode enough information from the sound to have an enjoyable experience.

But the audience is not just decoding information based on the performer's sound and gesture (which by itself is a multimodal phenomenon). Music is a social experience. In the case of a concert, there is an aspect of the experience that is similar to that of a sporting event. Will the performer(s) play the fast passage without a mistake? Will someone cough and ruin the quiet passages? Will the audience inspire the performer, or will the audience be antagonistic or apathetic, and thus ruin the night? All of these things become part of the multimodal information that the performer takes in during the course of a performance, and in the course of a career as a performer.

20.3 Fidelity to Reality vs. Flexibility in the Design of Reality in Interactive Sound Models

We understand that a performer has a musical idea and then initiates a gesture, shaping the gesture in such a way that the performer's instrument transforms that gesture into a sound that contains the musical idea. We also understand that there is a complex chain of communications that includes the performer, the instrument, the listener, and the audience. This communications chain is multimodal and it is multidirectional. Each part of the chain affects the other parts of the chain, so that the musical idea is affected by each part of this chain of communications, both in real time and outside of real time. For the instrument designer, then, the challenge is to create an instrument that maximizes the performer's ability to encode a musical thought through gesture using the interface of the musical instrument. There are many factors that go into this instrument design. The exact requirements of the instrument will vary according to the social and musical conventions of the performer and the music that the performer is playing.

In all cases, though, an instrument must behave in as intuitive a manner as possible. By the time that we are old enough to play an instrument, we have some kind of intuitive understanding of the laws of physics, and we expect that instruments will behave according to these principles. As we become more experienced, we find that the best instruments use the laws of physics to their advantage. While it is possible to design an interactive sound model that ignores many physical principles, we find that these instruments may behave in a manner that is counterintuitive to both the performer and the audience. It would be fair to guess that the best sound models will find some balance between fidelity to reality and flexibility in the use of reality.

The design of interactive sound models, then, offers both opportunities and challenges. One of the potential strengths of sound models is their ability to expand and question our concepts of reality—to create alternate realities. We have the opportunity to create truly unique instruments that would not have been possible without the aid of digital technology. So our initial intuition is to think that, since the laws of physics can be stretched and extended in the virtual world, we need not concern ourselves with the real world. The problem with this, of course, is that a real performer must play this instrument (even if indirectly, as is the case in automated composition); and a real audience will be listening to the instrument. Both the performer and the audience will have expectations, some of which may be hard-wired and some of which may be learned through experience with instruments, music making, and listening. This balance between reality and altered reality is a delicate balance to maintain, and it is this balance that we must always address in our design decisions. Just as in the design of "traditional" instruments, the more we know about reality, the more we become aware of ways in which it may be intuitively manipulated.

20.4 Examples of Traditional Mechanical Gesture Transformation Technologies

Different technologies will allow us to design different methods of transforming gesture into sound. The choice of technologies is driven by the results we seek. A clear example of this is the difference between the harpsichord and the piano. The harpsichord and piano are both keyboard instruments and both make sound using strings. They are played in more or less the same way, but they respond to the performer's touch in very different ways. The harpsichord creates sound through a mechanism that plucks the strings that are inside the instrument. When a key is pressed, a plectrum travels upward and across the corresponding string, plucking it and making sound. Because of the way the plectrum works, the harpsichord produces sounds of more or less the same volume each time a key is pressed. The piano, on the other hand, makes sound through a mechanism that strikes the instrument's strings with hammers. When a key is pressed, a mechanism causes a small, felt-covered hammer to rebound off of the strings. Pressing the key harder causes the hammer to strike the strings harder.

These two instruments interpret the same general gesture in different ways. The difference between piano and harpsichord is a difference in structure. The two instruments respond differently to the same gesture because of the way they are built. The choice of structural design was brought about by a musical choice and it will, in turn, influence subsequent musical choices. The piano "interface" was designed to afford a larger range of dynamics than the harpsichord (thus its full name, the pianoforte), and a greater amount of sound in general. A harpsichord has much in common with plucked string instruments such as the lute (and it even has a "lute stop"), while the piano has more in common with percussion instruments.

Those who want to hear or perform the music that is written for the harpsichord will by and large prefer the sound of the harpsichord and the feel of the harpsichord. Those who want to hear or perform the music that is written for the piano will by and large prefer the sound of the piano and the feel of the piano. As a result of the different design of the gesture transforming mechanisms, the action of each instrument is also different, so each instrument will provide different haptic feedback to the performer. The haptic feedback feels appropriate to the music performed on the instruments. All of these elements (the sound, the feel, the haptic feedback, and the visual differences between the instruments) combine to create very different overall presentations.

20.4.1 The Theremin as a Case Study of an Organic Coupling of Gesture and Sound

One of the earliest electronic instruments, the Theremin, is a good case study of a unique control mechanism and a unique synthesis engine that are brilliantly conceived and well matched. The Theremin is best known for its remarkable performance interface, of course, but a study of the instrument reveals that the sound production engine, though more subtle, is equally remarkable, and, in fact, the sound production engine is inseparable from the performance interface.

The performance interface is well known. The performer stands in front of two antennae and moves her hands closer to or further away from the antennae. One antenna (and, thus, one hand) controls the pitch of the instrument, while the other antenna (and, thus, the other hand) controls the volume of the instrument. The instrument is never actually touched—only the hands' proximity to the instrument is changed.

However, it is not just the interface that makes the Theremin a brilliant bit of engineering. It is also the mechanism that the Theremin uses to unite the gesture transformation and the sound synthesis engine into an integral unit. The Theremin uses a type of capacitive sensing in which the performer's arms and hands become part of the capacitive field. As the proximity of the arms and hands changes relative to the antennae, the capacitance of the field changes. Borrowing from radio technology (which Leon Theremin knew well), the capacitance affects the tuning of an LC oscillator, which is tuned above the frequency range of human hearing. This oscillator is combined with the output of another LC oscillator, whose frequency never changes. The tone that we hear is the result of the heterodyning of the two oscillators' frequencies.

The second antenna of the Theremin controls the volume of the instrument in much the same manner as the pitch of the instrument is controlled. Again, there is a pair of LC oscillators. In this circuit, however, the output of the oscillators controls the output of a bandpass filter. The filter output is then sent through an envelope controller, and this envelope controller controls the output of the VCA (voltage controlled amplifier) that ultimately controls the volume of the instrument (Moog 1996). Much of this may seem counterintuitive. Why not just attach the output of the first antenna to one oscillator tuned in the range of human hearing, and then attach the output of the second antenna directly to a VCA? The additional steps are needed because the changes in the capacitance field are tiny and must be scaled in some way to make the interface practical in a musical sense. These technical features enable both of the antennae to respond in a similar manner. For the performer, the feel of the response will be maximally similar in each hand. More to the point, these technical features fine-tune the transformation of gesture to sound, adding subtlety and detail to the result.

In this way, the Theremin is similar to traditional instruments such as the piano. One could simply make a stiff string and then strike it with a felt covered hammer, and the result would be similar to the sound of a piano. With the piano mechanism, however, one is able to use not just two hands, but in fact, all ten fingers of the two hands, transforming them into subtle and powerful percussive hammers. The piano mechanism allows for a more or less precise transfer of energy from the large muscles in the arm to the ten fingers. This mechanism has enabled and continues to enable a highly developed polyphony to be performed and invented on the piano.

While the Theremin has not yet fostered this level of repertoire development, the point here is that its construction is a good example of an electronic instrument that both points toward the future and yet also draws upon concepts from the past. While the Theremin disregards a traditional mechanical interface, it nonetheless requires and responds to gestural subtlety, and it has a refined and well-integrated mechanism for transforming those gestures into sound. This last point is the most important. The Theremin is an example of an early instrument that approaches the design of a nonmechanical instrument as an organic whole. While the unique interface is its most apparently striking feature, this interface is only the most obvious part of an integral system. It is this entire system, not just the interface, that serves to transform gesture into sound in a fascinating manner.

20.5 Decoupling Gesture and Sound in Virtual Instruments

Traditional instruments require specific means of gesture transformation, such as the mechanism of the piano, or the mechanism of the harpsichord (to cite the two examples that we have already discussed). These mechanisms make strategic use of physical principles to produce particular results. In these instruments, there is a physical connection between the performer and the instrument, and this interaction occurs according to well-known principles. Likewise, the range of sounds that each instrument produces through these mechanisms is the result of well-known principles.

With interactive sound models, however, there is no a priori need to design gesture transformation mechanisms that behave according to physical principles. With these sound models, both the interface between the performer and the instrument and the transformation of the performer's gestural input into sound may be designed in imaginative ways that may or may not reference the physical world. While this freedom gives our imaginations access to an enormous range of possibilities, it is this seemingly endless number of possibilities that may confound us.

When we are confronted by an enormous number of possibilities, the first thing we can do is to find ways of dividing the large task into smaller tasks. The way this has usually been done in the design of sound models is to divide instruments into two parts. The first part is the control mechanism (in the music programs of Max Mathews, this is the "score"), while the second part is the sound synthesis engine (in the music programs, this is the "orchestra"). While this is a somewhat arbitrary and problematic division, there is enough logic and history behind this division to make it useful until we can find other ways of thinking about sound model design.

In a classic digital synthesizer, for instance, the performer plays a keyboard (the control mechanism). The keyboard generates MIDI information that is sent to a sound synthesis engine (such as a sample playback module), and the synthesis engine translates the control information into sound. Using a typical classic synthesizer, the performer plays not only piano sounds with the keyboard, but also, for instance, saxophones, guitars, and drum sounds. The advantage of this is that one keyboard player may play many different types of sounds without changing instruments. The disadvantage is that a keyboard is a very different interface from, for instance, a saxophone, so that it is very difficult for the "saxophone preset" to truly sound or to perform or to feel (to the performer) like a saxophone. In order to compensate for this interface, an experienced keyboard synthesizer performer may add pitch bend and modulation wheel data into the data stream, and the performer may play signature riffs that are idiomatic to the emulated instrument (in this case, the saxophone). Nevertheless, the sound and the process, and especially the feel, are still a compromise noted by both the performer and the audience. Breath controllers were introduced for this reason. A breath controller attached to a keyboard synthesizer will allow the performer to introduce the breath control of volume, articulation, and/or modulation, and will make the sound and the feel more convincing. Again, though, this is a compromise, albeit one in the right direction. There is just one generic breath controller for all wind and brass instruments, and the keyboard is still used in conjunction with the breath controller to determine pitch and, to some degree, duration. The only haptic feedback that the performer will have is, again, from the keyboard, since the breath controller does not provide haptic feedback.

The choice of sound synthesis engine in the classic keyboard digital synthesizer is generally some form of sample playback, some variant of additive or subtractive synthesis, or some variant of nonlinear synthesis (such as frequency-modulation synthesis). There are, of course, many different types. Each synthesis technique has its strong points and its weak points. Certain synthesis techniques match well with certain control mechanisms while other techniques strain to work with other control mechanisms, and are in fact not well matched.

20.6 Physically Based Models

In the design of interactive sound models, we consider not only the way that the instrument sounds, but also the way the way that the instrument feels (including the degree and quality of haptic feedback) and the way that all of this information is communicated to the listener or audience. The fact that interactive sound models make no a priori demands on the designer is both an opportunity and a challenge. In this chapter, we are interested in looking at some fundamental principles of traditional instrument design so that we can make strategic choices in the way that we use the virtual world to transform the physical world. We are interested in the "plausible impossible."[1] It is for this reason that we are interested in exploring physically based sound models.

A *physically based* sound synthesis model produces sound by calculating the way a physical object might vibrate in response to some input force. This physically based response is a good basis for designing the coupling between gesture sensing and sound synthesis, creating instruments that have a traditional feel but that retain the flexibility of the virtual. Just as with other virtual instruments, physically based models may be used with a variety of controllers such as keyboards, breath controllers, sensors, and camera systems. The synthesis runs continually, responding to changes in its environment, and so is an especially good match for continuous input such as that from a breath controller or microphone. Physically based synthesis also opens up new possibilities for rich multimodal interaction involving synchronous motion, graphics, and sound.

The pairing of gestural sensing and sound synthesis plays a key role in the way an instrument responds to gesture, and, therefore, creates the way an instrument "feels." A performer plays the keys of a piano, and hears the strings and the resonant body, but it is the hammers and their mechanisms that interpret the performer's actions. Physically based models do not create this pairing automatically—in a sense, we wouldn't want them to—but they do give designers a familiar, physically rooted vocabulary with which they can adapt synthesis and sensing to one another. There are many physically based techniques, but here we will concentrate on one particular variety: Finite Difference Time Domain, or FDTD, models. Bilbao (2009) calls these "direct simulation" models, and he describes the mathematics of such models in depth.

Many sound models abstract away the physical form of a sounding object, but FDTD models retain this form, making, for example, input and output based upon location available at run-time. This allows for great flexibility in control. FDTD simulation is computationally expensive, but it is now possible to simulate significant models in real time using vector-unit CPU or GPU (Sosnick, 2010) techniques, allowing the use of FDTD models for interactive audio. An FDTD simulation is based on the way that sound waves move across some kind of material. Consider the way a guitar string moves after it is plucked: guitar strings vibrate in complex ways to create sound. (Figure 20.1 shows a snapshot of a simulated string just after it has been set into motion.) How can we determine the string's shape at any given moment?

To break the problem down into tractable parts, we can first divide the string into several discrete segments. This reduces the task of determining the string's shape to one of understanding how any given segment moves. If we understand how segments move, we can start from a known state and then calculate how each segment moves through a series of time steps. The resulting segment positions will describe the entire shape of the string. The movement of a string segment depends on how it reacts to forces around it. String motion is primarily due to tension in the string: each string segment pulls on its neighbors. External actions, such as a performer's touch, also exert force on a string. This situation can be described by equation 20.1.

$$y_{tt}(x,t) = \frac{T}{\mu} y_{xx}(x,t) + f(x,t). \tag{20.1}$$

FIGURE 20.1 A snapshot of a simulated string just after it has been set into motion.

Let's look at each term of this equation in turn. The equation as a whole calculates the acceleration, y_{tt}, of some part of the string. (We use the y dimension to denote the displacement of some part of the string from its normal, straight rest state, and the string's length is along the x dimension.) The acceleration is a function of space (x) and time (t). T is the tension in the string, and μ is its linear mass density. The term y_{xx} describes the curvature of some part of the string. So therefore this equation says that a string (under tension) with more curvature will move more quickly at any given time. Furthermore, a string under more tension will move more quickly, and a denser (heavier) string will move less quickly.

The curvature changes over time, but the tension and mass density normally remain constant. This suggests that a string under more tension will always move more quickly, and thus vibrate at a higher frequency and have a higher pitch. This matches our intuition about how guitar strings should behave. The term $f(x,t)$ describes external force being applied to the string at any given point. With an actual physical string, this might come from a performer's fingers, from a bow, or from a piano hammer. Virtual strings can be used to model these situations, given corresponding ideas about just how some object or another applies force to a string. The particular way in which applied force changes over time, both in amplitude and in the position where it is applied, can radically change the sound of a virtual string—just as is true for a real string. Models for different kinds of interaction can be found in the literature; for example, Cuzzucoli and Lombardo (1999) discuss a detailed model of a player's action in plucking a guitar string.

To write a computer program that runs the simulation, we can replace each term in the equation with an approximation that describes it in terms of differences between values at different time steps, or values in adjacent string segments. For example, we could replace y_{tt} with the approximation below, with Δt being the size of the time step used in our simulation.

$$\frac{y(x,t+1)-2y(x,t)+y(x,t-1)}{\Delta t^2} \tag{20.2}$$

Doing this for all the terms gives us an equation written in simple terms of values of y for various string segments at various time steps. We can then solve the equation at each time step for successive values of y, the displacement of the string. This is the so-called method of finite differences. We will not discuss many practical details of writing a simulation here, including choosing particular finite difference approximations, but many good discussions can be found in the academic literature (e.g., Chaigne 1994) and in Bilbao, mentioned above.

The equation above describes a so-called "ideal" string, one without any damping or motion due to stiffness. A practical string equation will add at least one damping term, and possibly two (equation 20.3). These work just as the terms in the ideal string equation, and can even be influenced externally due to force interaction (for example, fingers plucking a string both add force and damp the string as they move against it).

$$y_{tt}(x,t) = \frac{T}{\mu} y_{xx}(x,t) - b_1 y_t(x,t) + f(x,t). \quad (20.3)$$

The new term in equation 20.3, with coefficient b_1, describes a simple damping based on the velocity of the string. It causes all frequencies present in the string's vibration to fall off at the same rate. Additional terms can account for things like frequency-dependent damping and stiffness. Terms like these can be used to change the apparent sonic character of a model. For example, with FDTD plate models (which we won't discuss here in detail), changing the damping terms can make the same plate sound like metal, plastic, or wood.

A key strength of an FDTD model is its rich and varied response to interaction. Like a physical string (but unlike a sampled one), a model like the one above produces different sound depending on where and when force is applied. Furthermore, a synthesis program using a model like this can apply force anywhere along the string, in any imaginable way, even from multiple points at once. Since the model is virtual, this can be combined with novel kinds of input and all the flexibility of modern sensing techniques.

Plucking a real guitar string at different points lends a different tone to the sound produced by the string. Plucking it nearer to the middle will cause the sound to be clearer and more bell-like. Plucking it closer to the bridge will produce a more twangy sound. This effect can be difficult to reproduce in a synthetic model, but is reproduced naturally by an FDTD string: it follows from the way the string vibrates after being stretched to a particular shape.

Similarly, output can be taken from any point on the string or integrated over the entire string at once. Just as different placements of electric guitar pickups affect the guitar's sound, taking output from different places on an FDTD string model will emphasize different frequencies in the result. This follows naturally (in both cases) from the way that different points on the string vibrate: any given point on the string will move more for certain frequencies than for others. An FDTD string model also responds naturally to changes that happen when it is already sounding, such as transitions between notes, since these are just new forces added to a continuously sounding model. This doesn't mean that such transitions are effortless, but it does mean that an instrument designer or a performer can think about transitions in familiar, physically rooted terms, letting the model respond appropriately. These properties are invaluable to making an instrument that responds well to gesture. In particular, an instrument based on physical models is likely to respond well to novel gestural input. The model simply takes in the

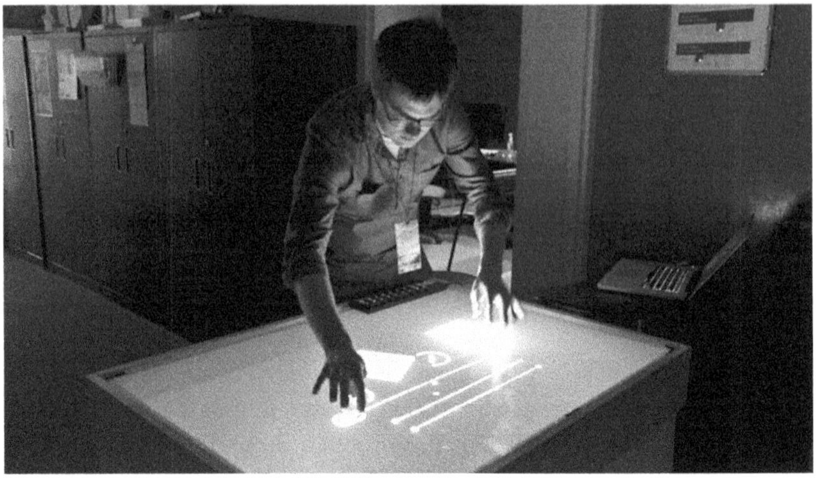

FIGURE 20.2 An interactive multitouch audio system. Courtesy of Jane Drozd.

new force, combines it with whatever else is already occurring, and continues to sound. The designer works to map gesture to the parameters of the model rather than directly to sounds, giving up some control but gaining a world of serendipity and fluid response.

As a nice corollary of the way the terms of the model are designed—in terms of forces applied at different locations—input doesn't need to be limited to a single point. At the same time that some input is adding energy to the string at one point, another input can be damping it somewhere else, or even stopping the string entirely to change its pitch. Alternatively, two or more performers can both play on the same string, letting their actions interact with one another. These capabilities act as a foundation for designing systems with interesting input and response. For instance, we could create a mapping between real-world positions and movement and virtual model positions and movement. A physically based model such as the FDTD string we have been discussing is an ideal candidate for direct representation on a multi-touch table. The spatial nature of the FDTD string design and response means that it is easy to draw an animated string and to situate it among other graphical objects. Figure 20.2 shows an interactive multitouch table audio system designed by the authors.

Performers can then either use their fingers to play the string directly or use active objects in the environment to do things like bounce balls off of the string. They can change the string's basic parameters while it is being played—perhaps one performer might adapt the environment while the other is playing. An extension of this idea is to use autonomous software agents as well as human performers. For example, agents might change the lengths (and thus the pitches) of a set of strings as a human plays them, or the agents might damp certain strings, encouraging the human performers to play elsewhere.

During all of this performance, the string and other objects in the environment can respond with synchronized sound and visuals. Because of the nature of the

FDTD simulation, the same data that drives sonic output can be used to draw the shape of the string as it moves. These things combined give performers a rich multimodal experience when working with the system, enabling them to create a multimodal performance, incorporating interrelated physical movement, visuals, and sound. This same idea about mapping real-world positions can be lifted off the touch table and extended to three-dimensional space through the use of a camera system. Dancers' movements and gestures can be mapped smoothly to forces applied to virtual models—either concretely, as with the touch table, or more abstractly. Compared with a touch table, an interface based on camera sensing is in a way less direct. On the other hand it simply offers a different kind of relationship between performer and instrument: one based more on whole-body movement than on virtual object manipulation. It is also of course possible to incorporate visuals for another layer of multimodal feedback.

Other input sensors such as buttons or levers, accelerometers and gyroscopes, and microphones may also be used to drive a physical model, given appropriate mappings. This last idea—using sound from another instrument—is especially interesting. The idea of filtering and transforming a sound with the input from a second audio source has a long history in electronic music, with the use of the vocoder and various convolution techniques.

Because physical models are driven by arbitrary force, at audio rates, they are well suited to respond to input created from audio input. A basic implementation of this is to use a sort of inverse pickup: an element that transforms sound directly into force applied at some position on a string. A performer can then use a microphone to interact with the string. This enables the player drive a realistic string in ways that would be difficult in the physical world, producing novel sounds. For example, the performer might make a sort of "blown string" instrument by blowing across the microphone, or speak into a bank of strings to produce a kind of vocoder effect.

We have used simple FDTD strings throughout our examples here. The literature also includes well-known models for objects such as plates, membranes, and tubes, and these may be connected with one another to form more complex instruments; all of them work on similar principles. It is presently possible to simulate many FDTD models, such as those discussed in this chapter, at interactive rates on commonly used computing hardware. At the time of this writing, it is not difficult, for example, to use a high-end laptop to simulate several strings and a plate at the same time. Even more complex objects, with arbitrary 3D shapes, require more advanced simulation techniques. Simulation based on finite elements (O'Brien, Cook, and Essl 2001) leads to promising results, but is computationally complex. Both finite-element models and large FDTD networks are presently beyond our ability to simulate in real time (at audio rates). Efficient methods for interactive simulation of such objects are a topic for future research.

We have not discussed the haptic aspect of FDTD sound models to any great extent, as this is a complicating factor that needs to be explored in greater depth at another time. It seems intuitive, though, that physical models could provide a good foundation for the modeling and mapping of haptic feedback. Physical models run continuously, responding to changes in their environment, enabling these models to provide feedback that varies with the input and with any subsequent changes anywhere in the system. Haptic feedback could be mapped in this same continuously and physically rooted manner. We will stress that we are speaking hypothetically here, though, and we do not have the same degree of practical experience with the haptic aspect of these models as we do with other aspects of these models.

Physical models provide a good foundation for musical objects that respond in realistic and flexible ways, combining the best of the physical and the virtual. Regarding the three themes that we stated at the beginning of the chapter, we can state the following:

1. Physically based models work naturally in multimodal environments, and are therefore well suited for these environments.
2. Physically based models provide performers with instruments that are physically rooted and, thus, interactive in intuitive ways. This intuitive interaction, in addition to the multimodal qualities of the models, provides the listener or audience with a familiar environment for "decoding" the performer's intentions.
3. While physical models allow us to describe the physical world in more or less precise ways, they also provide us with conceptually direct ways of transforming the physical world.

This last point is the greatest strength of physically based models. Physically based models provide us with a method for developing insights into the nature of reality. If we can understand the natural world, we can begin to develop interesting and meaningful ways to transform reality. The better we understand reality, the more we are able to create the "plausible impossible."

20.7 Conclusions

Digital sound technologies exist in a world of physical laws, yet they can seemingly extend and transform our conception of the physical world in ways that were not possible before their introduction. How, then, do we reconcile the physical world of sound and the many possible virtual worlds of sound? In order to begin to answer this question, we have discussed the essential role of gesture in music and in music making. We have observed how traditional mechanical instruments such as the piano and harpsichord—instruments that evolved over many hundreds of years—are able to transform gestures into sound in spectacularly rich and subtle ways. We have remarked that the introduction of digital technologies in the late twentieth century (and before digital technologies, the introduction of electronic technologies) has created entirely new possibilities for the design of musical instruments, or, more generally now, interactive sound models. The introduction of these new possibilities has brought with it new problems in design, since electronic and digital technologies behave according to a very different set of principles than mechanical technologies.

We have observed that gesture is a multimodal phenomenon. One could observe, for instance, that we do not use only our hands (or lungs, as another instance) to create sound. We also use our ears, and our ears depend upon our brain. We also use our eyes. After a while, we begin to understand that our entire being is involved in the making of music and in listening to music. We understand that gesture is not only a physical act, but also a cognitive and conceptual processing of a physical act. Given this, it is important to understand interactive sound models as integrated systems. Too often, we concentrate only on the gesture transformation interface. When we think of the piano, we think of the keys, maybe the hammers. We do not usually think of all of the many parts that work together in complex ways to finally create what we think of as "piano." This is also true of the Theremin. We think of the unique antenna interface, but we seldom think of all of the circuitry that works together to create the entity that we identify as "Theremin." It is this entire system that works together to create a multimodal system that gives us the sound, the feel, the look, the performance gestures, in short, the *gestalt* that we refer to as "instrument."

We also observed that the gestalt nature of what we call a "performance" is one that has many participants, including the music creator, the music performer, and the music audience. Each of these participants (whether they are embodied in one person, or in many) has their own set of expectations, all of which are rooted in the physical world. We noted that digital technologies allow us to design interactive sound models that exhibit varying degrees of fidelity (or nonfidelty) to the physical world, and we observed that it is desirable to be faithful to the physical world to at least some extent. We are searching for a balance between fidelity to reality vs. flexibility in the design of reality—we are searching for the "plausible impossible."

Speaking to the idea of the "plausible impossible," it is interesting to note that traditional mechanical instruments such as the piano already expand our concepts of physical reality. If you have ever heard or seen a very small person who knows how to play piano well, you may be surprised to note the "big sound" that they are able to produce. This is because the piano is well designed and the pianist knows how to maximize the efficiency of the piano design. We can still be surprised by this, but—perhaps because the piano has been around so long—we accept these transformations as logical.

Finally, in this chapter we proposed the use of physical modeling techniques as a logical candidate for use in the creation of interactive sound models that can be both rich and intuitive. Physical models are logical candidates because they work naturally in multimodal environments, because they are physically rooted, and because they provide conceptually direct methods for both describing reality and extending it. If we can understand the physical world, and if we can understand the way we as humans "process" the physical world, we can begin to develop interesting and meaningful ways to transform reality. The better we understand reality, the more we are able to create the "plausible impossible."

Further reading

We encourage the interested reader to consult Leman (2008) and Bilbao (2009) for in-depth discussions of many of the topics covered in this chapter. Cook (2002) also provides an accessible yet thorough introduction to working with physical models, including a discussion of different kinds of controllers. Smith (2012) discusses physical modeling from a signal-processing perspective. Fletcher and Rossing (1998) is a useful introduction and reference to the physics of many different kinds of musical instruments (though it does not address computational methods).

The annual NIME (New Interfaces for Musical Expression, www.nime.org) and DAFx (Digital Audio Effects, www.dafx.de) conferences discuss many topics related to the material presented in this chapter. We recommend their past proceedings to readers interested in learning more about interactive audio: synthesis, sensing, and the integration of the two.

Note

1. This phrase comes from a 1956 television special hosted by Walt Disney (*Disneyland*, 1956). He used the phrase in the context of explaining how animation creates worlds that seem real yet could not actually exist. It is from this idea that we take our inspiration in this chapter.

References

Bilbao, Stefan. 2009. *Numerical Sound Synthesis*. Chichester: John Wiley.

Chaigne, Antoine, and Anders Askenfel. 1994. Numerical Simulations of Piano Strings. *Journal of the Acoustical Society of America* 95 (2): 1112–18.

Cook, Perry R. 2002. *Real Sound Synthesis for Interactive Applications*. Natick, MA: A. K. Peters.

Cuzzucoli, Giuseppe, and Vincenzo Lombardo. 1999. A Physical Model of the Classical Guitar, Including the Player's Touch. *Computer Music Journal* 23 (2): 52–69.

Fletcher, N. H. and T. D. Rossing, 1998. *The Physics of Musical Instruments*, Second Edition. New York, NY, USA: Springer.

Godøy, R. I., E. Haga and A. R. Jensenius. 2006. Playing 'Air Instruments': Mimicry of Sound-Producing Gestures by Novices and Experts, in *Gesture in Human-Computer Interaction and Simulation*, Lecture Notes in Computer Science, vol. 3881, pp. 256-267.

Gritten, A. and E. King, 2006. *Music and Gesture*. Aldershot: Ashgate.

Leman, Marc. 2008. *Embodied Music Cognition and Mediation Technology*. Cambridge, MA: MIT Press.

Moog, R. 1996. Build the EM Theremin, *Electronic Musician*, 12/ 2, 86-99.

O'Brien, James F., Perry R. Cook, and Georg Essl. 2001.Synthesizing Sounds from Physically Based Motion. *Computer Graphics Proceedings, Annual Conference Series*, 529–36. http://graphics.berkeley.edu/papers/Obrien-SSF-2001-08/Obrien-SSF-2001-08.pdf.

"*The Plausible Impossible.*" *Disneyland*. ABC. October 31, 1956 [television program].

Schrader, Barry. 1982. *Introduction to Electro-acoustic Music*. Englewood Cliffs, NJ: Prentice-Hall.

Smith, Julius O. 2012. *Physical Audio Signal Processing*. W3K Publishing. http://ccrma.stanford.edu/~jos/pasp.

Sosnick, Marc, and William Hsu. 2010. Efficient Finite Difference-based Sound Synthesis Using GPUs. *Proceedings of the 7th Sound and Music Computing Conference (SMC 2010)*. http://smcnetwork.org/files/proceedings/2010/71.pdf.

CHAPTER 21

VIRTUAL MUSICIANS AND MACHINE LEARNING

NICK COLLINS

In an age of robotics and artificial intelligence, the music stars of tomorrow may not be human. We already see precedents for this in anime virtual pop stars from Japan like the Vocaloid icon Hatsune Miku, or cartoon bands from Alvin and the Chipmunks to the Gorillaz. These are all audiovisual fronts for human musicians, however, and a deeper involvement of artificial musical intelligence in such projects is projected. Could our concert halls, clubs, bars, and homes all play host to virtual musicians, working touring circuits independent of any human manager? The applications of such radical music technology extend from new art music concert works to mass music entertainment in games and education.

There is already a long and fascinating history to machine interaction in concert performance, from such 1960s and 1970s precedents as the analog machine listening pieces of Sonic Arts Union composers Gordon Mumma and David Behrman (Chadabe 1997) to the computerized online structure formation of *OMax* (Assayag et al. 2006), from George Lewis' many decades development of the computer improvisational system *Voyager* (Lewis 1999) to advances in musical robotics (Kapur 2005). Lessons from the creation of virtual musicians have an essential role to play in our understanding of interactive music settings in general, for such systems test the limits of engineering research and compositional ingenuity.

In order to work within human musical society, the machines need to be wise to human musical preferences, from the latest musical stylistic twists across human cultures, to more deep-rooted attributes of human auditory physiology. Creating truly adaptable virtual musicians is a grand challenge, essentially equivalent to the full artificial-intelligence problem, requiring enhanced modeling of social interaction and other worldly knowledge as much as specific musical learning (we will not attempt all of that in this chapter!). The payoff may be the creation of new generations of musically competent machines, equal participants in human musical discourse, wonderful partners in music making, and of redoubtable impact on music education and mass enjoyment. One vision of the future of musical interaction may be that of a "musical familiar" that adapts with a musician from childhood lessons into adult performance, developing as they grow.

Although such portrayals can be a great motivator of the overall research, we can also drift into more unrealistic dreams; the projects of virtual musicianship are bound up inextricably with the future of artificial intelligence (AI) research. Previously (Collins 2011a), I let speculation go unhindered. Herein, I shall keep things more tightly connected to the current state of the art and outline the challenges to come from technical and musical perspectives.

Key to the creation of enhanced autonomy in musical intelligences for live music is the incorporation of facilities of learning. We know that expert humans musicians go through many years of intensive training (ten years or 10,000 hours is one estimate of the time commitments already made in their lives by expert conservatoire students, see Ericsson and Lehmann 1996; Deliège and Sloboda 1996). A similar commitment to longer-term development can underwrite powerful new interactive systems. To go beyond over-fitting a single concert, and to move toward a longer lifetime for musical AIs, rests in practice upon incorporating machine-learning techniques as a matter of course for such systems.

There is an interesting parallel with tendencies in gaming toward larger game worlds, enhanced game character AI, and the necessity of being able to save and load state between gaming sessions. Interactive music systems need larger stylistic bases, enhanced AI, and longer-term existence. Where the current generation of musical rhythm games centerground motor skills over expressive creation, more flexible interaction systems may provide a future crossover of academic computer music to mass consumption.

We shall proceed by reviewing the various ways in which machine learning has been introduced in computer music, and especially to the situation of virtual musicians for live performance. We treat machine learning here above parallel engineering challenges in machine listening (the hearing and music-discerning capabilities of machines). For reviews of machine listening, the reader is pointed to Rowe (2001) and Collins (2007, 2011b).

21.1 Machine Learning and Music

The application of any machine-learning algorithm requires modeling assumptions to be made; music must be represented in a form amenable to computer calculation. In order to get to a form where standard machine-learning algorithms can be applied, the input musical data is preprocessed in various ways. Machine listening is the typical front end for a concert system, moving from a pure audio input to derived features of musical import, or packaging up sensor and controller data. The data points at a given moment in time themselves may be of one or more dimensions, taking on continuous or discrete values.

The treatment of time is *the* critical aspect of machine-learning applications for music. Whether denoted as time-series analysis (in the mold of statistics) or signal processing (in engineering), musical data forms streams of time-varying data. With respect to the time base, we tend to see a progression in preprocessing from evenly sampled signals to discretized events; AI's signal-to-symbol problem (Matarić 2007, 73) recognizes the difficulty of moving in perception from more continuous flows of input to detected events. Though signals and sequences may be clocked at an even rate, events occur non-isochronously in general. Where the timing is implicit in the signal case, events may be tagged with specific time stamps. In both situations, a window of the last *N* events can be examined to go beyond the immediate present, acknowledging the wider size of the perceptual present and the role of different memory mechanisms. For evenly sampled signals, the window size in time is a simple function of the number of past samples to involve; for discrete events, the number of events taken may be a function of the window size's duration (based on what fits in) or the window size in time may be a function of the number of events examined (in the latter case there would typically be a guarantee on the average number of events sampled per second, to avoid creating nonsensically massive windows, or checks in the code to avoid any aberrant scenario). Having gathered a window of data, in some applications the exact time ordering is then dropped (the "bag of features" approach, where the order of things in the bag is jumbled; see Casey et al. 2008) and in others it remains a critical consideration of the algorithm; some procedures may also concern themselves only with further derived properties of a window of data, such as statistical features across all the events.

Achieving some sort of representation which is musically relevant and yet compatible with an on-the-shelf machine-learning algorithm, a process of learning can take place over multiple examples of data following that representation. We should distinguish two sorts of learning tasks here. In supervised learning, the inputs always have directly associated outputs, and the mapping that is learnt must respect this function space, while generalizing to cope robustly with new input situations unseen in training. In unsupervised learning, the learning algorithm attempts to impose some order on the data, finding structure for itself from what was otherwise previously implicit. Learning algorithms can require a large amount of example data to train, and musical situations can sometimes not supply many examples on a given day. It will not always be practical to train on-the-fly in the moment of performance, instead it may require preparation steps; many machine-learning algorithms deployed in concert are not conducting the learning stage itself live, but were trained beforehand, and are now just being deployed. This mirrors the way human beings develop over a long haul of practice, rather than always being blank slates in the moment of need.

We cannot review all machine-learning theory and algorithms in this chapter. Good general reviews of machine learning as a discipline include textbooks by Mitchell (1997) and Alpaydin (2010), and the data mining book accompanying the open-source Weka software, by Witten and Frank (2005). Stanford professor Andrew Ng has also created an open machine-learning course available online, including video lectures and exercises (http://www.ml-class.org/course/video/preview_list). We will mention many kinds of machine-learning algorithm in the following sections without the space to treat formally their properties.

We also won't be able to review every musical application of every type of machine-learning algorithm herein, but will hopefully inspire the reader to pursue further examples through the references and further searches. As a rule of thumb, if an interesting learning technique arises, someone will attempt to apply it in computer music. Applications often follow trends in general engineering and computer science, for example, the boom in connectionist methods like neural nets in the 1990s, genetic algorithms over the same period, or the growth of data mining and Bayesian statistical approaches in to the 2000s.

21.2 Musical-learning Examples

Three examples of the sorts of musical task enabled by machine learning are:

- Learning from a corpus of musical examples, to train a composing mechanism for the generation of new musical materials.
- Learning from examples of musical pieces across a set of particular genres, to classify new examples within those genres.
- Creating a mapping from high-dimensional input sensor data to a few musical control parameters or states, allowing an engaging control space for a new digital musical instrument.

Although only the last is explicitly cast as for live music, all three could be applicable in a concert context; stylistically appropriate generative mechanisms are an essential part of a live musician's toolbox, and a live system might need to recognize the stylistic basis of the music being played before it dares to jump in to contribute! We review some associated projects around these three themes, knowing that the survey cannot be exhaustive.

Machine learning is intimately coupled to modeling of musical data, and many predictive and generative models of music that rest on initialization over a corpus of data have appeared in decades of research on algorithmic composition and computational musicology. The venerable Markov model, first posited by John Pierce in 1950 as applicable in music (Pierce 1968), is the premier example. Markov systems model the current state as dependent on previous states, with an "order" of the number of previous states taken into consideration (Ames 1989). To create the model, data sequences are analyzed for their transitions, and probability distributions created from counts of the transitions observed; the model is then usable for generation of novel material (new sequences) in keeping with those distributions. The popularity of Markov models and that of information theoretic variants has continued in literature on symbolic music modeling and pattern analysis in music (Conklin and Witten 1995; Wiggins, Pearce, and Müllensiefen 2009; Thornton 2011), as well as underlying the well-known (if not always clearly defined) work in automated composition of David Cope (2001). One famous interactive music system, the Continuator of François Pachet (Pachet 2003) is based on a variable-order Markov model. In its typical call-and-response mode of performance, the Continuator can build up its model on-the-fly, using human inputs to derive an internal tree of musical structure in what Pachet calls "reflexive" music making, because it borrows so closely from the human interlocutant. Begleiter, El-Yaniv, and Yona (2004) compare various variable-order Markov models, assessing them on text, music MIDI files, and bioinformatic data. Prediction by partial match is one such algorithm that has proved successful (the second best after the rather more difficult to implement context tree weighting in Begleiter, El-Yaniv, and Yona's study), and it has been extended to musical settings (Pearce and Wiggins 2004; see also Foster, Klapuri, and Plumbley 2011 for an application to audio feature vector prediction comparing various algorithms) (see also Chapters 22 and 25 in this volume).

The Begleiter, El-Yaniv, and Yona (2003) paper notes that any of the predictive algorithms from the literature on data compression can be adapted to sequence prediction. Further, any algorithms developed for analysis of strings in computer science can be readily applied to musical strings (whether of notes or of feature values). The Factor Oracle is one such mechanism, an automaton for finding common substring paths through an input string, as applied in the OMax interactive music system at IRCAM (Assayag et al. 2006). OMax can collect data live, forming a forwards and backwards set of paths through the data as it identifies recurrent substrings and, like a Markov model, it is able to use this graph representation for generating new strings "in the style of" the source. One drawback of this application of a string-matching algorithm is that its approach to pattern discovery is not necessarily very musically motivated; the space of all possible substrings is not the space of all musically useful ideas! As Schankler and colleagues (2011) note, the Factor Oracle tends to promote musical forms based on recurring musical cells, particularly favoring material presented to it earliest on in training (rondo-like forms); a human participant can cover up for some of the algorithm's deficiencies.

With the rise of data-mining approaches, an excellent example where the mass use of machine-learning algorithms occurs in computer music is the developing field of music information retrieval (MIR) (Downie 2003; Casey et al. 2008). Most of these algorithms operate offline, though there are circumstances, for example live radio broadcast, where classifications have to take place on-the-fly. There are certainly situations, for instance, the audio fingerprinting of the *Shazam* mobile service used to identify music in the wild, where as-fast-as-possible calculation is preferable. As for many interactive systems, MIR systems may have their most intensive model parameter construction precalculated in intensive computation, and they can then deploy the model on novel data much more easily. Nonetheless, newly arriving data may need to be incorporated into a revised model, leading to intensive parameter and structure revision cycles (e.g., as occurs if rebuilding a kD-tree). The gathering volume of MIR work is a beneficial resource of ideas to adapt for live performance systems.

Machine learning has also found its way into new musical controllers, particularly to create effective mappings between sense inputs and the sound engine (Hunt and Wanderley 2002). Applications may involve changes in the dimensionality of data, as in many-to-one or one-to-many mappings. For example, Chris Kiefer uses echo state networks (a form of connectionist learning algorithm) to manage the data from EchoFoam, a squeezable interface built from conductive foam, reducing from multiple sensors embedded in a 3D object to a lower number of synthesis parameters (Kiefer 2010). The MnM library for the graphical audio programming environment Max/MSP provides a range of statistical mapping techniques to support mapping work (Bevilacqua, Müller, and Schnell 2005); Rebecca Fiebrink has released the Wekinator software which packages the Weka machine-learning library into a system usable for real-time training and deployment (Fiebrink 2011). With increasingly complicated instruments, machine learning can help from calibration and fine tuning of the control mechanism, to making the sheer volume of data tractable for human use.

In Robert Rowe's taxonomy, there is a dimension on which interactive music systems move from more purely reactive instruments to independent agents (Rowe 1993). The production of increasingly autonomous interactive agents to operate in concert music conditions has increasingly drawn on machine-learning techniques. Examples range from the use of biological models such as genetic algorithms (Miranda and Biles 2007), through neural networks (Young 2008), to unsupervised clustering of antecedent and consequent phrases in the work of Belinda Thom (2003). Some of the most sophisticated work to date was carried out by Hamanaka and collaborators (2003), who modeled the interactions of a trio of guitarists (they applied such techniques as radial basis network mapping, Voronoi segmentation, and Hidden Markov Models). The litany of machine-learning techniques continues, though our survey must admit space limits; we might mention reinforcement learning (Le Groux and Verschure 2010;), case-based reasoning (Mantaras and Arcos 2002), or Bayesian modeling frameworks (Temperley 2007) as areas of interest for investigation.

21.3 Machine-learning Challenges

Whatever the machine-learning algorithm, there are issues in musical application that have repeatedly arisen in the literature. The problem of sparse data in any individual musical interaction was identified by Thom (2003) in her work on clustering. Although a common complaint of the contemporary composer is the lack of rehearsal time given by ensembles to their particular works, professional musicians have a lifetime of general practice to draw on and obtaining sufficient data to match this is a challenge. Methods equipped to work over large corpuses of musical data, whether audio files or symbolic data like MIDI files, can provide the extensive bootstrapping a given model may require. Rehearsal recordings can be taken, and passed over by a learning algorithm in multiple training runs (for example, as required in some reinforcement-learning approaches) or applied selectively in training an onset detector that has more negative examples than positive to learn from. Alternatively, algorithms may be preferred that need less data, or simply less data used in training at the cost of reduced performance effectiveness, as with some demonstrations of the Wekinator (Fiebrink 2011); the added noise of such a system can (charitably) be musically productive, as for example in the inaccuracies of a pitch tracker leading to a more unpredictable (and thus stimulating) response (Lewis 1999).

From a musician's perspective, minimal intervention in the training of a machine musician is preferable; humans are not renowned for patience with algorithms, and they certainly find it uncomfortable to play with others of a divergent standard. Even if the algorithm cannot turn up ready to play, unsupervised training in rehearsal or even during performance is beneficial. Smith and Garnett (2011) describe a "self-supervising machine" that provides an unsupervised guide process (based on adaptive resonance theory) above a supervised neural network; they claim benefits in avoiding costly pre-session training time as well as reduced cognitive load and increased flexibility. Few other projects have attempted the easy application of machine learning for musicians embodied by the *Wekinator*, though Martin, Jin, and Bown (2011) discuss one project to give live agents control of musical parameters, within an interactive machine-learning paradigm, where association rule learning is used to discover dependencies.

Machine learning in real applications forces various pragmatic decisions to be made. Musical parameter spaces show combinatorial explosions (for example, in considering increasingly long subsegments of melodies as the units of learning); keeping the dimension of the state space low requires compromises on the accuracy of the approximating representation. Without some simplification, the learning process may not be tractable at all, or may require too much training data to be practicable! A regression problem with continuous valued data may be reduced to discrete data by a preprocessing clustering or vector quantization step, at the cost of losing fine detail and imposing boundaries (this tension between continuous and discrete is familiar whenever we use convenient categories in natural language, which can distort the true distribution). Even when a musical representation is eminently sensible, the machine-learning algorithms themselves have differing inductive biases, with different performances in generalizing to unseen cases. It may be useful to train multiple models in parallel and select the best performing (there are technicalities here in holding back certain test data to measure this). Yet what works well as a small-scale solution to a particular concert task may prove less equipped to the vagaries of a whole tour!

A further issue for those researching the incorporation of learning agents in live music is evaluation of the effectiveness of these agents as musical participants, especially where we consider the longer-term existence of these systems. Even after building such systems, evaluating them through longitudinal studies is not easy. The attribution problem in machine learning notes the difficulty of assigning credit to guide the learning of a complex system, particularly when praise or negative feedback is itself scarce (Collins 2007). As well as confounding the application of algorithms such as reinforcement learning and the fitness functions of genetic algorithms, the lack of quality feedback undermines evaluation of system effectiveness. Human–computer interaction (HCI) methodologies for feedback from rehearsal or concerts are currently based around more qualitative methods of review such as postperformance interviews (Hsu and Sosnick 2009). In-the-moment quantitative evaluation methods (such as physiological measures from galvanic skin response or EEG) in HCI are at only a tentative stage (Kiefer, Collins, and Fitzpatrick 2008).

21.4 A Listening and Learning System

In order to illustrate a learning musical agent in more detail, we examine here the LL system, which was originally premiered in the summer of 2009 for a duet with the free-improvisation percussionist Eddie Prévost. The core unsupervised learning components of the system have subsequently been built into a freely available Max/MSP external object ll~ as a result of an AHRC-funded project by composer Sam Hayden and violinist Mieko Kanno on "Live Performance, the Interactive Computer and the Violectra." Sam's revised *schismatics II* (2010) and his newer *Adaptations* (2011) make use of the technology in works for laptop and electric violin (Hayden and Kanno 2011).

Figure 21.1 gives an overview of the whole system of the original LL. Ten parallel agents are associated with ten different musical states; the switching of state, and thus which agent is active, depends on machine learning from the human musician's inputs to the system. We avoid too much discussion of the machine-listening components, and the output synthesis models herein, instead concentrating on the learning aspects. The primary sites of machine learning in the system are:

- Feature adaptation (histogram equalization) to maximize feature dynamic range;
- Clustering of half second timbral feature windows;
- Continual collection of rhythmic data from the human performer for reuse by the machine, via a Markov model;
- Classification differentiating "free time" and more highly beat-based rhythmic playing.

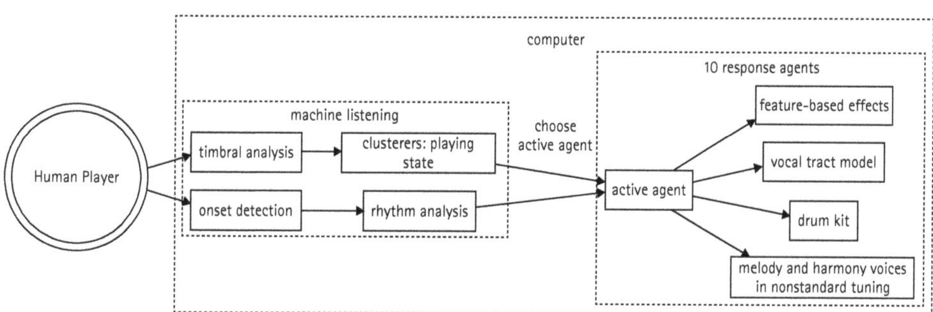

FIGURE 21.1 An overview of the whole LL.system.

The first three processes are unsupervised and automatic; the last involves training data collected in rehearsal.

The first process is a special low-level normalization step. Features provided in machine listening may have different parameter ranges, and some sort of max–min or statistical (mean and standard deviation) normalization is required for their ranges to be comparable. Histogram equalization is a further technique, lifted from computer vision (Bradski and Kaehler 2008, 188), where the area assigned between 0 and 1 in the normalized feature output is proportional to the actual feature distribution observed in the training data, through a histogramming estimation and linear segment model. This step then tries to make the different features maximally comparable in combined feature vectors of normalized values. The histogram equalization can be learned online (as values arrive), which can be especially useful where the distribution of data is not well known in advance (and may be an attribute of many musical situations, for example, microphones in unfamiliar acoustic environments or a system working across many types of musical input encountering bagpipes for the first time!).

In the second learning process, clustering operates on aggregated timbre points, constructed from an average of timbral feature vectors over a window of around 600 ms. In actual fact, the clustering is achieved by running multiple randomly initialized k-means (where k = 10 in LL) clustering algorithms, and taking the "best" with respect to an error condition (least total distance of training data to the cluster centers). Postprocessing is used on the clusterer output for stability; the majority state over the last ten checks (where checks occur around ten times per second as feature data updates) is taken as the output state. The best matching cluster is thus a result of feature data collected in the last 1.5 seconds, a reasonable figure for working memory and a good turnaround time for reaction to a shift in musical behavior from the musician being tracked. In application, multiple clustering units can be used, based on different combinations of source features as the data source; this keeps the dimensionality of input lower for an individual clusterer than using all features at once, making machine learning more effective with a smaller amount of input data (recall the discussion of tradeoffs above).

The third and fourth processes depend on event-timing data lifted from the human musician through onset detection, and machine-listening processes to assess the current metrical structure of performance (beat tracking). The classifier was constructed by observation as much as a fully supervised algorithm; indeed, when collecting materials in rehearsal, Eddie Prévost, when asked to provide examples of his freest playing, tended to still mix in short flashes of more beat-based material; given a smaller amount of data, human inspection of this was the most pragmatic solution (a future study may collect much more data across multiple drummers and investigate machine classification more rigorously). The classifier differentiates performance in a loose, highly improvisatory mode, from more highly beat-driven material. The response model for the musical agents' own timing then follows a Markov model of observed event timings collected during free playing, or works with respect to beat-boundaries discovered in the metrical analysis. The Markov model was constantly active in collecting data, and could develop from rehearsal through into the actual concert.

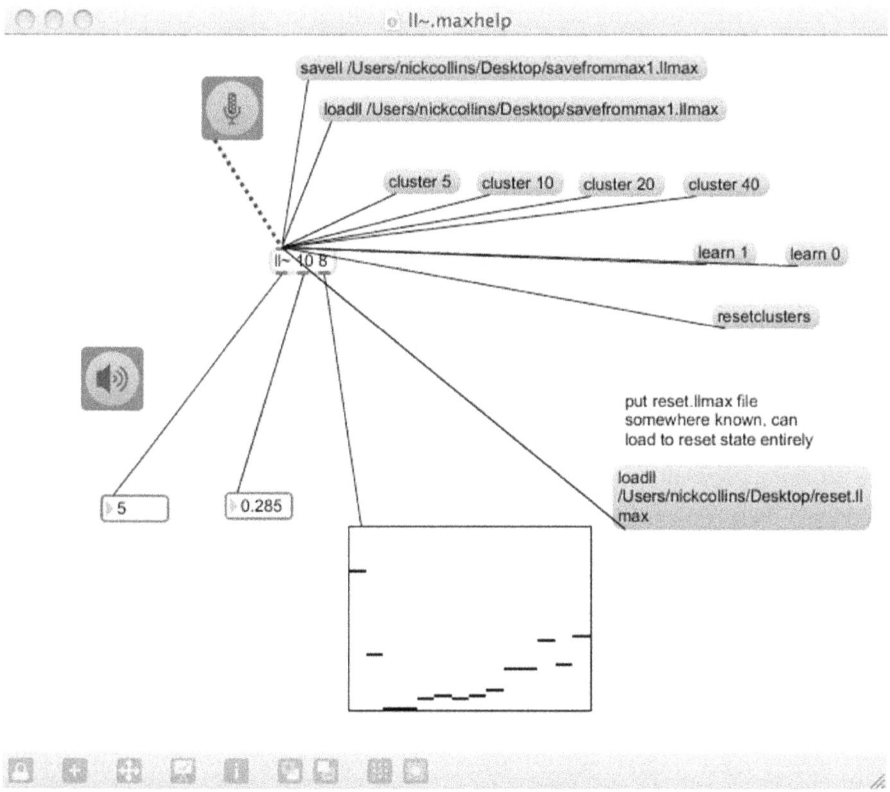

FIGURE 21.2 A screenshot of ll~ in action.

Shorn of the rhythmic analysis parts of LL, processes 1 and 2 were packaged into the more reusable ll~ external for Max/MSP. Figure 21.2 shows a screenshot of ll~ in action, illustrating feature collection and the classification of timbral states. The external's three outputs are the clusterer's current observed cluster number (the number of states, the "k" in k-means, can be chosen as an input argument), a measure of how full the memory is with collected feature data, and a list of histogram equalized and normalized feature values (which can be useful in further feature-adaptive sound synthesis and processing).

In practice, not all learning has to be online, adapting during a concert. For clustering, although online learning algorithms (such as agglomerative clustering) were implemented the most pragmatic technique was to run the k-means clustering at a predetermined (after enough data is collected) or user-selected point (this is the control mode of the ll~ external). This avoids transitory behavior of the clusterer particularly in the early stages of receiving data. While data collection is continuous, ll~ is quite flexible to being trained at chosen points, and particular clustering solutions can be frozen by saving (and later recalling) files. In practice, even if a system is learning from session to session, the hard work of reflection and learning may take place between rather than during sessions. The final system that performed with

Eddie Prévost in the evening concert had been trained in rehearsal sessions earlier in the day and the day before. The time constraints on available rehearsal had led me to train baseline systems on drum samples and human beat-boxing; we experimented in performing with systems trained on alternative musical input than Eddie's drum kit, perhaps justifiable as an attempt to give the system a divergent personality, and while some in-concert adaptation took place in the rhythmic domain, the feature adaptation and clusterers were fixed in advance, as well as the actual classification measure for free time versus highly beat-based.

Reaction to LL's premiere was positive from performer and audience, though in discussion after the event (a recording had been made), Eddie was more guarded in any praise. He was enthusiastic about the ideas of longer-term learning, though we both agreed that this system did not yet instantiate those dreams. Sam Hayden also kindly sent feedback on his own investigations of the ll~ object, noting: "I've been experimenting with using pre-trained ll~ objects and mapping the output values onto fx synthesis parameters then feeding the resultant audio back into the ll~ objects. Though the ll~ system is working as it should the musical results seem a little unpredictable... Perhaps the mappings are too arbitrary and the overall system too chaotic. I suppose the issue is of perception: as a listener I think you can hear that the system has some kind of autonomy. It is a question of how much you need to be able to follow what the system is doing for the musical interactions to be meaningful." In his *Adaptations*, Sam even feeds back the final output audio of the system, mixing into the input of the earliest ll~ object. Successive ll~ objects are introduced as the piece progresses over time, gradually increasing complexity; he writes "As a listener, you are aware of some kind of underlying controlling system, even if you're not quite sure what it's doing. It is this ambiguity that interest me."

These comments highlight the independent views of listener, critic, and composer, and a musician interacting with the system, and the need for further evaluation of such systems as new learning facilities are explored. The reader is invited to try the ll~ object, and consider the roles machine learning could play in their own work. Much remains to explore, as ever!

21.5 Virtual Musical Futures

Ultimately, artificial musical intelligence is a manifestation of the whole AI problem of interfacing machines to human society as full participants, and the learning capacity of human beings is of clear import here. Advances in the field of musical interaction employing machine learning can be of substantial potential impact to our understanding of human intelligence in general. This chapter has surveyed existing attempts to create flexible concert agents, the machine-learning technologies that may lead to future adaptive systems, and one modest attempt to work toward a longer-term learning agent for concerts.

Though our focus has been virtual musicians in concerts, developments in this technology interact with other media. Videogames include increasing amounts of AI, and where the 2000s craze for rhythm games has waned (perhaps as people have realized they are at heart rather linear piano-roll challenges, like specific musical technical exercises), future music games may embrace rather more open-ended worlds, where dynamic difficulty adjustment works over the lifetime of a player. Beyond touring AIs it is hard to resist the possibility of musical familiars, virtual-musician programs that act as lifelong musical companions, from tutors to partners in music making. Where fixed recording may falter after a busy twentieth century, the rise of gaming points to a return of adaptable music making for all.

Acknowledgments

With thanks to the editors, and Chris Thornton, for review feedback on the chapter, and Eddie Prévost and Sam Hayden for their highly musical input and careful reflection on the systems.

References

Alpaydin, Ethem. 2010. *Introduction to Machine Learning*. Cambridge, MA: MIT Press.

Ames, Charles. 1989. The Markov Process as a Compositional Model: A Survey and a Tutorial. *Leonardo* 22 (2): 175–187.

Assayag, Gérard, Georges Bloch, M. Marc Chemillier, Arshia Cont, and Shlomo Dubnov. 2006. OMax Brothers: A Dynamic Topology of Agents for Improvisation Learning. In *AMCMM '06: Proceedings of the 1st ACM Workshop on Audio and Music Computing Multimedia*. New York: ACM.

Begleiter, Ron, Ran El-Yaniv, and Golan Yona. 2004. On Prediction Using Variable Order Markov Models. *Journal of Artificial Intelligence Research* 22: 385–421.

Bevilacqua, Frédéric, Rémy Müller, and Norbert Schnell. 2005. MnM: A Max/MSP Mapping Toolbox. In *Proceedings of the International Conference on New Interfaces for Musical Expression (NIME05)*, Vancouver, BC.

Bradski, Gary, and Adrian Kaehler. 2008. *Learning OpenCV: Computer Vision with the OpenCV Library*. Sebastopol, CA: O'Reilly Media.

Casey, Michael A., Remco Veltkamp, Masataka Goto, Marc Leman, Christophe Rhodes, and Malcom Slaney. 2008. Content-based Music Information Retrieval: Current Directions and Future Challenges. *Proceedings of the IEEE* 96 (4): 668–696.

Chadabe, Joel. 1997. *Electric Sound: The Past and Promise of Electronic Music*. Englewood Cliffs, NJ: Prentice Hall.

Collins, Nick. 2007. Musical Robots and Listening Machines. In *The Cambridge Companion to Electronic Music*, ed. Nick Collins and Julio d'Escrivan, 171–184. Cambridge, UK: Cambridge University Press.

——. 2011a. Trading Faures: Virtual musicians and machine ethics. *Leonardo Music Journal* 21: 35–39.

——. 2011b. Machine Listening in SuperCollider. In *The SuperCollider Book*, ed. Scott Wilson, David Cottle, and Nick Collins, 439–460. Cambridge MA: MIT Press.

Conklin, Darrell, and Ian H. Witten. 1995. Multiple Viewpoint Systems for Music Prediction. *Journal of New Music Research* 24 (1): 51–73.

Cope, David, ed. 2001. *Virtual Music: Computer Synthesis of Musical Style*. Cambridge, MA: MIT Press.

Deliège, Irène, and John A. Sloboda, eds. 1996. *Musical Beginnings: Origins and Development of Musical Competence*. New York: Oxford University Press.

Downie, J. Stephen. 2003. Music Information Retrieval. *Annual Review of Information Science and Technology* 37: 295–340.

Ericsson, K. Anders, and A. C. Lehmann. 1996. Expert and Exceptional Performance: Evidence of Maximal Adaptation to Task. *Annual Review of Psychology* 47: 273–305.

Fiebrink, Rebecca. 2011. Real-time Human Interaction with Supervised Learning Algorithms for Music Composition and Performance. PhD diss., Princeton University. http://www.cs.princeton.edu/~fiebrink/Rebecca_Fiebrink/thesis.html.

Foster, Peter, Anssi Klapuri, and Mark. D. Plumbley. 2011. Causal Prediction of Continuous-valued Music Features. In the *Proceedings of the International Society of Music Information Retrieval Conference*, 501–506.

Hamanaka, Masatoshi, Masataka Goto, Hideki Asoh, and Nobuyuki Otsu. 2003. A Learning-based Jam Session System that Imitates a Player's Personality Model. *IJCAI: International Joint Conference on Artificial Intelligence*, 51–58.

Hayden, Sam, and Mieko Kanno. 2011. Towards Musical Interaction: Sam Hayden's *Schismatics* for E-violin and Computer. *Proceedings of the International Computer Music Conference*, 486–490.

Hsu, William, and Marc Sosnick. 2009. Evaluating Interactive Music Systems: An HCI Approach. In *Proceedings of the International Conference on New Interfaces for Musical Expression*, 25–28.

Hunt, Andy, and Marcelo M. Wanderley. 2002. Mapping Performer Parameters to Synthesis Engines. *Organised Sound* 7 (2): 97–108.

Kapur, Ajay. 2005. A History of Robotic Musical Instruments. In *Proceedings of the International Computer Music Conference*, 1–8.

Kiefer, Chris. 2010 A Malleable Interface for Sonic Exploration. In *Proceedings of the International Conference on New Interfaces for Musical Expression*, 291–296. Sydney, Australia. http://www.nime.org/proceedings/2010/nime2010_291.pdf.

Kiefer, Chris, Nick Collins, and Geraldine Fitzpatrick. 2008. HCI methodology for evaluating musical controllers: A case study. In *Proceedings of the International Conference on New Interfaces for Musical Expression*, 87–90. Genova, Italy. http://www.nime.org/proceedings/2008/nime2008_087.pdf.

Le Groux, Sylvain, and Paul F. M. J. Verschure. 2010. Towards Adaptive Music Generation by Reinforcement Learning of Musical Tension. *Proceedings of Sound and Music Computing*. http://smcnetwork.org/files/proceedings/2010/24.pdf.

Lewis, George E. 1999. Interacting with Latter-day Musical Automata. *Contemporary Music Review* 18 (3): 99–112.

Martin, Aengus, Craig T. Jin, and O. R. Bown. 2011. A Toolkit for Designing Interactive Musical Agents. *Proceedings of the 23rd Australian Computer-human Interaction Conference*, 194–197. New York: ACM.

Mantaras, Ramon Lopez de, and Josep Lluis Arcos. 2002. AI and Music: From Composition to Expressive Performance. *AI Magazine* 23 (3): 43–57.

Matarić, Maja J. 2007. *The Robotics Primer*. Cambridge, MA: MIT Press.

Miranda, Eduardo Reck, and John. A. Biles, eds. 2007. *Evolutionary Computer Music*. London: Springer-Verlag.

Mitchell, Tom. 1997. *Machine Learning*. Singapore: McGraw-Hill.

Pachet, François. 2003. The Continuator: Musical Interaction with Style. *Journal of New Music Research* 32 (3): 333–341.

Pearce, Marcus T., and Geraint A. Wiggins. 2004. Improved Methods for Statistical Modelling of Monophonic Music. *Journal of New Music Research* 33 (4): 367–385.

Pierce, John Robinson. 1968. *Science, Art, and Communication*. New York: Clarkson N. Potter.

Rowe, Robert. 1993. *Interactive Music Systems*. Cambridge, MA: MIT Press.

——. 2001. *Machine Musicianship*. Cambridge, MA: MIT Press.

Schankler, Isaac, Jordan B. L. Smith, Alexandre François, and Elaine Chew. 2011. Emergent Formal Structures of Factor Oracle-driven Musical Improvisations. In *Mathematics and Computation in Music*, ed. Carlos Agon, Moreno Andreatta, Gérard Assayag, Emmanuel Amiot, Jean Bresson, and John Mandereau, 241–254. Paris: IRCAM, CNRS, UPMC.

Smith, Benjamin D. and Guy E. Garnett. 2011. The Self-Supervising Machine. Proceedings of the International Conference on New Interfaces for Musical Expression, 30 May-1 June 2011, Oslo, Norway. http://www.nime2011.org/proceedings/papers/B21-Smith.pdf

Temperley, David. 2007. *Music and Probability*. Cambridge, MA: MIT Press.

Thom, Belinda. 2003. Interactive Improvisational Music Companionship: A User-modeling Approach. *User Modeling and User-adapted Interaction Journal* 13(1–2):133–177

Thornton, Chris. J. 2011. Generation of Folk Song Melodies Using Bayes Transforms. *Journal of New Music Research* 40 (4): 293–312

Wiggins, Geraint A., Marcus T. Pearce, and Daniel Müllensiefen. 2009. Computational Modeling of Music Cognition and Musical Creativity. In *The Oxford Handbook of Computer Music*, ed. Roger T. Dean, 383–420. New York: Oxford University Press.

Witten, Ian H., and Eibe Frank. 2005. *Data Mining: Practical Machine Learning Tools and Techniques*. San Francisco: Morgan Kaufmann.

Young, Michael. 2008. NN Music: Improvising with a "Living" Computer. In *Computer Music Modelling and Retrieval: Sense of Sounds*, ed. Richard Kronland-Martinet, Sølvi Ystad, Kristoffer Jensen, 337–350. Lecture Notes in Computer Science 4969. Berlin: Springer.

CHAPTER 22

MUSICAL BEHAVIOR AND AMERGENCE IN TECHNOETIC AND MEDIA ARTS

NORBERT HERBER

> We must arrange our music... so that people realise that they themselves are doing it, and not that something is being done to them. (John Cage in *Generation*. Cited in Ascott 2003, 123)

Music is made not only through a composer's particular arrangement of sounds, but by the listener's ability and willingness to appreciate these sounds and include them in that category. Cage's landmark "silent" piece, or as it is often called, *4′33″*, is the best example of this idea, in which the existence of a musical work is more dependent on the actions of the listener than those of the composer. Musicians working in ways complementary to Cage contrast the traditional compositional proposition, "I think this arrangement of sounds is interesting," with, "What would it sound like if...?" This difference in approach is the imperative of Experimental music and one of the foundational musical questions behind the thoughts in this chapter. Music that has an unknown outcome shares an ontological resonance with *technoetic* environments that possess similar uncertainties. Roy Ascott characterizes the *technoetic* as "a fusion of what we know and may yet discover about consciousness (*noetikos*) with what we can do and will eventually achieve with technology [*techne*]. It will make consciousness both the subject and object of art" (Ascott 2001).

This chapter takes the position that music can shape and transform consciousness; it can give rise to a new consciousness as it is experienced. As one transitions to an alternate or mixed reality using tools of mediation like the Internet, personal computers, mobile phones, or other devices, consciousness is altered. Music that operates in congruence with (rather than in parallel to) this reality becomes a more substantial ingredient in forming that new consciousness. As such, the artworks, projects, and systems of mediation to be discussed in this chapter will be referred to as *technoetic*, or as *technoetic environments*.

Amergent music is a generative style developed to complement the innate dynamics or ontology of technoetic environments. *Generative* means that the music is made in real time by algorithms that continuously vary the sonic output. Amergent music synthesizes the becoming and emergence of mediated interaction with generative processes and the aesthetics of Ambient music. What someone hears is not only the result of an algorithmic process, but the consequence of actions that have been taken in a technoetic environment. Whereas *effect* is a result, *emergence* is a behavior. The patterns of a cellular automaton or swarm algorithm are visually evident as an effect—or result—of a simple rule set. Where *affect* is a physical and mental sensation in the flow of becoming, *amergence* is a phenomenon of consciousness. It characterizes emergent behavior with an additional, affective dimension and works to bring forth subjective details of the emergent behavior that surrounds those immersed in a technoetic environment. Amergence evokes a qualitative behavior of potential. The act of music making involves seeded sound potential and presence in a mediated environment where sound, in the flow of interaction and generative processes, is experienced as a becoming of music. A work of Amergent music is rooted in the ontology and innate dynamics of a mediating technology. It recognizes the functioning order of the environment or platform that supports it. Sounds are layered in myriad combinations to spin a connective thread of musical experience that is brought forth by virtue of one's presence and engagement in a technoetic environment.

A more literal interpretation of the opening John Cage quote reveals one of the inherent tensions of Amergent music, and the relationship of music to technoetic environments in general. As one exists in these environments, his actions resonate throughout, potentially affecting or effecting every other person or element also within it. This kind of presence forms the basis of a relationship that includes not only the permeable sound–music boundary espoused by Cage, but a more literal version of the idea that "they themselves are doing it." The interconnectedness of these environments is not unique. The Dalai Lama reminds us that in our immediate reality, "Everything we do has some effect, some impact" (Dalai Lama 2001, 63). The difference is that in technoetic environments these a/effects can be sensed more immediately, or they can be used for exploration and experimentation as a simulation, and as the foundation of a mediated reality with the ability to transform consciousness.

This view of the world, in relation to music and art, has suggested a path of inquiry that follows in the steps of cybernetics. In its earliest years cybernetics was characterized as a science of regulation and control, and focused primarily on the communications that took place between people and machines. W. Ross Ashby discussed the homeostat (1956), or thermostat, as an example. Using this device, a human sets a threshold for the preferred room temperature and a mechanism within the device monitors that environment, introducing warmer or cooler air as needed. As the field matured, cybernetics became a useful means of conceptualizing and thinking about many different kinds of systems. Stafford Beer applied cybernetic principles to management and government, and referred to many of these systems as "too complex to fathom" (Beer 1972, 67). Beer encouraged others to apply cybernetic thinking to a variety of technical and artistic fields in which the complexity of interactions was often conceptually or creatively burdensome. In the realm of art, cybernetics was used as a means of redefining the relationship between the artist, the traditional notion of "viewer" and the work itself. As Roy Ascott originally suggested in 1967:

> It is necessary to differentiate between *l'esprit cybernétique*... and cybernetics as a descriptive method. Now, art, like any process or system, can be examined from the cybernetic point of view; it can also derive technical and theoretical support from this science—as in the past it has done from optics or geometry. This is not unimportant, since the artist's range can be extended considerably... But it is important to remember that the cybernetic vision in art, which will unify art with a cybernated society, is a matter of "stance," a fundamental attitude to events and human relationships, before it is in any sense a technical or procedural matter. (Beer 2003, 127)

In my research, cybernetics has provided models and a framework for structuring new ideas and techniques. It has facilitated the development of a fledgling practice and given voice to thoughts that were initially easier to execute as an artwork than explicate in a larger or more robust context. The work presented here is the culmination of a musical approach that draws on the theories and concepts of cybernetics but is not a literal manifestation of the circuits and wires one often associates with the field. This chapter looks at cybernetics as a means of coordinating the behavioral relationship between the art work and person engaged in it. Like Ashby's homeostat, music is regulated to be congruous with the dynamics of the environment and the behavior of those who exist within it.

22.1 Music as Behavior; Music as Movement

The idea of music as a behavior came not from discussions or writings about music but rather of biology. Cybernetically inclined biologists Humberto Maturana and Francisco Varela's research (1979, 1980, 1992) has contributed profoundly to informing this work. Their view of living systems as "systems" provides a framework for using computers to produce work with an organic feel. A final, technoetic work is not alive, but demonstrates some life-like characteristics. Structural coupling, Maturana and Varela's term for the relationship of mutual perturbations that binds adjacent autopoietic ("self-making" or autonomous) unities in a shared environment is reconceived as structaural coupling. In this model, two autonomous systems—a person and a system for generative music—are likewise bound in a continuous exchange of interactions within a mediated environment.

22.1.1 Autopoiesis and Organizational Closure

Maturana and Varela write that the defining element of all living things is their autonomy. Much of their work is based on their theory of *autopoiesis*. The term translates from two Greek words: αὐτὸσ, or *self*, and ποιειν, *to make*, in the sense of creation or production (Maturana and Varela 1980). Autopoiesis, simply put, states that the product of any living thing is itself; there is no separation between the producer and the produced (Maturana and Varela 1992). They define this functioning order as:

> An autopoietic machine is a machine organized (defined as a unity) as a network of processes of production (transformation and destruction) of components that produces the components which: (i) through their interactions and transformations continuously regenerate and realize the network of processes (relations) that produced them; and (ii) constitute it (the machine) as a concrete unity in the space in which they (the components) exist by specifying the topological domain of its realization as such a network. (Maturana and Varela 1980, 78–9)

The language that explains autopoiesis makes the connection between organic life and generative systems clear. Generative music systems have a sustaining order defined by their own processes. In such a system, an autopoietic "network of processes of production" (Maturana and Varela 1980, 78–9) could include computer code that produces random numbers, monitors timers, makes "if... else" decisions, and so on. A generative system is also a unity in the space that it exists. Either as a standalone work or a component in something larger, generative music systems have a discrete identity.

Where this comparison becomes less clear is in the regeneration of processes of production. A musical system that can, to use Maturana and Varela's words, "regenerate and realize the network of processes (relations) that produced [it]" (1980, 79), would be able to write additional rules that would be added to the generative procedures, or would be able to record itself and integrate those recordings into the body of sound material at its disposal. While such systems are likely to exist, their exploration is outside the scope of this chapter. Therefore *autopoietic* will be replaced by Varela's term *organizational closure* (1979, 55).

Organizational closure is related to autopoiesis. It includes some of the qualities and characteristics required in autopoiesis, but excludes the biologically focused idea of regenerating these processes. The generative music systems used in Amergent music are discrete in their environment and have the ability to continuously produce out of the network of components that comprise them. They are both autonomous and organizationally closed. They are not autopoietically alive, but "livinglike" (Varela 1979, 59) in their operation.

Living things are subject to disturbances, or *perturbations*, in their environment that present a threat or challenge, or simply a new set of circumstances that must be handled or overcome. Perturbations can be obstacles in the functional order of a unity and they can allow organizationally closed systems to interact, though their interactions are never tightly coordinated or specified between discrete unities. All interactions take place within an environment, which has an additional role to play in this mutual exchange. Perturbation, and the idea that living systems can both maintain and convey their autonomy within an environment, is a crucial component to the behavior of Amergent music.

22.1.2 Structural Coupling

When multiple unities coexist in an environment, there can be a relationship of *structural coupling*. This is a biological phenomenon described by Maturana and Varela as a history of "reciprocal perturbations" (1992, 75) between two or more living things, and these living things and their environment. The basic relationship is illustrated in Figure 22.1.

It is easiest to think of structural coupling as the relationship between two (or more) adjacent cells. Each is autopoietic and solely responsible for its own functioning; yet it is not isolated. Changes to the immediate environment will affect the cells just as changes within individual cells will have an impact both on their fellows and on the space in which they exist. Structural coupling is present "whenever there is a history of recurrent interactions leading to the structural congruence between two (or more) systems" (Maturana and Varela 1992, 75). This relationship of reciprocal perturbations triggers structural changes. These are never directed or specified, but remain congruent with the autopoiesis of the individual unities involved.

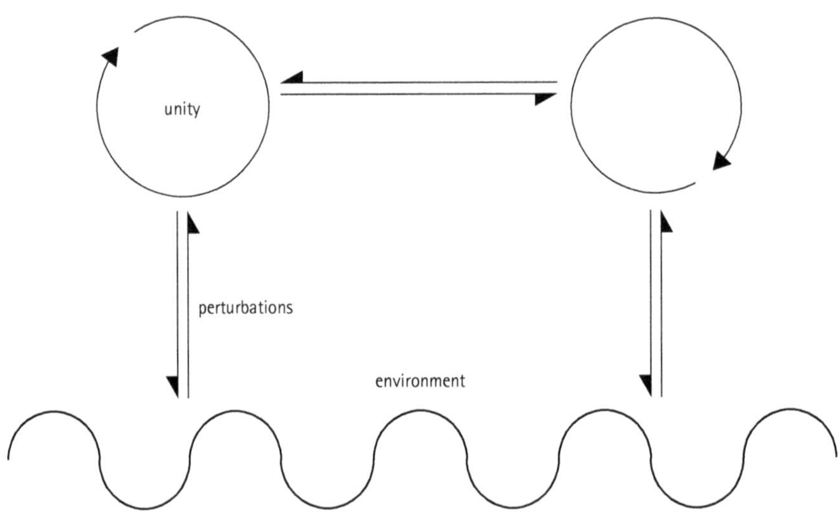

FIGURE 22.1 In Maturana and Varela's structural coupling (1992, 74), each unity is autonomous in its autopoiesis, and through its autopoiesis will make perturbations that are felt by adjacent unities and the environment in which they exist.

22.1.3 Structaural Coupling

Structural coupling belongs specifically to the domain of biological systems. It is a relationship that requires autopoiesis and, as such, should be discussed only as a mechanism of organic life. As Maturana and Varela have noted, autopoiesis applies to individual cells and should not be scaled or transposed to include higher levels of organization in an organism (1992). However, the concept is very powerful in the realm of transdisciplinary study and artistic creation. In the case of Amergent music, a human listener is one unity and a generative music system is another. Both are autonomous, organizationally closed, and struct*aurally* coupled (see Figure 22.2).

Struct*aural* coupling takes the same overall form as structural coupling in biological systems. There are mutual perturbations between organizationally closed—not autopoietic—unities. These perturbations characterize the kinds of interactions that take place between a generative music system, the listener within the mediated environment, and the environment itself. All interactions are recurring, which leads to continuous structural changes that are triggered, yet never specified. All changes remain compatible with the preservation of each unity's organizational closure.

A generative music system generally consists of computer code that manages random numbers, timers, and "if... else" decisions (to name a few examples) and sound resources (samples or a synthesis engine) and the rules or organization that define the

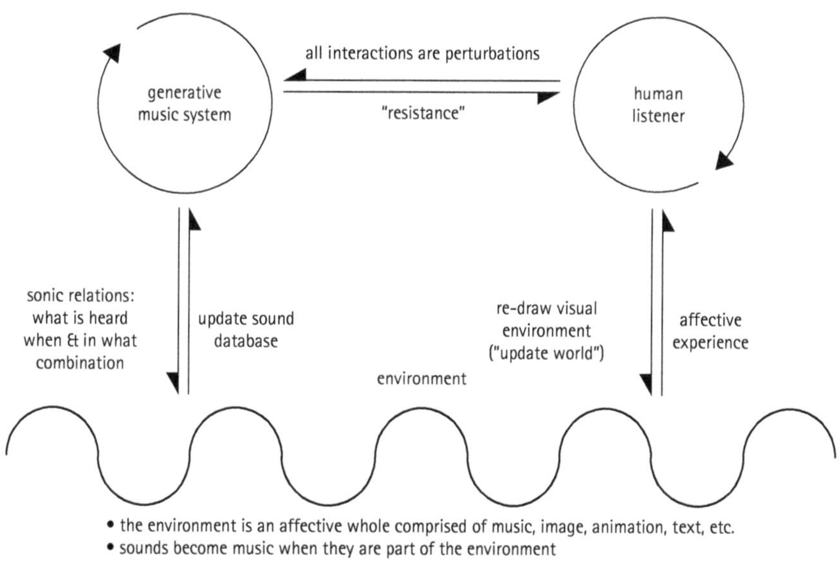

FIGURE 22.2 Struct*aural* coupling is the relationship of mutual perturbations between organizationally closed unities: a generative music system and a listener. The model (when in use or in context) creates a fluid stream of musical experience. Though it is often unclear how or where the perturbations that establish coupling begin, the listener develops a sense of congruence with the world through the music that comprises (a part of) it.

relationship between the code and its related audio assets. Together, these components comprise the organizational closure of the generative system as a unity.

The human listener is also an organizationally closed unity. Their biology defines them as such, but so does the process of mediation. Their unique abilities in the mediated world (as enabled by software) separate them from their environment. The environment is the mediated world that unifies an experience, binding listener to music. In addition to music, it can comprise images, video, animation, text, seeds of a narrative, and in some cases, other unities. Any perturbations made by the listener resonate with both the generative system and the environment. Similarly, the generative system perturbs the listener and environment, and the environment can perturb the listener and the generative system.

This is one of the most complex and important perturbations in the struct*aurally* coupled interaction model. Part of what it triggers in the listener is due to the affect of music—the perturbation that resonates from environment to listener. While the listener does not have direct or immediate control over what happens in the music, after a few reciprocal perturbations have passed, it becomes apparent to listeners that their actions have a congruence with the music. The arrangement of struct*aurally* coupled interaction makes it impossible to control anything directly, but a relationship becomes audible over time. It also becomes "tangible" in a sense. There is no direct contact, but through the same structural changes in the generative system that lead to new musical directions, there is a perturbation that pushes back, *against* the listener. This is a quality of musical instruments, something that Aden Evens refers to as "resistance." He explains:

> Defined by its resistance, the instrument does not just yield passively to the desire of the musician. It is not a blank slate waiting for an inscription....The instrument itself has a potential, a matter to-be-determined, and its use is always in relation to its own character as well as to the desire of the musician....Neither music nor instrument is predetermined, set in a specified direction from the beginning....The instrument's resistance holds within it its creative potential. (Evens 2005, 160–61)

The generative system pushes back to let the listener know its bounds and the possibilities it affords. The types of sounds that can be heard, overall texture and density, emergent melodies and introspective spaciousness—these are all sonic qualities under the control of the generative music system. In the biology of Maturana and Varela, the management of incoming perturbations causes this system to undergo structural changes that maintain organizational closure within its own "structural determination" (1992, 96). This means that the system of a cell will change but only within the range of possibilities afforded by its structure. Musical instruments have a similar structural determination. An FM synthesizer may not have a circuit board large enough to allow it to behave like a sampler. Without DSP or an additional reed mechanism, a trumpet cannot sound like an accordion. Synthesizers and brass instruments offer a wide range of sonic possibilities, but there are also limits set by their structure and materials. Similarly, within a work of Amergent music, there are many different sonic possibilities contained within (or limited to) the scope of the technoetic environment.

22.1.4 Perturbations and Behavior

Perturbation is the key concept in the struct*aurally* coupled relationship. All involved parties maintain their autonomy, organizational closure, functioning order, and so on, yet are still receptive to external forces. These forces (perturbations) cannot control them or specify changes in particular, but they trigger responses within the domain of the system's requisite organizational closure. This biological relationship is particularly compelling because it is so similar to interaction with digital, generative music systems.

In 1967, Roy Ascott imagined that such a practice would be possible. In his view, "The necessary conditions of behaviorist art are that the spectator is involved and that the artwork in some way behaves. Now it seems likely that in the artist's attempt to create structures that are probabilistic, the artifact may result from biological modeling. In short, it may be developed with the properties of growth" (Ascott 2003, 129). Clearly, even from this early perspective, a cybernetic view of biology that facilitated the modeling of living systems held great artistic and conceptual potential. It begins to get at the idea that any type of music—operating in an environment of mediated interaction—must change. Change how? When? And into what? Throughout the history of computer games, music has always changed in some way. Even *Space Invaders* (one of the earliest computer games made by Midway in 1978) would increase the tempo of a simple four-note melody as the player's situation grew more dire (Collins 2008, 12). It is important to draw a clear distinction between this early approach and the current directions of Amergent music.

The influence of biological models, my background as an improviser, and a guiding interest in developing music congruous with the ontology of contemporary technology pointed to a *behavior* of music. Music can be viewed as an unfolding process: What does it do over time? And how does it react in relation to one's use of the technology that supports it? Behavior is an ideal way to answer these general concerns and questions. It addresses the actions of music over time, and by viewing interactions as perturbations, it clarifies questions of change. This music doesn't just get slower, louder, or darker in relation to external events—it *behaves*.

Maturana and Varela write, "Behavior is not something that the living being does in itself (for in it there are only internal structural changes) but something that we point to" (1992, 138). Amergent music is built around musical systems that are capable of sending and receiving perturbations. These stimuli trigger in each system "internal structural changes" that produce the events interpreted as "behavior" to an observer. Consider the following statement from their book *The Tree of Knowledge*:

> Thus, the behavior of living beings is not an invention of the nervous system and it is not exclusively associated with it, for the observer will see behavior when he looks at any living being in its environment. What the nervous system does is expand the realm of possible behaviors by endowing the organism with a tremendously versatile and plastic structure. (Maturana and Varela 1992, 138)

Now replace all instances of *organism* and *living being(s)* with *music*, and *nervous system* with *generative system*:

> Thus, the behavior of **music** is not an invention of the **generative system** and it is not exclusively associated with it, for the observer will see behavior when he looks at any **music** in its environment. What the **generative system** does is expand the realm of possible behaviors by endowing the **music** with a tremendously versatile and plastic structure.

This transposition from the biological to the musical presents a welcome alternative to the standard notion that, in any work where music is coupled to interaction, "the music changes." Yes, there is change. But "change" and "change of state" can be more robustly described as dimensions of behavior. There is no deliberate action, no pre-planned response defined a priori within a database of all possible actions of the generative system, but a genuinely unique response given the conditions/perturbations the system confronts in the moment of action.

The distinctions between linear music and Amergent music can be further clarified with an additional example offered by Maturana and Varela. In *The Tree of Knowledge* they discuss the case of a particular plant (*Sagittaria sagitufolia*) that can transform between aquatic and terrestrial forms depending on the current water levels in its environment. This is behavior because there are "structural changes that appear as observable changes in the plant's form to compensate for recurrent disturbances of the environment" (Maturana and Varela 1992, 143). However, because the behavior happens so slowly, an observer is likely to cite these changes as part of the plant's development. It is much easier to think the plant grew that way due to the amount of water around it. Maturana and Varela argue that behavior is a structural response to external forces no matter what the tempo.

The case of behavior versus development in the sagittaria is much like the case of Amergent vs. linear music. Music that is composed in a linear model is told exactly what it must do to "develop" and meet the expectations of the situation for which it was composed. It operates in a prescribed way and conforms to a set of demands. This should not be misconstrued as a negative evaluation. However, when the situation in which the music is to be heard is *changing*, the music itself becomes less able to complement and support it. Much of the music that can be heard in contemporary mediated environments and art works is trapped in such a model of linear thinking. Alf Clausen, composer for the cartoon series *The Simpsons* recommends, "score the emotion not the action" (Chilvers 2004). This is appropriate for cartoons but not for environments of mediated interaction. Namely: what emotion? The emotional tenor is often unknown. Even if emotion could be surmised, it is not known what actions would produce it. It *is* known, however, what *ingredients* will be used to produce both action and emotion. That is the behavioral advantage of Amergent music.

Amergent music can, by comparison, act on its own accord. It is not "doing what it is told" nor is it predestined to purposefully connect with the events of its environment.

The generative systems that give rise to it simply respond to perturbations in the maintenance of their own internal functioning order. Compared side by side, an observer may hear a piece of linear music and a piece of Amergent music and think that both suit their expectations given the environment. But alter or transform that environment, and due to the lack of behavioral adaptation in a linear piece, its presence will be awkward or ill fitted when heard a second time. Like an organism, the Amergent piece is far more capable of behavior that responds to environmental changes and perturbations in the maintenance of its identity and functioning order.

22.2 First-, Second-, Third-order Cybernetic Systems

The cybernetic perspective of this research has served to inform a means of musical production that is ontologically congruent with the technoetic environments in which the music is created and heard. In the process of developing such a system, other factors surrounding the relationship between music, environment, and listener or interacter came to light. Throughout this development, it was necessary to study and compare various works of Experimental, Ambient, Generative, and my own Amergent music. These genres provided an excellent model in that they have a compelling mix of compositional control and freedom that lends itself to real-time musical behavior.

Cybernetics also plays a role (implicitly or explicitly) in each of these genres. And though they artistically distinct, there are commonalities that reveal a cybernetic relationship of a third order, in which the person engaged in interaction becomes part of the very system that gives rise to the work they are experiencing.

22.2.1 First-order Systems

Gordon Pask describes first-order systems (1°)[1] as "classical black boxes and negative feedback" (1996, 355). Heinz von Foerster refers to another of Pask's characterizations of first-order systems, stating that "the observer enters the system by stipulating the system's purpose" (2003a: 285). In short, 1° systems focus on autonomy and regulation. In a musical context, this is represented by instructions that lead to the autonomy and regulation (or organization) of sounds. Table 22.1 cites examples of relevant musical works and presents a simple 1° stipulation. These first-order stipulations do not represent any of these works in their entirety. All, except for those works of Amergent music by the author, are not complete until they reach the second-order stipulation. The Amergent pieces must reach the third-order stipulation to be complete.

The first order can be loosely described as various means of structural organization and algorithms that will lead to the production and performance of a musical work.

Table 22.1 First-order systems in Experimental, Ambient, Generative, and Amergent music

TITLE (GENRE)	*MUSICIAN*	*1° SYSTEM*
In C (Experimental)	Terry Riley	Elastic structure; sequential progression through the set of 53 phrases (Riley 1964)
Paragraph 7 of *The Great Learning* (Experimental)	Cornelius Cardew	Instructions for piece: "Do not sing the same note on two consecutive lines" "Sing any note that you can hear" Otherwise, "choose your next note freely" (Eno 1976: 3)
Music for Airports (Ambient)	Brian Eno	Tape-phasing structure at intervals of 21″, 17″, 25″, 18″, 31″, 20″, 22″ (Tamm 1995, 137)
Bloom (Generative)	Brian Eno and Peter Chilvers	Looping drone; melody generator
Dérive Entre Mille Sons (Amergent)	Norbert Herber	Generative timer, randomizer, sequencer and x-fader; spatial arrangement of audible zones

22.2.2 Second-order Systems

Again, von Foerster agrees with Pask and characterizes the second order (2°) as cases in which "the observer enters the system by stipulating his own purpose" (2003a, 285). The observer's purpose is frequently experimental: "what does (or could) this sound like?" This proposition calls to mind W. Ross Ashby's characterization that a system is "not a thing, but a list of variables. This list can be varied, and the experimenter's commonest task is that of varying the list... that gives the required singleness" (1956, 40). In these 2° musical systems (see Table 22.2), sounds are integrated with the system as variables in a musical experiment.

The system does not simply exist in some "final" form, but rather changes due to the role of the observer—the "composer" or musician who makes use of the system. In Generative and Amergent music, the system is a list of variables including the parameters of a generative instrument and a palette of sounds to which it is coupled.

22.2.3 The Third-order and Amergent Music

A third-order stipulation applies only to works of Amergent music, such as *Sound Garden* (2009), *Dérive Entre Mille Sons* (2010), and a pair of simultaneous installations called *I am Ai, We are Ai* and *Fields of Indigo* (*IAWA/FI*). These installations are a collaborative effort between the author and textile artist Rowland Ricketts (Ricketts 2013). These projects combine generative music with live field recordings to create an environment that reflects on the themes of tradition, interconnectivity, and processes of diminution and accretion both in sound and natural indigo dye.

Table 22.2 Second-order systems in Experimental, Ambient, Generative, and Amergent music

TITLE (GENRE)	*MUSICIAN*	*2° SYSTEM*
In C (Experimental)	Terry Riley	Phrases composed loosely in key of C; progression advances at performer's discretion (Riley 1964)
Paragraph 7 of *The Great Learning* (Experimental)	Cornelius Cardew	"[A]ccidents that are at work" such as "'unreliability' of a mixed group of singers," "beat frequency," "resonant frequency" of the room, "preference" or "taste" of the individual performers (Eno 1976: 4)
Music for Airports (Ambient)	Brian Eno	Pitched sounds are phased at various intervals to produce shifting tonalities over time (Tamm 1995: 137)
Bloom (Generative)	Brian Eno and Peter Chilvers	Drone plays in multiple keys; melodies constructed of pitches harmonically related to the drone
Dérive Entre Mille Sons (Amergent)	Norbert Herber	Sound palette assigned to generative instruments and linked to individual sonic zones within a spatial layout

IAWA/FI is an installation developed by two artists, but it is also a collaboration between two seemingly disparate geographic locations: Tokushima, Japan, and Champaign, Illinois. Tokushima represents both the history and current practice of indigo in Japan. Historically, this city was—and still is—the overwhelming source of indigo dye for the entire country. Installation locations are in the Tokushima prefecture at the Bandai Warehouse in Tokushima City (see Figure 22.3) and an indigo field in the mountains of Kamikatsu. The US location is based at both the Krannert Art Museum on the University of Illinois (Urbana-Champaign) campus and an indigo field in Bloomington, Indiana, where much of the plants used for the installation were sourced (see Figure 22.4). This site represents one contemporary expansion of the tradition and history of Japanese indigo. The visual portion of the installation consists of a variety of indigo-dyed textiles that explore the accumulative nature of the indigo-dyeing process, hand-cut indigo plants that will dry and oxidize (become blue) over time, and a time-lapse video of the indigo drying process.

The sound of the installation is built up in layers. Two field recorders create a sonic foundation. These are placed in both the US and Japanese indigo fields and continuously stream a real-time recording over the Internet to the installation location in the opposite country—those that visit in Japan hear the sound of the Illinois indigo field and vice versa. Additional layers comprise a digitally processed version of the live stream, as well as concrete and processed sounds related to indigo production and dyeing: winnowing, stomping dry leaves, stirring dye vats, dye running and dripping, and rinsing the dyed cloth with water. A final layer consists of voice recordings of people connected to the Tokushima indigo tradition through practices in agriculture, industry, and art.

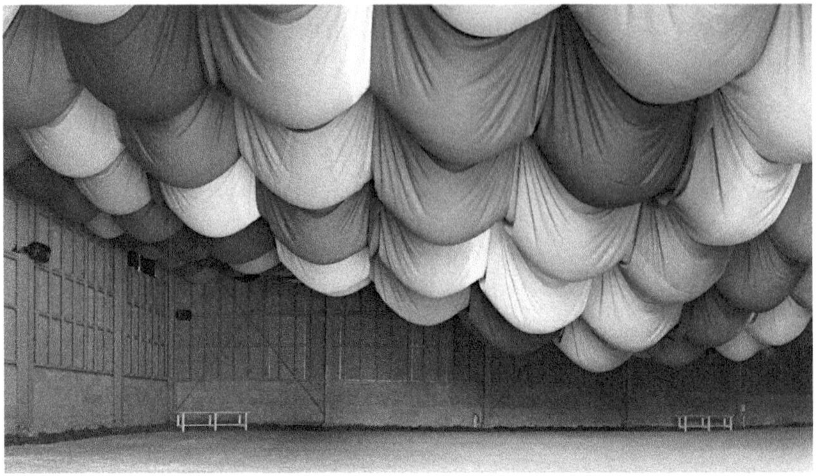

FIGURE 22.3 The Bandai Warehouse is an open space, approximately 44 feet wide, 80 feet long, and 16 feet to the ceiling. To realize the installation, the warehouse was filled with indigo-dyed textiles, speakers, and motion sensors.

FIGURE 22.4 The indigo field in Bloomington, Indiana where the field recording was placed.

The live audio streams play continuously and are subject to the weather conditions, flora, and fauna present at the site of recording. Prerecorded sounds related to indigo processes and production are part of a generative system. Both layers are autonomous and ever changing. The voice recordings are heard relative to the presence of people inside the installation space. As visitors move about the room, motion sensors trigger an additional generative system that plays these sounds. Their presence and engagement with the space connects them with sounds of a tradition that grew out of Tokushima and has spread across the globe. Recordings and images of the project can be found at http://iamai.jp/en/soundstreams.html.

What is heard is immaterial, or "not present in a physical state," much like the steps of the processes that leave their mark on the finished textiles. The sounds of *IAWA/FI* reflect on aspects of Japanese indigo such as connection to a place and cultural roots that will inevitably change and be influenced by each and any of us. Sound constructs a strong metaphor for the force of cultural influence and interaction on many traditions as one culture shapes and influences the other. Through a variety of generative techniques, the sounds of *IAWA/FI* are constantly heard in unique permutations. New sonic combinations and sequences regularly redefine the work, weakening the idea of what it *is* while giving strength to an overall sense of potential and possibility. Those that enter the installation space and engage with the work will make visual and aural contact with this tradition and its becoming.

22.2.4 Third-order Systems

The sonic portion of *IAWA/FI* is a third-order system. In the third order (3°), the observer and system have a shared purpose. The observer's purpose is an extension of the question posed in the 2°stipulation, asking "*why* does it sound this way and what does that say about the 'place I'm in'?" In the 3°, the observer is more technoetically oriented and coupled to an ever-changing 2° system. The reciprocal perturbations constitute both a question and an assertion of an unfolding, mutual purpose, as interactions indicate intent or desire and seek to draw out experience.

This "drawing out" in the 3° system demonstrates that both the generative system and observer are situated inside the work as an environment. However, as von Foerster states, "the environment as we perceive it is our invention" (1973, 1). The work of Amergent music does not exist without the dynamics that are created and sustained between the generative system and the observer. This is illustrated in Figure 22.5. It is the same struct*aural* coupling diagram as presented earlier in this chapter, but with an additional layer of information that reveals the presence of 1°, 2°, and 3° stipulations. The reciprocal perturbations exchanged between observer and generative system construct a mediated reality of emergence and becoming. Chris Lucas writes:

> The current "state-of-the-art" is in third-order cybernetics, where the observer is part of the coevolving system. This is a more intrinsic (embodied) methodology and shows the ongoing convergence of all the various systemic disciplines, as part of the general world paradigm shift noticed recently towards more integrated approaches to science and life. In 21-st century systematics, boundaries between systems are only partial and this implies that we must evolve with our systems and cannot remain static outsiders. Thus our mental beliefs echo our systemic behaviours, we co-create our realities and therefore internal and external realities become one. (Lucas 2001, XX)

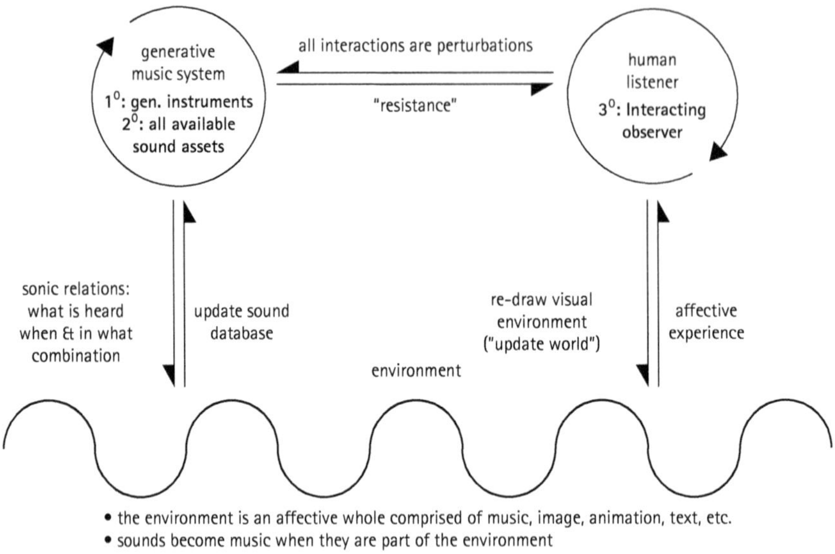

FIGURE 22.5 Struct*aural* coupling facilitates interaction within a 3° cybernetic system. The 1° is represented by the generative instruments, and the 2° by the system of sounds used by these instruments to create a complete generative system. The interacting observer constitutes the 3° as the reciprocal perturbations shared between them and the generative system give way to the environment out of which the affective experience emerges.

In technoetic environments this is a reality dominated by emergence, where the synergy of localized interactions churn endlessly, producing novelty in *this* moment, and in the next, and the next, and so on. There is an objective. These works produce a transformation of consciousness that is sustained by the artwork, not just a transformation of *any* consciousness. Stafford Beer thought of cybernetics as the science of exceedingly complex systems—of systems that *become* in an unpredictable manner—and a science that focused "on adaptation, on ways of coming to terms performatively with the unknown" (Pickering 2008, 129). For musicians and sound artists or designers that cultivate (or help to cultivate) these types of mediated experiences, *becoming* is always known. The ontology of that becoming will always be partly determined by the capabilities of the technical system that sustains the processes of mediation. But within those capabilities there is a great deal that is unknown. Struct*aural* coupling provides a 3°system that behaves so as to seamlessly integrate a musical becoming within the totality of the evolving, mediated reality.

In the context of business (strategic management) consulting, Vincent Kenny and Philip Boxer write, "We need to have a domain which contextualises the activities of, and relations among, the participant observer ontologies of the 2° domain...3° cybernetics must be a domain which allows us to come to contextualise this 'subject', with his 'ethical system' and his higher-order 'purpose.' We need to understand his phylogenesis as observer" (Kenny and Boxer 1990). While the work discussed here is substantially removed from the field of strategic management consulting, Kenny and Boxer express a shared need to characterize the overall dynamics and possible outcomes for situations in which an observer is coupled to another system and the pair have a shared purpose. What is most interesting is their reference to this person as a "participant observer," which implies he has both active and passive roles in this overall process. In a 3° stipulation, the system and individual evolve together. In works of Amergent music this partnership of transformation, continuous perturbation, and the tension of simultaneous (in)activity plays an essential role in shaping the experience of a technoetic environment.

22.3 Amergence and the Poiesist

The projects discussed in this chapter began as part of a research process. The objective was to answer questions relating to music and a coupled technological environment. But this inquiry additionally led to unexpected answers concerning the people involved in the interaction. The relationship described earlier makes it clear that these people are more than docile listeners. But they are also not involved to the degree that would engage them in any kind of "work." Specifically, these projects are not music production software or tools. The unique role of these people and the experience afforded by the technoetic environment was one of the more elusive and surprising outcomes of this research process.

With information technology and usability, the term *user* is common and effectively suggests the demand this person has for the utility of an object or the mediated environment (Norman 1989; Krug 2006). The potential of involvement and engagement with an interactive art work calls for the use of the term *participant* (Cornock and Edmonds 1973; Popper 1975). But whereas *user* has too much implied agency, *participant* has too little for the discussion at hand. *Player*, as used in games, conveys a more carefree sense of agency but it also connotes the hands-on act of playing music. This idea is of course related, but too specific to other realms of music making to be of use in this context.

In her book *The Utopian Entrepreneur* (2001), Brenda Laurel used the term *partner* to suggest a mutual agreement between artists or designers and the person engaged in their work. She favored the term because, unlike *participant*, there was clarity in the consensual nature of the agreement or relationship (Laurel 2001). There is also *vuser*, a combination of *viewer* and *user*, coined by Bill Seaman in 1998 (1999, 11), which encapsulates elements of surrender and agency inherent to these environments. In works such as those discussed in this chapter, a combination of *user, listener*, and *participant* is apropos, but none speak sufficiently to the ontology of technoetic environments.

Martin Heidegger's lecture "The Question Concerning Technology" argues that it is not important to ask what technology can do for us, but to become aware of what it can reveal about ourselves and the world in which we live. Technology is most beneficial in the long term when it is used to reveal and explore, not to exploit. If there is a question concerning technology, it is a question of *how*, and it focuses on a sustainable future. Technology itself challenges us to think about its essence: "what is that?" Heidegger discusses its tendency toward "revealing" and "enframing." Through enframing, "the subjugation of the world to already given human ends" (Pickering 2008, 131), technology provides resources, tools, and processes—a "standing-reserve"—that gives way to further technological developments. It has a recursive essence that, if not handled carefully, subjugates us to the service of technology at the expense of spiritual and other aspects of human development. Heidegger writes:

> So long as we represent technology as an instrument, we remain transfixed in the will to master it. We press on past the essence of technology. When, however, we ask how the instrumental unfolds essentially as a kind of causality, then we experience this essential unfolding as the destining of a revealing.... The question concerning technology is the question concerning the constellation in which revealing and concealing, in which the essential unfolding of truth propriates. (Heidegger 1977, 337–8)

Technology exists as a continuous cycle of "revealing and concealing" in which truth can be discovered. Through this process, "the essential unfolding of the essence of technology" should be approached with caution because the truth it offers is intertwined with demise. Pickering observes that Heidegger's notion of *revealing* "points us to a politics of emergence" (2008, 131). The tumult in a cellular automata creates a useful impression. Cells churning off and on, flickering in and out of coherent groups and patterns, appear similar to Heidegger's processes of revealing and concealing. Like order in any self-organizing system, truth is evanescent.

Heidegger's dynamics of revealing are discussed as an entangled network in which technology contains equal measures of interwoven "danger" and "saving power." He writes, "Human activity can never directly counter this danger. Human achievement alone can never banish it. But human reflection can ponder the fact that all saving power must be of a higher essence than what is endangered, though at the same time kindred to it" (Heidegger 1977, 339).

The danger is the effect of technology, the tangible results of enframing and standing-reserve. The saving power is affect; the unfolding of "ambiguity points to the mystery of all revealing, i.e., of truth" (Heidegger 1977, 337). Heidegger asserts that those who are attentive to the strand of revealing containing saving power are the ones who will become truly free. This dialectic of revealing is similar to the semantic tension between *effect* and *affect* that led to the term Amergent music. *Amergent* combines action and emotion. *Emergence* as a characterization of the action involved in reciprocal perturbation, and *Affect* as the emotional impact of this continuous exchange. Each dynamic is necessary to the processes that give rise to the musical experience.

While Amergent music has independence and autonomy within its environment, it does not unfold entirely of its own accord. The person who is simultaneously listening and engaged in the mediated environment is largely responsible for the totality of what is heard. This is the *poiesist*, the one who draws music out through the agency of their interaction. Heidegger writes:

> There was a time when it was not technology alone that bore the name *technē*. Once the revealing that brings forth truth into the splendor of radiant appearance was also called *technē*.
>
> There was a time when the bringing-forth of the true into the beautiful was called *technē*. The *poiēsis* of fine arts was also called *technē*. Heidegger 1977, 339)

Poiesis is a bringing forth. In works of Amergent music the person engaged in the experience, formerly known as the *participant, user, player*, and so on, is more appropriately called the *poiesist*. The experience of interaction facilitated by Amergent music is a poiesis—a bringing forth or drawing out—the catalyst to a becoming or emergence of sounds into music. The poiesist draws out sound to reveal music; the poiesist engages with "the constellation in which revealing and concealing, in which the essential unfolding of truth propriates" (Heidegger 1977, 338). This process and the experience of sound it engenders is amergent. As in our relationship with technology, we become aware of the things a sonic environment can reveal about ourselves and the technoetic places in which we inhabit.

22.4 Conclusions

In the *Biology of Cognition* (the first part of *Autopoiesis and Cognition*) Humberto Maturana tells a story that serves as a useful (and final) summary to the musical ideas presented in this chapter:

> Two groups of workers are assembled and each given the task of building something. In the first group a leader is appointed and he is given a book with drawings, measurements, and a discussion of the materials required to build a house. The leader dutifully follows the *descriptions* in the book and guides his team through all of the various tasks required to build their house to suit every last detail of the design. (Maturana and Varela, 1980, 53–5)

The second group has no leader. Instead each member starts in a single row and is given an identical copy of a book filled with a general set of instructions. In it there is no mention of *house*, no discussion of pipes or windows or electrical wires, and no drawings whatsoever. There are only instructions specifying what a worker should do given their starting position and all other possible positions they might encounter as the process ensues and their relations to the other workers changes.

An observer visits the worksite of the first group to see that they are in fact building a house. He clearly sees that it is a house and the workers know that it is a house they are building. They have seen the plans and discussed them to be certain that the finished product matches the description which they were provided.

The observer then travels to visit the site where the second group is working. There he finds that another house is in the process of construction, though if he were to ask the workers what it is they are building they could not give a definite answer, all they could do is point to individual steps within the process such as, "when the two-by-four is positioned like that, I put the nails in like this." In the second group there is no description to follow, only steps that constitute a process of changing relationships between the workers and available materials. Maturana writes:

> That the observer should call this system a house is a feature of his cognitive domain, not of the system itself. (Maturana and Varela, 1980, 54)

Performing a similar transposition from earlier in this chapter, the statement yields:

> That the observer should call this system **music** is a feature of his cognitive domain, not of the system itself.

The observer sees what he sees and hears what he hears. That it is a house or a piece of music is his construction and a function of his cognitive domain. The origin or defining order of what he hears is particular to the generating system and does not need to be known in advance for an observer to form his perception(s). Amergent music, like the working process of the second group in Maturana's story, *becomes*. It is emergent through a series of interactions based on changing relationships. How this is done is of little importance to the poiesist, yet he can hear transformations and accept them as part of his ongoing mediated reality. From a musical perspective this is not done to deliberately model what Maturana tells us about human cognition. It is not an attempt at making mediated reality *really* real. It simply offers a mechanism for creating music that is complementary to the flow of becoming in the human domain of perception, and for making that flow congruous to the perpetual emergence experienced in technoetic and media arts.

Note

1. The abbreviations for first order (1°), second order (2°), and third order (3°) are borrowed from Kenny and Boxer (1990).

References

Ascott, Roy. 2001. *When the Jaguar Lies Down with the Lamb: Speculations on the Post-biological Culture*. http://www.uoc.edu/artnodes/espai/eng/art/ascott1101/ascott1101.html.

——. 2003. Behaviourist Art and the Cybernetic Vision. In *Telematic Embrace: Visionary Theories of Art, Technology, and Consciousness*, ed. E. A. Shanken, 109–57. Berkeley: University of California Press.

Ashby, W. Ross. 1956. *An Introduction to Cybernetics*. London: Chapman and Hall.

Beer, Stafford. 1972. *Brain of the Firm*. London: Penguin.

Chilvers, Peter. 2004. The Music behind Creatures. *Gameware Development* http://www.gamewaredevelopment.co.uk/creatures_more.php?id=459_0_6_0_M27.

Collins, Karen. 2008. *Game Sound: An Introduction to the History, Theory, and Practice of Videogame Music and Sound Design*. Cambridge, MA: MIT Press.

Cornock, Stroud, and Ernest Edmonds. 1973. The Creative Process Where the Artist Is Amplified or Superseded by the Computer. *Leonardo* 6 (1): 11–16.

Dalai Lama. 2001. *The Dalai Lama's Book of Daily Meditations*. London: Rider.

Eno, Brian. 1976. *Generating and Organizing Variety in the Arts. Studio International* 984: 279–283. Reprinted in *Breaking the Sound Barrier: A Critical Anthology of the New Music*, ed. Gregory Battock. New York: Dutton, 1981. http://www4.ncsu.edu/~mseth2/com307S13/readings/EnoArts.pdf.

Evens, Aden. 2005. *Sound Ideas: Music, Machines, and Experience*. Minneapolis: University of Minnesota Press.

Harland, Kurt, 2000. Composing for Interactive Music. *Gamasutra*. http://www.gamasutra.com/features/20000217/harland_01.htm.

Heidegger, Martin. 1977. *Basic Writings: From "Being and Time" (1927) to "The Task of Thinking" (1964)*. Edited by David Farrell Krell. New York: HarperCollins.

Herber, Norbert. 2009, *Sound Garden*. http://www.x-tet.com/soundgarden.

——. 2010. *Dérive Entre Mille Sons*. http://vimeo.com/18756185.

Kenny, Vincent, and Boxer, Philip. 1990. *The Economy of Discourses: A Third Order Cybernetics?* http://www.oikos.org/discourses.htm.

Krug, Steve. 2006. *Don't Make Me Think: A Common Sense Approach to Web Usability*. Berkeley, CA: New Riders.

Laurel, Brenda. 2001. *The Utopian Entrepreneur*. Cambridge, MA: MIT Press.

Lucas, Chris. 2009. *Complexity Theory: Actions for a Better World*. http://www.calresco.org/action.htm.

Maturana, Humberto R., and Francisco J. Varela. 1980, *Autopoiesis and Cognition: The Realization of the Living*. Dordrecht, Netherlands: D. Reidel.

——. 1992. *The Tree of Knowledge: The Biological Roots of Human Understanding*. Boston: Random House.

Norman, Donald A. 1989. *The Design of Everyday Things*, New York: Doubleday.

Pask, Gordon. 1996. Heinz von Foerster's Self Organization, the Progenitor of Conversation and Interaction Theories. *Systems Research* 13 (3): 349–362.

Pickering, Andrew. 2008. Emergence and Synthesis: Science Studies, Cybernetics and Antidisciplinarity. *Technoetic Arts: A Journal of Speculative Research* 6: 127–133.

Popper, Frank. 1975. *Art-action and Participation*. New York: New York University Press.

Ricketts, Rowland. 2013. *Rowland and Chinami Ricketts: Indigo, Art, Textiles*. http://www.rickettsindigo.com.

Riley, Terry. 1964, *In C, Other Minds*. http://imslp.org/wiki/In_C_(Riley,_Terry).

Seaman, William C. 1999. *Recombinant Poetics: Emergent Meaning as Examined and Explored within a Specific Generative Virtual Environment*. PhD diss., Centre for Advanced Inquiry in the Interactive Arts, University of Wales.

Tamm, Eric Enno. 1995. *Brian Eno: His Music and the Vertical Color of Sound.* New York: Da Capo.

Varela, Francisco J. 1979. *Principles of Biological Autonomy*. New York: Elsevier.

von Foerster, Heinz. 2003a. Cybernetics of Cybernetics. In *Understanding Understanding: Essays on Cybernetics and Cognition*, 283–286. New York: Springer.

——. 1973. *On Constructing a Reality*. http://ada.evergreen.edu/~arunc/texts/InventingSystems/readings2.pdf.

SECTION 5

TOOLS AND TECHNIQUES

CHAPTER 23

FLOW OF CREATIVE INTERACTION WITH DIGITAL MUSIC NOTATIONS

CHRIS NASH
AND ALAN F. BLACKWELL

PRACTICE-BASED research into digital audio technology is the source of many new and exciting interactions, instruments, and sonorities. However, the nature of the technologies used raises significant challenges for traditional conceptions of musical practice. The disjunctions between composition, performance, and improvisation, between the use of common score notation and other graphical representations, and between discrete and continuous expressive scales can be compared to long-standing debates in human–computer interaction (HCI) regarding direct manipulation (e.g., mouse-based point-and-click, drag-and-drop, etc.) and abstract programming (e.g., keyboard-based notation editing), graphical user interfaces (GUIs) and command lines, and visual (e.g., Max/MSP) versus textual (e.g., SuperCollider) programming languages (see also Chapter 24 in this volume).

Our HCI research group has a long-standing program of work understanding the characteristics of notational systems in the broadest sense (Blackwell and Green 2003). We consider any visual, textual, or symbolic user interface to be a notation, which can be treated as directly analogous to music notation in the sense that it guides the future operation of the computer, just as music notation guides the "operation" of a performance. Performances can be more or less literal, more or less improvised, more or less edited and so on. All of these variations are found in both digital music systems and other digital systems, and raise theoretical challenges for computing as they do for music.

Nevertheless, the tools provided by traditional HCI theories and usability techniques have found only limited utility in catering for musicians (Paradiso and O'Modhrain 2003), especially in guiding the design of notation-based interactions (Church, Nash, and Blackwell 2010). In music, these debates are often framed in terms of the personal style of artists and practitioners, or within broad traditions and communities of practice (for example, individual preferences for SuperCollider or Max/MSP). However, this approach to analysis can obscure useful commonalities. In this chapter, we therefore combine research perspectives from HCI with those of digital music production. Our intent is to document the theoretical considerations and issues that emerge when designing and evaluating interfaces for musical expression and creativity. Drawing from other fields, such as psychology and programming practice, we discuss models of the creative process, notation use (Green 1989), skill development (virtuosity) (Nash and Blackwell 2012), flow (Csikszentmihalyi 1996), and the "liveness" (Tanimoto 1990) of musical feedback (Church, Nash, and Blackwell 2010), to highlight limitations in the use of HCI models and theories for music. We propose design heuristics for the support of virtuosity in music systems, to complement those more generally used to provide *usability* (Nielsen 1993), and present a modeling framework for considering these issues within the creative user experience, in the context of real-world music applications.

The concepts, themes, and theories behind the models and recommendations presented in this chapter are the product of a large-scale, two-year study of over one thousand sequencer and tracker users, using a variety of HCI techniques, including interaction logging, video studies, and user surveys. Our findings, which are presented elsewhere (Nash and Blackwell 2011, 2012), complement the theoretical work presented here. Wider applications of the model and details of flow and liveness in programming activities, which may be relevant to live coding practices, have also been published (Church, Nash, and Blackwell 2010).

23.1 The Creative Process

Most theories of creativity attribute the creation of novel ideas to the unconscious mind, where an individual's experiences and stimuli are aggregated into new forms, ultimately surfacing into conscious awareness (Sternberg 1999). Wallas's *stage theory* (1926), based on the earlier reflections of Helmholtz and Poincaré, forms the basis of many recent descriptions of the creative process, describing distinct stages in this process (Csikszentmihalyi 1996; Sternberg 1999) (Table 23.1).

Table 23.1 Overview of the creative process

preparation	conscious, active work to thoroughly familiarize oneself with the problem or task
incubation	unconscious processing of the problem, often over time, away from the task
intimation	where the individual becomes aware that a solution is close at hand
illumination	the moment when a solution emerges into conscious thought
evaluation	a period of critical, conscious work, to verify the suitability of the solution
elaboration	a final period where refinements are made to an otherwise verified solution

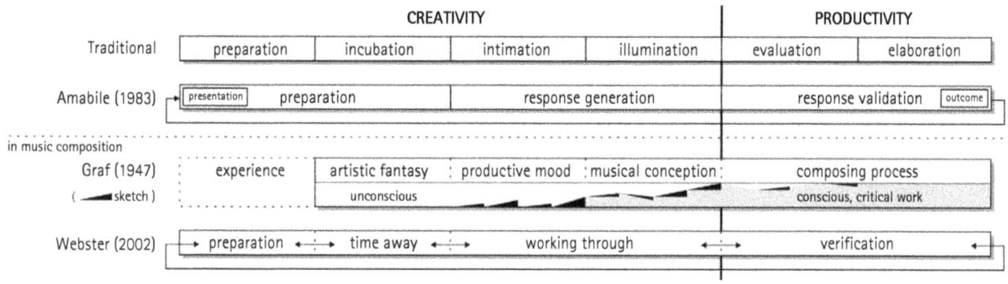

FIGURE 23.1 Stage-based theories of the creative process (Csikszentmihalyi 1996, Wallas 1926), and two descriptions of the music composition process (Graf 1947; Webster 2002), in the context of broader "creativity" and "productivity" phases of "innovation," as characterized by Amabile (1983). See references for detailed descriptions.

Stage theory's linearity and apparent focus on goal-oriented, creative *problem solving*, rather than the more exploratory examples of creative self-expression found in art and music (Sternberg 1999), have encouraged recent theorists to consider more iterative, recursive, parallelized, and less directed forms of the model, as shown in Figure 23.1. In this way, artistic expression, such as music composition, is often characterized as an ill-defined creative problem, where the creativity rests as much in finding problems, as solving them (Amabile 1983).

Amabile's *componential theory of creativity* (1983) expanded stage-based accounts to reflect the ongoing iterative process within creativity, as well as the crucial roles of expertise and intrinsic motivation, which enable an individual to progress and persevere within a domain. In music, Webster's model (2002) echoes this cyclic process, but also accounts for the tendency to jump between stages, observable in many composers' less formally structured, sometimes erratic, working practices.

Graf's review of composition practices (1947), a rare example of the limited canon of composition research, describes the stages more as moods, and emphasizes the importance of the *musical sketch*, as a tool composers use to probe and elicit musical ideas from their unconscious. Sketches, by virtue of their low-fidelity and exclusively personal use, enable the artist to very quickly experiment with novel ideas, without more formal verification or external oversight, economically trialing a more involved creative process. They allow an individual to explore more ideas, which can be accepted or rejected without significant penalty; facilitating creativity through greater ideation (Sternberg 1999), as illustrated in Figure 23.1.

23.2 Performance-based Music Production

While the score was once the only method of distributing music, the introduction of recording technologies allowed live performances to be captured, thus partly obviating the need for formal notation and literacy. The audio-processing model of music production became even more widespread when computer technology brought the digital studio to the desktop, in the form of the sequencer and *digital audio workstation* (DAW). These programs used visual metaphors (Blackwell 2006), drawing analogies to pianos, mixers, tape recorders, and even dangling wires, to support and preserve the working methods of studio musician, allowing the recording of live performances from acoustic or digital (MIDI) musical instruments (Duignan et al. 2004).

Though these packages offer a multitude of editing and postprocessing tools, the sequencer user interface is principally designed around the manipulation of recorded data, reflecting a division in the creative process—the creativity supported by the live performance of musical instruments, and the productivity supported by subsequent *windows, icons, menus*, and *pointer* (WIMP)-based editing, which is considerably less live (Nash and Blackwell 2011). Consequently, studies have observed a tendency for music software to support only the final, refinement stages of the creative process (Blackwell and Green 2000), and not the generation of new ideas (Smith, Mould, and Daley 2009).

23.3 Feedback and Liveness

In Marc Leman's compelling argument for more engaging *embodied cognition and interaction* in music technology (2008), he cites inherent limitations in any attempt to interact with music *indirectly* through an abstract layer of notation such as a score, piano roll, waveform, or graphic user interface. His perspective implicitly rationalizes the focus on live, real-time performance (and its discrete capture) and the peripheral role of computer editing, in the use of software such as sequencers, DAWs, Max/MSP, and the like to create music. The process of sketching, however, illustrates how notations can be used to support creativity and encourages us to think with greater optimism about the opportunities afforded by notation-mediated music interaction. A central element of Leman's thesis centers on supporting fast *action-reaction cycles* between the individual and music, replacing abstract visual modes of feedback (notation) with more direct real-time modes, such as haptics and sound itself.

In other work (Church, Nash, and Blackwell 2010; Nash and Blackwell 2011, 2012), we explored the role of feedback and interaction rates, looking at the specific interaction issues resulting from the use of *direct manipulation* and *WIMP* interfaces (e.g., sequencers, DAWs), which focus on continuous visual representations of musical parameters in real-time, in comparison to *programming-like* notation-based interfaces, like soundtracking (MacDonald 2007), which revolve around the very fast keyboard editing of scripts for future events, similar to live coding (Blackwell and Collins 2005). Borrowing from programming, we adapted Tanimoto's concept of "liveness" (1990), which describes the level of availability of feedback about the end product (the program or piece of music) from within the development environment (a code editor, sequencer, or tracker).

We found that although the sequencer architecture supported the highest level of liveness, through live performance capture (Level 4, *stream-driven*: continuous, real-time manipulation of the domain, e.g., sound), subsequent visual and mouse-based editing activities were significantly less live (Level 2, *executable*: interaction with a visual specification of the domain). By comparison, the rapid interaction rate and broad availability and prominence of musical feedback during editing in the tracker provided greater overall liveness in the user experience (Level 3, *edit-triggered*: feedback from the domain is available after any user input).

The speed with which the tracker user interacts is aided by the ergonomics and motor-learning supported by the computer keyboard, leading some to describe "the art of tracking" as "some sort of musical touch-typing" (MacDonald 2007). A tight edit–audition feedback cycle is possible because the keyboard is used not only for note entry, but also music editing, program navigation, and playback control. At the same time, the focus provided by the editing cursor provides an implicit playback marker, from which edits can be quickly auditioned, without having to consciously move a song pointer. The motor and keyboard skills learned by the user mean that, with practice, many interactions become *ready-to-hand*, and can be executed without reflecting on the physical action. In this sense, at least part of the interaction becomes *embodied*.

23.4 Virtuosity in Computer Music Interaction

Much of the speed advantage demonstrated in the tracker user experience is enabled by the development of expertise; motor skills and program knowledge learned and practiced over an extended period of time. Supporting expert use in a program can introduce learning curves that conflict with the goals of natural and intuitive usage by novices that dominate mainstream approaches to design for usability (e.g., Nielsen 1993).

Usability approaches are prominent in the sequencer and DAW, and their use of visual metaphor, which allows the user to apply knowledge learned elsewhere, thus minimizing the need for further learning (Duignan et al. 2004). However, controlling virtual representations of physical devices allows only a limited transfer of the associated procedural knowledge learned with the original device: motor skills, built on the learning of spatial schemata and haptic feedback, cannot be transferred, nor easily redeveloped using the mouse (Smyth et al. 1994). Moreover, dynamic layouts and windowing can impede learning of the interface, requiring a visual search before most interactions to locate the window, icon, menu, or pointer.

Many principles of usability design are outlined by Nielsen (1993), in his set of *usability heuristics*, used in the design and evaluation of user interfaces. While advocating minimizing a user's memory load ("recognition rather than recall"), he also suggests "shortcuts" for experienced users ("unseen by the novice user"). Similar design principles, which treat the computer as a fundamentally visual medium, are evident in most modern consumer software, including audio software like sequencers and DAWs, in contrast to those for hardware audio interfaces, which focus on skilled interaction, motor learning, and nonvisual feedback modes, such as haptics and sound (Paradiso and O'Modhrain 2003). Consequently, in the next section, we propose design heuristics for computer music interfaces, which specifically account for virtuosity and nonvisual feedback, and which are designed to aid the development of user experiences supporting the creative process, drawing on concepts of feedback, liveness, and direct involvement.

23.5 Design Heuristics for Virtuosity

Following the principles presented above, we suggest design heuristics for interfaces to support virtuosity. Designing *multilayered* interfaces that suit both novice and expert users presents design challenges (Shneiderman et al. 2005). A distinction is made in the targeting of expert users; a virtuosity-enabled system enables a novice user to become expert. It does not rely on domain expertise learned elsewhere (e.g., music literacy), but should consider the transferability of skills learned.

Some of these heuristics draw upon and develop the recommendations of a recent workshop report on creativity support tools (Resnick et al. 2005). Various aspects of computer-based notations are also discussed in the context of the *cognitive dimensions of notations* (CD) framework (Green 1989), which has been previously used to highlight interaction issues in music software (Blackwell and Green 2000).

> Heuristic$_1$ (H_1): Support learning, memorization, and prediction (or "recall rather than recognition")

Expert methods can be enabled by the use of memory (Smyth et al. 1994). Although some interface widgets allow both novice and expert interaction (e.g., the use of mnemonics, in menu accelerators), provisions for usability (e.g., "recognition rather than recall"; Nielsen 1993) can hamper experts (Gentner and Nielsen 1996) and their impact should be considered carefully in systems designed for virtuosity. Using memory, interaction is no longer mediated through visual metaphors fixed by the interface designer, but by schema derived from physical interaction and personal experience.

Notations should not aim solely to be "intuitive," rely heavily on domain-specific knowledge, or otherwise devalue the learning experience. Instead, they should provide a rewarding challenge that scales with user experience (Csikszentmihalyi 1996). Shneiderman and others (2005) describe a similar requirement that creative support systems should have "low threshold, high ceiling, wide walls," respectively offering: a minimal initial learning barrier to support novice use (see H_3); a maximal scope for advanced and more complex edits to facilitate the greater ambitions of experts; and the opportunity for users to define their own paths and working processes, without being constrained to established systems or practices.

Unfortunately, HCI methodologies provide limited account of "learnability" (Elliot, Jones, and Barker 2002), either assuming prior user expertise or explicitly obviating the learning requirement. Although the CD framework (Green 1989) reserves judgment as to the desirability of various aspects (dimensions) of a notation, the presence of *hard mental operations* is invariably viewed as a negative quantity, in HCI. In the context of virtuosity, perhaps we have found a context in which such mental challenges are actually beneficial.

H_2: Support rapid feedback cycles and responsiveness

To master a system, its behavior must be "transparent" (Holtzblatt, Jones, and Good 1988; Kitzmann 2003), allowing the user to easily equate cause with effect in their interactions. Reducing the delay between action and reaction is an effective way to achieve this (Leman 2008).

In computer interaction, basic control feedback should be provided within approximately 100 ms (Nielsen 1993) to appear instantaneous. Complicated operations should complete within roughly 1s (~300 ms to 3 s), or otherwise risk interrupting the flow of thought. After 10 s of idleness, users actively become restless, and will look to fill the time with other tasks. As such, longer delays, especially those requiring wait cursors or progress meters, should be avoided; and are "only acceptable during natural breaks in the user's work."

To support live performance and recording, there are even stricter criteria for a music system, which must respond within a few milliseconds (Walker 1999). Dedicated low-latency sound drivers (e.g., ASIO, WDM) have been developed to provide such latencies, typically confining delays to under 25 ms, and potentially as low as 2 ms. Even below this threshold, musicians and professional recording engineers are sensitive to jitter (the moment-to-moment fluctuations of clock pulses, measured in nanoseconds), but the impact is perceived in terms of sound quality (the introduction of noise and enharmonic distortions, and deterioration of the stereo image), rather than system responsiveness. While less "live" interactions such as playback control and general UI responses tolerate higher latencies, longer delays nonetheless affect the perceived directness of the user experience. Table 23.2 summarizes these requirements for interaction in a musical system. A relationship between timing and control emerges; the finer the required control, the tighter the demands on responsiveness.

As much as the timing, the quality of feedback also affects perceived "liveness" of a system (Church, Nash, and Blackwell 2010; Nash and Blackwell 2012). Liveness, in the context of notation use,[1] is a quality of the design experience that indicates how easy it is for users to get an impression of the end product during intermediate stages of design. UI designers should apply the timing constraints in Table 23.1 to both visual and musical feedback, delivering them in synchrony, where possible. At the same time, increased liveness can reduce the opportunity for useful abstraction and increase the skill required

Table 23.2 Timing of feedback in a music system, listing the changing perceptions of delays at different timescales, and consequences for interaction if they are exceeded (Nielsen 1993; Walker 1999).

Timing	Perception	Consequence if violated
< 1 ms	sound quality ("tightness," "jitter")	user hears noise artifacts, inharmonic distortions, muddied stereo image
< 25 ms	real-time audio ("low latency")	user has difficulty keeping musical time, maintaining sync during performance
< 100 ms	"instantaneous" UI response	system feels slow and unwieldy, harming user's performance and sense of control
< 1 s	noticeable delay	user has difficulty planning ahead or maintaining "flow of thought" and continuity of action
< 10 s	tolerable delay	user loses focus, attention wanders to other tasks

to interact with a program. For example, music programs usefully abstract time, allowing the editing of events at arbitrary points in the music, whereas real-time music input requires performance skills.

In programming, promoting liveness is an example of the push to accelerate the feedback cycle in software design, complimenting the philosophies of similar moves toward *rapid application development*. In rapid application development, the early availability of testable prototypes allows more flexible targets, and supports experimentation and ideation (Resnick et al. 2005), both of which facilitate creativity (Sternberg 1999).

Authoring software, including music production software, has seen the provision of increasingly complex functionality, which can not only be difficult for the user to understand, but can also take time to execute, reducing the perceived responsiveness of the system (Resnick et al. 2005). An interaction designer, in automating trivial yet laborious tasks, must also take note of the period a user must then spend idle (see Table 23.1). In both cases, the goal is to keep the user active and engaged, and avoid interrupting the flow of action.

H_3: Minimize musical (domain) abstractions and metaphors

In HCI, UI designers try to reify the user's "mental model" to represent and operationalize a task domain (Norman 1993), using formal abstractions and metaphor (Duignan et al. 2004; Blackwell 2006) for processes, properties, states, relationships. The formalisms of any notation determine the expressive flexibility it allows, shaping a user's perspective of the domain, or even a culture's at large (Sloboda 1985). It is difficult for an interface designer to match the user's internal representation of musical expression without inadvertently shaping or even limiting the creative output (Blackwell and Collins 2005; Kitzmann 2003). This is particularly the case when designing a unified interface for an artistic audience, who define themselves by their uniqueness and innovation.

Although classically trained or musically literate users will share many perceptions of musical structure, there are few widely accepted formalisms encapsulating the full gamut of computer music capabilities, and nondigital abstractions (such as wires, pots, pedals, mixers, or other metaphors from electronic and acoustic music; see Duignan et al. 2004) can be confusing, confining, and cumbersome. In such cases, "the *closeness of mapping* to conventional audio processing equipment... is indicative of a corresponding reduction in potential for creative exploration" (Blackwell and Collins 2005).

Resnick and his colleagues (2005) advocate avoiding the use of higher-level abstractions, in favor of low-level primitives that can be layered and combined to produce equivalent or greater functionality by the user. At the same time, the simpler functionality of each primitive makes it easier to understand and learn. As more primitives are layered and combined, the challenge increases, providing a scalable learning experience, toward the development of broader mastery. Turkle and Papert (1990) call this approach "soft-mastery," observing that it encourages "closeness to the object," and that such bottom-up perspectives are common in fine artists and musicians.

A visual UI presents explicit abstractions of the music, but the interface designer should also not overlook the implicit abstractions a user will make by simply listening to the sound. In this capacity, simply the broad availability of such musical feedback may be the best scaffolding enabling users to form their own abstractions. In this sense, such "audibility" may be seen as a correlate of the *visibility* dimension in the CD framework, which concerns the degree to which systems "bury information in encapsulations" (Blackwell and Green 2003).

H_4: Support consistent output and focused, modeless input

An interface that remains consistent, from moment to moment, can be more easily remembered and predicted. Fixed, static layouts enable the development of not only spatial schemata, but also motor learning (Smyth et al. 1994), both of which allow a degree of interaction to be handled unconsciously (see H_1). Changeable, dynamic screen layouts, such as floating windows, require conscious reflection, interrupting thought processes and hampering the performance of experienced users. Users should not have to visually search through different windows, modes, or other views, to locate information or effect minor edits, but "should be able to perform any task at any time" (Gentner and Nielsen 1996).

Most programs present a primary notation, which is kept in view for the majority of the time, and which constitutes the focus of activity, for example the source code in an integrated development environment (IDE), the document in a word processor, or the waveform in a sound editor. Such a focal point allows a user to maintain concentration.

By contrast, a decentralized interface split across multiple subdevices hides functionality, complicating learning and use of the program.

Hardware controllers (e.g., MIDI devices, control surfaces, digital mixers) can make visual metaphors tangible and enable peripheral interaction (Paradiso and O'Modhrain 2003), often presenting fixed, physical layouts that can aid motor learning. These devices support more direct involvement with music in a live context, but potentially draw attention away from the software environment, and may have limited use in non-real-time edits.

H_5: Support informal interaction and secondary notation.

Blackwell, Church, and Green (2008) noted that formalism is an unavoidable consequence of encoding processes in the digital domain, but also that UI conventions have, at times, restricted the use of design software. Even when using digital tools, users often support more complex interaction with pen and paper, to make notes, reminders, calculations, or sketches of representations not easily or quickly executed in the UI (Sellen and Harper 2001). Earlier, we discussed the benefit this medium brought to music, in the composer's use of the musical sketch. However, there is no reason why similar informal, low-fidelity representations can't be supported digitally (Blackwell et al. 2008).

In the CD framework, this provision is known as a secondary notation and specifically describes those aspects of a programming environment that allow a user to record information without using the formalisms in the main notation. In a score editor, for example, this might involve allowing the user to annotate any part of the score with text or drawings. Because such notes needn't be interpreted by the program in any way (e.g., as music), users are not limited in what they can write or draw, allowing them not only to quickly scribble down ideas while fresh in their mind, but also record information and ideas that the program's designers never thought to support explicitly.

23.6 Systems of Musical Flow

Similar to stage theory (see section above), many HCI methodologies (e.g., GOMS/KLM, Cognitive Walkthrough) address scenarios characterized by a well-defined problem or task, entailing established interactive procedures and known solutions (e.g., "correct actions") (see Sharp, Rogers, and Preece 2007). Artistic creativity, by comparison, can be likened to a "wicked problem": ongoing, ill-defined, with no "correct" solution or definitive end condition (Rittel and Webber 1973). Consequently, the creative user experience can be difficult to model using standard HCI methods. In this section, we propose a modeling approach to the creative user experience, which combines our previously discussed theories of feedback, liveness, and virtuosity, with Csikszentmihalyi's theory of *flow*. This theory describes a mental state, where a delicate *balance of challenge and ability* leads to a *feeling of control* and a *loss of self-consciousness*, engendering a working environment that can benefit creativity (Csikszentmihalyi 1996). Over time, ability increases, requiring greater challenges to maintain flow, ultimately leading to the development of mastery in a domain. In this context, the flow concept describes an intrinsically rewarding path to building ability, through enjoyable and fulfilling challenges matched to the individual, engendering a scalable learning experience.

Earlier sections identified the central role of feedback in creative user experience, as recognized in flow theory's call for *direct and immediate feedback* (Csikszentmihalyi 1996). Timely, unambiguous modes of feedback, in interfaces that favor fast audio feedback rather than abstract visual representations, also help a user to identify *clear goals* during interaction, and the actions taken to meet them, enabling them to learn the system through tinkering (Scripp, Meyaard, and Davidson 1988). Further components of flow have also been touched on, such as *concentration and focus* and the *feeling of control* (*autonomy*). More generally, feedback enables users to track their progress, allowing them to reflect on their musical creativity, and providing either a source of satisfaction or prompting further editing. In this way, interaction with the music system can be *intrinsically rewarding*, if progress is maintained and distractions or interruptions are avoided.

This section presents a model of interaction within interactive systems designed for authoring or creative design, such as music. Rather than explicitly identifying goal-oriented, stepwise, or linear processes, the model seeks to characterize the moment-to-moment feedback, flow, and liveness properties of the system to highlight issues in the user experience that might be detrimental to the performance and motivation of users, and thus also their creativity.

23.6.1 Overview

We model the creative user experience as a combination of closed feedback loops (Figure 23.2), between the *user, domain*, and a system's *notation* (i.e., interface). Musical performance is modeled as *direct involvement* with the domain, where the user's expertise with the performance device (e.g., instrument) supports *embodied interaction* (Leman 2008). At the same time, a user can interact indirectly with the domain, through the notation. In this latter scenario, the speed of notation manipulation and the speed of audio realization and musical feedback determine the liveness of the experience. A system that enables fast manipulation of the notation and fast audio feedback, synchronous with visual feedback, involves the user more directly in the domain (see Section 23.3), without sacrificing the utility of a visual notation.

Systems are modeled by evaluating the emphasis of each arc or loop in a given music interaction or interactive system. In addition to the three basic loops indicated in Figure 23.2, larger cycles emerge within the combinations of arcs and loops, notably defining broad user-interaction paradigms that span the user, notation, *and* domain: a *recording* paradigm (*performance* → *transcription* → *visualization*) and a *composition* paradigm (*manipulation* → *realization* → *audition*). Combinations of

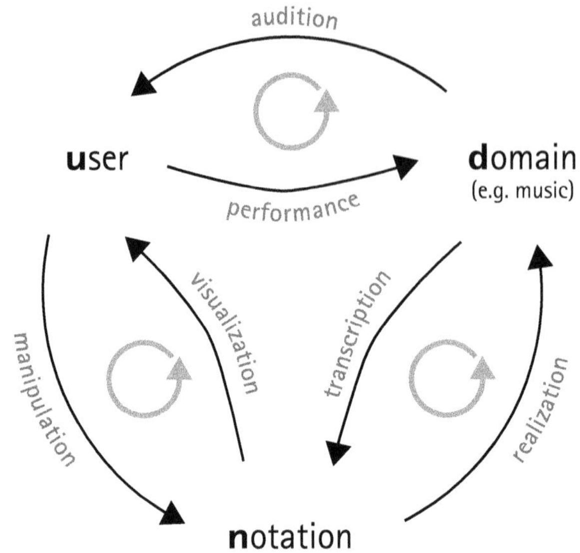

FIGURE 23.2 The systems of musical flow model of interactive music, characterizing the user experience as a network of feedback loops based on interactions with and between the task domain (e.g. music) and notation.

these cycles with the smaller loops can be used to define basic system archetypes, which can be used to characterize most real-world music systems, such as sequencers, score editors, and trackers. Figure 23.3 shows the synthesis of two popular archetypes, but space constrains us from exhibiting the full (super)set in this chapter, which extends the model to encapsulate applications such as augmented instruments, algorithmic composition, Max/MSP, or even *Guitar Hero*-like programs (see Nash 2012) and thus may form the basis for a taxonomy of computer music interaction.

23.6.2 Application

The capacity for flow is indicated by the level of liveness of each interaction loop, as annotated in Figure 23.5, but is also influenced by the basic configuration of the program. Real-world systems may be many times more complicated than their underlying archetype, involving more than one notation, or even domain. Features of the modeled system can highlight interaction issues that may impact the associated user experience, as illustrated and briefly described in Figure 23.4.

For example, a digital audio workstation (DAW) (Figure 23.5a) provides many features and interfaces beyond the basic sequencer model. The system is split between

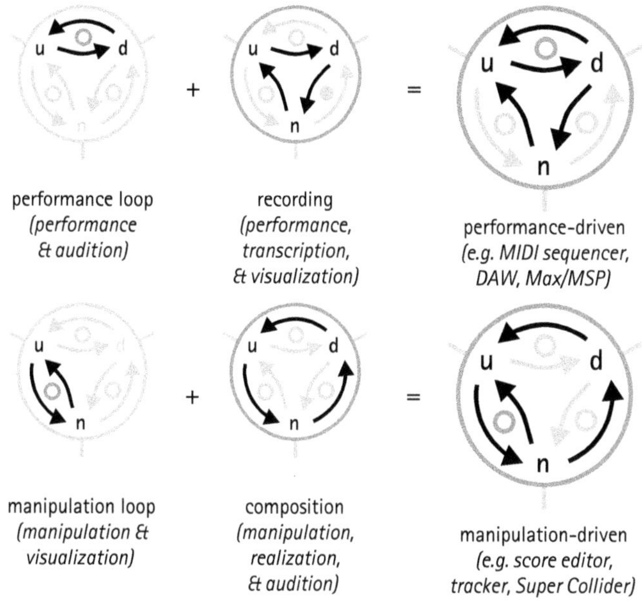

FIGURE 23.3 Example system archetypes, characterized as the synthesis of feedback loops coupling notation and domain, with real-world examples. See also Nash (2012) for further examples and discussion.

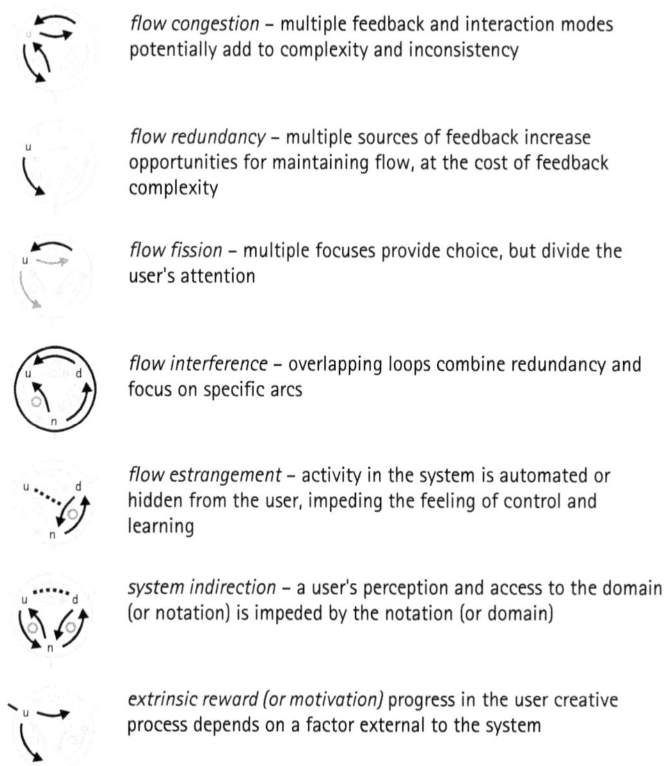

FIGURE 23.4 Interaction issues highlighted by system features.

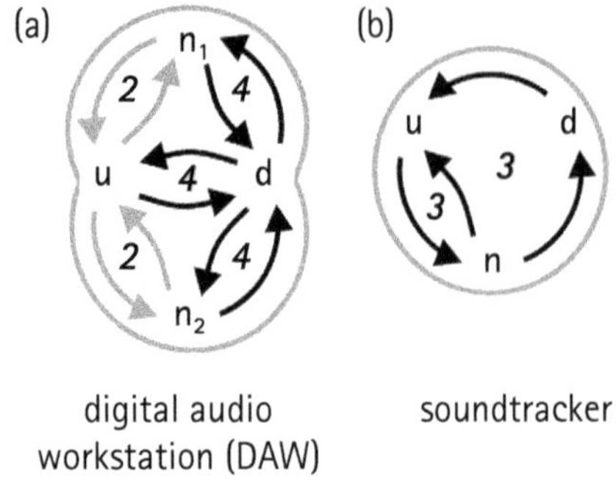

FIGURE 23.5 Real-world system examples, annotated with respective liveness levels for constituent feedback loops.

multiple visual notations ("devices," in the program), arranged around a central performance loop. As established in our earlier discussions, Level 4 liveness is offered during performance, but interaction with the various notations is more cumbersome. No interaction scenario that includes the user, a notation, and the domain exhibits liveness above Level 2. Moreover, the congestion of arcs intersecting the user, caused by multiple partial-notations (or "subdevices"), highlights issues with divided attention and focus switching (Flow Congestion and Flow Fission in Figure 23.4), also affecting the consistency and complexity of the user interface or interaction.

By comparison, the simpler *soundtracker* (Figure 23.5b) model supports rapid edit–audition cycles through a single, more-encompassing notation. Naturally, this unified representation of the domain is likely more complicated than any one of the DAW's subdevices, and architects of a system must balance usability and virtuosity considerations; for example, using standard usability techniques (e.g., Nielsen 1993) and the heuristics proposed in Section 23.5. Naturally, there is nothing to prevent designers integrating a faster, more powerful, expert-oriented notation as an *extra* subdevice in the DAW, with minimal impact on the current usability provided by existing notations, which could be used by novices and ignored by experts.

23.7 Conclusions

Musical creativity relies on exploration, expertise, and intrinsic motivation, contrasting goal-based usability approaches in mainstream HCI practice and analysis. In music, creative individuals not only develop virtuosity with instruments, but also notations, and while new performance devices provide new modes of real-time musical expression, relatively little research has looked at the composer's use of notation, and how it can be supported by the computer, as a tool for abstracting, sketching, and exploring creative ideas. Moreover, the linear recording process predominant in existing performance-based music software, such as sequencers and DAWs, can be seen to serve only an ancillary role in creativity, focusing on the transcription or refinement of a musical idea, and not the earlier, more exploratory stages of creativity (see Blackwell and Green, 2000; Duignan 2004; Smith, Mould, and Daley 2009).

Integrating psychologies of expertise and intrinsic motivation, this chapter introduces flow theory (Csikszentmihalyi 1996) as a framework for identifying and analyzing the properties of a notation-based music system that supports learning and creativity, integrating it with HCI theories of liveness (Tanimoto 1990) and the cognitive dimensions of notations (Green 1996), working toward a theoretical foundation for HCI in digital music.

Supporting flow in notation use entails a design shift from usability to virtuosity; advocating user interfaces for composition that facilitate a rapid edit–audition cycle and the development of skill (motor and memory) with input devices. By improving liveness and virtuosity (e.g., motor learning), programs can support embodied interaction through notation and levels of immersion and flow in music that are comparable to those found with live performance devices, as evident in studies of interaction with tracker notation (Nash and Blackwell 2011, 2012). Expert tracker users, for example, trade off editing complexity to maintain faster feedback cycles that preserve the liveness of interaction, favoring shorter editing episodes and more iterative editing styles, with more frequent playback of material (Nash and Blackwell 2012), facilitated by fluency with the QWERTY keyboard (Nash and Blackwell, 2011).

Another critical component of flow is the balance of challenge and ability. Through moving away from real-time music, users also gain control of the pace of interaction, enabling them to self-regulate the challenge involved, and preserving a sense of autonomy. For novices, this relaxes the virtuosity required to engage in the musical domain, lowering the threshold for creativity (Scripp, Meyaard, and Davidson 1988). At the same time, it gives experts the time to consider and experiment with more complex, advanced, and original musical solutions (compared with what is solvable in real-time performance or improvisation, see Sloboda 1985), ultimately raising the ceiling of creativity.

The goal of this chapter was to provide the designers of interactive audio systems with a new perspective on the musical interaction engendered by their systems, emphasizing aspects of the user experience that are not easily encapsulated using traditional HCI methods, nor widely supported in UIs for authoring music. Our hope is that the concepts, recommendations, and theoretical models presented here—including virtuosity, liveness, and flow—will encourage wider discussion and development of the HCI issues and theory facing interactive audio practitioners.

Acknowledgments

Our research was funded by the Harold Hyam Wingate Foundation. In the course of this research, Nash spent several months as a research intern with Steinberg and is grateful for input, support, and assistance from colleagues at Yamaha, Steinberg, and Cakewalk.

Note

1. Distinct from other uses of the term "liveness" in music, used to describe the immersive quality of a live performance viewed from the social perspective (e.g., that of the audience; Auslander 1999). See Nash and Blackwell (2012).

References

Amabile, Teresa M. 1983. *The Social Psychology of Creativity*. New York: Springer-Verlag.

Auslander, Philip. 1999. *Liveness: Performance in a Mediatized Culture*. Abingdon, UK: Routledge.

Blackwell, Alan F. 2006. The Reification of Metaphor as a Design Tool. *ACM Transactions on Computer-Human Interaction (TOCHI)* 13 (4): 490–530.

Blackwell, Alan. F., Luke Church, and Thomas. R. G. Green. 2008. The Abstract is "an Enemy": Alternative Perspectives to Computational Thinking. *Proceedings of PPIG 2008* 20: 34–43.

Blackwell, Alan. F., Luke Church, Beryl Plimmer, and Dave Gray. 2008. Formality in Sketches and Visual Representation: Some Informal Reflections. In *Sketch Tools for Diagramming*, ed. Beryl Plimmer and Tracy Hammond, 11–18. Workshop at VL/HCC 2008. Washington, DC: IEEE Computer Society.

Blackwell, Alan F., and Nick Collins. 2005. The Programming Language as Musical Instrument. *Proceedings of PPIG 2005* 17: 120–130.

Blackwell, Alan F., and Thomas R. G. Green. 2000. A Cognitive Dimensions Questionnaire Optimised for Users. *Proceedings of PPIG 2000* 12: 137–152.

——. 2003. Notational Systems: The Cognitive Dimensions of Notations Framework. In *HCI Models, Theories and Frameworks: Toward a Multidisciplinary Science*, ed. J. M. Carroll, 103–134. San Francisco: Morgan Kaufmann.

Church, Luke, Christopher Nash, and Alan F. Blackwell. 2010. Liveness in Notation Use: From Music to Programming. *Proceedings of PPIG 2010*, 22: 2–11.

Collins, David. 2005. A Synthesis Process Model of Creative Thinking in Music Composition. *Psychology of Music*, 33: 193–216.

Csikszentmihalyi, Mihaly. 1996. *Creativity: Flow and the Psychology of Discovery and Invention*. New York: HarperCollins.

Duignan, Matthew, James Noble, Pippin Barr, and Robert Biddle. 2004. Metaphors for Electronic Music Production in Reason and Live. In Computer Human Interaction, *Lecture Notes in Computer Science (LNCS)*, ed. M. Masoodian, S. Jones, and B. Rogers, 3101: 111–120. Berlin: Springer.

Elliot, G. J., E. Jones, and P. Barker. 2002. A Grounded Theory Approach to Modelling Learnability of Hypermedia Authoring Tools. *Interacting with Computers* 14: 547–574.

Gentner, Don, and Jakob Nielsen. 1996. The Anti-Mac Interface. *Communications of the ACM* 39 (8): 70–82.

Graf, Max. 1947. *From Beethoven to Shostakovich*. New York: Philosophical Library.

Green, T. R. G. 1989. Cognitive Dimensions of Notations. In *People and Computers V*, ed. Alistair Sutcliffe and Linda Macaulay, 443–460. Cambridge, UK: Cambridge University Press.

Holtzblatt, Karen A., Sandy Jones, and Michael Good. 1988. Articulating the Experience of Transparency: An Example of Field Research Techniques. *SIGCHI Bulletin* 20 (2): 46–48.

Kitzmann, Andreas. 2003. Transparency, Standardization and Servitude: The Paradoxes of Friendly Software, In *Perspectives on Multimedia: Communication, Media and Information Technology*, ed. Robert Burnett, Anna Brunstrom, and Anders G. Nilsson, 41–54. Chichester, UK: John Wiley.

Leman, Marc. 2008. *Embodied Music Cognition and Mediation Technology*, Cambridge, MA: MIT Press.

MacDonald, R. 2007. Trackers! *Computer Music* 113: 27–35.

Nash, Christopher. 2012. Supporting Virtuosity and Flow in Computer Music. PhD diss., University of Cambridge. http://research.nashnet.co.uk.

Nash, Christopher, and Alan Blackwell. 2011. Tracking Virtuosity and Flow in Computer Music. *Proceedings of ICMC 2011*, 575–582. Huddersfield, UK 31 July–5 August.

——. 2012. Liveness and Flow in Notation Use. *Proceedings of NIME 2012*, 28–33. http://www.eecs.umich.edu/nime2012/Proceedings/NIME2012WebProceedings.html.

Nielsen, Jakob. 1993. *Usability Engineering*. Cambridge, MA: AP Professional.

Norman, Donald. 1993. *The Psychology of Everyday Things*. New York: Basic Books.

Paradiso, Joseph A., and Sile O'Modhrain. 2003. Current Trends in Electronic Music Interfaces. *Journal of New Music Research* 32 (4): 345–349.

Resnick, Mitchel, Brad Myers, Kumiyo Nakakoji, Ben Shneiderman, Randy Pausch, Ted Selker, and Mike Eisenberg. 2005. Design Principles for Tools to Support Creative Thinking. In *NSF Workshop Report on Creativity Support Tools*, ed. Ben Shneiderman, Gerhard Fischer, Mary Czerwinski, Brad Myers, and Mitch Resnick http://www.cs.umd.edu/hcil/CST/.

Rittel, Horst W. J., and Melvin Webber. 1973. Dilemmas in a General Theory of Planning. *Policy Science* 4: 155–169.

Scripp, Larry, Joan Meyaard, and Lyle Davidson. 1988. Discerning Musical Development: Using Computers to Discover What We Know. *Journal of Aesthetic Education* 22 (1): 75–88.

Sellen, Abigail J., and Richard H. R. Harper. 2001. *The Myth of the Paperless Office*. Cambridge, MA: MIT Press.

Sharp, Helen, Yvonne Rogers, and Jenny Preece. 2007. *Interaction Design: Beyond Human-Computer Interaction*. Chichester, UK: Wiley.

Shneiderman, Ben, Gerhard Fischer, Mary Czerwinski, Brad Myers, and Mitch Resnick. 2005. *NSF Workshop Report on Creativity Support Tools*. http://www.cs.umd.edu/hcil/CST/.

Sloboda, John A. 1985. *The Musical Mind*. Oxford: Oxford University Press.

Smith, Jeff, David Mould, and Mark Daley. 2009. Constructures: Supporting Human Ingenuity in Software. *Digital Creativity* 20 (1–2): 79–94.

Smyth, Mary M., Alan F. Collins, Peter E. Morris, and Philip Levy. 1994. *Cognition in Action 2nd edn*. Hove, UK: Lawrence Erlbaum.

Sternberg, Robert, ed. 1999. *Handbook of Creativity*. Cambridge, UK: Cambridge University Press.

Tanimoto, Steven L. 1990. VIVA: A Visual Language for Image Processing. *Journal of Visual Languages and Computing* 1: 127–139.

Turkle, Shelley, and Seymour Papert. 1990. Epistemological Pluralism. *Signs* 19 (1): 128–157.

Walker, Martin. 1999. Mind the Gap: Dealing with Computer Audio Latency. *Sound On Sound*, April. Cambridge, UK: SOS. http://www.soundonsound.com/sos/apr99/articles/letency.htm.

Wallas, Graham. 1926. *The Art of Thought*. New York: Harcourt-Brace.

Webster, Peter. 2002. Creative Thinking in Music: Advancing a Model. In *Creativity and Music Education*, ed. Timothy Sullivan and Lee Willingham, 16–34. Toronto: Britannia.

CHAPTER 24

BLURRING BOUNDARIES

Trends and Implications in Audio Production Software Developments

DAVID BESSELL

THE development of music software is a continuously evolving area that is subject to a wide array of developmental pressures. Often these pressures have competing priorities, which can lead to a variety of desirable, and not so desirable, features in the user experience of that software. Examination of the impact and implications of these software-development priorities is interesting in itself but can also, by association, throw some light on the wider technological, commercial and theoretical forces at work in music of the early twenty-first century. In the words of Miller Puckette, "Computer music software can also encode ideas that reach beyond any individual implementation of them" (2002, 39).

Music software development has some overlapping similarities in common with other media software such as that used in graphics, animation and film, and therefore some of the observations noted here will have relevance in those wider fields. Why does any of this matter? As Brown and others have noted, "the computer... like any other medium, effects the information that is stored in it or passes through it. The medium is not neutral; it has an effect on the music. When we are aware of this transforming nature of a medium, we can either compensate or use it. Only when we ignore it, or deny it, we risk the transformational change taking us by surprise or undermining our true intention" (Brown 2007, 9). Although Brown couches this idea in terms of the computer, it is equally applicable to the software that mediates our interaction with the hardware: "The design of the software cannot help but affect what computer music will sound like" (Puckette 2002, 31) (see also Chapter 23 in this volume).

24.1 Software Categorization

At first sight, it may seem that software for musicians could be placed into broad categories. The type of categories that might be envisaged include the following:

- Digital audio workstations
- Audio editors
- Music programming languages
- Score writers
- Algorithmic composition aids
- Software instruments

These are all manifestations to a greater or lesser extent of the way in which musicians themselves conceptualize their relation to music through software.

However, an attempt at this kind of neat categorization immediately throws up issues of the type that form the focus of this chapter (for example, see the comments on feature creep discussed below). An alternative more analytical approach might take a step back from the software itself and make distinctions related to the medium in which the software exists. Brown has suggested such a model based on categories such as "the computer as a musical tool, the computer as musical medium and the computer as musical interface" (Brown 2007, 6). Although neither of these two approaches fully captures the complexity and fluidity of the relationship between musician and software (compare the more philosophical approach based on notions of play, e.g., Dixon 2006), nevertheless these types of concept and category will be used in this discussion for pragmatic reasons. The focus here will be primarily on audio production software of the type usually referred to as a digital audio workstation (DAW), as it exemplifies many of the key trends in music software development. However, mention of other categories of software will be made periodically to illustrate aspects of the subject.

24.2 The Development of the Digital Audio Workstation

An examination of aspects of the development of the DAW raises some interesting factors. Firstly, even the concept of the DAW as a multipurpose workbench for audio production was not clear in the original form that the software took. At the present time (2013) leading DAWs might include ProTools, Cubase, and Logic in Europe with the addition of Digital Performer in the United States. All of these pieces of software began life with considerably more limited and focused functionality than the current versions, and the ongoing trend of cumulative creeping additional functionality shows no sign of abating. To give some specific examples, Cubase, Logic, and Digital Performer all originated in the mid-1980s as sequencers (that is programs designed to record, playback, and edit MIDI information only), without any audio recording, notation, or instrumentation functions. This MIDI pattern-based approach is probably clearest in the early versions of Logic, then called Creator.

ProTools has a somewhat different lineage to the other three DAWs. Also appearing in the mid-eighties, ProTools' first version, called Sound Designer, is almost unrecognizable when compared with the current DAW iteration. Initially, Sound Designer was intended as an audio editor for Emu sampling keyboards. It was not until the late eighties that a two-track audio editor–recorder, now called Sound Tools, was introduced, which subsequently went on to add multitrack audio recording in the early nineties, when it also acquired the name ProTools.

24.3 Commercial Pressures

To delve somewhat deeper into the development of this type of software, Cubase will be used as a representative case study. To disentangle the competing pressures on the development of DAW software, the following general categories will be employed: commercial pressures, usability, and technological advances.

The commercial pressures impacting on the developers of DAW software come in various forms. There is the basic competition model, where software developers are inclined to imitate popular software features implemented by rival manufacturers in order to increase their user base: "A larger network of software users usually means compatibility with more users" (Cheng and Tang 2010, 437). Or as Puckette has noted, "the good, perennial ideas will jump from implementation to implementation" (Puckette 2002, 40). An early example of this type of expansion can be found in the adoption by Cubase of some aspects of the pattern-editing approach seen in early versions of Logic. This spread of features from one proto-DAW to another can be seen working in the opposite direction in this case as well. Early versions of Logic soon added aspects of the track-based approach to the display of MIDI information that were found in the Cubase interface. Almost immediately, we have a situation in the development of the DAW where two alternate paradigms for the editing and display of the same type of musical information are supported side by side. This duplication of function, particularly notable in the commercial professional and semiprofessional area of the market, is a key trend in audio production software development that has implications for usability, described below. Lyon has observed that "now that we have software models that are very clear and palpable, it also seems that we are opening them up and trying to grab ideas from other (software) sources" (Lyon 2002, 14).

It is quite obvious when looking at the development of the major DAWs that when one developer introduces a feature that is perceived to be well received by the end user, it is not long before the other developers either copy the feature directly or implement it in a similar manner. This aspect of DAW development can be seen in the introduction of audio recording and editing alongside existing MIDI functionality, or in the case of ProTools, for example, the implementation of the host-based approach to plug-in architecture first introduced by Steinberg in Cubase version 3.2 in 1996. Audio plug-ins are small pieces of self-contained software that can be added by the user to the DAW host. Initially these dealt with audio signal processing only, but Steinberg expanded the concept in 1999 to allow the creation of plug-in virtual instruments known as VSTi (virtual studio instruments) plug-ins. Also in 1999, ProTools introduced its own host-based plug-in format, RTAS (real time audio suite). The concept was obviously borrowed, but the plug-in format was not compatible with Steinberg's VST (virtual studio technology). Logic soon followed and when the company was taken over by Apple in 2002, and then also developed its own incompatible Macintosh-specific plug-in format called Audio Units. Likewise, Digital Performer also introduced native plug-in functionality in version 5. Like Logic and ProTools, the developers of Digital Performer (Mark of the Unicorn) also decided to implement a proprietary noncompatible plug-in format called MAS (Motu audio system). This proliferation of incompatible formats is another key feature of music software development and seems likely to have arisen because of commercial pressures to compete, combined with copyright protections (Cockburn and MacGarvie 2011; Bainbridge 2009) and a desire to prevent a competing software developer acquiring dominance in the potentially lucrative plug-in market.

Further examples of features migrating from one software DAW platform to another include: the introduction of the ability to correct the timing of live audio (audio quantizing) and the related introduction of the ability to time-stretch and compress audio without affecting the pitch; the ability to detect the tempo of an audio clip; the addition of video synchronization and playback functionality from within the DAW; and a long list of other features.

Another key driver for this type of feature creep is the commercial desirability of adding new "improvements" to a piece of software. This drive to continual updating of software has become a key part of the marketing strategies of the DAW manufacturers, all of whom have a rolling system of paid major updates for existing users alongside their free minor updates service. Usually the manufacturers have built into the paid update model a pricing structure that penalizes users who do not update regularly. Although manufacturers do not publish figures, it is reasonable to assume that they may derive a significant proportion of their income from charging existing customers for updates that are marketed on the basis of their added functionality.

A further notable but less obvious commercial driver can be seen in the way that the various DAWs position themselves in relation to the concept of professionalism. ProTools is particularly interesting in this regard as it has traded heavily on trying to promote itself as the "industry standard" for audio recording in professional recording studios. Indeed, the concept of professionalism is even expressed in its title, indicating that this may have been a conscious aspect of its market positioning even before it achieved dominance in commercial recording studios in the late 1990s. The impact of this can be seen in various ways. Firstly, there is some correlation between the perceived professionalism of a DAW and the soberness of the color scheme graphics used in the interface. Secondly, ProTools has for a long time been considerably more expensive than its rival DAWs, partly justifying this price differential on its "professional" status. Interestingly, ProTools' commercial dominance in the music studio and film postproduction markets has been somewhat eroded since the year 2000, and simultaneously the price differential with the other DAWs, while still appreciable, has also been reduced. Thirdly, it is noticeable that in comparison to Cubase, Logic and Digital Performer, ProTools has somewhat resisted the pressure to overcomplicate its interface and still relies primarily on a two-window approach to editing and display of musical information as opposed to the proliferation of multiple window environments common in other DAWs. This can be read as a utilitarian approach that might appeal to professionals more interested in workflow functionality than superfluous features and distracting colorful graphics. The downside of this Spartan approach to the interface and features is that ProTools has been criticized for its slowness in adding functions commonplace on the other DAWs, such as advanced MIDI functionality, plug-in delay compensation and 64-bit support: "ProTools 10 is, to the dismay of many, still a 32-bit application" (Wherry 2012). This relation between perceived professionalism and interface design extends beyond the DAW software area. There is a broad continuum between the relatively colorful and graphically elaborate interfaces of the more consumer-orientated music software such as Propellerhead's Reason and the minimal graphical interface of a text-based programming language such as C++.

Graphically, Cubase has in the past taken a slightly different path regarding professionalism. In the late 1990s Cubase adopted a pseudo-3D graphical appearance that was characterized by a worn and distressed graphical appearance to its virtual transport controls. This is an interesting example of a software trend that has recurred in many aspects of music software but is particularly associated with the synthesizer plug-in software that was first introduced by Steinberg in 1996. The graphic emulation of hardware interfaces in software has several economic factors that may be driving it. There is the association with the professional hardware of the past, and the related issue of perceived value. The plug-in's visual association with a previously expensive piece of hardware intimates that the cheaper software is a good value path to the same functionality. A good example of this might be the Arturia Minimoog plug-in software emulation, which closely resembles the original hardware Minimoog analog synthesizer in its graphic appearance. This imitation of vintage hardware is not just confined to the interface. It is also a notable trend, particularly in plug-in software, that developers have tried to emulate the audio behavior and functionality of previously existing hardware. While this has some obvious commercial implications there are also technological issues which may be influencing this drive toward hardware emulation in software.

A more subtle influence of commercial pressures can be detected in the manner in which software features are presented to the end user. An interesting comparison can be made between the closely related programs Max/MSP and Pure Data. These two applications are rather similar graphical-based approaches to a music and video-specific programming language. They share quite a degree of compatibility in design philosophy, components, and file formats. Max/MSP is primarily authored by David Zicarrelli and Miller Puckette (Puckette 1988, 420–29). While Pure Data (Pd) is authored by Miller Puckette alone (Puckette 1996, 37–41). In short, the main difference between these two programs is that one, Max/MSP, is marketed commercially while the other, Pd, is free.

When comparing the two similar programs, it is apparent that there is a large difference in the number and functionality of the small modular pieces of graphically represented atomized code that these dataflow programming languages use as their main user interface. Max/MSP has over 500 and rising code-based objects, which range from quite fundamental and primary programming functions through to some higher-level objects embodying a collection of related functions. By contrast, the basic distribution of Pd currently has nearer 200 such objects and these are heavily concentrated on the more primary coding functions, with far fewer of the higher-level objects. In terms of functionality it is perfectly possible to do everything with the Pd set of objects that can be done with the larger set of Max/MSP objects, but this use does require a considerably higher level of programming and theoretical knowledge from the end user. As one of the software's authors has noted himself, "The marketplace forces commercial software vendors to load their products with features" (Puckette 2002, 42).

It would seem that the commercial forces driving the marketing of Max/MSP have influenced the need to introduce more high-level prebuilt audio functions into the program in an effort to widen the potential user base beyond those with relatively in-depth audio programming skills. This key distinction between high- and low-level expertise required from the end user can be seen reflected across music software and correlated in different ways with the perceived professionalism of the software, the desired breadth of its user base, and the degree of technical explicitness with which features are presented to the end user. There is not always a neat dividing line between commercial drivers for software trends and technical considerations.

24.4 Technical Constraints

The technical nature of music software and the computing platforms on which it relies have of course been heavily influenced by the rapid developments in computing hardware, which are still ongoing. As previously mentioned, the early DAW software was functionally tightly focused and MIDI-oriented. ProTools, the earliest commercial software dealing with audio in a way that we might recognize as related to the current DAWs, had to use additional add-on hardware processing in order to be able to manipulate audio. So even if the early software developers of the DAWs had intended to implement something related to the extensive multifunctionality typical of their later versions, then technical constraints would have prevented them from doing so.

More likely though, it would appear that increases in available computing power have been followed by the developers with the addition of new features on a somewhat ad-hoc basis. In other words, the software has expanded its functionality to soak up any expansion in computing power that has become available: "we want software to do everything, and our notion of 'everything' grows broader every year" (Puckette 2002, 40). To give a concrete example of this tendency in action, Logic was one of the leaders in introducing the ability to time-stretch audio. Initially this function was achieved using a process related to granular synthesis (a sound-synthesis technique based on the manipulation of small digital samples of sound known as grains) that has the advantage of fairly modest processor (CPU) demand. With the further expansion of computing power it soon became possible to perform frequency domain processing based on the Fast Fourier Transform[1] in something close to real-time. Thus, when Cubase later adopted the time-stretch function, it did so using frequency domain techniques, providing better quality audio results at the expense of higher CPU load. Perhaps inevitably, Logic then eventually went on to adopt the frequency domain approach to audio stretching, as did ProTools and Digital Performer. Thus, another key driver for trends in audio software development can be seen at work. There is also likely to be a reinforcing effect here, where commercial and technical imperatives align and drive the software development in the same direction.

It is interesting to compare the technical development of the DAW software approach with that of some other types of music software. In the case of programs that began as dedicated two-track audio editors such as Sound Forge, Wavelab, Audition, and Bias Peak, it is significant that the function creep noted in the development of the DAW software has also been present. Indeed, the addition of multitrack audio recording, plug-in hosting, and video synchronization facilities has seen a large-scale convergence between so-called audio editors and the DAWs. This development is particularly curious in the case of Steinberg's Wavelab which now shares so much functionality with Cubase that there is a question as to whether they actually are directly competing with each other in terms of technical features.

For other categories of software this blurring of functionality associated with technological advances in computing power is expressed in slightly different ways. For example early version of the program Max dealt only with MIDI information, as it was subject to similar technical constraints to the DAWs. With the increase in computing power, this category of graphic programming language also added native audio functionality in 1997 and acquired the MSP part of its name (Max Signal Processing). Similarly, a video extension for Max/MSP called Jitter was introduced in 2003. This process has parallels with that of the spread of functions in the DAW software already noted.

The expression of this aspect of the response to technical advances in score-writing software follows a path that has recognizable similarities with that of the other programs already examined. The score-writing software Sibelius, for example, also started life with limited playback functionality, but has since added more advanced MIDI features, plug-in hosting, video synchronization, and other features identifiable from the history of DAW development.

The main exceptions to this type of convergence of function across audio software include text-based programming languages, algorithmic composition tools, and instrument plug-ins. Text-based programming languages commonly used for music such as C/C++ have always had a somewhat protean nature and so perhaps the concept of feature creep due to commercial or technical pressures is not a particularly good fit in this case. In addition, many of the other music applications are written in these text-based languages and so the relation between them in terms of features is complex and far from straightforward. At the present time, algorithmic composition tools are still more often found in an academic context than a commercial one. Perhaps for this reason the type of expanding multifunctionality noted in other areas of software is not so apparent. The more commercially oriented examples of this software paradigm, such as Microsoft's Songsmith, are relatively recent developments so perhaps it is too early to expect to see the kinds of development trends that are apparent over larger timescales with more established software. Some earlier attempts at similar commercial algorithmic approaches such as Band in a Box show some signs of a similar developmental path to the DAWs, with the appearance of additional audio functionality and limited plug-in hosting, but the trend is not quite so clearly defined as in other software categories.

Instrument and signal processing plug-ins are an interesting exception to this general trend of increasing multipurpose functionality. Despite the technical advances in computing power, these plug-ins have largely resisted the wider trend to expand into other related audio areas. Indeed the very conceptual nature of the plug-in architecture is bound up with a single defined function and may be an example of a newly emerging contradictory software trend focused more on modular usability than extensive functionality.

Issues such as programming priorities sometimes combine with hardware technical advances to produce additional features in software that are driven at least partially by extramusical pressures. To return to the Cubase case study, it is notable that some of the additional functions that have appeared as the program has developed seem to be driven by a combination of increased processing power and the fact that it is relatively straightforward from the programmer's point of view to provide such features. An example might be the Cubase logical editor, which takes a Boolean approach to specifying MIDI parameters for editing. While its potential power has been acknowledged, this Boolean interface has also been criticized as being rather an unfamiliar way for musicians to conceptualize selection of parameters and yet is closely related to the types of logical structured architectures seen in text-based programming languages. This comment is typical, "The Logical Editor window is a powerful way to process MIDI data in *Cubase*, although it can often appear daunting to new users" (Wherry 2003).

24.5 Usability in Music Software

The issue of usability in audio production software is an interesting one. On first consideration it would appear self-evident that user friendliness would be among the highest priorities for software developers. However, on closer examination it is clear that many drivers of trends within the software development environment have aspects which have the potential to conflict with usability for certain groups of potential end users. So of what would the priorities for software driven primarily by usability consist? Raskin has suggested that in order to be "humane," or, in other words, user-friendly, an interface should have the following attributes:

- It should be modeless; user-initiated actions should have the same result whatever the current state of the application.
- Functions and features should be obvious to most intended users and make use of existing skills in the target user community.
- It should be monotonous; there should be just one way to accomplish a task.
- Features should be visible; the relevant features of an application should be visible at any given time and users should not be required to memorize that a function exists.

For most users, the interface is their experience of the application, so although these points are expressed in relation to the user interface they still hold when applied to the totality of the application. Similarly Nielsen (1993) has attempted to define usability along the following lines:

- Satisfaction; the application should be subjectively pleasurable to use.
- Learnability; the application should not have such a steep learning curve that productivity with it is excessively delayed.
- Efficiency; the application should enable a high level of productivity once it has been learned.
- Errors; in addition to the application being catastrophic error-free, end users should experience few operator errors when using the application, and those errors should be easily recoverable.
- Memorable; the application should be easy to remember so that users can return to the application after a period without having to relearn features.

Taken together, these suggested criteria provide a basis on which to assess the usability of audio production software. These criteria are chosen as examples and this is not to deny that other factors such as the cognitive style of the end user may be important.

Bearing these criteria in mind, it can immediately be seen that there is a potential conflict between many of the drivers for software development already discussed and notions of user friendliness. Indeed it may be suggested that almost every criterion for usability suggested by Nielsen (1993) and Raskin (2000) has a contradictory pressure at work. To give just a few representative examples, the previously noted accretion of ever more widespread features in many audio applications mitigates against many aspects of Raskin's criteria. With increasing complexity the danger of actions becoming state-dependent as opposed to the modeless ideal is increased and features are likely to become less obvious to end users.

An example of the duplication of functionality that highlights the potential for conflict with usability criteria, the environment is a Logic-specific feature not shared by any other DAW and it operates on a building-block paradigm which does however share some features with software such as the previously mentioned Max/MSP and Pd. It allows the end user to reconfigure and, to a degree, add functionality to Logic by connecting together graphically represented small, limited-function objects. Historically, the environment has been the source of some confusion in the Logic user community. "Making the most of your virtual instruments in Logic can involve some complex Environment trickery... Using multi-channel and multi-timbral plug-ins in *Logic* can often lead to confusion. As ever with DAW software, there are several solutions to the problems encountered" (Moores 2007).

It has been particularly noticeable since the takeover of Logic development by Apple that a great number of functions previously accessible only from the environment have since been duplicated in a more intuitive (for most musicians) graphical manner in other windows. This has resulted in quite widespread duplication of function. It is a feature of this type of duplication that the original interface, the environment, has remained in place as a kind of historical legacy feature for established users who have spent time learning how to use it. This term "legacy" is something that is referred to specifically in audio production software manuals and can over time lead to a kind of archeological stratification of old software features, which obviously has implications for clarity of function and usability. This archeological software stratification can be seen to a greater or lesser extent in the majority of the audio applications mentioned so far.

Beyond a certain level of functionality it becomes impossible to display all options on a single computer screen simultaneously, so the strategy of multiple window operation gradually became widely adopted in the DAW software. This pragmatic feature for managing burgeoning complexity is in opposition to Raskin's criterion for visibility, as is the accompanying proliferation of hidden menus, submenus and sub-submenus that have appeared in response to the creeping functionality already noted.

However, not all aspects of usability have been undermined by competing design imperatives. It is possible to observe trends in audio software development where usability can be identified as the dominant force; for example, in the introduction of plug-in functionality in all of the major DAWs. The plug-in concept revolves around relatively small self-contained pieces of code which typically perform well-defined and bounded functions. These bounded software representations usually add extra functionality to the host DAW and thus provide a way for the end user to configure the features that they wish to see in the software. This modular approach, if carried beyond the rather limited current realm of plug-in instruments and effects, has the potential to reverse aspects of the burgeoning complexity of audio production software. The modular concept can also be seen at work in other types of software in addition to the DAWs. Max/MSP and Pd both allow the end user to implement additional functionality by creating their own original code objects and similarly the score writer Sibelius has additional features implemented through plug-in scripts that can be added to by the end user. "Max allows a proliferation of individual elements (species) that never overwhelm the overall environment" (Zicarelli 2002, 46).

It can be suggested that the increasing implementation of what might be called intelligent assistance, which is a particularly notable aspect of the recent development of DAWs, can also be viewed from the usability perspective: "the expert's musical knowledge and control have been virtualized, that is, abstracted and disjoined from the person and moved into the tools" (Keislar 2009, 39).

An example of this idea might be the automation of time-stretching of an audio sample loop so that it conforms to the tempo of the current project. Again we can look at Cubase as a representative example. Once the ability to time-stretch audio independently of pitch was introduced, the complexity of functions and controls surrounding it multiplied, arguably to the point where for less-experienced users it became difficult to achieve tasks common in popular music, such as matching a sample loop to the prevailing tempo. In response to this situation, in version 5, the developers introduced a feature in which an audio sample loop containing appropriate metadata is automatically time-stretched to conform to tempo as soon as the sample loop is imported into the DAW. Although it is possible to disable this automatic function, the onus has been shifted from higher-level professional knowledge needed to implement the feature, to higher-level knowledge needed to disable the automatic engagement of the feature. For some users this has a positive impact on usability, while for others it may have the reverse effect. Professional users in some styles of music may wish to avoid the quality degradation associated with time-stretching for instance, and thus may experience additional complexity in the need to disable this unwanted intelligent assistance.

In the score writer Sibelius we can see a greater degree of intelligent assistance that has a less ambiguous impact on usability for end users. Perhaps in this case these features seem more unambiguously useful because the end-user group for music score-writing is more tightly defined than for a DAW. For instance, there are many hidden automated layout functions implemented in Sibelius which are generally very well orientated toward assisting the user to create legible scores without having to engage with a lot of low-level positioning detail for things such as notes, accidentals, and bar lines. Usually this intelligent assistance works well and only really impacts on usability when nonstandard scoring is attempted, such as that associated with classical avant-garde styles. It is also interesting to note that in Sibelius the type of intelligent assistance that happens automatically by default is largely confined to areas that do not have implications for the use of musical skills, whereas intelligent assistance with more musical tasks, such as the creation of a retrograde version of a musical motif, is confined to user-elected functions such as those found in the plug-ins menu.

This concept of intelligent assistance, or expert systems (Cowan 2001) as they might also be thought of, has a number of implications for the use and future development of both the software itself and the skills needed to produce and record music with it. Originally, features such as automatic time-stretching and tempo conform were seen in programs such as Sony's Acid and Ableton Live. Initially these programs seemed to be targeted more at the amateur hobbyist and semiprofessional user interested in the popular electronic dance music styles of the 1980s and 1990s. It is interesting to note the trend for such features to migrate from semiprofessional software such as those mentioned to the traditionally more professional-orientated DAWs such as Cubase. As an aside, it should also be noted that Ableton Live has acquired a good number of more professional features since its early development, such as the tight integration of Max/MSP.

24.6 Emergent Trends

All of these developments suggest the question, what themes and trends can we identify in these seemingly myriad and often contradictory cross-currents in audio production software development? Key trends identifiable across a number of software and hardware platforms in the history of recent developments in this field might include:

- The tendency for the number of features in a software application to increase over time.
- An increase in the number of different ways in which features can be accessed and executed.
- The introduction of modular architectures, as exemplified by plug-ins.
- The development of increasingly elaborate multiwindow interfaces over time.
- The tendency for software to acquire additional functionality in areas not strictly related to its original function.
- The introduction of increasing amounts of expert assistance by default.
- A blurring of the boundaries between so called "professional" features and non-professional features.
- The accretion of historical levels of legacy features.
- The proliferation of a variety of incompatible audio standards.

This provisional list has a number of implications for the current and future use of music software. One of the more important emerging features with possible long-term implications is that of the introduction of automated expert assistance. This feature is one of the key trends responsible for the blurring of the boundaries between the historically well-defined segregation of consumer and professionally orientated software. In a situation where there is a clear distinction between automation of low-level, low-skill tasks and user-invoked higher-level assistance, then there are potential end-user benefits. However, in other cases where more skilled and sometimes even creative tasks are automatically performed by routines embodying expert knowledge, there is potential for the deskilling of future end users and a stifling of original creativity, because the expertise encoded within these functions by definition draws from existing practice.

One of the more curious features to emerge from this short survey of aspects of audio software development is the growing tendency for applications to converge in functionality. As we have seen over a period of time, MIDI sequencers became audio recorders, audio editors acquired audio effects, developed instrument capabilities, and so on. Software score writers became multitrack audio capable, added instrument plug-in hosting and developed Internet publishing features. Loop-based DJ tools such as early versions of Ableton have added programming language functionality in Max for Live. Max/MSP objects have been developed to allow integration of other programming languages, such as Java, C, Python, Csound, and so on. It may not be overstating the case to say that each piece of audio software is gradually taking on the functions of all of the other audio applications. In a discussion in 2002, Lyon states, "The distinction between experimental music and what might be termed normative music—that is, music based on accepted stylistic norms—is mirrored in our software." (2002, 13). From the examples given above it would seem this distinction is showing signs of breaking down, at least in relation to the software.

24.7 Possible Futures

Perhaps one of the possible futures we could extrapolate from these emergent trends is that of a multipurpose generic music software framework into which the end user could plug in just those features that were required for the musical tasks at hand. Software developers have commented on this type of approach in relation to programming languages such as Max/MSP, "Because most people build Max programs for their own use... they only create the features they need for a particular task." (Zicarelli 2002, 45)

Could the popularity of single-function music apps for the iOS mobile operating system from Apple point in this direction? When one of these music apps is loaded on to an iPad, then in one sense the iPad transforms itself into a synthesizer or other single-function music device. The nearest currently existing to this kind of modular model in software might be some of the VST instrument-hosting applications that have been developed specifically to access the features of small, single-task plug-ins without the need for running the large-scale, multipurpose host DAWs for which they were originally designed. An example of this might be the software Plogue Bidule, whose developers state on their website, "The new paradigm of computer audio can be summed up in two words: real-time and modular" (http://www.plogue.com 2012).

On the subject of multipurpose DAWs, one of the pioneers of computer music, Max Mathews, has commented: "having a properly modularized program seems like one of the very important things" (Lyon 2002, 16). Of course, the history of technology tells us that tomorrow something may come along that will completely change the game, but until that moment these speculations may help to make us just a little more aware of the transforming nature of our medium, to paraphrase Brown.

Note

1. Fast Fourier Transform, a mathematical algorithm used in digital audio as part of a process to describe and manipulate the frequency components of a sound. For further information on real-time audio applications of this algorithm, see Settel and Lippe (1994).

References

Brown, Andrew R. 2007. *Computers in Music Education: Amplifying Musicality*. New York: Routledge.

Bainbridge, David I. 2009. *Intellectual Property*. Harlow, UK: Pearson Longman.

Cheng, Kenneth Hsing, and Qian Candy Tang. 2010. Free Trial or No Free Trial: Optimal Software Product Design with Network Effects. *European Journal of Operational Research* 205(2): 437–447.

Cockburn, Iain M., and Megan J. MacGarvie. 2011. Entry and Patenting in the Software Industry. *Management Science* 57 (5): 915–933.

Cowan, Robin. 2001. Expert Systems: Aspects of and Limitations to the Codifiability of Knowledge. *Research Policy* 30 (9): 1355–1372.

Dixon, Martin. 2006. Echo's Body: Play and Representation in Interactive Music Software. *Contemporary Music Review* 25 (1–2): 17–25.

Keislar, Douglas. 2009. A Historical View of Computer Music Technology. In *The Oxford Handbook of Computer Music*, ed. Roger T. Dean, 11–43. Oxford: Oxford University Press.

Lyon, Eric. 2002. Dartmouth Symposium on the Future of Computer Music Software: A Panel Discussion. *Computer Music Journal* 26 (4): 13–30.

Moores, John. 2007. Using Multi-channel & Multitimbral Plug-ins: Logic Notes and Techniques. *Sound on Sound*, September. http://www.soundonsound.com/sos/sep07/articles/logictech_0907.htm.

Nielsen, Jakob. 1993. *Usability Engineering*. Boston: Academic Press.

Puckette, Miller. 1988. The Patcher. *Proceedings of the International Computer Music Conference*, 420–429. San Francisco: International Computer Music Association.

——. 1996. Pure Data: Another Integrated Computer Music Environment. *Proceedings, Second Intercollege Computer Music Concerts*, 37–41. Tachikawa, Japan.

——. 2002. Max at Seventeen. *Computer Music Journal* 26 (4): 31–43.

Raskin, Jeff. 2000. *The Humane Interface: New Directions for Designing Interactive Systems*. New York: Addison-Wesley Professional.

Settel, Zack and Cort Lippe. 1994. Real-time Musical Applications using FFT based Resynthesis. *Proceedings of the International Computer Music Conference*, 338–343. San Francisco: International Computer Music Association.

Wherry, Mark. 2003. Logical Editor Window, Steinberg Cubase Tips and Techniques. *Sound on Sound*, May. http://www.soundonsound.com/sos/may03/articles/cubasenotes0503.asp.

——. 2012. Product Review: Avid Pro Tools 10. *Sound on Sound*, March. http://www.soundonsound.com/sos/mar12/articles/pt-10.htm.

Zicarelli, David. 2002. How I Learned to Love a Program that Does Nothing. *Computer Music Journal* 26 (4): 44–51.

CHAPTER 25

DELIVERING INTERACTIVE EXPERIENCES THROUGH THE EMOTIONAL ADAPTATION OF AUTOMATICALLY COMPOSED MUSIC

MAIA HOEBERECHTS,
JEFF SHANTZ,
AND MICHAEL KATCHABAW

FIXED compositions of music are often sufficient for noninteractive uses, such as film and television. In these instances, there is a singular experience delivered to the audience, with no opportunity for them to have any real say in what is going on. Interactive applications such as videogames, on the other hand, are quite different. The user is now a full participant, and their actions help to shape the experience that they are taking part in. Such experiences, however, tend to fall flat, seem inconsistent, or lack harmony when musical accompaniments remain static and unchanging while the rest of the experience is dynamic and responsive to the actions of the users. Music often creates and supports mood, tone, tension, pacing, and emotional impact of an experience: if it is not adapted or adjusted as a result of user interactivity, it will only lessen the experience.

Fortunately, most interactive entertainment applications are software based, and thus are flexible. Software can be instrumented to monitor and sense various aspects of user experience and to analyze this data to determine when musical accompaniments need to be changed in response to user activity and how they should be changed to reinforce the experience and emotion being delivered. Software can then implement these changes in music to avoid the divergent problems discussed above and achieve a truly, fully interactive experience for the user. This is the approach taken in our work, the Algorithmic Music Evolution Engine (AMEE) (Hoeberechts, Demopoulos, and Katchabaw 2007, 2009; Hoeberechts and Shantz 2009), and is the focus of this chapter.

This chapter will provide a thorough and detailed treatment of algorithmic interactive music, delving into both theoretical and practical aspects of automated music adaptation to improve the user experience delivered by modern interactive applications. We have divided the chapter as follows:

Introduction to music and emotion: A brief overview of the relevant aspects of music psychology and perception, and how these elements impact user experience.

Affective algorithmic generation: A discussion of the current state-of-the-art and literature in this area.

Designing for interactivity and adaptation: An examination of how to structure a software system to enable music adaptation and support for interactivity; the basis for our AMEE software system.

Implementation guide: Practical aspects of implementing this design into AMEE.

Case studies: The concepts, techniques, and tools discussed throughout this chapter are applied to the creation of a novel game, *Pop Tones* (Hoeberechts et al. 2012); a social music authoring environment, *AMUSE*; and a mobile ring-tone generator, *ART-G*, which uses AMEE to create music that adapts emotionally in response to user interactivity.

25.1 Introduction to Music and Emotion

Before exploring technical aspects of algorithmic generation, it is important to first discuss some of the theoretical background in music psychology that underlies the concept, and in particular, emotional adaptation. Our perspective on emotional expression in music comprises several key points: There is a set of moods that can be convincingly expressed in music. Listeners with a similar cultural background will identify the moods conveyed by a given piece of music relatively consistently. Perceived moods in a composition can be linked to specific characteristics of the piece such as mode, articulation, tempo, and volume, among others. Changes to these characteristics alter perceived mood.

These statements are not without controversy. That music is capable of evoking and/or expressing emotions is not a universally accepted idea. Stephen Davies summarizes some of the problems as follows: "Given that music is non-sentient, how could emotions be expressed in it? Are musical phrases understood as utterances about emotions? Or are they interpreted as gestures betraying the composer's or performer's emotions? Or are they expressive insofar as they move the listener?" (Davies 1994, x). Supposing we accept that there is an association between music and emotions, we also face the question of whether the emotions are merely *perceived* in the composition by the listener, or whether they are *felt*—that is, experienced more fundamentally than by simple cognitive identification (Juslin and Västfjäll 2008).

The major theoretical positions regarding emotion and music can be distinguished as *absolutist* versus *referentialist*, and *formalist* versus *expressionist* (Abeles and Chung 1996; Radocy and Boyle 2003). An absolutist holds that the meaning of a musical work is contained within the work itself and can be conveyed through the perception of the work in isolation. A referentialist maintains that music's meaning involves more than the notes itself; for example, a story or spiritual state is necessary as background context for understanding the music. As an illustration, consider the *Star Wars* title theme by John Williams (Williams 1977). An absolutist would hold that the meaning of the piece is derived entirely from the music itself: if you feel excited while listening to the music, it is because there are characteristics of the composition that evoke this emotion. A referentialist, on the other hand, would attribute some or all of the emotions you feel to your knowledge of the plot of the movie or associations with cultural understanding of the film.

The absolutist viewpoint can be further classified as either formalist or expressionist. A formalist holds that the meaning of a piece of music is primarily intellectual and is derived from perception of the structural relationships between elements in a composition (for example, the musical form or harmonic progressions). An expressionist believes that the principal sources of meaning of a composition are the emotions evoked in a listener as a result of listening to (or perceiving the structural relationships within) the piece. Thus, to an expressionist, musical meaning is primarily emotional.

For our purposes, we adopt an expressionist viewpoint, since we are designing a means of interaction that allows the user to choose characteristics that the generated music should be capable of expressing. Both the absolutist and referentialist perspectives are useful: We can aim to generate music that "stands on its own" in the sense that the generated music contains elements designed to evoke specific emotions in the listener. However, in the context of a computer or videogame where there is a narrative, we might also use a referentialist perspective in which the game players' associations between the narrative and the background music serve to evoke emotions.

25.1.1 Types of Emotions Expressed in Music

In the context of the psychology of music, emotional response to music is divided into several categories. These include, for instance, the following (definitions are based on those in Abeles and Chung 1996 and Radocy and Boyle 2003). *Affect/emotion/mood*: This is the "usual" interpretation of emotional response. While listening to a given piece, a person feels an emotion such as joy or anxiety. *Aesthetic*: An aesthetic response has both an affective and intellectual component. If we think a piece of music is "beautiful" or "well structured," these would be aesthetic evaluations of it. *Preference*: This term encompasses "like" or "dislike" of a particular piece of music, or preference of one over another. To distinguish preference from the previous two categories, imagine a musical piece that makes you sad, that you think is beautiful, yet something you would rather not listen to because of the sadness it brings. *Taste*: Musical taste reflects long-term attitudes toward music, developed from experience. The content of your CD collection would reflect your musical taste.

25.1.2 Emotion Perception Studies

The methodology used in studies on music and emotion varies widely, due in part to the lack of a standard theoretical framework (McMullen 1996). Juslin and Västfjäll's recent work (2008) shows promise in addressing this problem in arguing that musical emotions must be investigated with regard to their underlying mechanisms, such as brain response and musical expectancy. They claim that some disparities in experimental results can be explained by the researchers' failure to take these mechanisms into account. In other studies of musical emotion, there is variation in the music or musical stimuli that are used. Some analyze whole pieces of classical music or popular music songs (Gabrielsson and Juslin 1996; Hevner 1936), segments of pieces (Terwogt and van Grinsven 1991), altered pieces (order of sections, ending, tempo, harmony, and so on changed) (Hevner 1936), pieces composed specifically for the study to represent a desired musical quality (Gabrielsson and Juslin 1996; Hevner 1935), and even single-line melodies or single intervals (Costa, Bitti, and Bonfiglioli 2000). The studies that consider a musical element in isolation are perhaps the most useful for developing an engine to reflect musical meaning, since they state a result about one clearly defined musical element. However, it is unclear whether this methodology produces musically meaningful results since they are devoid of musical context (Hevner 1936).

In addition to the perspectives on musical meaning, there are different perspectives on the appropriate metric for measuring emotional response. Physiological approaches test physiological reactions to musical stimuli, such as changes to heart rate, respiration rate, blood pressure, and muscle tension. Some researchers maintain that music is capable of evoking emotions by causing a physiological reaction, which is in turn perceived by the subject as an emotion, as discussed in (Bartlett 1996). The advantage of studying emotional response in this way is that the response is quantifiable and easily measured. However, the link between physiological reaction and subjective emotion is not clear-cut.

Positron Emission Tomography (PET) scans, Functional Magnetic Resonance Imaging (fMRI), Electroencephalography (EEG), and other brain-imaging techniques have been used to visualize neurological response to musical stimuli. Studies have compared responses from nonmusicians and musicians, between music listening and playing, and from perception of harmony, rhythm, melody, and so on (Peretz and Zatorre 2003).

Adjective descriptors are a traditional means for evaluating emotional response to music. Normally, subjects are asked to check off from a list adjectives that they feel best describe their feelings while listening to music or samples. Hevner's adjective circle (Hevner 1935, 1936), is an early example of this research method and has served as the basis of much subsequent research. There have been many other attempts to compile complete and appropriate adjective lists; for example, in Terwogt and van Grinsven (1991) an experiment was conducted in which 100 mood indicators were rated on their usefulness in capturing musical meaning. However, one problem with using adjective descriptors is that they do not account for degrees of response.

Semantic differentials are pairs of bipolar adjectives for which a rating can be chosen by the subject. An example would be a ten-point scale labeled "sorrow" at one end and "joy" at the other. This technique overcomes the main limitation of adjective descriptors in that one can specify the degree to which a given emotion is experienced. However, there are implied semantics in the choice of opposite terms for each differential. To illustrate, consider the pair "excited–calm." Suppose you find a musical sample not to be exciting—with this differential, you would be forced to rate it as calm, despite the fact that you might feel that some other adjective, such as "stately," better describes it.

The approach taken in our research is to combine adjective descriptors and semantic differentials. Adjective descriptors are used, but a degree of intensity of each desired emotion between 0 (emotion not present) and 100 (extreme emotion) can be specified. There are two significant challenges in determining the relative contribution of musical characteristics to perceived mood. First, characteristics affect more than one mood. For example, increased tempo is associated with increased perception of happiness, excitement, and anger. Second, combinations of desired moods pose problems since the expected changes to characteristics sometimes conflict. For instance, tempo should be increased to evoke happiness, but decreased for serenity. Hence, we encounter a conflict when attempting to produce a mood that is both happy and serene.

25.2 Affective Algorithmic Generation

A number of approaches for affective composition have emerged in the literature that can be roughly divided into the following categories: stochastic processes, which incorporate random choice or probability in composition; rule- or knowledge-based methods, where input parameters are mapped to musical features based on well-established rules; evolutionary or genetic techniques, which create new patterns through mutation; recombinatorial approaches that combine short, prearranged music segments to produce longer pieces; state machine or grammar driven approaches in which composition is controlled though state or symbol transitions respectively; and hybrid approaches.

MySong (Simon, Morris, and Basu 2008) uses a stochastic, machine-learning approach, and serves as the underlying technology for Microsoft Research *Songsmith* (Microsoft Corporation 2009). *Songsmith* allows a user to sing a melody track, and it will automatically generate a full band accompaniment according to one of a number of user-selectable styles, such as 1950s Rock or Big Band. Users can tune the emotional quality of generated music by adjusting a "Happy slider," which alters the mode of the piece, and a "Jazzy slider," which affects how closely selected chords fit the notes of the melody. The system employs a Hidden Markov Model (HMM) that is trained using a database of melodies, along with their associated chord progressions.

Livingstone and his colleagues make use of the rule-based Two-Dimensional Emotion Space (2DES) (Schubert 1999), which maps the input parameters of valence (positive/negative charge) and arousal (energy level) to a set of both structural and performative musical features, such as tempo, mode, accents, and slurs (Livingstone et al. 2007). Emotions can be expressed by selecting appropriate values for these parameters. For instance, selecting high values for both valence and arousal might produce a happy composition with increased tempo and a major mode employed. A comprehensive treatment of 2DES and its capacity for musical feature transformations is given by Schubert (1999).

Evolutionary approaches have also demonstrated promising results in adaptive music composition. Legaspi and his colleagues asked human subjects to rate a set of musical scores according to several bipolar emotional adjectives using a web-based tool (Legaspi et al. 2007). These ratings are used by their system to generate rules to identify musical structures that express various emotions. Composition occurs by means of a genetic algorithm that employs a fitness function accounting for both adherence to musical structure and desired emotional expression when evaluating mutations.

Eladhari and colleagues use a neural network approach to generate music tailored to suit individual characters in videogames (Eladhari, Nieuwdorp, and Fridenfalk 2006). The system employs a "mind module" that consists of a spreading activation network, comprised of interconnected nodes modeling traits, emotions, moods, and sentiments. The mood of a character is measured in two-dimensional space, with one axis representing an inner mood scale (depressed to happy) and the other representing an outer mood scale (angry to exultant). Inner mood influences the harmonic qualities of the composition, such as the scale employed, while outer mood affects the time signature of the piece. Real-time affective changes occur when a sentiment felt by a character exceeds a given threshold, triggering an emotional response in the activation network and influencing the values of its inner and outer mood scales.

Comair and colleagues combine a finite state machine with a recombinatorial approach to produce adaptive audio for videogames (Comair et al. 2004). Precomposed blocks of music are associated with the states of the machine. For instance, one state might be assigned a relaxed music block, while another state might be assigned a more excited block. An interactivity parameter modeling some aspect of gameplay determines when to transition between states through comparison to numeric entry and exit values assigned to each state. Although composition in the system does not occur in real time, each player experiences a different composition since gameplay changes the order in which blocks of music are played.

Affective audio has also penetrated the commercial videogames market, with an increasing number of games adopting at least some form of emotional adaptation to enhance gameplay. A modification developed for the popular *Unreal Tournament 2004* changes music based on events occurring within the game as well as a player's status (Dobson 2006). For instance, entering a battle increases the intensity of the music played, while dwindling player health will shift the music to a more ominous feel.

25.3 Designing for Interactivity and Adaptation

Creating a software system capable of supporting interactive applications and emotional adaptation requires careful planning and a solid framework. In this section, we examine the design goals used in the creation of such a system, and provide the architecture of our system, AMEE.

25.3.1 Design Goals

In designing our AMEE system, we strove to meet several design goals, including flexibility, reusability, extensibility, adaptability, accessibility, portability, and enablement of collaboration. The engine must permit maximal flexibility in the composition process. No undue limitations should be placed on what music can be generated by the engine, to give users as much control over their music as possible. Furthermore, the engine must either generate music without any restrictions or permit a human composer to guide musical choices. Moreover, it must be possible to reuse musical elements, such as note sequences and harmonic structure, from both existing human-composed pieces and computer-generated material. These reusable elements should be logically organized into various databases or libraries that could be loaded or unloaded at will, and could be easily updated with additional content from various sources, including the engine itself.

It is also important to provide an extensible architecture that can be easily integrated with other software, such as music tools, computer and videogames, mobile applications, web services, and so on. Reusability, as discussed above, can help enable extensibility; additional musical styles can be readily added to the engine's repertoire by creating pattern databases based on existing pieces. This is in contrast to machine-learning approaches, for instance, in which extensive training is necessary to adopt a new style with palatable results, and composition quality is highly dependent on the selection of sufficiently representative training data.

To support interactive music and applications, it is critical to allow music to be adapted both offline and online, in real time alongside the composition process. These adaptations should include a wide variety of parameters governing the composition process and the various attributes of the piece itself, including for example alterations based on various emotional characteristics such as happiness, sadness, anxiety, liveliness, and so on.

By incorporating a multilevel application-programming interface (API), engine functionality can be made accessible to users with varying levels of musical and/or programming knowledge. For instance, changes to the mood of a given composition can be made by selecting intensities for specific emotions, such as triumphant or scared. Working directly with emotional adjectives is more intuitive to end users and programmers who may not be sufficiently familiar with the theory of emotion to use more abstract parameters such as valence and arousal. Users with such experience, however, can access a lower-level API that permits access to these parameters, as well as other music parameters such as mode, tempo, and so on.

To ensure that the engine can be used in a variety of applications across a great number of platforms, the engine itself must be highly portable. Great care must be taken to minimize programming language dependence and to decouple the engine from the underlying music output technology. While MIDI is a logical choice at its output production layer, it must be feasible that the engine could be used with other technologies, such as those based on recorded samples.

Lastly, it is important for the engine to encourage collaboration between composers who might wish to use the engine to generate ideas and share interesting pieces with others. The engine should facilitate this process by allowing users to create and export musical styles, musicians, and generated compositions, which can then be shared with other composers using e-mail, web services, or social platforms such as Facebook or Twitter.

25.3.2 Framework

With these design goals in mind, a design framework was created for AMEE (Figure 25.1). The High Level API exposes elements of the engine modeled after real-world entities such as Musician, Instrument, Performer, Piece, Style, Ability, Emotion, and Mood. In doing so, the engine becomes more accessible to more novice users, such as those with more limited music or programming experience. For example, suppose that an application requires a piano accompaniment. The user creates a Musician that plays an Instrument and has a Mood. A Musician also has an Ability and knows a number of Styles. We can thus model a real musician, who can play one instrument at a time, but who has the ability to play that instrument in different styles with varying ability. The Musician could be assigned a piano as an Instrument, and a Style appropriate to the application, such as classical or improvisational jazz. The Musician can then be instructed to play their Instrument using their current (default) Mood, which can be adjusted in real time in response to user interaction. For instance, the user can instruct the Musician to be sadder in their playing.

The Low Level API exposes access to the underlying complexity and detail of the music and the composition process and, as such, is primarily intended for more advanced users who possess more extensive music or programming experience. Continuing the example from the previous section, suppose that the user wants the pianist Musician to play sadder music. When done through the High Level API, this in turn results in the adjustment of various parameters through the Low Level API, such as mode, pitch, and tempo. Alternatively, an expert user could bypass the High Level API entirely and directly adjust these parameters themselves to achieve the same effect. The Core Engine encapsulates the music composition and production facilities of the engine as a whole. It is the Core Engine that delivers the functionality exposed through the High and Low Level APIs (see below).

With the goals of reusability and extensibility in mind, Pattern Libraries provide the Core Engine with the ability to load, save, and organize various musical elements, such as note sequences and harmonic structures. Storage could be local (in files or a database) or remote (web-based, cloud-based, and so on). Libraries can be created and populated in several ways. Authoring tools allow users to write and produce their own content for the libraries. Extraction tools, such as MUSIDO (Demopoulos and Katchabaw 2007), allow users to pull musical elements from existing pieces for insertion into the libraries. Lastly, the Core Engine itself can store elements of composed music in the libraries for future use. When needed by the Core Engine, these elements can be loaded and used. Musical elements stored in the libraries can be organized in various ways; for example, different elements can be collected together to define a variety of musical styles.

As the Core Engine engages in composition and production, the results of its efforts are sent to one of many possible Output Devices. These devices could either be for

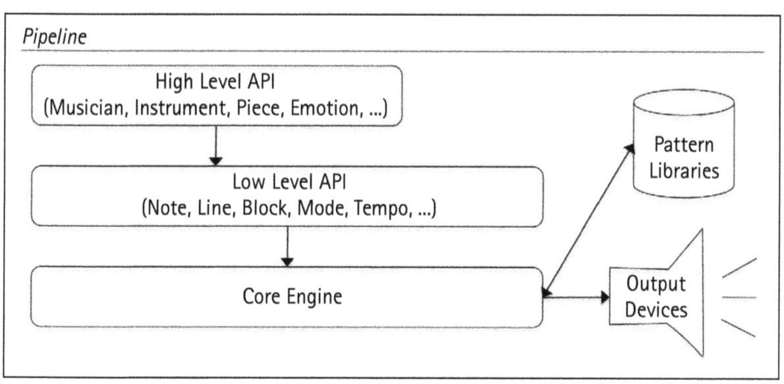

FIGURE 25.1 AMEE Framework.

playback of music (such as MIDI devices) or for capture of music for later playback (such as MIDI, WAV, or MP3 files).

25.3.3 Implementation Guide

In this section, we refine the design and framework from the previous section, and discuss how we implemented our adaptive music composition system, AMEE, based on these concepts. The Core Engine of AMEE is divided into several main groups of classes. High-level classes implement the real-world entities modeled in the engine, including Musician, Instrument, Performer, Piece, Style, Mode, Meter, and Mood. The pipeline classes control the flow of the music-generation process and are responsible for calling methods on the generating classes. This pipeline is depicted in Figure 25.2 and its operation is discussed in detail in the next section. The producer classes create high-level musical elements that comprise the composed piece (sections, blocks, and musical lines). The generator classes create the lower-level musical elements (harmonic patterns, motif patterns, modes, meters) from which the products are assembled.

The pipelined architecture illustrated in Figure 25.2 has several significant advantages over other potential architectures. The generation process can be viewed as an assembly line for constructing musical blocks. Each of the producers along the pipeline fills in the elements that its generators create, until the finished block is eventually passed to the Output Producer for playback. The producers all work independently, which means that there is potential to parallelize the generation process. Furthermore, to dynamically alter a composition, a different generator can be substituted in a producer without affecting the rest of the pipeline.

25.3.4 Process of Music Generation

In order to use the engine, producers and generators must be created and loaded in the pipeline. The first step is to use a GeneratorFactory to create the generators. Five types of generators are necessary for the engine: a StructureGenerator, which creates the overall sectional structure of the piece (e.g., ABA form); a HarmonicGenerator, which creates a sequence of chords for each section (e.g., I–IV–V–I); a MotifGenerator, which creates short sequences of notes (e.g., a four-note ascending scale of sixteenth notes beginning on the tonic); a ModeGenerator, which returns modes for the piece (e.g., start in F major, progress to C major, divert to D—and return to F major), and a MeterGenerator (e.g., 4/4 time). Every generator contains at least one pseudo-random number component that is used for decisions it needs to make.

The generator might also make use of a PatternLibrary, a class representing Pattern Libraries as depicted in Figure 25.1, which provides the generator with musical elements it can use directly, or as starting points for musical element generation. As discussed earlier, these libraries are created in advance, and are intended to embody

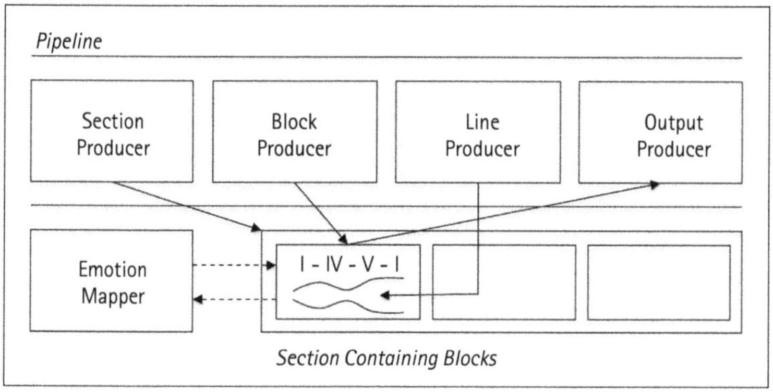

FIGURE 25.2 AMEE Core Engine Pipeline.

musical knowledge. For example, in our earlier work (Hoeberechts, Demopoulos, and Katchabaw 2007), we created a "Bach" MotifPatternLibrary, containing motifs from Bach's Invention No. 8 and a HarmonicPatternLibrary, based on the same piece.

Once the generators have been created, the producers must be created using a ProducerFactory. There are four producers necessary. The SectionProducer uses the StructureGenerator to produce sections, where a section is a chunk of the piece with an associated length in seconds. Each section contains a number of "blocks," which are short segments of the piece (for example, four bars) composed of a musical line played by each Musician. The BlockProducer is responsible for deciding how to coordinate the musical lines using a HarmonicPattern created by a HarmonicGenerator associated with the BlockProducer. The musical lines themselves are composed by the LineProducer that uses a MotifGenerator to create the actual note sequences in the musical line played by each Musician. Finally, an OutputProducer must be initialized, which will convert the generated piece to an appropriate form and output it (for example, as MIDI played as audio).

When the producers have been initialized, they are loaded into a pipeline as shown in Figure 25.2. The pipeline oversees the generation process and calls on each producer in turn to assemble the piece. An example of the process is as follows: the SectionProducer is called on to get a new section of the piece, then the BlockProducer returns the first block of that section, which in turn is passed to the LineProducer and filled in with notes, and lastly the completed block is sent to the OutputProducer for sound output. At each step in the pipeline, if desired, the products can be sent to the EmotionMapper for Mood-dependent adjustments. Music generation continues until the duration of the piece, specified by the user or determined by the engine, is completed.

25.3.5 Using Pattern Libraries

As discussed earlier, the purpose of the PatternLibraries is to allow new music to be generated by reusing compositional elements: either elements from existing pieces or elements that have been previously generated by the engine. As such, a PatternLibrary can be thought of as a repository of musical ideas that can be used to guide the composition process; instead of generating new elements for a piece, elements can be taken from the PatternLibrary. The four libraries that are currently used in AMEE are the HarmonicPatternLibrary, the MotifPatternLibrary, the ModePatternLibrary, and the MeterPatternLibrary. Our motivation in creating the PatternLibraries was twofold: to give the system the ability to compose in different styles and to provide a mechanism for exchanging and storing musical ideas. These concepts will be discussed in turn.

To answer the question, "What style is this piece of music?" one would listen for clues among the musical elements of the piece to determine its classification. The instruments being played are often an initial hint (the "sound" of the piece). Further to that, you would attend to the rhythmic structure, the harmony, the melody, the mode, and the meter, and transitions between elements. Consider the knowledge of a jazz trumpet player, for instance Louis Armstrong. He knows typical harmonic progressions that will be used in a piece, and has many "riffs" in mind that he can use and improvise on. In our system, this knowledge would be captured in a HarmonicLibrary and MotifLibrary respectively.

How the libraries are used is determined by the implementation of the generators. A generator implementation could use the library as its only compositional resource, that is, all the musical elements returned by the generator are those taken from the library, or it could use the library patterns as starting points, which are then modified, or it could return a mixture of library patterns and generated patterns (and naturally, it could ignore the library contents entirely and return only generated patterns). Thus, the libraries are rich musical resources that can be flexibly used depending on the desired musical outcome. The more heavily the elements from a PatternLibrary are used in a new composition, the more that new composition will adhere to the style defined by the elements in the PatternLibrary. Proper organization and structuring of a collection of PatternLibraries is consequently crucial to defining the various styles supported by the engine; improper blending of elements may result in the muddling of styles and the loss of their distinctive qualities.

Regarding the exchange of musical ideas, consider a situation where you meet someone using AMEE who, in their collection of Musicians, has a very dynamic guitar player (let us call him Jesse Cook). The knowledge of the guitar player is contained in the libraries the guitar player is using for music generation. One could allow the Louis Armstrong player from above to learn how to make his trumpet sound like a flamenco guitar by incorporating the riffs from Jesse Cook's MotifLibrary into Louis Armstrong's existing library. A different possibility is that both Musicians could be added to the same band. What would it sound like if the two Musicians jammed together? What if they collaborated in an ensemble? Could they influence each other while playing? All of these functions are directly supported by the idea of PatternLibraries as embodiments of musical knowledge.

25.3.6 Inside the MotifGenerator

The MotifGenerator, although it is only one small component in AMEE, contains the code that one would normally think of as the main element of a music-generation system: it generates sequences of notes and rests with associated timing. One very important design decision differentiates AMEE from most other music-generation systems: the notes (Motifs) that are generated are *independent of both the mode and the harmony*. This is best illustrated by example.

Consider the opening phrase of Mozart's Sonata in A major, K331 as shown in Figure 25.3.

The right-hand melody in the first bar begins on C♯, the third of the scale, and is played over the tonic chord in root position in the bass (harmony I). In the second bar, in the left hand part, we see the exact same melodic pattern, this time starting on G♯, played over the dominant chord in first inversion (harmony V_6).

In a motif, we would encode this musical idea as:

Pitches	2	3	2	4	4
Locations	0.0	1.5	2.0	3.0	5.0
Duration	1.5	0.5	1.0	2.0	1.0

where "pitches" are positions in the mode relative to the root of the harmonic chord (with the root as 0); the "locations" indicate an offset from the beginning of the bar (location 0.0); and the "duration" specifies the length of the note. Locations and durations are expressed in terms of number of beats (in 6/8 time, an eighth note gets one beat).

The purpose of encoding motifs in this way is to capture the musical pattern associated with the sequence of notes, without restricting ourselves to a specific mode or harmonic chord. The approach we have chosen for motif storage allows composed pieces and musical lines to be transposed and reinterpreted in any mode. Moreover, as illustrated in the above example, a particular pattern of notes often appears more

FIGURE 25.3 Mozart Sonata in A major, K331, opening.

than once during a piece, but serves a different function depending on the underlying harmony.

The StandardMotifGenerator in AMEE operates in both the library-based and pseudo-random manner. When a motif is requested, it first checks whether a PatternLibrary has been loaded. If it has, it attempts to retrieve a motif of the specified length (number of beats), and the desired type (end pattern, which is one that could be found at the end of a section, or regular pattern, or either). If any suitable patterns are found, they are returned. Otherwise, a new pattern will be generated.

Two pseudorandom generators are used in motif creation: a number generator and a pitch generator. A loop continues generating notes/rests as long as the desired number of bars has not been filled. The motifs are generated according to musically plausible rules and probabilities.

As previously mentioned, the motif is encoded in a mode-independent and harmony-independent manner. Thus, the "pitches" that are generated and stored are actually relative pitches to the tonic note in the mode. Consider the following example: Suppose the motif contains the pitches values 0–1–0. If that motif is eventually resolved in a position where it appears as part of the I chord in C major, it would be resolved to the actual MIDI pitches for C–D–C. However, if that same motif were used in a position where the harmony required the V chord in C major, the motif would now be resolved to G–A–G. Suppose now that the motif contains the pitch values 0—0.5—1—0. The value 0.5 indicates that a note between the first and second tone should be sounded, if it exists (this will produce a dissonance). Thus, in C major for chord I, 0–0.5–1–0 would be resolved as C–C♯–D–C. If an attempt is made to resolve a dissonant note where one does not exist (for example, between E and F in C major), one of the neighboring notes is selected instead.

25.3.7 Harmonizing between Musicians

The purpose of generating a harmonic structure for a piece is to permit groups of Musicians to perform a piece together that will sound musically coordinated. The LineProducer is the entity in the pipeline that is responsible for resolving motifs into notes that fit into chords within a mode. When multiple Musicians are playing together in an Ensemble, the harmonic structure of the block being generated is determined (a HarmonicPattern is chosen), and then each Musician's line is generated *based on this same HarmonicPattern*. The result is that even though each Musician is playing a different line from the others (and possibly different instruments, each with its own range), at any given time each Musician will be playing a motif that fits into the current harmonic chord and mode for the piece. Of course, this does not imply that every note each Musician plays will be consonant with all other notes sounding at that time—that would be musically uninteresting. Musicians might be playing passing tones, ornaments, dissonant notes, and so on, but the harmonic analysis of each of their musical lines will be the same.

In addition to playing in an Ensemble, using one global style, each Musician's line can be created by the LineProducer to reflect a personal style for the Musician. This effectively enables a Jam Session mode, in addition to the more coordinated Ensemble.

25.3.8 Choice of Mode

Music can be generated in AMEE in *any* mode that can be supported by MIDI. This is a very flexible implementation that allows music played to be played in any major or minor key, or using a whole-tone scale, chromatic scale, blues scale, Japanese scale, and so on. The restrictions imposed by MIDI on the scale are that the period of repetition for the scale is an octave, and that the smallest distance between two notes is a semitone. Thus, an octave is divided into twelve semitones, designated by consecutive whole numbers in the MIDI format (i.e., 0, 1, 2, 3, 4, 5, 6, 7, 8, 9, 10, 11). Note that these are "normal" restrictions for Western music. The pattern of the scale is stored using offsets from the starting note of the scale. These offsets specify the number of semitones between steps in the scale. For example, every major scale has the following step pattern: 0, 2, 4, 5, 7, 9, 11. To specify which major scale is desired, we also store the starting note of the scale, which is a whole number between 0 and 11 that corresponds to an offset from MIDI pitch 0, which represents C.

Motifs can be resolved into any mode, regardless of the mode in which they were originally specified. An unavoidable consequence of this flexible implementation is that the motifs might sound strange if the mode into which they are resolved contains fewer tones than the mode in which they were designed. The resolution method that is used in this case is that the notes are "wrapped around" using a modulus-style computation (i.e., the sixth note in a five-note mode becomes the first note one octave higher).

25.3.9 Enabling Emotional Adaptations of Music through the Emotion Mapper

A key feature of AMEE is the incorporation of Emotion and Mood as factors that can affect music generation. Mood is considered in two contexts: individual Musicians have a Mood which can be adjusted independently of other Musicians, and a piece can have a Mood as well. To illustrate these different contexts, consider an example of an orchestra with twenty-seven members. Suppose that the bassoon player is depressed because she just learned that she could no longer afford her car payments on her musician's salary. Suppose that the orchestra conductor is trying to achieve a "happy" sound at the end of the piece currently being played. Depending on how professional the bassoon player is, she will play in a way that reflects her own "sad" mood, and the desired "happy" mood to varying degrees. These are the two contexts in which we have considered musical mood.

The Emotion Mapper (as shown in Figure 25.1) acts on low-level musical elements (volume, tempo, mode, harmonic patterns, and motif patterns). As discussed earlier in this chapter, motif patterns are sequences of harmony and mode-independent notes: for instance, a descending sequence of three whole tones. Within a composition, the motif will be resolved to actual mode tones such as B, A, G, or F♯, E, D, depending on the current mode and the harmonic function of the motif. The Emotion Mapper adjusts motif patterns through alterations to pitch, consonance, and articulation.

There are three factors that contribute to the adjustment made to a specific musical characteristic: the desired mood, the initial value of the characteristic, and the degree of effect of the mood on that characteristic. For the ten supported emotions in AMEE (happy, sad, triumphant, defeated, excited, serene, scared, ominous, angry, and crazy), the *emotional intensity*, a percentage ranging from 0 (emotion not present) to 100 (extreme emotion), can be specified through the High Level API. Any combination of the supported emotions is permitted. The Mood is the collection of emotion intensities. As mentioned earlier, Mood is considered in two contexts in the system: global mood, which represents the overall feeling of the composition, and individual moods, which characterize the mood of each musician in the performance. Depending on the desired outcome, a mood from either of these categories could be sent to the Emotion Mapper.

The process of emotional adjustment of a musical element is initiated by one of the producers in the engine pipeline from Figure 25.1, and therefore could occur at each stage through the generation process. To complete an adjustment, if necessary, the producer uses a generator to create a low-level element. Then, the producer calls the Emotion Mapper to adjust the element according to the desired mood. Following this, the Emotion Mapper determines adjustments to each dimension (mode, tempo, pitch, consonance, volume, and articulation) based on mood. Finally, the relevant features of the low-level element are adjusted.

The relative contributions of each emotion in the desired mood are combined to determine the requisite effect on the musical characteristics. As an example of relative contributions, consider the influence of happy, sad, and scared on consonance as summarized in Table 25.1. The references used to establish each result are also shown.

The "contribution" of each emotion is a real number between −1.0 and 1.0 that reflects the polarity and strength of the influence of that emotion on the respective musical characteristic. In Table 25.1 we see that happy has a strong positive impact on consonance, whereas scared has a very strong negative impact on consonance (therefore producing dissonance). The dimensional contributions are combined using a formula, as discussed in Hoeberechts and Shantz (2009), to produce a final adjustment number for each musical characteristic.

As the final step, the Emotion Mapper uses the calculated dimension adjustments to make changes to the low-level musical elements. These changes could be, for instance, discrete alteration, continuous alteration, or effect alteration. Discrete alteration is used for characteristics with discrete values (e.g., mode and consonance). Depending on the final dimension adjustment and thresholds for change, the characteristic may be altered to a different value. For example, the number of dissonant notes in a motif might be increased or decreased. Continuous alteration is used for characteristics with a range of possible values (e.g., tempo, pitch, articulation). The existing characteristic is increased

Table 25-1 Relative contribution example

Emotion	Effect	(References)	Contribution
Happy	0+	(Costa, Bitti, and Bonfiglioli 2000)(Hevner 1936)	+0.75
Sad	–	(Costa, Bitti, and Bonfiglioli 2000; Hevner 1936)	–0.75
Scared	–	(Costa, Bitti, and Bonfiglioli 2000; Hevner 1936; Nielzén, and Cesarec 1981)	–1.0

0: no effect; –: more dissonant; +: more consonant

or decreased by a percentage depending on the dimension adjustment. For example, the tempo is reduced to 75% of its present value. Finally, the effect alteration is used to portray the desired emotion; the characteristic should vary in time (e.g., volume). For example, for a defeated feeling, we gradually decrease the volume, and for crazy, there are sudden changes in volume in the composition. This model could support other types of alterations, such as pitch bend effects, distortion, and rhythm alterations, amongst others.

25.4 Case Studies

Using the AMEE software system, numerous interactive music applications have been developed. In this section, we examine a few of these applications, highlighting their use of AMEE and its ability to support interactive applications and adaptive music.

25.4.1 Pop Tones

Pop Tones (Hoeberechts et al. 2012) is a match-3 style game with a focus on music and emotion. The objective of the game is to catch color-coded balls that float down the screen, with each color corresponding to a different emotion. By lining up and matching three balls of the same color or emotion together, the balls disappear and a bar associated with the emotion fills. The more an emotion bar fills, the more that emotion influences the music being played. Each level has quotas for the various emotions and the level is completed only when the quotas are all met. Levels are grouped into different worlds of various themes, and the player advances from world to world based on the completion of levels. The player loses the game if too many balls are not caught, or if the stack of caught balls accumulates to reach the top of the screen.

A screenshot of the gameplay of *Pop Tones* is given in Figure 25.4. On the left of the screen are the emotion bars that fill with matches; those that are grayed out do not need to be filled on the current level. On the right of the screen are indicators of the current level and world. The middle of the screen contains the main play field. At the bottom of the play field is the core ball that the player controls and uses to catch the colored emotion ball as they fall down the screen. Balls stick to one another, causing them to stack and accumulate if matches are not made, as shown in Figure 25.4.

Pop Tones uses AMEE to compose, generate, and play back all music. One Pattern Library is used per world, to provide a unified style of music on each world, while providing variety between worlds. Pseudo-random generators are used to provide different music for the various levels of a world, although keys are carefully selected and reused to ensure that the same level plays the same piece on each play through. Naturally, AMEE's Emotion Mapper handles all emotional adaptations in response to matches made in the game. When the player matches three balls together, filling the corresponding emotion bar further, *Pop Tones* uses AMEE's High Level API to signal the Emotion Mapper to adjust the corresponding emotion for the piece being played. These adjustments can have interesting implications for player choice and strategy, as aspects of the music impact gameplay. For example, the tempo of the music influences the speed of the game, including both ball spawning and drop rates; the faster the music, the faster the game. As a result, when filling emotion bars through matches, the player needs to be careful to choose the matches made to regulate the emotion and attributes of the music. Otherwise, the player could very quickly lose control of the balls, the music, and the game, bringing an early loss to the player.

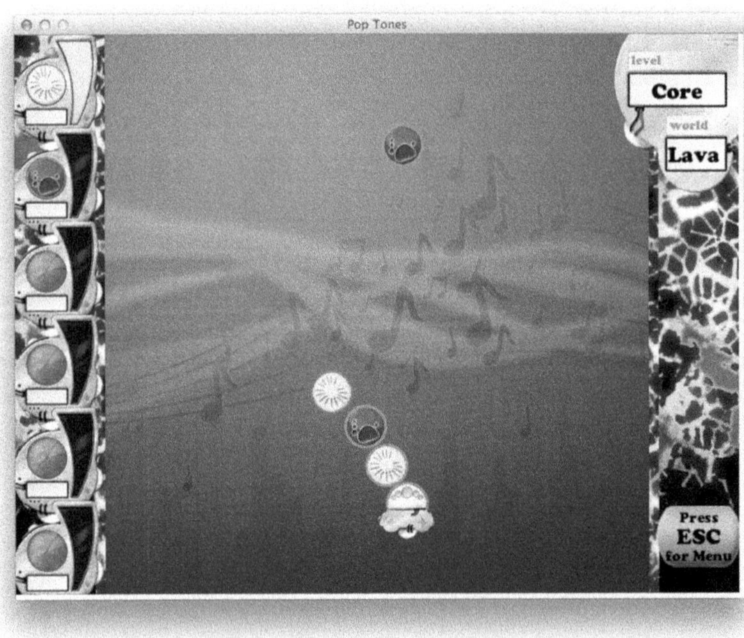

FIGURE 25.4 Screenshot of Pop Tones Gameplay

25.4.2 AMUSE

AMUSE is the AMEE Music Sharing Environment. It is a music-authoring tool that gives users access to the functionality of the AMEE engine for generating music, and is based on the application prototypes discussed in Hoeberechts and Shantz (2009). In addition to providing access to AMEE, *AMUSE* facilitates the sharing of concepts and musical ideas using Facebook as a social platform and cloud-based storage of content.

A screen shot of *AMUSE* is shown in Figure 25.5. As can be seen, the interface for *AMUSE* is organized using standards found in typical audio and music software. In edit mode, the user can create and work with multiple tracks and can assign instruments and edit attributes of tracks independently. The user can also tune attributes of the piece as a whole, including global mood settings that are used by the piece by default. The edit mode can be thought of as an offline mode, where all the necessary configurations and adjustments are made prior to the generation of the piece; any run-time manipulations of the piece are carried out in play mode.

Figure 25.6 shows *AMUSE* in play mode. In this mode, not only can the piece be played, but the player can also interact with the music, making various real-time adjustments to the piece. Typically, this involves splitting tracks into segments, with the various segments having some attributes changed between them during a transition period between the segments. This can be as simple as turning a track off and back on again or something more complex, like tuning the mood or emotion of the piece at various points, which is more relevant to the chapter at hand.

Such emotion adaptations can be specified using the sliders shown in Figure 25.6. When an adaptation is required, *AMUSE* signals AMEE's Emotion Mapper through its High Level API to ensure that the adaptation occurs when and as needed. Emotion

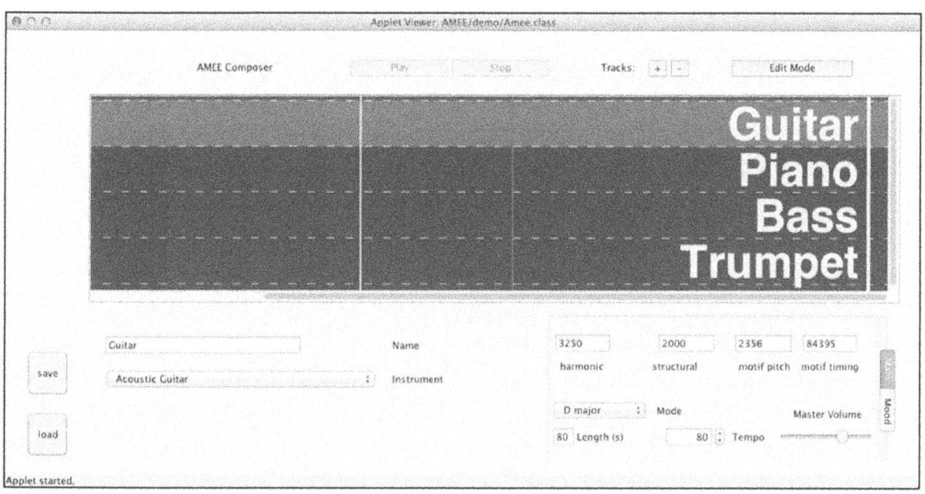

FIGURE 25.5 Screenshot of AMUSE in Edit Mode

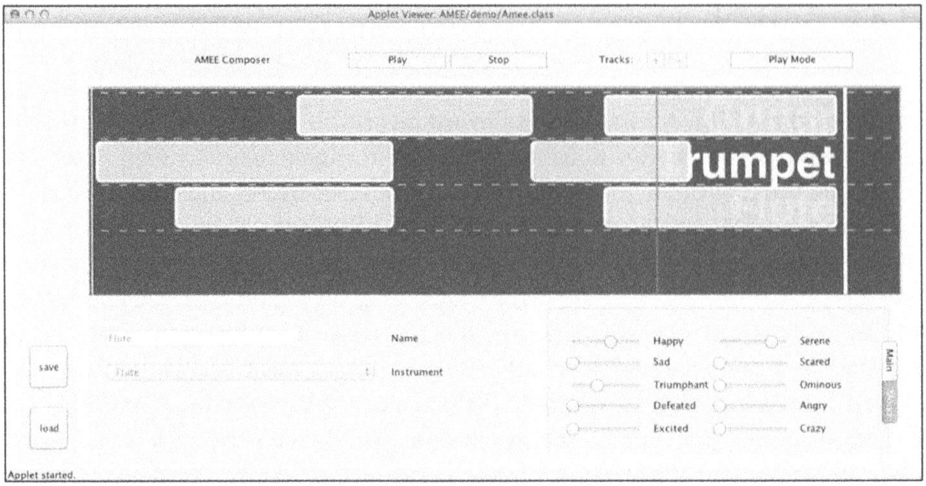

FIGURE **25.6** Screenshot of *AMUSE* in Play Mode.

adaptations are thus specified in an interactive fashion, applying to either individual tracks or the piece as a whole. In doing so, we can create a variety of emotionally engaging music in an interactive and adaptive fashion.

Since *AMUSE* is implemented as a Java applet, it can be executed in any web browser with an appropriate Java Virtual Machine. All storage of work is done remotely in a cloud, with access controlled through Facebook credentials to allow sharing of work, commentary, and so on, thus enabling a highly collaborative approach to music generation using *AMUSE*.

25.4.3 ART-G

ART-G is the AMEE Ring Tone Generator for mobile devices using the Java ME platform. The purpose of this software is to provide facilities to mobile phone users to create ring tones using AMEE that can be both personalized to the caller and emotionally adapted using AMEE's Emotion Mapper. *ART-G* is conceptually simple, but limitations built into most Java ME implementations on smart phones introduce many challenges. Generating a ring tone is as simple as providing to *ART-G* a phone number, which is used to seed the pseudo-random generators in AMEE, which in turn allows the same music to be generated every time that phone number is seen. Other attributes of the piece can be configured, including the emotional context to be used in the ring tone.

One issue limiting the usefulness of *ART-G* is the implementation and support of the Java telephony API in most smart phones. The original hope was that *ART-G* could run as a service on a phone, effectively replacing the standard ringer. When a call is received, the incoming phone number would be passed into *ART-G*, and the resulting ring tone would be played on demand. The user could set and periodically update their current mood, which would cause the ring tones to be adapted accordingly. Unfortunately the telephony API is not supported on most devices, and there is no way for a Java MIDlet to intercept the ringing of the phone. (While this is done as a reasonable security precaution, it limits *ART-G* to being a tool to generate ring tones for later use, as opposed to a background service for smart phones.)

Another issue is the sandboxing of the Java Virtual Machine on most smart phones. For security reasons, running Java MIDlets cannot access much of the local storage on a phone, and in particular the standard location for storing ring tones on the device. Consequently, *ART-G* cannot directly add generated ring tones to the phone's collection, and so any ring tones created by *ART-G* must be pushed off the phone first and then manually added to the phone's ring tone collection afterwards.

Lastly, real-time playback of generated music through AMEE or *ART-G* on a smart phone is complicated by the implementations of the Java mobile media API found on most phones. A Java Virtual Machine on a smart phone is required to support the playback of MIDI files according to the API standards. Providing access to the MIDI synthesizer on the phone, however, is optional, and most smart phones choose not to provide it. As a result, *ART-G* often lacks the direct access to MIDI controls that are necessary to generate and play back music in real time.

This issue was solved in the Java ME version of AMEE by creating a new MIDI control that leverages the fact that the Java Virtual Machine must support MIDI playback. Instead of directly accessing the MIDI synthesizer when it is not allowed, AMEE instead renders generated music into a collection of file buffers (music is typically double or triple buffered for performance reasons). The MIDI playback feature is then used to play these file buffers in sequence in real time, with specialized threads handling the timing of hand-offs between buffers to avoid skips in the music playback. With such an implementation, the Java ME version of AMEE creates the illusion of music being generated and played back in real-time through the MIDI synthesizer on the phone when this actually is not the case. *ART-G* takes advantage of this, allowing ring tones to be played back and adapted in real time (for example to change emotional context) or by saving them directly to files for later transfer and use.

25.5 Conclusions

Throughout this chapter we have explored the creation and use of emotional adaptation from automatically composed music. The basis for this is our AMEE software system, which has been used to create numerous engaging and interactive user experiences in the form of computer and videogames, music-authoring software, and so on.

While we have enjoyed great success on many fronts, this work is not without controversy of its own. When first exposed to the notion of automatically composed music, a common reaction is to consider the implications and impact on human composers, with the concern often being that technological advancements will displace human talent.

From our experiences, this can be a significant barrier to the adoption and development of this sort of technology. To the contrary, we see automatic composition as assisting human composers, enhancing and extending their capabilities. Experienced composers can encode their musical ideas into the Pattern Libraries of AMEE, allowing their own work to be resequenced and reimagined in new ways or embedded into interactive software and emotionally adapted in real time. To users with little or no musical background or expertise, automated systems such as AMEE present an opportunity to create music for the first time. In either case, the users are empowered by the technology and are not minimized or marginalized by it.

Content producers are also empowered by automatic composition systems like AMEE. Variety and adaptability allow for more engaging and immersive experiences for users. These systems also allow content producers to better leverage their in-house music talent, as discussed above. For small to medium-sized enterprises that cannot cost-justify their own in-house talent, systems such as AMEE provide access to affordable, royalty-free music content.

The implications of automatically composed music for end users of interactive applications are also substantial. Systems such as AMEE allow the creation of applications like *Pop Tones* that are adaptive and free of endless repetition and looping. Users of such applications can enjoy more engaging and more immersive experiences for longer periods of time. End users of authoring tools such as *AMUSE* are also empowered in creating their own music and collaborating with other users from around the world.

While much progress in this area has been made, there are still tremendous research opportunities to explore in the future. Numerous researchers, ourselves included, are continuing to study various strategies for improving the musicality of automatically composed music. Empirical studies in the area must continue to better understand the interplay between music, emotion, and user engagement and interest in interactive applications such as computer- and videogames. Such studies figure prominently in our ongoing work with AMEE. Automatically composed music also introduces many opportunities in games for using music as a gameplay mechanic, instead of just an accompaniment to the experience. We plan to explore these possibilities with AMEE going forward as well.

References

Abeles, H., and J. Chung. 1996. Responses to Music. In *Handbook of Music Psychology*, ed. Donald Hodges, 285–342. San Antonio, TX: IMR.

Bartlett, Dale. 1996. Physiological Responses to Music. In *Handbook of Music Psychology*, ed. Donald Hodges, 343–385. San Antonio, TX: IMR.

Comair, Claude, Rory Johnston, Lawrence Schwedler, and James Phillipsen. 2004. Method and Apparatus for Interactive Real Time Music Composition. US Patent 6,822,153, filed May 14, 2002.

Costa, Marco, Pio Bitti, and Luisa Bonfiglioli. 2000. Psychological Connotations of Harmonic Musical Intervals. *Psychology of Music* 28: 4–22.

Davies, Stephen. 1994. *Musical Meaning and Expression*. Ithaca, NY: Cornell University Press.

Demopoulos, Ryan, and Michael Katchabaw. 2007. MUSIDO: A Framework for Musical Data Organization to Support Automatic Music Composition. In *Audio Mostly 2007 Conference Proceedings*. Ilmenau, Germany: Fraunhofer Institute for Digital Media Technology.

Dobson, Jason. 2006. Product: Creative, dSonic Release UT2004 Audio Patch. Appeared in Gamasutra Webzine on July 27, 2006. Accessible online at: http://www.gamasutra.com/view/news/101200/Product_Creative_dSonic_Release_UT2004_Audio_Patch.php.

Eladhari, Mirjam, Rik Nieuwdorp, and Mikael Fridenfalk. 2006. The Soundtrack of your Mind: Mind Music—Adaptive Audio for Game Characters. In *Proceedings of the 2006 ACM SIGCHI International Conference on Advances in Computer Entertainment Technology*, 54. New York: ACM.

Gabrielsson, Alf, and Patrick Juslin. 1996. Emotional Expression in Music Performance: Between the Performer's Intention and the Listener's Experience. *Psychology of Music* 24: 68–91.

Hevner, Kate. 1935. The Affective Character of the Major and Minor Modes in Music. *American Journal of Psychology* 47 (1): 103–118.

——. 1936. Experimental Studies of the Elements of Expression in Music. *American Journal of Psychology* 48 (2): 246–268.

Hodges, Donald, ed. 1996. *Handbook of Music Psychology*. San Antonio, TX: IMR.

Hoeberechts, Maia, Ryan J. Demopoulos, and Michael Katchabaw. 2007. A Flexible Music Composition Engine. In *Audio Mostly 2007 Conference Proceedings*. Ilmenau, Germany: Fraunhofer Institute for Digital Media Technology.

——. 2009. *A Flexible Music Composition Engine*. PCT Patent WO 2009/036564.

Hoeberechts, Maia, D. Kerr, R. Pacheco, C. Eineke, B. Wheeler, Jeffrey Shantz, N. DiZazzo, J. Bindra, M. Katchabaw, and M. Park. 2012. Pop Tones. *To Be Published*, Q1 2014.

Hoeberechts, Maia, and Jeffrey Shantz. 2009. Real-time Emotional Adaptation in Automated Composition. *Proceedings of Audio Mostly 2009*, September 2–3, 2009, Glasgow Caledonian University, Scotland.

Juslin, Patrick, and John A. Sloboda. 2001. *Music and Emotion: Theory and Research*. Oxford: Oxford University Press.

Juslin, Patrick, and Daniel Västfjäll. 2008. Emotional Responses to Music: The Need to Consider Underlying Mechanisms. *Behavioral and Brain Sciences* 31 (5): 559–621.

Legaspi, Roberto, Yuya Hashimoto, Koichi Moriyama, Satoshi Kurihara and Masayuki Numao. 2007. Music Compositional Intelligence with an Affective Flavor. *Proceedings of the 12th International Conference on Intelligent User Interfaces*, 216–224, New York: ACM.

Livingstone, Stephen R., Ralf Mühlberger, Andrew R. Brown, and Andrew Loch. 2007. Controlling Musical Emotionality: An Affective Computational Architecture for Influencing Musical Emotions. *Digital Creativity* 18 (1): 43–53.

McMullen, Patrick. 1996. The Musical Experience and Affective/Aesthetic Responses: A Theoretical Framework for Empirical Research. *Handbook of Music Psychology*, ed. Donald Hodges, 387–400. San Antonio, TX: IMR.

Nielzén, S. and Cesarec, Z. 1981. On the Perception of Emotional Meaning in Music. *Psychology of Music* 9: 17–31.

Peretz, Isabelle, and Robert J. Zatorre, eds. 2003. *The Cognitive Neuroscience of Music*. Oxford: Oxford University Press.

Radocy, Rudolf E. and J. David Boyle. 2003. *Psychological Foundations of Musical Behavior.* Springfield, IL: Charles C. Thomas.

Schubert, Emery. 1999. *Measurement and Time Series Analysis of Emotion in Music.* PhD diss., University of New South Wales, Sydney.

Simon, Ian, Dan Morris, and Sumit Basu. 2008. MySong: Automatic Accompaniment Generation for Vocal Melodies. In *Proceeding of the Twenty-Sixth Annual SIGCHI Conference on Human Factors in Computing Systems*, 725–734. New York: ACM.

Terwogt, M., and F. van Grinsven. 1991. Musical Expression of Mood States. *Psychology of Music* 19: 99–109.

Williams, John. 1977. *Star Wars* (Main Theme), London Symphony Orchestra, conducted by John Williams, recorded March 1977, 20th Century Records, 2T–541, 33⅓ rpm [film later retitled *Star Wars Episode IV: A New Hope*].

CHAPTER 26

A REVIEW OF INTERACTIVE SOUND IN COMPUTER GAMES

Can Sound Affect the Motoric Behavior of a Player?

NIELS BÖTTCHER
AND STEFANIA SERAFIN

> We used to use MIDI and operational code to basically control FM synthesis in realtime on the super Nintendo, because it had actually an FM synth in it. (Gordon Durity, EA games Vancouver, from an interview in April 2012)

Procedural or real-time synthesized sound is not a new phenomenon. Consoles as early as the PlayStation and the Nintendo 64 had real-time FM synthesis built in, and were controlled by MIDI and operational code. With the introduction of the CD and sampling techniques, real-time synthesis was removed from the consoles and substituted instead by high-quality samples loaded into the RAM. This development was a major advancement in terms of sound quality, since game audio turned toward the direction of film sound in relation to sound quality. However, this big step forward could also be seen as a step backward in relation to the exploitation of the interactivity, dynamic narrative, and structure that a computer game has to offer. This idea is exemplified with the game *Myst* where Collins (2008, 66–7) describes how different minimal compositional approaches are used on the looped musical motifs in order to make them less memorable. This game is a very good example of how sound designers and composers had to think alternatively to avoid repetition of the music in the games as the more dynamic MIDI techniques were abandoned in favor of more linear tracks and loops.

Game sound includes music, background atmospheres, reward sound, navigational sounds, dialog, and much more. However, all of these categories cannot be dealt with in one chapter. Accordingly, this chapter will explore the current status of procedural or real-time synthesized audio in computer games in general, with a focus on interactive sounds related to the kinetic gestural movements and actions of the player. Such actions are most often found in fighting games, family entertainment, first-person shooters, sports games, and music games.

26.1 Why Is Procedural Audio Interesting?

When discussing procedural audio, it is sometimes argued that it is not yet possible to implement real-time procedural models that have the same sound quality as high-quality samples. Procedurally generated audio still has a lot to offer that has not yet been fully explored. This chapter will among other issues discuss some of the advantages as well as disadvantages of procedural audio for computer games (see also Chapter 32 in this volume).

In a typical game production, game audio is allocated a very small portion of the total CPU and memory of the console. The rest of the CPU and memory is used for graphics, AI, networks and similar tasks. Some sound designers describe the process of having available even 10 percent of the CPU to be a hard task, which needs good arguments when facing the producer. The very limited memory for samples is often a major problem for sound designers, as they have to compromise with the number of samples to stream or load into memory.

In order to avoid repetition on sounds such as footsteps, shotguns, swords, or similar, sound designers often record many different variations of the same sound that can be randomized. In addition, different randomizations on parameters such as amplitude and pitch, among others, are also applied to the sound. The task of recording many different versions of the same sound as well as the programming of the randomization can be a very time-consuming process. If tools for applying randomization on the frequency spectrum and characteristics of the sound samples could be used, sound designers could save a lot of time in recording and producing several different versions of the same sound. Some tools already exist for this purpose and will be discussed later in this chapter.

Within the last few years, new interfaces to control computer games have started to appear on the market. These products include the Nintendo Wii controller, Sony Eyetoy, PlayStation Move, Microsoft Kinect, Novint Falcon, Apple iPad, and many others. These devices contain a variety of sensors such as microphones, accelerometers, and cameras that make it possible to analyze sound input, detect motion, and retrieve sophisticated information about acceleration, 3D orientation, and more in relation to the user's motion. These new means of controlling games open up the possibility of engaging the motor behavior of the players in a more varied and active way, compared with the traditional joypad, joystick, mouse, or keyboard. These new interfaces also provide alternatives for interactive sound. Today, in most computer games designed for sensor-based controllers, prerecorded samples are still commonly used to simulate sounds produced when users interact within a scenario by moving parts of their body. When using a more traditional game controller this interaction seems to be satisfying. However, with the much more detailed gestural information retrieved from the newer game controllers this interaction tends to seem somewhat limited, especially in games reliant on user movement: sword fighting games, sports games, and the like.

In most movement-based computer games a prerecorded sound may be triggered whenever the player moves a body part or controller above a certain threshold. But what would happen if these sounds were generated in a purely procedural way that reflects the user's movements and other variables from the game engine? Could this perhaps affect the player's perception of sound quality, engagement, motor behavior, or other elements of gameplay?

26.2 Where to Apply Procedural Audio?

Procedurally generated audio will of course require CPU processing and in this context the sound designer must compete with all the other processes in the game. However, procedural audio can actually lead to memory savings in the RAM, as either no or very few samples have to be loaded into memory. Until procedural audio can produce the same sound quality and level of detail as a well-processed sample, it may not make sense that all of the audio within a game should be made procedurally. But in the cases where audio requires dynamic changes and where the interactive aspects become important, it certainly makes sense. When physics are involved in a game, the use of real-time synthesized audio becomes extremely relevant, as it would normally require a large number of samples and involve time-consuming programming to make the sound realistic and nonrepetitive.

In a typical first-person shooter (FPS) game one will hear the weapons fired by the main character and the enemies close to the player. These sounds often have to be of as high quality and as impressive as possible, as this is what the user expects from generations of FPS games using high-quality Foley recordings of real weapons (Foley sound is reproduction of everyday sounds, which are added in the postproduction stage of typically a film). Presently, it would be difficult to generate these kinds of sounds purely through procedural models in a convincing way without losing many important microdetails. But in the background, the player will also hear a lot of shooting from secondary characters and action happening in the distance. These sounds are often variations of the same samples, being used in the front just with a filter applied or lower amplitude, and they still take up RAM or streaming space. A possible alternative could be to apply the same kinds of thought to procedural audio as with layered 3D graphics: differentiating the level of detail and implementing these sounds by simple noise-based procedural models not taking up space in the RAM. Noise-based source-filter models would most likely not be convincing enough for the front sounds in terms of quality, but for the background the user would potentially not notice it, as they would be unlikely to focus on residual audio and microdetails while engaged in gameplay. Inexpensive computational models could produce these kinds of sounds and procedural variations could easily be applied to them, in the same way that less-detailed 3D graphics are typically found in the layered background of a game. This approach could be highly relevant in games where the player can see several distant background actions. An example of this could be FPS games similar to *Far Cry 3* or the series of *Call of Duty*.

Another approach is to incorporate a Level Of Audio Detail (LOAD) directly in a procedural audio model, as described in by Andy Farnell (2012). The same concept is also referred to as the sonic level of detail: Eoin Mullan (2012) discusses the sonic level of detail for modal synthesis designed for computer games. Modal synthesis is a sound synthesis technique using modal analysis to describe the vibrations of an object. The acoustic mode is found in the frequency domain looking at the resonances in an audio spectrum.

Another use of procedural models could be in environmental sounds. Anyone with experience of microphones and Foley recordings understands the difficulties of recording environmental sounds such as wind, rain, and stormy weather. Environmental sounds often require longer samples in order to avoid repetition, which also takes up a lot of the memory. By contrast, procedural models offer a significant reduction of memory usage because they are most likely generated by synthesis in real time and therefore don't need any samples to be stored in the memory.

26.3 Procedural Audio in Commercial Games and Available Tools for the Production

Very few commercial games actually use procedural audio today. What may be the reason behind this will be discussed later in this chapter. In the following are a few examples of commercial games using procedural audio and the available tools for its production. The game *Spore* incorporates procedurally generated music, including different algorithms for increasing and decreasing musical density and other musical parameters, based on the interactions of the user in the game. The visual programming language Pure Data was among other elements used to generate the interactive sound in *Spore*.

In addition, some racing games have also used procedural techniques, mostly techniques similar to granular synthesis. One example is the game *Need for Speed*, where the car engine sound is based on recordings of parts of the sound, such as the gearshift, among others. A model was then developed for simulating the engine acceleration and de-acceleration, based on offline analyses of engine recordings. The RPM of the engine were then simulated and different modulations performed in real time based on the variables in the game.

In general, many racing or vehicle games tend to be forerunners in the development of creative interactive audio, and in some cases also procedurally (or almost procedurally) driven audio. One reason may be that sound plays a very central role in these games, but these kinds of racing games do not contain many other types of audio content compared to other types of games. The game *SSX* employs real-time digital signal processing (DSP). In *SSX* various music and sound analysis techniques are used on the personal music that the user loads into the game.

In particular, the discrete Fourier transform (DFT), look-ahead fast Fourier transform (FFT), and transient analysis of the music are applied in order to perform tasks such as real-time beat detection and other techniques. Using in-house production tools, different effects and modulations can then be applied to the music, including the application glitches, sample and hold, filters, and more.[1] Today, most companies use audio middleware solutions to deal with the sound in the game. Audio middleware is an interface between software, whereby the interface design is intended to reduce programming requirements for the sound designers, while allowing for connectivity with the game engine. By using audio middleware the sound designer can link sound to different game objects, such as specific areas, variables, or event triggers.

One of the important aspects about audio middleware is that a part of the tool is often designed with an easy-to-use graphical interface, specifically created for sound designers. In this way designers can implement sounds directly and test them with the game running in real-time. Two of the main middleware solutions of today are Fmod and Wwise. Both Fmod and Wwise allow for third-party plug-ins and effects, as well as a few additional plug-ins for applying procedural audio. Wwise has a plug-in called SoundSeed, which has been used in a few commercial games. SoundSeed is a parameter-based synthesis plug-in consisting of an air model for generating wind as well as "whooshy" aerodynamic sounds like a sword in motion. The plug-in also includes a modal synthesis-based impact module for generating variations on audio files, typically for impact sounds, footsteps, and similar.

The French company audioGaming has developed a set of plug-ins for the audio middleware Fmod. One plug-in is called AudioWeather, which was developed for procedurally generating weather sounds such as wind and rain, where the user can change the amount of wind, rain drops, fluctuation, and different timbres of the sound. Another plug-in called AudioFire is designed for generating fire sounds. AudioGaming has also developed a middleware solution named audioGesture, specially designed for sonifying gestures procedurally. This middleware includes different premade weapon models and other objects.

Sony Computer Entertainment Europe (SCEE) has an in-house middleware tool called SPARK (SCEE Procedural Audio Real-time Kernel), which was developed by the audio programmer Nicolas Fournel. The *SPARK* tool is an analysis and resynthesis tool, specifically designed so the sound designer can load in a static sample, analyze it, and then automatically create a dynamic model from the original sample. The tool uses many different analytical elements, including envelope followers, transient analysis, spectral analysis, and others. Subsequent modeling uses more refined techniques, including curve fitting, distribution finding, and modal analysis. Different materials and variations can afterwards easily be applied to the sound in the tool. The basic idea behind this software was to make a tool specifically designed for the nontechnical sound designers using a simple GUI, where the focus should not be on developing technical sound models or researching audio production mechanisms or physics, but rather on the creation of samples where the software will create dynamic models automatically. With this technique the sound designer is not limited to premade models designed by an audio programmer, but can use his own recorded or processed sample and preserve the personal artistic characteristics of this sound.

Premade procedural models could for some sound designers seem to be limited, if they do not follow the overall sound aesthetics of that specific sound designer or game or if the model is completely fixed with a few possibilities for changing different parameters or timbre.

26.4 Why Is Procedural Audio not already Implemented?

It has been possible to employ both physical sound models and more perceptual-based sound models for decades. But why have those models not already been integrated in computer games? When the early 3D graphics were introduced, they were slow to be integrated into computer games because of their poor quality. Over time and through the evolution of processors, graphic cards, and programming techniques, the quality of 3D graphics improved substantially. As mentioned earlier in this chapter, the integration of sampling and of the CD to consoles slowed the use of real-time synthesis, and game developers started using sample-based prerecorded audio, which has much better fidelity. It was now possible to use the same sound quality in videogames as in movies, and from this point forward it would be difficult to envision a return to real-time synthesized audio.

Much research shows that the quality of audio has a strong influence on the visual perception. In Storms and Zyda (2000), it was shown that high-quality audio could increase the perception of high-quality visuals in audiovisual applications. It was also shown that low-quality visuals could decrease the quality perception of the audio. It was not shown however that high-quality visuals could increase the quality perception of audio. Many other experiments have shown different influences of audio on graphics. Skalski and Whitbred (2010), describe how sound quality (in this case Dolby 5.1 surround sound vs. Dolby Stereo) has a great impact on the perceived presence and enjoyment in a computer game, whereas high-definition image quality has no effect compared to standard image quality.

Similarly, Gaston and others (2010) showed that MP3-quality audio enhances the perception of pixilation in video, and Hulusic and colleagues (2011) describe how movement-related sound effects can improve the perceived quality of the animations in a computer game by proving it is possible to lower the frame rate of an animation when movement-based sound effects are used. The above examples illustrate how it is possible to improve the perceived quality of graphics by the use of good-quality audio. But it does not seem to be as easy to do it the other way around. Concerning procedural audio it seems to be critical that bad-quality audio can decrease the perceived quality of the visuals. Therefore, if real-time synthesis has to be applied it is not necessary to go through the same extended development period as 3D graphics had. If procedural audio is to be introduced, it simply has to sound as good as a well-made Foley recording, maybe even better, or in some way increase the entertainment value significantly for the player in order to be accepted.

The main reason for procedural audio not being used more in games is the sound quality of the models. It is not yet possible to compete with high-quality Foley recordings in terms of microdetails and sound quality. It is an extremely complex procedure to create models that sound as promising as a rich sound recording. In some cases precise and very computational heavy physical models have been developed (James, Babic, and Pai 2006; Zheng and James 2011). These models contain some of the richness of the real samples, but they are far too heavy to compute in real time with a game running 3D graphics, AI, and other network tasks.

As the time for producing sound in computer games is often very limited, sound designers often simply do not have the time to program or develop procedural models for each new game. Not many tools currently exist for applying procedural audio in games using audio middleware. Until this situation is resolved, it is likely that not much procedural audio will be seen in commercial computer games, particularly not in the bigger commercial computer games with large budgets and tight time schedules.

Another complication from using real-time-synthesized audio could be the task of testing the game for eventual errors. When testing a game based on sample playback it is always fairly easy to locate an eventual error in the sound. Then one often has to discover if it is a wrong sample not corresponding to the specific asset or perhaps a wrong sample rate that is being used. The error can be corrected quickly and the problem is solved. When using an interactive procedural audio model there could in some cases be many different elements that have caused the error: there could be wrong input parameters or many other things could be wrong. In some cases, it could become a time-consuming process to locate errors within the model, and again the time factor would become a critical issue for the game developers.

26.5 How Realistic Should Procedural Models Be?

> In many ways the realism aspired to in games is not a naturalistic realism in the sense of being a simulation of reality, but a cinematic realism that relies on established motion picture convention. (Collins 2008, 134)

Reality is one thing, and its transposition into audiovisual two dimensionality (a flat image and usually a monaural soundtrack), which involves radical sensory

> reduction, is another...there is really no reason for audiovisual relationships thus transposed to appear the same to us as they are in reality, and especially for the original sound to ring true. (Chion 1994, 96)

Much has happened since the book *Audio-Vision: Sound on Screen* by Michel Chion (1994) was written. We now have advanced 3D sound, 3D graphics, and in some cases even 3D stereo graphics—and as input we have interfaces with several different sensors. But still this idealized state is very far from the realistic sensorial input we get from our every day environments.

People working with real-time synthesis or procedural audio for games often tend to simulate reality to a hyperrealistic degree. Classical sound designers educated in film schools or similar tend to have a completely opposite approach. Perhaps the reason for this division is that many of the academic people working with procedural audio, real-time synthesis, or physical modeling often have a very technical background, coming from fields such as computer science or music technology. Here we often see an attempt to use physics to reproduce reality in a very naturalistic way. In the case of an experiment that aims to achieve complete immersion in a virtual-reality environment it makes more sense to attempt to reproduce a naturalistic reality, but with a computer game, it does not make sense to try to reproduce exact realism from nature.

Currently, there are two key theories regarding procedural audio. People such as Nicolas Fourne argue that the essence of sound aesthetics should be with the sound designer and that the tools for generating the procedural content should be an automated process of well-produced samples, presently seen in tools such as SPARK. Conversely, people like Andy Farnell argue that "procedural audio is a philosophy of understanding the physical basis of the sound, and that sound designers should work with the physical behavior of the objects rather than just a manifestation."[2]

Many interesting physical sound models have already been implemented. The aerodynamic model from Dobashi, Yamamoto and Nishita (2003) is a good example of a very complex and realistic aerodynamic fluid model for simulating among other thing aerodynamic sword sounds for games. Many advanced impact models have been developed (Rath et. al. 2003: Van den Doel, Kry, and Pai 2001) simulating contact between different materials with different textures, different force, weight, and more. For the interested reader Eoin Mullan gives an overview of the physical models relevant for games and describes some of the problems and benefits that these models has to offer in Grimshaw (2010, chapter 16). All of these sound models are extremely interesting and important knowledge can be gained from them, but in a game context, they seem to miss the "game realism," and for an experienced gamer the sounds may even be perceived as unrealistic.

In the book *Audio Vision: Sound on Screen* (1994), Chion discusses the phenomenon of "added value": if you hear the footsteps of an evil villain in a movie (or in a game), the sound of the footsteps should carry the information that this character is a villain. Similarly, if you hear the sound of a wooden stick held by a murderer, the sound of the stick should express the character's intention with this object—this is the so-called added value of the sound. So the "meaning" and "intention" (affecting emotional connotation) might be the element that many of the hyperrealistic sound models lack in order for them to be perceived as "real," in the context of a computer game. It does not seem to be that important that the sound generated is based on correct physical parameters, but more that it is understood as a killer's weapon, a friendly stick, or something else.

When designing procedural audio we should on one hand look at the physics in order to retrieve as many parameters for the interaction as possible (or at least try to get inspiration from the physics), but we should also consider the emotional aspects the player should carry after having interacted with the object. Chion (1998) introduced another relevant concept, the "ergo-audition," the experience of hearing oneself acting within an acoustic manifestation of the world. According to Chion, the effect of the sound on the listener will be more joyful if the sound is meaningful, nuanced, and a bit surprising. Again this idea backs up the need for procedural nonrepetitive sound as well as procedural audio with meaning, intention, and added value.

Another interesting question to consider is what happens when we raise the level of interaction and use these new interfaces capable of tracking precise velocity, acceleration, and similar elements? Does our perception of game realism change? And could an interactive sound model with lower sound quality be perceived as better, when the user interacts physically with the model and understands all the varying aspects of the sound model? It has been shown that active participation in a computer game can decrease the rating of audio correctness compared with a passive participation, where players only see and listen to a game (Grimshaw 2010). One could imagine that the more the movements with their body reflect the sound generated, the more the player will accept the quality derived from procedural audio.

26.6 An Example of an Interactive Procedurally Generated Sound Model

In May 2012, at Aalborg University in Copenhagen, Niels Böttcher designed an interactive sound model of a sword intended for use with the Wiimote. The model was designed based on the following four specifications:

- The model should be as interactive as possible and should be designed in close correlation to the kinetic movements of the player, simulating the physics that are happening when an object like a sword is swung in the air.
- The model should sound as "high quality" as possible in relation to what is normal standard in a typical commercial action computer game.
- The model should give the user a feeling of holding a killer's metal sword in the hand.

- The mapping between the gestures and the sound generated should be designed in such a way that it would open up the possibility to force the player to use his body in a more expressive or exploratory way than usual.

The sword sound was implemented in the visual programming language Max/MSP and combined with a running sword game implemented in the game engine Unity3D. The data retrieved from the Wiimote was sent via the OSC (Open Sound Control) protocol from Unity to Max/MSP. The sound of the sword consisted of three different main sound modules. Two granular modules and a subtractive-synthesis-based module. In order to simulate the aerodynamic turbulence between the movement of the sword and the air a subtractive-synthesis-based method was used. A simple bandwidth filter was applied to white noise and the acceleration of the Wiimote was mapped to the cutoff frequency of the filter, going from 400 Hz to 2000 Hz exponentially mapped so the user would achieve the higher frequencies only if the Wiimote was swung with a strong acceleration.

The Q-factor of the filter was also mapped to the acceleration, so that it would increase with acceleration, in the same way the bandwidth of the filter became narrower in case of a strong acceleration. The Q-factor describes the width of the bandpass filter as a function instead of a static bandwidth and keeps the proportion of the frequencies passed independent of its center frequency. In addition, the volume was mapped exponentially to the acceleration, so that one would have to swing the sword fairly hard to be able to hear the effect of this filter. The result was not a precise simulation of aerodynamic fluids, but a very simple and easy-to-understand swoosh effect similar to an object swung in air.

The first of the two granular modules was based on a Foley recording of a thin bamboo stick attached to a flat plastic sword being swung through the air. The flat plastic sword gave a low frequent massive sound and the bamboo stick gave a swooshy more high-frequency sound. The process of finding the right materials to do the Foley recording was one of trial and error, using a combination of many different sticks, wires, and materials. The main aim was to find a sound that was similar to what one would hear in a typical sword-fighting game on the market. The main feature of the granular synthesizer was simply to scrub synchronously through the sample.

Synchronous granular synthesis with rather large grains is typically good for reproducing the original sound. The acceleration of the Wiimote was mapped exponentially to the position of the scrubbing through the sample. The mapping was designed in a way that the player would never reach the end point of the sample, where the sound had an emphasized high-frequency swoosh-sound, if the wiimote was not swung hard enough. The pitch of the granulized sound was also mapped exponentially to the acceleration of the Wiimote, so the harder you accelerated the Wiimote, the more the pitch would increase. The amplitude was exponentially mapped in a similar way, where the amplitude would increase the harder the Wiimote was swung.

Ten voices were used for the granular synthesizer to make it sound wider, but fewer could also be used in order to save CPU. The distance between the grains was set to 20 ms, and the duration of the grains was set above 100 ms, to keep as many of the transients as possible and to lower the clicks when sliding into a new grain. The second granular synthesizer was applied in order to give the total sound a characteristic of metal on top of the wooshy sounds. The sample used for this part was obtained by recording a large kitchen knife being slid across an iron pot. The scrubbing position was mapped in the same way as the acceleration of the Wiimote and the amplitude and pitch were designed in the same way as the other granular modules.

In the end the three different sound modules were mixed in such a way that the first granular module of the main wooshy sword sound was placed highest in the mix. The metal sound was kept very low so that the sound would be only just audible for the player. Finally the wooshy aerodynamic sound performed by the filter-based subtractive sound module was mixed slightly higher than the metal sound, so one would just about notice the sound.

The samples in the granular modules can easily be changed and as a result there would be a completely different timbre and style of sword, which fits with the exact type of computer game and chosen aesthetics. This example shows the flexibility of the model's nature, which requires only that a high-quality sample is used. The same subtractive-synthesis model can be reused as it is used only as a subtle effect of aerodynamics and kept low in the total mix. Still this subtractive-based sound has a significant effect in giving the user a feeling of actually interacting with a real object and swinging it in the air. By changing the Q-factor of the filter, different sizes and shapes can also be simulated. By changing the pitch of the granular synthesizer, the perceieved size and shape of the object can also be changed very easily. Finally by changing the metal sound sample one could simulate a wooden sword or any other material.

26.7 Can Sound Affect the Physical Behavior of the Players?

In Böttcher and Serafin (2009) different interactive sound models were compared in a simple audio-only game. The main purpose of the test was to compare a set of different interactive sound models and evaluate the perceived sound quality as well as entertainment level between the different models. The sound models compared were modal synthesis, sampled sound, granular synthesis, and subtractive filter-based synthesis. The most interesting findings in the test were not the result of the test, but an unexpected observation made during it. The simplest of the four models, a simple subtractive-synthesized model similar to the one used in the above example of the sword design, made a huge impact on how the players performed while swinging the Wiimote.

The test showed that a very simple subtractive-synthesis-based sword sound model was rated as the most entertaining of the different models, and the test persons seemed to behave in a much more engaged way when using this model. Observing how the users were playing with the Wiimote during the game, there was a strong difference in body gestures, depending on which sound model was mapped to the controller. When playing with the model based on sampled sounds, users demonstrated a tendency to make the same gestures again and again. In only a short amount of time they learned exactly what they had to do in order to trigger the swing of the sword. The subtractive-synthesis patch was very interesting in relation to this observation. This sound seemed to cause the players to behave more exploratively and with an increased variety of gestural movement patterns.

This development was one of the most interesting observations of the test, and it was rather surprising how significant different gestural expressions were related to different sound models. It must be noted that these observations cannot be directly applied to other types of computer game, as this experiment was conducted using an audio-only game, which is likely to have had an influence on the players' perception of sound. The above test led to a second experiment, also performed at Aalborg University in May 2012. A prototype of a sword game controlled by the Wiimote was implemented to include AI enemy animation and graphics, as seen in a standard sword-fighting game.

The sword game was a simple FPS sword game, where the player would control an arm holding a huge metal sword. In order to keep the focus on the sound, premade animations were triggered according to a set of thresholds in the game, as is also the case in most contemporary sword-fighting games. The basic task in the game was to fight against a simple AI knight and hit the enemy more times than you were hit. In the test two different sound implementations were compared. A sample-based version and a version based on the above-mentioned interactive sword sound model. The sample-based version used a recording of the interactive sound model from above, with applied randomization on pitch, amplitude, and applied lowpass filter, in order to avoid repetition. At the end of the test the participants were asked if they had noticed any difference between the two games. Surprisingly, most of the test participants did not mention any difference in the sound, but some of them replied that they believed the gameplay was different in some situations. For example the test participants answered: "I could hit the knight harder in the last game," or, "The swing was more forceful in the second version." These responses led to an interesting observation, as the gameplay offered in the two games was completely identical. The design of the sound made the test players believe that the game mechanics were actually different in the two games. The sample-based sound was a recording of the interactive model, based on the exact same sound, just not interactive in the same way. The more interactive kinematic self-sound made the players believe that the game mechanics were actually different, and that they could affect the damage on the enemy in another way.

26.8 Conclusions

Procedural and real-time synthesis for computer games is still in its very early stages. Much work has to be done before we will see this type of audio as standard in computer games. There are many arguments that promote the use of procedural and real-time-generated audio. First of all there is a great potential for an improved game experience on the player's side. Second, the player could develop a more explorative and less repetitive motor behavior. By using procedural audio the programmers could save memory, and finally the sound designers could potentially retrieve new tools and ideas for applying less linear and more dynamic sound design. Currently most of the sound design implemented in contemporary commercial computer games seems to have its inspiration from the many Hollywood film productions. As an example, most weapon sounds are produced in the same way as the Foley artists have done for decades. Sounds are to be embedded in games in different variations in order to avoid repetition. But why not explore the media and technology and try to do something completely differently? There is so much potential for creativity and exploration in computer games with the technology available today.

There is a need for more procedural models or tools capable of being embedded directly into the audio middleware. Yet before this is the case we will not see much procedural audio implemented in commercial computer games. As always we will most likely not see innovative and experimental sound design from the large commercial game companies, mainly due to tight budgets and little time for creative exploration. The first real use of procedural audio in a commercial computer game will most likely come from one of the smaller independent companies who can risk taking a chance in doing something different.

Notes

1. Interview with Gordon Durity, EA Games, Vancouver in March 2012
2. This argument is a fragment of an interview with Andy Farnell conducted in December 2011 by Niels Böttcher.

References

Böttcher, Niels, and Stefania Serafin. 2009. Design and Evaluation of Physically Inspired Models of Sound Effects in Computer Games. *Proceedings of the 35th AES International Conference on Audio for Games*. New York: Audio Engineering Society.

Chion, Michel. 1994. *Audio Vision: Sound on screen*. New York: Columbia University Press.

——. 1998. *Le son*. Paris: Paris Editions Nation.

Collins, Karen. 2008. *Game Sound: An Introduction to the History, Theory and Practice of Video Game Music and Sound Design*. Cambridge, MA: MIT Press.

Dobashi Yoshinori, Tsuyoshi Yamamoto, and Tomoyuki Nishita. 2003. Real-time rendering of Aerodynamic Sound Using Sound Textures Based on Computational Fluid Dynamics. *ACM Transactions on Graphics* 22 (3): 732–740.

Farnell, Andy. 2010. *Designing Sound*. Cambridge, MA: MIT Press.

——. 2012. Behaviour, Structure and Causality in Procedural Audio. In *Game Sound Technology and Player Interaction*, ed. Mark Grimshaw, 313–339. Hershey, PA: Information Science Reference.

Gaston, Leslie, Jon Boley, Scott Selter, and Jeffrey Ratterman. 2010. The Influence of Individual Audio Impairments on Perceived Video Quality. In *Proceedings of the Audio Engineering Society 128th convention* (AES 128), London, UK.

Grimshaw, Mark. 2010. *Game Sound Technology and Player Interaction: Concepts and Developments*. Hershey, PA: IGI Global.

Hulusic, Vedad, Kurt Debattista, Vibhor Aggarwal, and Alan Chalmers. 2011. Exploiting Audio-visual Cross-modal Interaction to Reduce Computational Requirements in Interactive Environments. *Visual Computer* 27 (1): 57–66.

James, Doug L., Jernej Babic, and Dinesh K. Pai. 2006. Precomputed Acoustic Transfer: Output-sensitive, Accurate Sound Generation for Geometrically Complex Vibration Sources. *ACM Transactions on Graphics* 25 (3): 987–995.

Mullan, Eoin. 2012. Physical Modelling for Sound Synthesis. In *Game Sound Technology and Player Interaction*, ed. Mark Grimshaw, 340–360. Hershey, PA: IGI Global.

Rath, M., F. Avanzini, N. Bernardini, G. Borin, F. Fontain, L. Ottaviani, and D. Rocchesso. 2003. An Introductory Catalog of Computer-synthesized Contact Sounds, in Real-time. In *Proceedings of XIV Colloquium on Musical Informatics* (XIV CIM 2003) Firenze, Italy, May 8–10.

Skalski, Paul, and Robert Whitbred. 2010. Image Versus Sound: A Comparison of Formal Feature Effects on Presence and Video Game Enjoyment. *PsychNology* 8 (1): 67–84.

Storms, Russell L., and Michael J. Zyda. 2000. Interactions in Perceived Quality of Audio-visual Displays. *Presence: Teleoperators and Virtual Environments* 9 (6): 557–580.

Van den Doel, Kees, Paul G. Kry, and Dinesh K. Pai. 2001. Foley Automatic: Physically-based Sound Effects for Interactive Simulation and Animation. In *Proceedings of the 26th Annual Conference on Computer Graphics (SIGGRAPH 2001)*, 537–544. New York: ACM.

Zheng, Changxi, and Doug L. James. 2011. Toward High-quality Modal Contact Sound. In *Proceedings of SIGGRAPH 2011*. New York: ACM.

CHAPTER 27

INTERACTIVE SPECTRAL PROCESSING OF MUSICAL AUDIO

VICTOR LAZZARINI

THE spectral processing of digital audio signals has been an area of active research interest for over four decades. Since the major developments in computing capacities over the recent decades, it has become a realistic proposition for practical applications in real-time[1] audio processing with general-purpose computing devices (such as desktops, tablets, mobile phones). Established analysis-synthesis methods such as the phase vocoder, as well as a variety of transformation techniques, have provided sound designers with an arsenal of audio manipulation processes.

In this chapter, we will explore the main principles of spectral processing from a practical viewpoint. The discussion will include an exploration of the most useful methods of sound processing that can be employed in interactive applications for different scenarios. Code examples are provided in the high-level language Python, and as Csound instruments, which are ready to be deployed in a variety of platforms using the Csound API and the Mobile Csound Platform (MCP), a software developer's kit (SDK) for ubiquitous musical applications.

As discussed by Moorer more than a decade ago (Moorer 2000), the computing power that is available for audio processing in general-purpose platforms has been increasing constantly, not only in terms of processor clock speed, but also in memory, storage, and networking capacity. In conjunction with these developments, the revolution brought about by ubiquitous computing puts significant resources at our disposal. These can be harnessed to create applications that could provide interesting opportunities for musical expression. The fine-grained control afforded by the spectral processing methods discussed in this chapter, in conjunction with the new methods of control and interaction present in modern hardware (touch surfaces, accelerometers, cameras, etc.), gives us great tools for developing and using new computer music instruments.

Various use cases of spectral processing stand out in musical applications. The capacity of stretching or compressing sound without the side-effect of pitch transposition is an attractive prospect, as we can manipulate the tempo of recordings to suit our needs. With fine independent control of pitch in real time, we can correct mistunings on the fly. For vocal sounds, we can add harmonizing background voices without the distorting effects of formant shifts, which are responsible for the infamous "chipmunk" songs. Hybrid textures and instrument can be created with cross-synthesis and interpolation of spectra. A whole collection of unusual and striking effects can be created with freezing and delaying different parts of the spectrum.

These are but a few of the various applications of the methods discussed in this chapter. The examples provided in the text will cover some of these possibilities, building a foundation for further exploration of spectral processing techniques. Composers and sound artists should be able to employ these examples directly or modify them for their purposes, either for real-time performance or offline composition and sound design.

For general-purpose interactive audio applications, spectral processing offers extensive forms of transformations. Media application designers can take advantage of this to add capabilities such as sound morphing or room models to their work. Likewise, game developers will be able to enrich the audio interaction features currently supported in their products. The Csound synthesis engine can be embedded in all of these application scenarios (Walsh 2010), enabling comprehensive support for spectral processing.

27.1 From the Beginning: Spectral Analysis Fundamentals

What does "spectral" mean, in relation to audio? There is a wide range of concepts that are linked to this word, but they all point to looking at or considering the *components* of a sound waveform, and the principle of breaking down an audio signal into a set of such components. So, from this perspective, any audio signal is made up of simpler, pure, sound waveforms, which, combined, provide us with the various sounds we hear. Such simple waveforms are seldom heard in isolation in the natural environment, but the sound of a tuning fork offers a good approximation to them.

This is an idea with long pedigree, going back to Helmholtz's pioneering studies in the nineteenth century (Helmholtz 1954), which applied the mathematical principles defined by Fourier (Fourier 1873) to soundwaves. The simple, pure, components in this analysis are sinusoidal (sine, cosine) waves. We can call the set of such waves, which, when combined make up an audio signal, the *spectrum* of that sound.

Designing sounds from a spectral perspective is a very established principle (Fischmann 2000). From the early electronic instrument designers to the latest computer software, this approach has been prevalent, and provides the basis for the development of a number of techniques of synthetic audio generation (Dodge and Jerse 1996), such as additive, subtractive, and nonlinear distortion. All of these methods consider

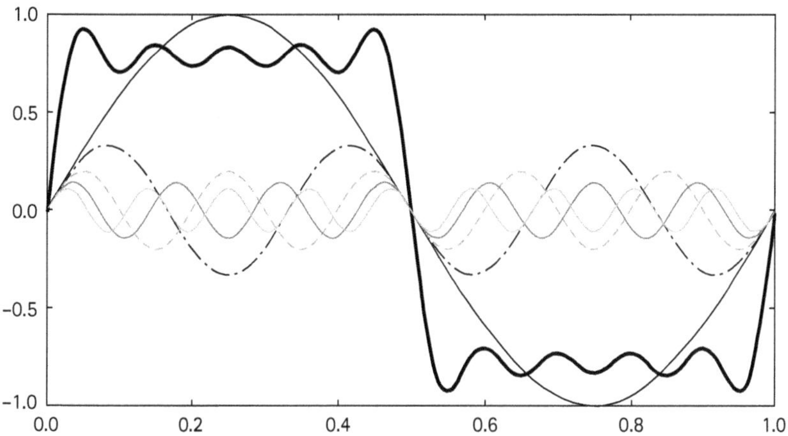

FIGURE 27.1 A square wave (thick line) as a sum of ten sinusoidal components via additive synthesis.

the target sound as having a certain spectral makeup and then employ different methods to recreate it, including: adding components together, sculpting them out, and generating them as a result of nonlinear operations.

Of these techniques, additive synthesis, which is simply based on mixing up carefully controlled sinusoidal components, is particularly relevant to the discussion in this chapter (Figure 27.1). Our goal is to be able to determine what the spectrum of a sound is, manipulate it and then create an output signal based on this manipulation.

27.1.1 The Fourier Transform

In order to obtain the "recipe" for a given "mix" that will reproduce a target sound, we can apply a number of analytical methods, all of which are based on a single principle, the Fourier Transform (FT). What we will try to find out is this: Given a target sound, what are its sinusoidal components?

These components will be identified by three parameters: frequency, amplitude, and phase (Figure 27.2). Frequency is related to how high or how low the component is, in perceptual terms (think, in a very loose way, of the pitch of a tone), and, physically, to how small or large a single sinusoidal wave cycle is, the inverse of its duration (or period). Amplitude relates to how prevalent the partial is in the mix (how much energy it has, relatively). Phase is more subtle: it indicates whether the component is either a sine, a cosine, an inverted sine, an inverted cosine, or something in between these four waveform shapes. The difference between a sine and a cosine wave is that the latter starts a quarter cycle later than the former. Inverted forms of these are half-cycle delayed in relation to the original waves. Simply put, as shown in Figure 27.2 phase can be summarized as the start position of a sinusoid at time 0. It is then possible to see that phase has to do with the synchronization (time delay) between the start positions of the various components.

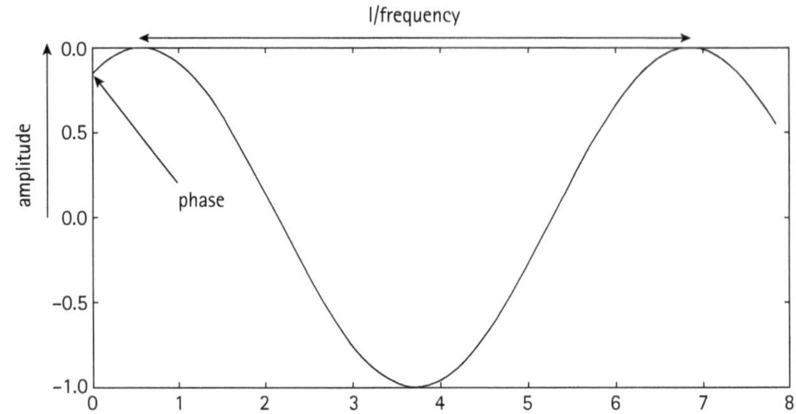

FIGURE 27.2 The parameters of each spectral component: frequency (represented in the figure as the waveform duration *T=1/frequency*), amplitude and phase.

The idea is straightforward, although its mathematical formulation might be seen as complicated. The FT uses a pair of sinusoid functions (a cosine and a sine), which make up what we call a *complex* sinusoid, as a component "detector." Such detectors are placed at all possible frequencies from −infinity to +infinity, and at each frequency the FT multiplies the signal being analyzed (which is taken to extend from −infinity to +infinity in time) with the respective sinusoid. This multiplication is point-by-point (every infinitesimal value of each signal) and the results at all of these points are then added together (an integral sum). The FT output is then the amplitude of cosines and sines detected in each frequency, called a spectral coefficient. If there is no component at that given frequency, this amplitude is zero and nothing is detected. If a component exists, then the result of the multiplication and integral sum will yield a non-zero coefficient. At all partial frequencies, the FT will produce these coefficients that will describe the spectrum of the input sound.

This operation gives us all the data we need: frequency of the component (which is the frequency of the "detector" sinusoid), its amplitude (spread out in the sine and cosine parts, obtained from the *magnitude* of the coefficient), and the phase (which can be worked out from the balance of energy in the sine and cosine parts, or the *argument* of coefficient). There is only a minor problem: The FT cannot be implemented in a digital computer. It is based on continuous time and continuous frequency (Steiglitz 1994), which is not ideally suited to such an implementation. It nevertheless remains a useful mathematical tool.

27.1.2 The Discrete Fourier Transform

In order to accomplish a practical implementation, we need a discrete form of this operation. A formulation of the transform that is adapted to sampled signals is called the Discrete Fourier Transform (Lazzarini 2010a). From the original FT, we keep the basic method, and we will use a complex sinusoid as a detector. Instead of using sinusoids at all possible frequencies, however, we will set these in a grid at discrete steps. The input signal will also be time-constrained, using a *window* of a given duration. Finally, the output will also be constrained, just a set of amplitudes for cosines and sines at the given step frequencies, as shown in Figure 27.3.

The operation works in a similar fashion to the FT, as we multiply the signal by the given complex sinusoid and obtain the spectral coefficient for the given frequency point. We repeat the process for each successive point, resulting in a sequence of sampled spectral coefficients. To determine frequencies, we start at 0 Hz, then proceed at integer multiples of a base (or fundamental) frequency, which is equivalent to the inverse of the window duration. For instance, if we analyze 4410 samples recorded at 44.1 kHz, then the window length is 0.1 seconds and the fundamental of the analysis is 10 Hz (1/0.1). The frequency points will be 0, 10, 20, 30, 40 . . . 22040, 22050 Hz.

We stop at half the sampling rate, 22.05 kHz (also known as the *Nyquist frequency*) because it is the limit for discrete-time signals. The next issue to determine is the number of spectral coefficients. If we started with *N* discrete time-points, we will have *N* discrete frequency points. But if the points listed above make up only 2206 coefficients, where are the other 2204? Beyond the Nyquist frequency, we fold back to the *negative* spectrum and continue incrementing. So the following points will be −22040, −22030, . . . −20, −10. The halfway point actually represents both the *positive* and the *negative* Nyquist

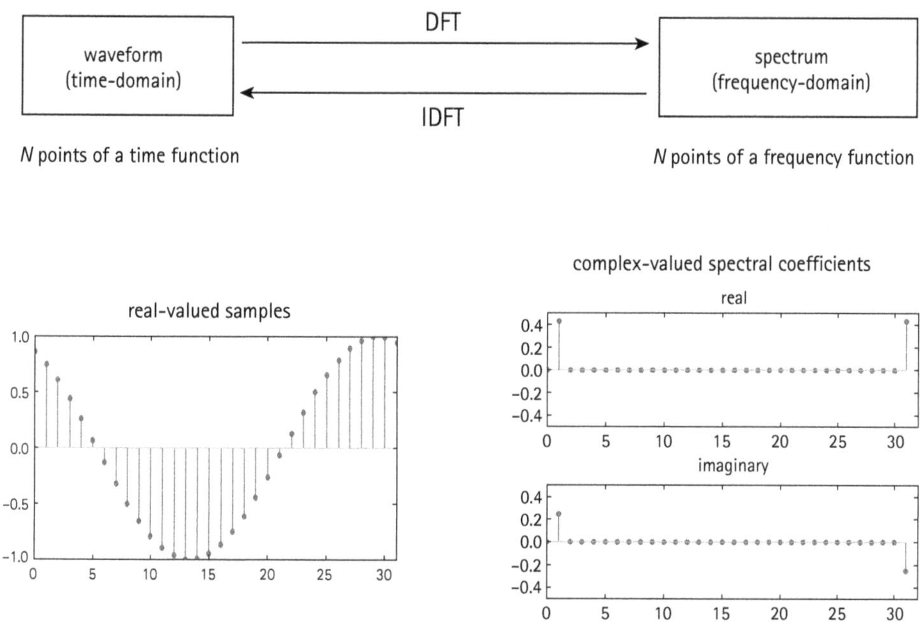

FIGURE 27.3 The DFT process, producing a sampled spectrum (right) from a sampled waveform (left). The spectral coefficients are amplitudes for cosines (real) and sines (imaginary) at each frequency point. The example uses N=32 points of a sampled waveform.

k	0	1	2	...	$N/2$	$N/2+1$	...	$N/1$
f_c	0	f_s/N	$2[f_s/N]$	...	$f_s/2$ AND $-f_s/2$	$-[N/2+1] \times [f_s/N]$	...	$-f_s/N$
C_k	a_0 0	a_1 b_1	a_2 b_2	...	$a_{N/2}$ 0	$a_{N/2+1}$ $b_{N/2+1}$		a_{N-1} b_{N-1}
	positive frequencies					*negative frequencies*		

FIGURE 27.4 The DFT frequency points: k is the point number, f_c the frequency of point k, c_k its spectral coefficient (a_k is the cosine part and b_k is the sine part), f_s the sampling rate and N the number of points in the analysis.

frequency (Figure 27.4). For a more mathematically involved description of these ideas see Lazzarini (2010a).

27.1.3 Understanding the DFT spectrum

Clearly, the DFT is not the FT, so we do not get the same fine-grained detail in frequencies that we would perhaps have liked. However, there is plenty of good information about the spectrum, which can be used in various applications.

One important point that should be made at this point is that the DFT decomposes the signal being analyzed in N components, which are multiples of the fundamental analysis frequency. This guarantees that if we add these N components together, we will get the back the original signal. The analysis might not be too meaningful, but it is still correct. In the special case where we have a periodic signal, where a full period of a waveform is captured by the analysis window, we will be able to recuperate the exact frequencies, amplitudes, and phases of that signal (as shown in Figure 27.3).

Another way at looking at the DFT results is to think of each frequency point as the centre of an analysis band, and the *magnitude* of its coefficient as the amount of energy in that band. This is how spectral displays often work, plotting the magnitude of these coefficients. A fast algorithm for calculating the DFT is normally used, the *Fast Fourier Transform* (FFT), so operations of spectral analysis can be very computationally efficient. Our first programming example, in the Python language, demonstrates a spectral display application, using a radix-2 FFT (based on power-of-two sizes, the most common type):

```
import pylab as p
from scipy.io import wavfile
import sys
# read the soundfile to be analyzed
(sr,sig) = wavfile.read(sys.argv[1])
```

```
if len(sig.shape) > 1:
    print("stereo files not supported")
    exit()
N = 2048
x = p.arange(0,N/2)
# setup the frequency points
bins = x*sr/N
# extract the window to be analyzed
window = sig[0:N]
# take the magnitudes of spec coeffs using abs()
spec = abs(p.fft(window))
# plot the positive spectrum only
p.plot(bins,spec[0:N/2]/max(spec), "k-")
p.ylabel("magnitudes," size=16)
p.xlabel("freq(Hz)," size=16)
p.show()
```

27.1.4 Processing Applications

The DFT has one outstanding application, which is in the implementation of spatial, reverberation, and related effects. It involves a digital signal processing operation called *convolution*. In situations where we have the response of a system to an impulse, the appropriately named *impulse response*, we can simulate how this system responds to an arbitrary signal (Kleiner, Dalenbäck, and Svensson 1993). For instance, if we have the impulse response of a room, and we want the result of a dry sound played in that room, we combine the two using the convolution operation.

The impulse response is the record of all the reflections of a signal that consist of a short burst, a single discrete value ("sample") followed by zeros, in a given system (say a room). The convolution operation takes the input signal, copies it to the time position of each reflection in the impulse response, scales it (i.e., boosts or attenuates) by the level of the reflected impulse at that position, and then mixes all these copies together. In other words: delay, scale, and mix. If we have an impulse response of T seconds, at f_s samples per second, we will have T f_s delay, scale, and mix operations for every output value. For some applications, this can be quite costly in computational terms.

Thankfully, there is a spectral way of implementing convolution. It uses the principle that this operation in the time domain (i.e., a convolution of two waveforms) is equivalent to multiplication of spectral coefficients. By applying an efficient DFT algorithm (the FFT), we can reduce the computational complexity of the above to two transforms, a block multiplication and an inverse transform (Figure 27.5). Moreover, if the impulse response is much smaller than the input signal (which is normally the case), we can break down the operations in blocks that are relative to the size of the impulse response and then reconstitute the signal back via overlap-add. The following programming example in Python implements this principle:

```
import sys
import pylab as pl
from scipy.io import wavfile
# read impulse and signal input
(sr,impulse) = wavfile.read(sys.argv[1])
(sr,signalin) = wavfile.read(sys.argv[2])
if len(signalin.shape) > 1 || len(impulse.shape) > 1:
    print("stereo files not supported")
    exit()
S = len(impulse) # impulse length
L = len(signalin) # signal length
# findfftsize as the next power of 2
# beyond S*2-1
N = 2
while(N <= S*2-1): N *= 2
# signal blocks for FFT
imp = pl.zeros(N)
sig = pl.zeros(N)
# output signal & amplitude
sigout = pl.zeros(L+S-1)
amp = max(signalin)
# copy impulse for FFT
imp[0:S] = impulse
spec_imp = pl.fft(imp)
for i in range(0, L/S):
    p = S*i
    sig[0:S] = signalin[p:p+S] # get block from input
    # perform convolution and overlap-add
sigout[p:p+2*S] += pl.ifft(spec_imp*pl.fft(sig))[0:2*S].real
# write file to output, scaling it to original amp
wavfile.write(sys.argv[3],sr,pl.array(amp*sigout/max(sigout), dtype="int16"))
```

27.2 Beyond the DFT: The Phase Vocoder

The limitations of the raw DFT, such as the fact that it is a one-shot, static process, make it difficult to use in a flexible way for general-purpose audio signal processing. However, it forms the basis for a more useful method of spectral analysis, the phase vocoder (PV). Initially designed as a filter bank algorithm (Flanagan and Golden 1966), where a number of parallel heterodyne filters were employed to break down the spectrum into a set of sub-bands, this technique soon was employed in conjunction with the FFT algorithm in

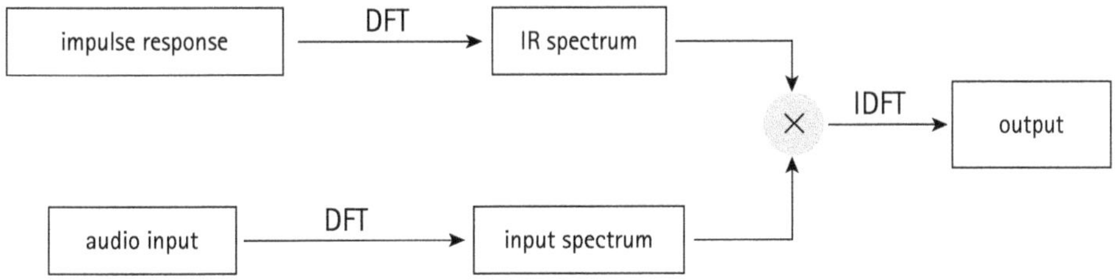

FIGURE 27.5 The DFT-based convolution process.

its practical implementations (Dolson 1986). Instead of describing it using the original layout, we will try to see it as an enhancement to DFT analysis (Lazzarini 2010b).

The first step toward the PV is the use of smoothing envelopes to shape the analysis signal, so that we eliminate any harsh cuts or discontinuities at its ends (where the signal was cut for the DFT operation). These envelopes are also known as *analysis windows* (the case we have been using so far is the *rectangular* window, which just extracts the audio block, without any smoothing). The net effect of this is to reduce artifacts: that is, spurious information that the DFT process has generated. Commonly used windows are the inverse-raised cosine ones, such as the Von Hann (also known as Hanning) and Hamming. These and other window shapes are shown in Figure 27.6.

Secondly, we need to make the operation time-varying. Originally, the DFT was a single-shot analysis, the "picture" of the spectrum at a given time. Now we want to make it so that we have a dynamic analysis of the sound as it changes through time. To do this, we have to take a sequence of windows, which can be minimally spaced by one sampling period, but generally take a larger time interval. The space between the analysis frames is known as the *hopsize*. Generally speaking, the largest practical hopsize used is a quarter of a window length. Now we have a series of spectral analyses, frames, containing N coefficients, each ordered in time.

In addition to using the FFT, we can also save space and processing by considering that, for audio signals, there is a duplication of information between positive and negative spectra. In this case, we can employ transforms that carry only the nonnegative spectrum (half the points plus the Nyquist one). These are often called "real" FFTs, as they are designed to work with real-valued signals (as in the audio case).[2]

With these three modifications (windowing, time frames, and real FFT operation, also known as the Short-Time Fourier Transform, or STFT, for real signals, as shown in Figure 27.7), we can start to define the phase vocoder analysis. The principle is this: for every DFT frequency point, which we will now call a *frequency bin*, we will be able to obtain time-varying frequency and amplitude data. The latter is simply obtained as before, by calculating the magnitude of each coefficient. The frequency data are extracted by first taking the phases of coefficients, then bin by bin, finding the phase differences between consecutive frames. As the phase is a periodic quantity (with period 2π), we need to "unwrap" it (i.e., bring it to the base range of $-\pi$ to π). The phase difference is going to be equivalent to the frequency detected at that bin (in radians per

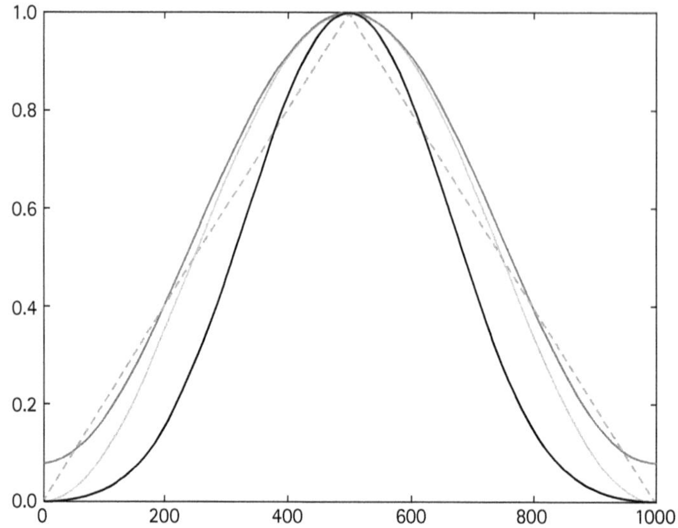

FIGURE 27.6 Four window types: Von Hann, or Hanning, Hamming, Bartlett, or triangle, and Kaiser.

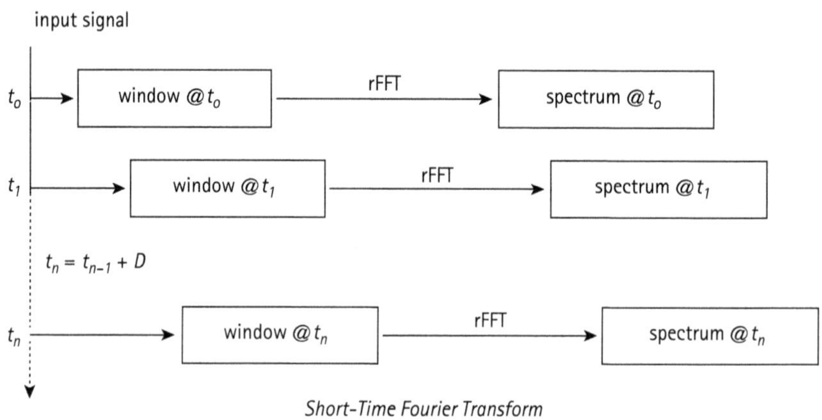

FIGURE 27.7 The STFT as a time-varying enhancement of the DFT.

hopsize samples): how much the sinusoid has "moved" between analysis snapshots. Finally, depending on how we want to store and manipulate this, we can convert it to Hz (Figure 27.8).

For the resynthesis of PV data we can use straightforward additive synthesis, since we have frequencies and amplitudes, we can drive a bank of sinusoidal oscillators with this data. In this case, the frequency and amplitude in each bin are used as control signals for each oscillator, which will model each spectral component analyzed. Alternatively, and often more efficiently, we can reverse the operations used to obtain the amplitude

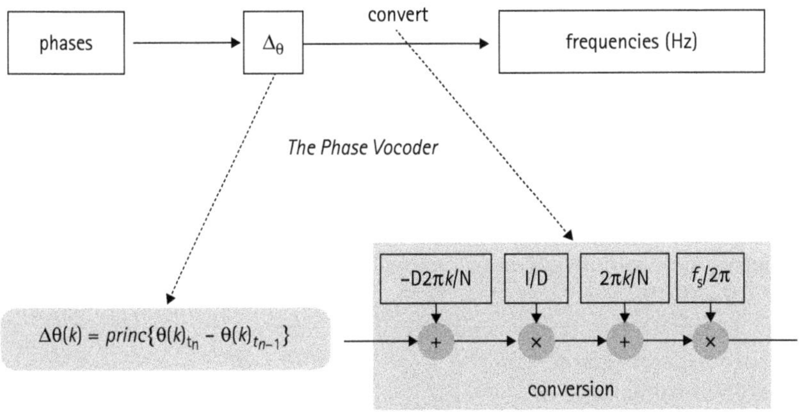

FIGURE 27.8 Frequency estimation in the Phase Vocoder and the optional conversion into Hz, where k is the bin number, t_n the frame number, f_s the sampling rate, D the hopsize, and N the analysis size (number of points).

and frequency data from the spectral coefficients and apply the Inverse Discrete Fourier Transform (implemented using the FFT algorithm) to recreate each analysis frame. If the frequencies are in Hz, they are converted to radians per hopsize, then integrated (i.e., added) to get the current phase. These, together with the amplitude, are then converted into the complex form of the coefficients (as real and imaginary pairs, in the same formas produced by the DFT). The analysis window is then reapplied to each frame and we overlap-add them in sequence to produce the output waveform.

The advantage of the PV representation is that now we have more meaningful analysis data, instead of a block of coefficients, we have time-varying amplitudes and frequencies in each bin. This is much more amenable to further manipulation than raw spectral data. For instance, we can perform timescaling (contraction, stretching) that is independent of frequency.

The following Python example implements phase vocoder analysis-resynthesis for simple timescale modifications. It uses the phase differences of two consecutive frames to obtain frequency estimates for each bin, and in sequence these are integrated to get a running phase. From the magnitudes and phases, we obtain the spectral coefficients to use in the inverse DFT. This allows us to hop the read point by any rate we choose, either compressing or stretching the sound. The output write position always proceeds at the analysis hopsize, controlling the overlap-add operation and ensuring the signal is reconstructed correctly:

```
import sys
import pylab as pl
from scipy.io import wavfile
N = 2048 # window size
H = N/4 #hopsize
```

```
Pi = pl.pi
# read input and get the timescale factor
(sr,signalin) = wavfile.read(sys.argv[2])
if len(signalin.shape) > 1:
   print("stereo files not supported")
   exit()
L = len(signalin)
tscale = float(sys.argv[1])
# signal blocks for processing and output
phi =pl.zeros(N/2+1)
out = pl.zeros(N/2+1, dtype=complex)
sigout = pl.zeros(L/tscale+N)
# max input amp, window
amp = max(signalin)
win = pl.hanning(N)
p = 0.0 # input read position in samples
pp = 0 # output read position in samples
while p < L-(N+H):
   # take the spectra of two consecutive windows
   p1 = int(p)
   spec1 = pl.rfft(win*signalin[p1:p1+N])
   spec2 = pl.rfft(win*signalin[p1+H:p1+N+H])
   # take their phase difference (to get freq) and
   # then integrate to get the running phase phi
   phi += (pl.angle(spec2) – pl.angle(spec1))
   # bring the phase back to between pi and – pi
   for i in range(0, N/2+1):
   while phi[i] < – Pi: phi[i] += 2*Pi;
   while phi[i] >= Pi: phi[i] – = 2*Pi
   # convert from mags (abs(spec2)) + phases (phi
   # tocoeffs (real,imag
   out.real, out.imag = abs(spec2)*pl.cos(phi),
   abs(spec2)*pl.sin(phi)
   # inverse FFT and overlap-add
sigout[pp:pp+N] += win*pl.irfft(out)
pp += H
   p += H*tscale
# write file to output, scaling it to original amp
wavfile.write(sys.argv[3],sr,pl.array(amp*sigout/max(sigout), dtype="int16"))
```

27.2.1 Streaming

In Csound, the phase vocoder is implemented as a streaming operation (Lazzarini, Lysaght, and Timoney 2006): it produces an output signal that is a sequence of frames, spaced by a given hopsize and containing frequency and amplitude pairs for all N/2+1 bins (the non-negative spectrum plus the Nyquist frequency bin). The output signal is a special *f* type that is self-describing and can be used as an input to several unit generators. Such data can be analyzed on-the-fly from an input signal or from memory (function table), or obtained from preanalyzed PV data stored in disk files.

27.3 Spectral Manipulation

A number of transformations can be applied to spectral data in the phase vocoder format. In addition to the timescale modifications introduced above, a number of frequency, pitch, amplitude, filtering, delay, cross-synthesis, and morphing processes are possible.

27.3.1 Frequency and Pitch

Frequency can be altered in a number of ways: We can transpose signals, which will scale up or down the frequency of all analysis components, causing a pitch shift. We can also shift linearly or nonlinearly the frequency data, which will generally not preserve harmonic relationships (if these exist) in the spectrum, rendering it inharmonic.

Pitch shifting can be performed in two basic ways: (1) we can shift the pitch in the time domain by resampling (i.e., reading the input data at a different rate) and then using the timescaling capacity of the PV to keep the signal at the original duration; (2) we can scale the frequencies found in each bin by the pitch shift factor, taking care to reallocate them to new bins that reflect their new value.

The former is usually applied to stored data (say on disk file or on memory), as the resampling process is facilitated in this scenario. The following Python example demonstrates this idea, using a very simple transposition method, which does not employ any interpolation (in practice, most applications will employ at least linear interpolation in the process):

```
import sys
import pylab as pl
from scipy.io import wavfile
N = 2048 # window size
H = N/4 #hopsize
Pi = pl.pi
# read input and get the timescale factor
(sr,signalin) = wavfile.read(sys.argv[2])
if len(signalin.shape) > 1:
   print("stereo files not supported")
   exit()
```

```
L = len(signalin)
pitch = float(sys.argv[1])
# signal blocks for processing and output
phi =pl.zeros(N/2+1)
out = pl.zeros(N/2+1, dtype=complex)
sig1 = pl.zeros(N)
sig2 = pl.zeros(N)
sigout = pl.zeros(L)
# max input amp, window
amp = max(signalin)
win = pl.hanning(N)
p = 0.0 # read position in samples
if pitch <= 1: end = L-(N+H)
else: end = L-(N+H)*pitch
while p < end:
   # transpose the audio (no interpolation)
   # and copy into frames for FFT analysis
pos = p
   for i in range(0, N):
posi = int(pos)
frac = pos – posi
   sig1[i] = signalin[posi]
posi = int(pos + H*pitch)
   sig2[i] = signalin[posi]
pos += pitch
spec1 = pl.rfft(win*sig1)
spec2 = pl.rfft(win*sig2)
# take their phase difference (to get freq) and
# then integrate to get the running phase phi
phi += (pl.angle(spec2) – pl.angle(spec1))
# bring the phase back to between pi and – pi
for i in range(0, N/2+1):
while phi[i] < – Pi: phi[i] += 2*Pi;
while phi[i] >= Pi: phi[i] – = 2*Pi
# convert from mags (abs(spec2)) + phases (phi)
# tocoeffs (real,imag)
out.real, out.imag = abs(spec2)*pl.cos(phi),
   abs(spec2)*pl.sin(phi)
# inverse FFT and overlap-add at the original timescale
sigout[p:p+N] += win*pl.irfft(out)
p += H
# write file to output, scaling it to original amp
wavfile.write(sys.argv[3],sr,pl.array(amp*sigout/max(sigout), dtype="int16")
```

The second approach can be used more flexibly with streaming PV data. In Csound, the pvscale unit generator implements this type of frequency scaling. It can be used in conjunction with a pitch detector to implement autotune and harmonizer effects.

27.3.2 Amplitude and Filtering

One of the problems of pitch shifting, especially of vocal sounds, is the side-effect of formant distortion, which is caused by formants being transposed alongside with the frequency. This leads to the well-known "chipmunk" or "Donald Duck" effect. With the PV, it is possible to retain the original formants while transposing the signal. This is a prime example of amplitude and filtering process in the spectral domain.

The principle is this: we scale the frequencies as described in the previous section. Following this, we apply the spectral envelope of the original signal to these scaled frequencies, to combine the modified frequencies with a different set of amplitudes. Because the PV gives us separate control of these two parameters, this effect is possible.

The spectral envelope can be extracted from the original amplitudes via a variety of methods. In Csound there are two alternatives, a simple cepstrum method and the true envelope estimation (Imai and Abe 1979). While the latter is more precise, the former is much more efficient and generally yields good results. The example below shows its use in the pvscale unit generator for a harmonizer application:

```
<CsoundSynthesizer>
<CsInstruments>
chnset 1.2, "pitch"; pitch control bus init
chnset 1, "mode"; mode control bus init
instr 1
a1 in; input audio
kpitch chnget "pitch"; pitch control
iform chnget "mode"; preserv mode: 0—none, 1—ceps, 2—true env
fsig pvsanal a1,2048,256,2048,1; PV analysis
fharm pvscale fsig,kpitch,iform; pitch shift
fmix pvsmix fsig, fharm; mix
a2 pvsynth fmix; PV synth
out a2
endin
</CsInstruments>
<CsScore>
i1 0 36000; activate instrument
</CsScore>
</CsoundSynthesizer>
```

27.3.3 Spectral Delays

A novel transformation enabled by the PV is the *spectral delay*. This effect allows users to apply separate delay effects to specific bands in the spectrum, which can be made up of one to several bins. In Csound, this process is enabled by a unit generator that sets up and writes PV data to a circular buffer. This data can then be accessed by a number of

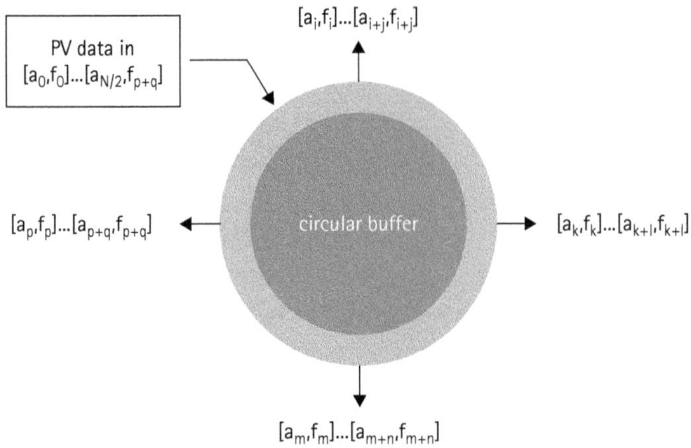

FIGURE 27.9 Spectral Delays, with taps that select specific ranges of bins.

instances of a buffer reader, setting up taps that are able to select specific spectral bands (Figure 27.9).

The example below demonstrates the spectral delay effect. It implements four taps covering different ranges of the spectrum with increasing relative delay times. It also employs feedback to reinsert the output signal (scaled) back into the delay line.

```
<CsoundSynthesizer>
<CsInstruments>
chnset 1, “delay”; relative delay control init
chnset 0.9, “feedback”; feedback control init
instr 1
iws = 2048; window size
ihs = iws/8; hopsize
kdelchnget “delay”; delay control
kfdbchnget “feedback”; feedback control
a1 in; input audio
fsig pvsanal a1,iws,ihs,iws,1;PV analysis
fmix pvsinit iws,ihs,iws,1;initfdb signal
fsig1 pvsmix fsig, fmix;mix input + fdb
iha,ktr pvsbuffer fsig1,5; circ buffer, 5s long
fsig2 pvsbufread ktr-0.23*kdel,iha,100,800,1; 3 oct above 100Hz
fsig3 pvsbufread ktr-0.31*kdel,iha,500,4000,1; 3 oct above 500Hz
fsig4 pvsbufread ktr-0.54*kdel,iha,1500,1200,1; 3 oct above 1500Hz
fsig5 pvsbufread ktr-0.87*kdel,iha,4000,22050,1; 4000Hz—Nyquist.
fmix1 pvsmix fsig2, fsig3; mix taps
fmix2 pvsmix fsig3, fsig4
fmix pvsmix fmix1, fmix2
a2 pvsynth fmix; PV resynthesis
fmix pvsgain fmix, kfdb; apply fdb scaling
```

```
out a2
endin
</CsInstruments>
<CsScore>
i1 0 36000
</CsScore>
</CsoundSynthesizer>
```

27.3.4 Cross-synthesis and Morphing

Another class of PV transformations is linked to the combination of features of two or more spectra. There are two basic forms in which these normally appear: cross-synthesis, where the amplitude features of one signal are combined with the frequencies of another; and morphing, where an intermediary sound is obtained by spectral interpolation.

The first type of effect is very similar to the effects found in the original *channel vocoder*. In that algorithm, using a filterbank analysis and synthesis, we obtain the energies at several bands across the spectrum. Then we impose these onto another sound, which would generally be component-rich with a flat spectral envelope. The PV version of this algorithm is very similar, except that in place of the filterbank we have our analysis bins (which are filterbank of a sort) and we will extract the spectral envelopes using one of the methods outlined above in the discussion of formant preservation. We will then combine this spectral envelope with the frequencies obtained from another source, cross-synthesizing a hybrid sound. In Csound, the pvsvoc unit generator provides the support for such operation.

Morphing on the other hand is a more difficult and subtle effect (Wishart 1995). There have been a variety of approaches to this problem. In the PV environment, one straightforward way of achieving morphing effects is to find intermediary spectra via separate interpolation of amplitudes and frequencies. In this case, the user will provide an interpolation factor, which will be used to combine the spectral data on a bin-by-bin basis. The most successful applications of this transformations rely on two factors: (1) the choice of spectra that have some common elements; (2) morphing over time, starting from one sound and then slowly interpolating toward the other. In Csound, the pvsmorph unit generator implements this transformation and it can be used to create interesting transitions between sounds, as shown in the example below:

```
<CsoundSynthesizer>
<CsInstruments>
nchnls = 2
chnset 0.5, "ampmorph"; amp morphing init
chnset 0.5, "freqmorph"; freq morphing init
instr 1
iws = 2048; window size
ihs = iws/8; hopsize
```

```
a1,a2 ins; inputs
kam chnget "ampmorph"; amp morphing
kfm chnget "freqmorph"; freq morphing
fsig1 pvsanal a1,iws,ihs,iws,1; chn 1 analysis
fsig2 pvsanal a2,iws,ihs,iws,1; chn 2 analysis
fmorph pvsmorph fsig1,fsig2,kam,kfm; morph
am pvsynth fmorph; PV resynthesis
outsam,am
endin
</CsInstruments>
<CsScore>
i1 0 36000
</CsScore>
</CsoundSynthesizer>
```

27.4 Conclusions

In this chapter, we set out to introduce and explore interactive spectral processing of audio. We have provided a straightforward, nonmathematical, introduction to the principles that underpin the area. As noted in the introduction and demonstrated in the chapter, spectral processing provides a fine degree of control in sound transformations. Interactive audio manipulation can benefit a lot from these methods.

Common applications are found in situations where independent control of tempo and pitch is required. For instance, this can be used in an audiovisual scenario where a given audio sample needs to synchronize perfectly with visual cues happening at specific points in time and no change in pitch is desirable. The reverse situation is also often required, where a sample needs to be transposed in pitch without loss of synchronization.

Spectral transformations can also be used to produce expressive dynamic soundscapes for various composition and sound design applications. As well as modifying pitch and time, a great variety of timbral effects can be achieved, through fine-grained filtering of spectral components, transitions and morphing, and delays of specific frequency bands, to list but a few possibilities.

In this chapter, we offered a number of example programs in the Python language, which demonstrated the implementation of basic spectral processes. This software can be applied directly in frequency-domain manipulation, or modified and translated to other languages and systems to suit the reader's needs. In addition to these, we provided various samples of Csound code that can be readily applied in a variety of interactive desktop or mobile application projects. All of these examples attempted to illustrate some very useful effects and transformations that can be implemented with frequency-domain processes. It is hoped that this introduction to the field will help provide the basis for further exploration of these and other related processes.

Notes

1. Real-time processing is understood here as the opposite to offline rendering, where the user can hear the results of their actions instantly, with little latency.
2. In the case of real-valued signals (as opposed to ones that are defined by complex numbers), the positive and negative spectra are Hermitian symmetrical, which means that the negative part can be inferred from the positive, and thus it does not need to be kept.

References

Dodge, Charles, and Thomas A. Jerse. 1996. *Computer Music*. New York: Schirmer.

Dolson, Mark. 1986. The Phase Vocoder Tutorial. *Computer Music Journal* 10 (4): 14–27.

Fischman, Rajmil. 2000. A Survey of Classical Synthesis Techniques in Csound. In *The Csound Book*, ed. R. Boulanger, 223–260. Cambridge, MA: MIT Press.

Flanagan, J. L., and R. M. Golden. 1996. Phase Vocoder. *Bell System Technical Journal* 45, 1493–1500.

Fourier, Jean Baptiste. 1878. *The Analytical Theory of Heat*. Cambridge, UK: Cambridge University Press.

Helmholtz, Hermann. 1954. *On the Sensations of Tone*. London: Dover.

Imai Satoshi and Y. Abe. 1979. Spectral Envelope Extraction by Improved Cepstral Method, *Electronics and Communication* 62 (4): 10–17.

Kleiner, Mendel, Bengt-Inge Dalenbäck, and Peter Svensson. 1993. Auralization: An Overview. *Journal of the Audio Engineering Society* 41 (11): 861–875.

Lazzarini, Victor. 2010a. Spectral Audio Programming Basics: The DFT, the FFT, and Convolution. In *The Audio Programming Book*, ed. Richard Boulanger and Victor Lazzarini, 521–538. Cambridge, MA: MIT Press.

——. 2010b. Programming the Phase Vocoder. In *The Audio Programming Book*, ed. Richard Boulanger and Victor Lazzarini, 557–580. Cambridge, MA: MIT Press.

Lazzarini, Victor, Thomas Lysaght, and Joseph Timoney. 2006. Spectral Signal Processing in Csound 5. *Proceedings of the International Conference on Computer Music 2006*, 160–163. New Orleans: ICMA.

Moorer, James A. 2000. Audio in the New Millenium. *Journal of the Audio Engineering Society*, 48 (5): 490–498

Steiglitz, Ken. 1996. *A Digital Signal Primer*. Menlo Park, CA: Addison-Wesley.

Walsh, Rory. 2010. Developing Audio Software with the Csound Host API. In *The Audio Programming Book*, ed. Richard Boulanger and Victor Lazzarini, [on DVD]. Cambridge, MA: MIT Press.

Wishart, Trevor. 1995. *Audible Design*. York: Orpheus the Pantomime.

SECTION 6

THE PRACTITIONER'S POINT OF VIEW

CHAPTER 28

LET'S MIX IT UP

Interviews Exploring the Practical and Technical Challenges of Interactive Mixing in Games

HELEN MITCHELL

GAME audio has come a long way since the simple electronic beeps of the early 1970s, when significant technical constraints governed the scope of creative possibilities. Recent years have witnessed technological advancements on an unprecedented scale; no sooner is one technology introduced than it is superseded by another, boasting a range of new refinements and enhanced performance. Those working creatively within the sphere of audio for games have been faced with a growing array of creative possibilities resulting from significant technological advancements and changing game platforms, fuelled by increased market demand. While these are all positive and exciting developments, they are driven by a rate of change that allows little time for the absorption and exploration of these developments and the new creative opportunities that they offer.

Certain parallels might be drawn between this history and the early days of sound film. The development of film was also driven by the introduction and progress of technology and by market pressures (Lastra 2000). Following the success of the Warner Brothers film *The Jazz Singer* (1927), the film studios realized that synchronized sound would be central to the future of the film industry. During subsequent years the studios sought to answer box office demand for sound films by producing films that showcased the new technology, such as musicals. However, to remain competitive the studios had to find more subtle ways of exploring the possibilities of sound film while maintaining tight production schedules. Consequently, films became vehicles for the creative exploration of new and emergent technologies, resulting in a fast succession of film firsts. Some of these film firsts proved more successful than others and would be imitated in future films; less successful approaches were discarded. Over many years of trial and error, a sophisticated and standardized approach to sound for film emerged.

In the same way that early sound films explored new creative possibilities, modern videogames offer the potential to explore new creative approaches as technical constraints are gradually eroded. There is one essential difference however; film sound emerged with few points of reference other than radio and stage performance and so had to forge a new path: games, on the other hand, emerged into a world familiar with a variety of audiovisual media. It is not surprising therefore, that film sound should become just one point of reference for the game industry, particularly since the tastes and expectations of potential gamers and game developers have been shaped by current conventions of sound for film and television. Game sound designer Jerry Ibbotson, for instance, states that he has often been asked for specific film sound styles in his work for game developers, and traces this trend to the formative influence of film and television from an early age:

> Games have definitely been influenced by film... it's subconscious because you've grown up watching films in the cinema and on television; you know what things are supposed to sound like. You know what guns are supposed to sound like from watching films... [but] they don't actually sound like that in real life (we're based opposite a firing range and guns don't sound like that at all).... Everything you think you know about things sounding is based on film, so even subconsciously it's carried through into what you do... the designer and the producer will tell you how they want it to sound, and nine times out of ten... they will use sound analogies like, "can you make it sound like the drop-ship from *Alien*" or "something from *The Fast and the Furious*" [or] "that has to sound Hollywood." ... What they mean is that it's got to be like... watching a film, so it runs through everything.

The desire to achieve the Hollywood sound as expressed by some game developers is reflected in much of the literature relating to game audio today. Even the briefest foray into game sound literature (academic and practice based) soon reveals references to film sound and to examples of best practice in particular. Without question the film industry's experience of sound, acquired over many years, represents an invaluable resource within various spheres of creative endeavor, but within the context of game audio there is one essential flaw in the film sound model: linearity.

Dynamic mixing, defined as the "real-time manipulation of audio replay parameters and inline effects treatments for organizational, technical, or creative purposes" (Broomhall 2010) distinguishes game sound from film sound. With the exception of some linear sequences such as cut scenes, sound in games is dynamic and changes in response to the interaction of the player(s). By contrast, film sound is mixed to footage, which once the final cut has been made will not change. The linearity of film sound allows tremendous creative and technical control over every aspect of the final mix. Such levels of control become quite challenging within the context of dynamic mixing, in which the mixing console is replaced by computer code that governs the ways in which sounds are triggered and controls changing audio parameters in real time. This chapter seeks to explore some of the practical and creative challenges of dynamic mixing for videogames, focusing on sound effects in particular. The discussion will be based on a series of interviews by the author and Jamie Maxwell in 2010 with leading audio specialists who are actively involved in the creation and standardization of audio for games.

28.1 Defining a "Good" Game Mix

Before outlining some of the challenges of dynamic mixing for games, it is necessary to consider what is generally meant by "a good game mix" in the industry. Nick Peck defines this as "a mix in which the appropriate layers of material are the ones which are the most audible to the listener at the particular time that they need to be," and suggests that this definition is equally applicable to film and television mixes. In refining the definition still further it is possible to identify three general classifications of sound that are present in most game and film mixes: music, dialog, and sound effects. These three basic categories may be further deconstructed into subdivisions that give some indication of the ways in which audio functions and is applied. Peck identifies five categories of sound:[1] ambience or background sounds, namely, "location-based material that is used to set the scene and give a context to the narrative"; Foley sounds, defined as "footsteps, cloth rustling, prop manipulation, and other sounds made by people"; hard sound effects, or "primary sounds that are often signature elements of the game"; music, which conveys the emotional content of the material, telling the viewer "what to feel"; and finally, "the most important layer," dialog, the function of which is to "tell the story." Of paramount importance is the ability to hear and understand what people are saying, although, as John Broomhall points out, one should not discount the "storytelling power of sound."

Having outlined the individual layers making up the mix, it is necessary to consider what is meant by the word "appropriate" in this context. This word is the linchpin of the definition and also the most problematic element of it since it suggests a subjective assessment. If dialog is to be heard clearly throughout game play then it would be logical to assume that dialog should always take precedence when balancing different elements within the mix. However, imagine a scenario in which a military commander is shouting instructions to his men during a combat sequence. A huge explosion takes place, deafening all those close to it. In such circumstances it could be argued that it would be more *appropriate* to duck the dialog below the sound of the explosion or to filter out the high frequencies of the dialog to mimic the effects of temporary hearing damage, making the game feel more real and believable but potentially compromising the intelligibility of the dialog. Alternatively, the clarity of the dialog might be preserved by ensuring that the frequency range of the explosion remains well below that of the dialog, possibly dropping the volume of the explosion in relation to the voice as well. This approach would prioritize the dialog at the expense of other sonic elements within the mix. Broomhall points out that some game developers today have taken this approach one stage further by introducing the concept of "critical dialog." He defines this as "dialog which must be heard come what may" because it provides important information and feedback to the player. He suggests that the technical treatment and handling of "critical dialog" differs from approaches applied to other types of in-game dialog. Perhaps the treatment of dialog in games seems an obvious example, but achieving a good game mix involves making many decisions relating to the prioritization and technical treatment of sound but on a much greater scale, involving multiple audio layers responding fluidly to interactive game play.

There are, however, some generally accepted and objective hallmarks of a good game mix, hallmarks that might indicate that the right and most "appropriate" decisions have been made. Ibbotson suggests that a good game mix is one that "you shouldn't really notice." A mix that brings together individual, high-quality "elements" in a "balanced" way, preventing what might otherwise become "a complete cacophony" while avoiding "irritating" repetitions of individual elements and providing informative "feedback." A good mix therefore is as much about omission as inclusion. The result however, according to Ibbotson, is a mix no one notices; one that draws the player into the game. In this regard Peck suggests that a good game and film mix share the same goal:

> The best soundtrack is the one that completely melts away in the perception of the viewer so that they're not thinking about the sound; they're letting the sound wash over them as they are immersing themselves in the experience and they are engaging in the story itself... to that end the aims of a good film mix and a good game mix should really be the same.... The aim of both of them is to engage the player, engage the viewer; to take them away from their life for a little while... [placing] them into the middle of another story.

John Broomhall makes a further addition to this appraisal of a good game mix, suggesting that there may be moments when it might be appropriate for a game mix to draw attention to itself by highlighting particular sounds that facilitate game play by, for example, helping a player to navigate within the game world or to "understand their next objective," and so on.

In summarizing the overriding objective of a good mix, we might therefore define this as the creation of a more "interesting and entertaining gaming experience," one that enables players to "suspend [their] disbelief" as they become immersed and absorbed in game play. The extent to which this is achieved must surely be the quality benchmark for dynamic mixing in games today.

28.2 Creative and Technical Challenges: Integration

For those working toward the achievement of successful game mixes there are many creative and technical challenges. One of the key challenges concerns the integration of sound into the game mix itself. The extent to which sound designers are involved in this process can vary significantly. An in-house sound designer may have considerable input when determining how the finished sounds might be integrated into the mix and processed during game play. Having in-house status provides opportunities to facilitate creative collaboration. In addition, being onsite as the game is developed provides the potential to become involved in the development of the audio much earlier in the production process. Clearly when sound design work is outsourced, creative collaboration can become more difficult and time constraints can be intensified as production schedules slip; something an in-house sound designer would be better positioned to monitor and potentially counteract.

It should be noted that neither of these models offers standardized approaches in terms of integrating sound into games. In some cases sound designers might be given the opportunity to work with audio programmers and other in-house personnel, directly influencing the ways in which their audio samples are integrated and processed within the game. In others they will simply be asked to deliver audio assets and will have little or no involvement in the integration process. Ibbotson's experience of working on games in which the sound design is outsourced, testifies to these differences of approach:

> We're working on a PS3 game at the moment with someone in-house who we ship the samples to; ... [he asks] our opinion, but he's the one who is laying ... [the audio assets] out. ... In some projects we've worked on, we've had completely bespoke tools. In *Broken Sword: Angel of Death* we had tools where you could load up every level and move in through the level, placing sound emitters where we wanted. ... Likewise we have worked on games where we've had a lot of control but then it's been the world builders instead who are placing the sounds and, when you see it, you're banging your head against a wall saying "No, no, no, I want that moving!" They don't necessarily understand because, even though they've placed it, they haven't even had headphones on or played the game. They're just placing little dots where they think a sound should be.

Thankfully Broomhall suggests that the experience Ibbotson describes is not as common as it used to be, although he does acknowledge that it is still largely true of "handheld" games. On the other hand, Ibbotson's experience of nonstandardized approaches and integration is echoed by that of Peck:

> [The] type of integration process you have ... to get your sounds working in the game ... changes on a game-to-game basis. There are many games that use middleware solutions ... in which, if there is enough time, the sound designers can ... sculpt and change some of the interactive parameters of the sounds as well as the sounds themselves. However there are many other situations ... in which ... the sound designer spots the game, figures out what it is they want to happen, and then generate(s) the assets. [The sound designer] ... hand(s) off the sound effects to the client who takes them ... to an engineer, who wires them up to the game itself. In reality that often happens, particularly when you're operating in a consulting ... realm ... without very much input from the sound designer after the fact.

Clearly these differences of approach can have a profound effect on the success or failure of the audio mix. Distancing the sound designer from the integration process and from decisions relating to the audio mix and real-time processing are less likely to yield satisfactory results. However, it is not simply a case of collaboration; successful audio integration hinges upon the expertise of the audio programmers and the dynamics between individuals, whether sound designers, audio programmers, audio directors, audio editors, producers, or game designers, as Ibbotson describes:

> We've worked on games with award-winning sound and a large part of that is down to the sound coder. I sometimes think that producers don't realize just how massively important a sound coder is. When you look at really successful games, one of the reasons why they sound so good is because they have dedicated audio programmers who really know how to get stuff to work, and when it works on a project it's normally because they are on the same wavelength as you. They are just as creative as any of the artists or graphics people.

According to Nick Peck, the best audio programmers are musical and have an inherent interest in sound. However, he also suggests that the context in which personnel such as programmers and audio directors work is also central to the success or failure of the finished audio:

> There is a need for game designers to be able to have trust in their audio directors; to be able to let them go and do what it is that they think is best for the game. The best game audio programmers that I have worked with are the ones that are also musicians and that love manipulating sound on their own at night when they're not doing it during the day for a job. The communication with them becomes effortless. There's no question... that audio directors often face an uphill battle in terms of being able to educate the... people around them as to the importance of sound... at LucasArts sound was always considered to be of utmost importance because that was the aesthetic that was laid down by George Lucas across his entire company.

Encouraging a creative culture in which audio programmers and other personnel recognize the potential power of sound and value creative exploration as well as the expertise of sound specialists, whether in-house or out of house, offers a number of distinct advantages during the integration process. Integrating sounds into a game mix involves making a number of decisions regarding the prioritization and processing of certain sounds, as outlined earlier. It could be argued that distancing sound specialists from these core decisions limits the scope to use the full potential of sound within games. John Broomhall suggests that, in his experience, most game developers now have in-house sound integrators, even when the creation of some sounds is outsourced. Developments of this type are already helping to resolve some of the issues surrounding the integration of audio into games.

Conversely, one might assume that audio samples might be integrated within the game and mixed or processed according to principles of physics without reference to sound specialists. Individual volume levels might be governed by distance measurements for example. Reverberation times might be set according to typical acoustic principles and absorption rates. Sound-emitting objects traveling at speed might be subject to pitch processing to mimic the Doppler effect. Frequency filtering might be linked to changes in distance, obstruction, occlusion, and so on. Such attempts to recreate the real sound world in miniature are relatively standard but do not fully address the sonic needs of games today, perhaps because very few games truly represent reality. Nick Peck asserts that "we need to understand that a video game is not the real world," and goes on to suggest that "while they're rooted in reality," this is supplemented with "artistic license," the object of which is "to heighten the dramatic aspect... of the story." He adds that "it is more important for a game to be able to succeed aesthetically than to succeed realistically." He illustrates his point by referencing "phenomenal examples of sound design" in films such as *Apocalypse Now, The Conversation*, and *Saving Private Ryan*. With reference to game sound, he rightly points out that the emphasis on scientific reality could yield less than entertaining results; for example, "it would be a mighty boring outer space science fiction game if we tried to mimic the real world since there is no sound in a vacuum."

Just as representations of sounds in space are not scientifically accurate, so it might be argued that the very act of recording sounds and diffusing them between stereo or surround speakers creates an artifice, a removal from the real world, altering the ways in which the human brain perceives those sounds. As anyone who has ever done any recording could testify, the results achieved from the placement of a single microphone in a room tend not to recreate the auditory reality of standing in the same position and simply listening. The effect and filter of human perception may not operate in the same ways when applied to recorded sounds as opposed to real-world sounds. For example, John Broomhall highlights the ability of the human brain to filter out certain sounds, and points to the use of Foley sounds in the game *Operation Flashpoint* as an illustration of one way in which this facility might be recreated in games. The Foley sounds are kept very low within the mix unless there is a dramatic reason for highlighting them; for example, if it becomes important for the game character to move stealthily, the volume of the footsteps increases, adding to the dramatic tension of the game play and focusing the attention of the player more directly on their movement within the game environment. If scientific modeling governs the ways in which sounds are integrated within game play without taking into account the effect of human perception upon those sounds then the model is incomplete and therefore flawed.

28.3 Creative and Technical Challenges: Nonliteral Sound

If the scientific model is incomplete or inadequate when applied within the unreal realms of fantasy or hyperexaggerated reality, it must be supplemented by something else, namely nonliteral sound. The requirement to consider nonliteral approaches to sound, mixing, and processing makes the involvement of sound designers and other sound specialists desirable much earlier in the development of the game and particularly throughout the integration process. It could be argued that if this approach were standardized, the potential quality of interactive game audio, and its ability to entertain, manipulate, and immerse players, would be significantly enhanced.

It should be pointed out that an exploration of past and current game titles would reveal many examples of nonliteral sound successfully integrated into dynamic game play. However if one considers the use of nonliteral sound within the context of pioneering films such as *Apocalypse Now*, it seems logical to consider whether similar approaches might be possible when translated from the linear sphere to the nonlinear. For example, in thinking about one function of sound in *Apocalypse Now*, namely the subtle heralding of future events, we might consider to what extent future events might be anticipated or heralded in game play when those events have not yet been determined by the player. One application of this technique in games would be the introduction of dark or menacing sounds in the mix to suggest danger, implying, almost subliminally to the player, that if they were to pursue their current course of action the outcome would not be good. For example, as a player approaches a door, a sound might be added to the mix, or the real-time processing of the existing game sounds might be manipulated, indicating danger behind the door. Alternatively, sounds might be pulled from the mix, as if to focus the attention, creating a sense of apprehension and internalizing the action. To use a real example; in an unreleased game project John Broomhall worked on some years ago, nonliteral sounds and music were linked to game states such as "exploring, enemy not present," "enemy present not sighted," "enemy sighted," and "combat." When "exploring," the music and sound combined to create unease, suggesting the enemy might be present; when the enemy was present but not sighted, changes in the music, such as greater use of dissonance and altered instrumentation, suggested greater danger. Additional sounds, such as whispering, were also added to the audio mix to intensify the sense of growing danger and fear. Once the enemy was sighted, the music and sound combined to add even greater tension and drama, anticipating the combat to follow. Interestingly once the combat game state was triggered, all music was pulled from the mix, allowing the focus to transfer to the sounds of combat, effectively shifting the focus of the player from the tension of anticipation to the reality of combat, while also preventing a cacophony of competing sounds. In this way the music and sound combined to "lead the player by the nose to the point of combat," and once there, the sounds of combat held the player in the moment, right in the heart of the action. Tying the design of the music and sound to specific game states as the dynamic mix responded in real time to changing game conditions helped to herald danger, and direct game play.

In a more recent example, the audio team for the game *Dead Space* created a game object called the "fear emitter,"[2] which attaches to other game objects, reacting dynamically in response to the location of the avatar. As the avatar approaches the fear emitter, the scaled response of the emitter adjusts the mix and the real-time processing in order to ratchet up the sense of tension and fear. By manipulating the emotional response of the player as they move within the game environment it becomes possible to manipulate and direct the player by heralding possible dangers. The player's response to these object-focused "fear emitters" is intensified by a dynamically changing RAM-based "creepy ambi patch,"[3] designed to recreate something of the player's internal and subconscious mind. The sound samples making up this patch may not correspond to a visible onscreen source since they represent the internalized world of the mind, or "in your head sounds." The use of "fear emitters" and the "creepy ambi patch" also help to create a more engaging gaming experience, one that offers other layers of interactive control and emotional manipulation.

However, such approaches raise other questions, such as to what extent should the mix seek to help the player by warning them of danger, or should such sounds actively mislead players to make game play more challenging? Jerry Ibbotson suggests that the use of sound to mislead or confuse would be appropriate in some games, and highlights *Bioshock* as an example of one that draws near to this approach, singling out the game as significant for its use of sound as "motif" and for its focus on sound design:

> If you think of the start of *Bioshock*... you have no idea what's going... in minutes it... runs into double figures before you have any clue what's going on... Sound design is such a bandied-around phrase, often it's just sound effects, but then sometimes it just steps over into something else. The whole game had... [a] motif of sound.

It is interesting to consider to what extent such approaches might work during protracted game play, since players might inevitably become acclimatized to such effects and therefore anaesthetized to them. Answers to such questions will always be subjective, but the early and active involvement of sound specialists in seeking to address these issues, given their knowledge of the scope and potential of sound, should be at the forefront of game sound development today. Unfortunately, all such explorations and innovations take time and within the pressurized environment of game development, time is a rare luxury.

28.4 Creative and Technical Challenges: Production Schedules

Game production schedules, unlike film schedules, tend not to incorporate an official postproduction stage, although John Broomhall suggests this is now being addressed through the work of people such as Rob Bridgett and Marty O'Donnell. Production schedules for those involved in the creation and integration of game sounds are extremely tight and this can have a significant impact upon the quality of the finished sound and the mix, according to Peck:

> It's an invariable aspect of game production that audio always comes at the end of the line. Often what ends up happening is that the people that are upstream from the audio... will continue tweaking and changing their materials past the point where the audio people need to be doing their work, but the deadline, the ship date itself,

> doesn't slip on the other end. As a result, the audio people are often pinched... there's a time crunch right at the very end of development when they're trying... to get their work in. So as a result, during the time when the sound designers and audio directors should be mixing their game, they may very well be altering sounds to be able to fit revised animations or fielding change requests from their creative directors and other people on the project; the result being that there's less time to mix than you would ever like... There's no question about the fact that having less resources in terms of less people... [and]... less time to be able to work on the game, means that the mix, which is... the very last step in the production chain of audio, is going to be affected.

Peck asserts that allowing more time for the creation and integration of audio by bringing in audio personnel right at the beginning of game development yields the most successful results:

> The games that succeed the most as far as audio goes, are the ones in which audio is brought in at the very beginning of the process just as in film. If audio is left until the end... you will often have a "slapdash," stress[ful]... race... to shovel sound effects in as quickly as possible. Sadly this happens all too often, however there are... many superb exceptions... particularly in the case of some of the larger videogame developers.

Sometimes scheduling pressures are intensified by the availability of key personnel, such as audio programmers; as Ibbotson reflects, "sometimes you work on a game with a coder who's doing many other things and has got Thursday alone to do the sound." Broomhall suggests that situations of this type are less common today but does acknowledge that leaving the mixing to the end of the process is "problematic." He points out that there might be a place for mixing "as the game progresses through production milestones," so avoiding problems of this type. Scheduling pressures compound the creative and technical challenges of audio creation and dynamic mixing. In addition, the challenges outlined so far are also impacted by the fact that the games industry lacks standardization, which of itself presents something of a challenge for audio personnel.

28.5 Standardization: Tools

Steve Martz likens the games industry to the "Wild, Wild West." It is an interesting analogy, which conjures up images both good and bad (and perhaps ugly) in equal measure. Given the proliferation of game genres and platforms, coupled with the fast pace of technological change, it is easy to see why standardization has not been a priority. For some the lack of standardization provides greater creative flexibility and more opportunities for experimentation when developing and mixing sound. Jerry Ibbotson likens "game sound people" to problem solvers; "we are inherently tinkerers... people in games development are always... trying to find different ways of doing things. Even if the first method works... we'll look at another method." He contrasts this with the more set approaches of colleagues in film and television, but makes no judgment between game and film industry methods.

However, when it comes to mixing, the film model does have real appeal, whether or not it is achievable in a nonlinear context. Peck suggests that "when you can get to the point when you can mix a game [in] the way that you can mix a film, on a digital console, then I will say that we're finally getting somewhere." A point that is echoed by John Broomhall, who argues for accessible technical facilities that enable "the sort of freedom a film mixer/sound editor would have to manage and exploit dramatic [and] narrative... opportunities." Goals such as these have encouraged the development of various middleware tools for the development and integration of game sound.

Recent years have seen the move towards greater standardization of these tools as they have been developed by companies such as Audiokinetic (Wwise) and Firelight Technologies (Fmod). Middleware solutions have had a significant and positive impact and are addressing some of the creative and technical challenges facing sound designers and other audio personnel. Nick Peck states that "Wwise and Fmod... allow you to be able to spend more time working on fine-tuning your sounds and less time having to... shovel the sounds into the game." He goes on to suggest that these tools have "allowed mixes to sound better."

The contribution of middleware tools cannot be overestimated. However, as middleware providers operate in a competitive market place, it is not surprising that different game developers use different middleware tools. Sound designers have to work with different middleware tools and also with custom software tools in some cases. Ibbotson's experience reflects this:

> Sometimes coders will design tools for you. For example, we've got vehicle tools; if you want an engine to backfire, it will dip everything else down underneath it, because you haven't got limitless headroom... So you've got fairly complicated... [things] going on in real time using these tools. The tools used will vary from developer to developer of course; some will just ship software in and other times they'll get things written for that particular project.

If one compares this with the more standardized preference for ProTools software found elsewhere in the film and music industries, the contrast appears quite marked. The use of custom tools would be unimaginable in this context. Even within the context of game audio, the development of custom tools has to be carefully considered, as Michael Kelly points out:

> I think designing and supporting custom tools for one title is expensive. To make it cost effective, it's likely you would want to reuse it for many titles. However, even very similar titles (such as sequels) will have different requirements, so rework will be necessary. Further, technology changes so fast that even the most flexible tool will date if it is customized to a particular technology level. My view would be that bespoke technology appears to become unsustainable for anything but the largest

> developers. The analogy in the music industry would be if each record label were to write their own mixing software and build their own mixing desks. As technology gets more complicated, I expect we'll see more from off-the-shelf tools which people can easily adapt to their own requirements; plug in their own extensions, or integrate with their own custom tools by exporting and importing files between them.

Adaptation of software to suit individual requirements might offer a move away from custom software to "generalized frameworks for software which are heavily customizable to bespoke requirements," but such a development would require agreement regarding "plug-in frameworks and... standardized data formats so the plug-ins... could communicate with each other." Clearly anything that simplifies the process of creating and integrating game sound would be a step forward, as Peck reflects:

> I'm a huge fan of simplicity; I want to put the power into the hands of the creative audio professionals... What is still to be achieved is the notion of making it easier... to be able to get sounds into the game interactive environment quickly and simply and to be able to mould and manipulate those materials with the minimal amount of hassle and number tweaking.

One of the most significant steps taken towards achieving this was made by the MIDI Manufacturers Association in October 2001, when they published plans for an interactive eXtensible Music Format (iXMF):

> Interactive XMF is intended to be an open-standard, cross-platform means for audio artists to bundle audio content files with general information and audio implementation instructions. Initially, it will be highly beneficial for the game industry, and since it may be used in any interactive audio application a potential to expand or create markets in other areas exists. This new file format will put artistic control into the hands of the artists, keep programmers from having to make artistic decisions, eliminate rework for porting to new platforms, and reduce production time, cost, and stress. (MIDI Manufacturers Association 2010)

The file format is being developed by the iXMF Working Group (IXWG) of the Interactive Audio Special Interest Group and work is ongoing. Just as MIDI revolutionized music making when it was first introduced, a universally recognized and interactive file format would be likely to have a similar impact and would offer a number of advantages, as Michael Kelly points out:

> In terms of more linear audio tools (like ProTools) there's a clear advantage to game developers if pro-audio (linear) tool developers were able to provide data in a format that was game-ready. Currently many developers must implement their own conversion tools, or worse still reinvent very similar tools which can never do the job as well as off-the-shelf tools which can afford investment in advanced interfaces. Of course, linear tools don't directly meet the interactive needs of game studios, but if pro-audio tools have a means to better fit into the game audio pipeline this has to be a good thing for everyone.

Imagine a situation in which sound designers could create and mix their sounds in ProTools before saving their work to a standardized, interactive file format for use with any interactive software, middleware, or hardware, perhaps including an embedded scripting layer holding detailed information for implementation and integration by any game engine. This would give tremendous creative freedom to sound designers and other audio personnel, offering them more direct control over the integration of audio into the mix. Implementing this level of control and access to a scripting layer would represent a challenge, as Kelly describes:

> The requirement for a scripting language, and ultimately the tool that makes this visible to the sound designer, is that it shouldn't limit the imagination of the sound designers. One way or another it should allow them to derive mix values from whatever they want, whether this is something high level like the number of players on screen, the weather in the game, or something low level like the current gain value of a compressor on a parallel bus. Similarly it should not limit what they can use to mix, whether it is volume, pitch or center frequency of a filter, or another scripted variable that links to any parameter on any other voice or bus. When you have flexibility like that, the team can decide what mixing system meets their specific game requirements. Such flexibility does come at a cost, but the proportional cost is lower and lower all the time. Ultimately it should be an agreement between the sound designer and the programmer... Traditionally it was dictated by the programmer; most likely because it was dictated by the hardware. Since the hardware restrictions are reducing, it is important that tools offer this flexibility and therefore it is important that any scripting language supports that, both for now and in preparation for the future.

If such tools were available, it is likely that the general public would also be interested in exploring the possibilities offered by such standardized interactive formats, something that Kelly recognizes: "I think there is a need for tools the public can get their hands on to create interactive audio; it is not just the games industry who need them." Above all he stresses the need for flexibility: "standardized file formats and standardized connectivity (such as plug-ins systems) between interactive and noninteractive tools mean that the user can choose the right combination of off-the-shelf and bespoke tools for their needs—that is what we lack at present." With reference to plug-ins, Kelly suggests that the range of available plug-ins is limited but adds that third-party plug-ins are gaining some ground in the games industry:

> we're beginning to see some use of third party plug-ins in the games industry. It will be interesting to see how it pans out. If the aim is flexibility and customization then plug-ins are definitely useful. For a developer who works on a cross-platform title it is more difficult because not all of the plug-ins will be available on all platforms. There may be technical reasons why a plug-in cannot be supported on all platforms, or more commercial reasons. Whilst these constraints exist (and I expect they will for some time to come), the tools can help by finding the best way to author data by making the best of each platform (and available plug-ins) but prevent the need to duplicate design work for different platforms with different abilities.

All of these developments would be to the advantage of game developers and content creators, but the same might not be said of middleware providers, who have invested time and money in the development of the tools they offer; standardization would inevitably involve redevelopment and could provide competition from creators of linear audio tools. Steve Martz suggests that there might also be some resistance to a standardized game data format on the part of publishers, developers, and some larger studios:

> They spend a considerable amount of their internal resources customizing... game engines to accommodate new features and advancements for the next title as well as streamlining project workflow. [That is] not to say that... resistance is unwarranted. They do what works best for their production pipeline;... third-party software or [standardized game data] format(s) may or may not be able to meet their needs in a timely manner.

For those more loosely associated with game development, the advantages of an open standard would be significant, but for now, despite various proposals and the establishment of working groups, Steve Martz notes that "their progress in creating a ubiquitous standard has achieved limited success. They may be successful at some point, but just not now."

28.5.1 Toward Standardization

Questions of standardization apply not only to tools for interactive audio and interactive data formats but also to technical specifications for game audio; a full discussion of which would be beyond the scope of this chapter. However, if we consider one example, namely reference levels, questions of standardization soon become apparent. In a paper written for the Audio Engineering Society's 35th International Conference in 2009, Masataka Nakahara, Kazutaka Someya, and Steven Martz surveyed differences in playback levels for in-game sound, game start-up sounds, and reference levels of game production studios. In the opening statement of the paper the authors state that:

> The audio experience can vary greatly when playing a video game. After establishing a playback level for a reproduction device, the game may still be too loud or too soft. Average reproduction levels can fluctuate from title to title and even within a game. It is important for game production studios to carefully consider how they configure their systems and the effects it may have on consistency in their games. (Nakahara, Someya, and Martz 2009).

In addition to suggesting useful reference levels for games based on average figures resulting from the survey, the authors of the paper conclude that "there are large differences among production companies, and... that the level of start-up sounds which are implemented into game consoles are approximately 10dB lower than the level of in-game sounds in opening cut-scenes" (Nakahara, Someya, and Martz 2009). If one compares such differences with the standardized specifications found in the film industry, the contrast appears quite marked. The work of the Audio Engineering Society and its Technical Committee on Audio for Games seeks to address issues of this type while furthering educational goals. Conferences, such as those organized by the AES and also the games industry itself, also have a role to play in establishing greater standardization, while also sharing best practice and promoting education.

The lack of standardization in relation to various technical specifications has a direct impact on mixing for games. Consider one example: game audio, unlike film audio, is mixed primarily for home media systems and mobile media devices. The speakers of these systems vary hugely in quality and specification, ranging from the cheapest in-built speakers through to high-end, audiophile systems designed for the well-heeled connoisseur. While equipment bearing the THX logo indicates a certain standard of maximized performance for audiovisual components and systems based on standards and parameters adapted from the cinematic context, not all manufacturers follow THX guidelines. Despite some general improvements in the quality of cheaper speakers for the home, it could be argued that the growing market for surround-sound systems has proved somewhat counterproductive and has not facilitated quality enhancement. For those on a budget, the money that would once have secured a reasonably decent pair of stereo speakers is now expected to secure anything up to seven speakers plus a subwoofer. In some cases surround-sound systems are simply thrown in as an extra feature accompanying the purchase of large-screen television sets, arguably resulting in a dilution of the quality of these systems. Even the best surround-sound systems may not be calibrated correctly for the domestic spaces in which they are placed. Consequently, those mixing audio in preparation for release to the home entertainment and mobile device markets must consider the variable quality and specifications of this equipment. Jerry Ibbotson accommodates such considerations in the final stage of the mix process:

> One of the key problems people have is that if they have a 5.1 setup, the majority of them will be setup incorrectly. They've probably got the sub up way too loud, drowning everything else out and many people won't know what the center speaker is for... With games that are played in the home it is grimly inevitable that someone has their speaker system setup oddly... With a game you always try and work on it in a studio environment and have that benchmark, but then you need to know what it's going to sound like in a domestic environment. When we worked on a PSP game we had a PSP development kit in the studio. At first we had the sound coming out of the studio speakers but as we got further towards the end of the project we were unplugging it from the mixer in the studio and either plugging in the worst headphones we could find or just playing it through the inbuilt speakers, so you know what it's going to sound like to the punter on the bus. For us, the logical process would be to work on the samples and the overall implementation at the highest possible quality you can manage, and then towards the end, to get the overall mix right, listen to it on what the consumer would listen on. Most home systems are pretty good these days and will pick up on most things but on a handheld like the PSP with tiny speakers, some things will just get completely lost. We have done stuff in the past, say a menu sound, where we have added a frequency element on top so that you'll be able to hear it better through smaller speakers.

Broomhall also suggests that great care is taken within the industry to consider different home setups. For example, depending on the playback configuration selected, adjustments can be made to the settings of multiband compressors strapped across the final mix, optimizing the sound for a particular setup. He also points to a growing trend towards the use of external, premium studios in which game teams set up their game development equipment and "tweak their mix in a high-end reference listening environment, possibly with a consulting mix engineer on hand to advise." It is clear that those within the games industry are concerned about the integrity of dynamic game mixes when transferred to the home environment and it is also clear that steps are being taken within the industry to move towards greater standardization in terms of tools and technical specifications. Just as standardization developed over many years within the film industry, it appears as if the games industry is at the start of a similar journey towards greater standardization.

The limitations of game audio technology, which were once so apparent within the games industry, have faded into the background, being replaced by rapidly evolving and unique creative and technical challenges. In addressing these challenges lies real potential for innovation and the exploration of "the power of ideas" (Broomhall 2010). The rich resource offered by the best examples of film audio and practice provides a valuable source of inspiration, but the interactive nature of game audio opens up new avenues waiting to be explored by the power of the imagination.

28.6 Conclusions

From very simple beginnings, game sound has progressed rapidly and continues to evolve. Film sound has provided an important referential resource in the development of game sound and continues to be of relevance moving forward into what promises to be a future of further rapid technological change and innovation. However, despite some common ground, there is one crucial point of divergence at which the paths of film and games separate, namely interactivity.

The creative and technical challenges presented by interactivity, as outlined in this chapter, fall into a number of key areas: the integration and mixing of sound within dynamic, real-time environments and the involvement of audio creators in this process (determining the prioritization, placement, real-time processing, and functionality of audio); and the impact of production schedules and associated budget constraints, coupled with nonstandardized approaches, tools, and audio specifications.

What might be the outcomes if these challenges are successfully tackled? It seems likely that as the challenges of integration and dynamic mixing are gradually eroded, audio creators will have far greater input in determining the ways in which sound functions and is prioritized within games. If this were to occur we might expect to see a shift away from literal, "see it, hear it" approaches, toward nonliteral explorations of the "storytelling power of sound." Within this context, the old adage "sound keeps it real, and music tells you what to feel" is likely to be challenged as the full potential and functionality of sound are explored within games, matching the very best examples of nonliteral sound on film but, more importantly, exploring the unique possibilities of sound in interactive environments.

Ironically such nonliteral explorations of functionality and precedence within game mixes might create games that feel far more real and immersive. For example, the prioritization of dialog in games mixes today can conjure up the impression of disembodied voices floating above the action, recreating anything *but* reality. The primary reason for the development of this approach is the need to communicate information to the player to facilitate game play. Perhaps this is appropriate in the first levels of a game, or in games for children, but should dialog be prioritized in this way as a rule of thumb? Might there be other, more subtle ways of communicating information to the player through sound rather than dialog? Might there be a case for depriving the player of information, or even misleading them, in order for game play to become more real or challenging?

If production schedules become a little more accommodating and consideration is given to the use of audio right from the beginning of the game development process, there should be more time for experimentation of this sort, posing and addressing questions such as these. Technological advances, in terms of integration, development, and so on, might "free up" additional time within production schedules, allowing more time for creative experimentation while also providing audio creators with easier ways of integrating audio assets, enabling them to try out different ideas and approaches. If, in addition, a dedicated postproduction phase of development becomes standardized we should also see improvements in sound production and mastering.

In terms of answering the technological challenges of dynamic mixing and games development more specifically, we should expect to see further developments in middleware tools and dedicated game development technologies. However, given technological trends already outlined, we are also likely to witness developments in "off the shelf" and customizable software and plug-ins. If the move toward standardization progresses, the features of standardized connectivity and multiplatform use will become more important, as will standardized data formats and audio specifications. Developments such as these would certainly be of interest to the general public and to providers of pro-audio (linear) tools, and perhaps to the film industry as well. At this point we might witness a reversal in the tide of influence as the film industry adopts technologies originally developed for games. More specifically, technological advances developed for the purposes of integration, dynamic mixing, and real-time processing, might be of use when producing, pre-dubbing and mixing audio stems for film.

In addition to technological advances supporting the development of games, we should expect to see further developments in gaming hardware and user interfaces. It seems likely that the human body itself and associated physiological or psychophysiological responses might well become the focus of interface development. Nintendo's ongoing development of a "vitality sensor" (designed to measure pulse response) and Microsoft's proposed "Joule" heart-rate monitor for the Kinect Play Fit, might be indicative of future trends. A patent application for biometric controls (designed to measure galvanic skin resistance, electrocardio data, and electromuscular data), filed by Sony in 2010, would suggest that the company is at least thinking along similar lines. The introduction of interfaces with biosensors of this type could be quite significant for developments in dynamic mixing. For example, galvanic skin response, heart rate, respiration, and so on, might be linked to real-time processing and changes in the layering of the mix; changes designed to intensify physiological and also emotional responses. Technological advances such as these will no doubt raise new technical and creative challenges, not to mention ethical and medical considerations. As the games industry addresses these challenges in the interests of interactive and immersive entertainment, outcomes might result that could be applied in other fields, such as medicine and music therapy, perhaps offering new approaches in the treatment of specific physical and mental health conditions.

To date, innovation and a pioneering spirit have characterized the games industry. The technical and creative challenges of interactivity have fuelled rapid and diverse developments. If moves toward standardization progress, with all the potential benefits this might afford, it will be interesting to see whether these traits continue to flourish. A standardized games industry might not nurture innovation, experimentation, and diversity to the same degree. The exploratory zeal of the games industry might be stifled in less fertile, homogenized ground.

Moving into the future, the advantages of standardization might well outweigh concerns of this kind, paralleling the approach taken by the film industry in its own historical development. But do the parallels end there? It seems likely that the influence of the film industry will lessen in coming years. As the games industry grapples with the challenges of interactivity, the relevance of linear film as a referential resource will continue to diminish. From its first tottering steps the games industry continues to mature towards greater independence. Like a stroppy teenager with a growing sense of their own identity, the games industry will become more self-sufficient and the balance of influence will shift full circle.

Acknowledgments

The author wishes to thank the contributors, all of whom gave so generously of their time, expertise and contacts in the preparatory interviews for this paper:

John Broomhall (2010): Broomhall Projects Limited; audio director and consultant (see http://www.johnbroomhall.co.uk for further details), BAFTA Video Games Committee and Chair of BAFTA Interactive Entertainment Audio Awards, Columnist for *Develop Magazine*. Previous roles: formerly Head of Audio (and Game Producer) at MicroProse, Spectrum Holobyte, Hasbro Interactive, and Infogrames UK Internal Studios. Interviewed February 2010.

Jerry Ibbotson (2010): founder and Creative Manager of Media Mill (a game audio production company founded in 2000; see http://www.mediamill.co.uk/ for further details). Writer and reviewer for *Audio Media* magazine. Guest lecturer in sound design at Hull University, York St. John University, and York University. Consultant and beta tester for Adobe Systems. Previously a BBC radio journalist and freelance radio producer. Interviewed February 2010.

Michael Kelly (2010): Senior Audio Engineer, SCEE R and D, Sony Computer Entertainment Europe and Co-chair of the AES Technical Committee on Audio for Games. Previous roles: Audio Research Engineer at Creative; Lead Sound Designer at Revolution Software. Interviewed by telephone and email, July and August 2010.

Steve Martz (2010): Senior Design Engineer THX Ltd and Co-chair of the AES Technical Committee on Audio for Games. Interviewed by telephone and email, June to August 2010.

Jamie Maxwell: Research Assistant.

Nick Peck (2010): Perceptive Sound Design (owner). Previous roles include: Audio Director, Activision; Sound Supervisor, LucasArts Entertainment Company Games; Game Sound Supervisor, Skywalker Sound (see http://www.underthebigtree.com/perceptivesound/ for further details). Interviewed August 2010.

Notes

1. Academic literature tends towards identifying a greater number of audio types or classifications. These are sometimes linked to terminology more commonly associated with film studies, such as diegetic and nondiegetic sound or music. See Collins 2008, 125–7.
2. For further details, see Original Sound Version 2010.
3. See Original Sound Version 2010.

References

Collins, Karen. 2008. *Game Sound: An Introduction to the History, Theory and Practice of Video Game Music and Sound Design*. Cambridge, MA: MIT Press.

Lastra, James. 2000. *Sound Technology and the American Cinema: Perception, Representation, Modernity*. New York: Columbia University Press

Nakahara, Masataka, Kazutaka Someya, and Steven P. Martz. 2009. An Analysis of Differences in Playback Levels Among In-game Sound Levels, Game Start-up Sounds, and Reference Levels of Game Production Studios. *Proceedings of the 35th AES International Conference on Audio for Games*. New York: Audio Engineering Society.

MIDI Manufacturers Association. 2010. *The Interactive Audio Special Interest Group and the iXMF Working Group* (IXWG). http://www.iasig.org/wg/ixwg/.

Original Sound Version. 2010. *Don Veca Interview: Dead Space Audio*, http://www.originalsoundversion.com/?p=693.

CHAPTER 29

OUR INTERACTIVE AUDIO FUTURE

DAMIAN KASTBAUER

In the not too distant future . . .

I'm standing on the bank of a river. It's late and the heat of the afternoon sun is fading. The sounds of cicadas are all around me. Their placid chirruping voices, accompanied by the nearby burbling of a stream, sets a tone of peaceful tranquility. The simulation in here is really good. As I kneel next to the shore, splashing water on my face, the sound erupts in a perfectly modeled and synchronized symphony of harmonic fluids (see Kurachi 2010). Each drop of water from my fingertips rippling on the surface is reflected in automatic sound synthesis (see Moss et al. 2010) and tied so tightly to the visuals that I'm struck by the fact that it was just a short time ago that we were locked into a sample-based methodology. It didn't happen overnight but, standing on the shore of this digital river, it's clear that audio technology has finally made good on the promises of transparent interactivity.

It should be said that I'm inputting this log in real time while working inside the beta release of our current simulation based on a historical recreation of Earth *circa* 2012. The simulation has been developed using an industry-standard authoring application by specialists spread out across the galaxy. I'm currently reviewing the work done by our technical sound synthesis specialists who have recently made a pass at updating some outdated sound models. It seems like ages ago that we began working with our library of reference materials to synthesize different aspects of Earth's soundscape into the living, breathing, and sounding world that is represented today.

I am lost in thought, and the sound of rushing water catches my memory. My mind is transported to a sunny day from my past. A reunion has brought family members together within a branch of an older beta version of a simulation located by a nearby stream in the countryside of Earth 2012. It was during this time in simulation technology that our industry was just beginning to iron out several inconsistencies inherent within the burgeoning field of procedural audio, synthesis, and the advanced manipulation of sound: baby steps toward the expansive fully realized simulation I'm testing today. As we laugh and carry on inside my mind's eye, the children play and chase butterflies along the edge of the rushing water. From the corner of my eye I can see my young daughter rushing in and out of cattails twice her height with her favorite doll. The look of freedom and wild abandon as the fronds whoosh back and forth while she is chased by cousins brings a smile to my face. I move to speak, and urge her to stay clear of the treacherous stream edge, but they are gone before the words leave my lips.

I become lost in debate with various relations over the use of sound as a storytelling device. It seems that we have finally found a way to effectively use all five senses to convey an emotional arc as part of an interactive experience. Of course, there continue to be new ways to channel and leverage our past through simulations. New ways of combining what has already happened with new technologies as a way of making sense of the future. Through the lens of creativity we may finally understand who we are and where our civilization is headed.

Back at the reunion, the sun is beginning to set in the distance and it's time for us to take leave of this place. I head toward the pack of children collapsed by the riverside, in search of my daughter. My steps echo coldly in the grass. Her smiling face goes unseen, so I ask the random gathering of nieces and nephews where she might be. No one has seen her for some time, they say. In a moment I feel my every fiber stiffen toward a heightened awareness of my surroundings. Nothing has changed, and yet in that instant it's as if all sound has been removed but for the deep churning of water. I look around frantically, my eyes darting between reeds as I begin calling her name. People are gathered around me; I can see their mouths moving but can no longer hear their questions. My mind is sharply tuned to the sound of dramatically frothing whitecaps as I desperately attempt to keep my thoughts above water.

I wade into the sea of cattails, moving further and further upstream. Their fronds brush past me in a noiseless hush of extreme focus. In an instant I'm conscious of others who have crossed a nearby bridge and begun combing the bank on the other side. My voice becomes louder and more urgent, punctuating the ebb and flow of water to my left which sounds far away. A curious rocky outcropping comes into view, as the taste of saltwater reaches my lips. I see my daughter playing noiselessly with her doll in the shade of a dark black stone column. I call to her, but she doesn't look up. I quickly close the gap between us and suddenly recognize that the sound of the stream has disappeared.

Suddenly, my daughter becomes startled by the sound of my voice calling her name. She looks up as I rush to her and hold her in a deep embrace, my footsteps resonating with an alien quality. The sound, clothing brushing against her white dress, explodes within the vacuum of soundlessness I find myself enveloped in. I look into my daughter's eyes as she smiles and holds out a ring of flower stems that she's woven into a ring. She looks at me and smiles. When asked by friends later, she'll say, "I wasn't lost, I just found a quiet place to play." We emerge from the noiseless, interstitial space, call off the search party, and return to the sound of the world around us.

These early simulations weren't perfect, but the feelings generated by them still resonate. The difficulty of authoring propagation for every explicit sound once meant that the process was subject to gaps in the handwoven audio fabric. Every inch "painted" by hand with sounds. It's easy to see, looking back on the technology, how these "dead spots" could exist—where no extraneous sounds can reach. Thankfully, work continued to be done toward an accurate representation of sound propagating within a simulation (e.g., Schissler and Manocha 2011). These advances were, over time, able to leapfrog the manual process that had been bogging down the development (e.g., Raghuvanshi et al. 2010). I have to say that the radical sound-mix change brought on by my alerted state operated perfectly, in retrospect. It's hard to believe the leaps and bounds that technology has achieved, even during my short life span. Especially so, as I have found myself deeply embedded in driving the current state of the art in our current recreation of Earth.

Standing up from splashing water on my face, I dry my hands on my coarse cotton pants. This action elicits a similar soft explosion of sound as that day so long ago. I'm reminded of how long it took us to get the subtlety of movement and material interaction to accurately reflect the complex dance of physical sound. Tying into the cloth simulation was the easy part. Once we had all of *that* data on our hands it was purely a process of sculpture, mostly subtractive but surprisingly influential on the resulting sound. Things became more complex with the addition of liquids and temperature... but at the end of the day, our flexible synthesis models stood up to the barrage of parameterization from each of the systems.

I step back onto the path and continue to a nearby watermill. I'm reminded at each footfall how long it took to get these footsteps right. Everything is now handled inherently by the procedural models in place, which take into account things like mass, velocity, materials, viscosity, among a variety of other things. What began as a purely academic pursuit swiftly blossomed into a creative endeavor, thanks to some early frontrunners in the field. Once the underlying mathematics were in place, the tools quickly followed. This development allowed early sound designers to move beyond hard science running behind the scenes. Enabled by smart and creative authoring applications, the sound of footsteps and movement sounds were able to transcend the synthetic quality of early experiments and emerge as a realistic representation.

As I approach, I note the churning sound of water on the undershot-wheel paddles grows louder with each step. I'm struck by the quality of dynamic frequency filtering over distance. The simple 3D-point sourced, volume-only attenuation curves of yesteryear have long ago been replaced by a complex matrix of real-time comb filtering, taking into account everything from air density to temperature and humidity. The spinning wheel splashes on effortlessly while some unseen turbine labors to convert the power of nature into energy. Of course, this one is just for show. However, it wouldn't surprise me to find that it powers all of the lights inside. In such a complex woven fabric of interaction, you can count on everything being designed with a purpose.

Approaching the door I reach for the handle with trepidation; the last time I tested it, it didn't accurately convey the weight and material type, so I sent the sound designer back to the drawing board. Turning the knob I hear the rusty mechanism come to life with a sharp snap of metal, hollow-wooden resonance, and deep unsettling creak. Each individual aspect of the sound takes into account the compositional breakdown occurring: damp location (as evidenced by the deep rust flaking off each hinge), wood density and thickness, in addition to the age and tenacity of the steel (James, Barbič, and Pai 2006). The door closes with a deep thud that reverberates as I let it close behind me (Taylor et al. 2009). That was much better, I think, as I send off a friendly note to the designer who is a million miles away. Within moments I've received a response which reads like a sigh of relief.

With so much to do, it's hard to get caught up on the little things. Thankfully the work we've been doing on the underlying physical modeling propagates throughout the entire world. Imagine if we had to place sounds by hand for every river or door; the worlds we're creating are just too large for such a labor-intensive process. While we do try to take a *couture* approach to sounds that require something a bit unique, the expectation that every nook and cranny sound convincingly Earth-like in this simulation involves more than just a few sound files spread across the face of the planet.

Back inside, there is the pervasive, low-end rumbling sound of rushing water, coupled with an oppressive clattering. I'm standing in front of a succession of wooden gears locked together in spiraling groove of perpetual motion. The hollow "thonking" sound they make as each tooth locks into place is a direct result of the latest wood models that have recently been shared by the foremost modal synthesis historians (e.g., Bonneel et al. 2008). Piggybacking on their research into old technologies and material types has brought a satisfying level of detail to the rich characteristic of wood. With their research in place we were able to further embellish the machinery using several creative tools running within the simulation.

The whole contraption is shuddering with an unbelievable rumble that I can feel in my gut. With a gesture I engage the authoring tools: an interface for interacting with the world appears in front of me. As it springs to life, ready for action, I quickly navigate the controls and begin slowing down the water flow outside in order to hear the change in sound. As the internal mechanism begins to wind down, there are no artifacts in either the pitch or timbre as the gears begin slowing to a halt—this isn't a simple sound file representation. The only remaining samples of an actual watermill were recorded at the low sample rate of 192 kHz/24-bit. We were able to extract some relevant data from these recordings, which were used to inform the creative application of various processes. These samples were critical, since we had never seen a working watermill in person, and ended up affecting the overall sound presentation.

As things grind to a halt I notice a gentle whistling finding its way through the cracks in the thatched roof overhead. Wind was the first and easiest dynamic synthesis that was applied to these simulations. Pitched and filtered noise of different colors was applied in combination with reflectors and deflectors, both abstracted within authoring toolsets and programmatically based on geometric representations within the different environments. The technology was very futuristic at the time. What it lacked with regard to its ability to sound natural, it made up for in its ability to be modified in real time using parameters from the simulations. As the technology progressed, the natural randomness of nature swiftly took the place of consistently sloping pitch curves and unreal sounding representations.

My footsteps echo on the hollow wooden floor convincingly as I resume my test and calibration procedures. I find the watermill synthesis and modeling holding up well under these extreme circumstances. I stop my testing just short of flooding the entire mill in order to listen to the resulting forces of nature on the tiny primitive structure. Safe within the confines of the simulation, the properties of the debug shard that instantiated when I enabled the authoring tools, I would have a unique perspective on the resulting mayhem. Distanced from the confines of the simulated physical world within this unique instance, I'm free to run amok. I'll save the creative joy of destruction for another day. I start the river gently flowing again and exit the tiny building and continue my testing.

Outside the sky has blossomed into a majestic eruption of purple and pink at dusk. The late afternoon cicadas have all been replaced with a chorus of crickets and occasional bird chatter. It's been so long since we were restricted to a single looping audio file that I don't even notice how diverse the soundscape is. While there continues to be a place for field recording as the basis for building a library of granular models and anomalies, it's no longer possible to capture these sounds in their natural habitat. Whereas these sounds may have existed in nature a long time ago, that time has since past. Luckily we have access to a wide variety of artifacts from the twenty-first century, including: field notes, recordings, and skeletons.

Some of the environments initially proved difficult to recreate. Isolating elements amid the noise of what was then modern culture proved easy enough at first. When it became more difficult was when we began to note peculiar behavioral changes between records taken from different decades in many of the vocalizing birds and animals. As industrialized society began to take hold, so too did the sound of a new era of machinery and technology. These sounds become an inexorable part of the complex auditory fabric of the earth and, over time, completely modified the speech and frequency range of vocalizations wherever nature overlapped with industrialized society. The difficulty became understanding the complex interaction that developed over time and how to realistically represent the ensuing soundscape. The result is a blending of both natural and manufactured sounds in a complex cacophony quite unlike anything heard since.

I watch the sun slowly slipping behind the rolling hills off in the distance and I'm struck by the true beauty of the 2012 Earth embodied in our simulation. Everything represented with a rhythm that can be found resonating throughout—from the tall grass gently swaying in the breeze to the waterwheel working in ceaseless syncopation. It seems that in this moment, there was a balance between the elemental forces at work and the swiftly encroaching hands of progress. It's impossible not to judge the years that have transpired in the interim as tragic when faced with such beauty. It is the hope of all of us that this experience can serve as a future road map for how to proceed as a society, now that the damage has been done.

Back to work, I quickly navigate a user-interface terminal and instantiate a new area of the world to test in. A transition opens soundlessly in front of me and floods my senses with the sound of the city. I'm overwhelmed by the density of the experience. All oppression and intensity as drones weave their way in and out of a sympathetic embrace. This humming metropolis is a hive, a thrumming and humming frenetic audio vibration that typified 2012 Earth in all its glory. In a moment: speeding cars, trickling fountains, and skyscrapers resonating in an orchestration of the, then, modern age.

Between the insistent air-conditioner rattle and intermittent elevated train clatter lies the golden age of information technology; an endless stream of activity amid the rapid acceleration of interconnectivity. It's not long, as I cross the threshold of modernity, before I'm confronted with the endless saturation of sensual information. These simulated molecules caress my battered brain-stem in a dance of orchestrated input toward an inevitable overflow of stimulation. Whereas moments ago I was adrift in the tranquility of the rural countryside, I'm now firmly frenetic in the no-hold-back-all-attack modern age.

I engage a frequency analyzer that immediately projects a rainbow of visual analysis across every corner of the world. I can visualize the density of the frequency spectrum based on color as it emanates from every object, radiating a kind of ethereal cloud of color based on the sounds of the city. As cars speed by I see their unique voicing characteristics reflected in a wash of frequency analysis, the hue and saturation accentuating a buildup of tonality across the spectrum. I'm looking for an anomaly reported by early adopters of this simulation, some sort of unnatural resonance.

I load a restore point and the world around me snaps into its prerecorded routine. It must have been a familiar scene: buildings reaching to the sky, vehicles treading familiar paths, a swarm of people vibrating with energy. Each building resonating with a hum. From the depths of their mechanical rooms to the vibrations of people cascading throughout the interior, each building is like an oscillator in this polyphonic synthesizer of a city. Vehicles zipping around like oversized insects buzzing with momentum, intent on reaching their destination through a complex mechanical ballet of belching exhaust and the endless drone of so-and-so revolutions per minute.

It's the same on the sidewalks that flank every thoroughfare. Everybody on the street moving, seemingly, in sync toward an unseen destination. Moving lockstep in time with the heartbeat of the city; like a million tracks of kick drum mixed below the threshold of hearing, but undeniably propelling each step. The individuality of their concrete footfalls lost in a sea of clear density (see Murch 2005).

Aside from the occasional amplitude increase, so far, the playback seems normal. I scrub through the capture to the moment specified in the report and all at once the world is saturated in a haze of pink, representing a frequency pileup. I switch to manual control and reverse to the origin of the anomaly. I peruse the data streaming in the debug window in an attempt to isolate the source and immediately see the problem. I navigate to the source; it seems at first glance to be a discrepancy in the modeling. As I trace along a sinuous pulsating line that connects the extreme data-point to a mass of overlapping frequencies, I find myself standing low to the ground somewhere in the middle of the sidewalk.

I switch off the frequency analyzer and find myself face to face with a little girl who is sobbing uncontrollably. This explains it, the quality of our procedural voice synthesis always begins to break down during moments of extreme emotional response. Our models take into account the simulated size, shape, and composition of the simulated vocal chords and we've captured and recreated the entire range of motion, but there are extremes in every case that still require tuning.

Of course we began by nailing down celebrity vocalizations first: these were the most profitable. As soon as an actress or actor achieved a certain status we rushed in to take a vocal snapshot which could be perpetuated beyond the peak of their profession. Coupled with a full body scan, procedural animations, and phoneme-based lip-syncing, we've had digital doubles as part of our experiences for a long time. There still exist moments when even our best technology cannot yet achieve the artistry present within a live body. At the end of the day it is still the performance, whether acted or authored, that has the potential to connect with the audience. It remains a testament to the human body-instrument that there are still secrets held within.

The girl seems to have quieted down and her face begins to scan the crowd frantically. My heart goes out to her as I remember that day so long ago when, as a parent, I lost someone dear to me, if only for a moment. Of course this is a simulation, none of this is happening, but I'm still here having a real reaction that I can feel in my chest as my breathing accelerates. It's more a result of my experience than any belief in this as "real." Regardless, I kneel down on the sidewalk in front of her and ask if she is lost. She looks in my face, not knowing exactly what the answer is, but eventually nodding in acknowledgment. I begin to ask who she is with when, mid-sentence, the message is broken by a mother quickly lofting the little girl into her arms. Amid a volley of hugs and exasperated condolences, she's quickly whisked away through the bustling parade of lunchtime foot-traffic. I can see her smile through the sea of people as I return to the task at hand.

Scrubbing back in time to the original anomaly, now identified as the little girl crying, I utilize the controls attached to the data-point for the girls vocalizations. Soloing the sound of her voice, I begin to apply various smoothing algorithms. Without losing any of the ferocity or emotion, I methodically constrain the parameters that have caused things to become unnatural. The process takes just seconds, but in a simulation of this scope it could take hours to address every similar case. Instead, I apply the adjustments to the vocalization models to be used in the event of other anomalies. While the changes don't go as deep as the simulation itself, they can be applied in real time if the same behavior is found. If it happens often enough, the solution will be used to inform the next update to the vocalization model.

This period of Earth has come to be known as a "great turning point" in the evolution of life on the planet. The majority was beginning to wake to the changing environment. Even amid the abstracted nature of the city, people are beginning to take note of the fact that we're all trying to share this planet. The recognition spreads across the planet and embraces the population in a warm realization that we are all part of something bigger than ourselves.

I exhale and return to the task at hand, flipping to the next restore point that needs investigating. I'm faced with a wall of water, an undulating sea, and wind set to tear the roofs off the nearby houses. The howl and moan as the waves increase their amplitude approaching shore are unhinged in a moment of sheer sonic terror. I bring up my display, and prepare to orchestrate the pure power of nature.

Epilog

Through a combination of fictional story and factual reference the idea of this story is to inspire the work being done to help envision the way forward for interactive audio. While some of these workflows and methodologies live strictly in the realm of science fiction, there are aspects that can be found running in simulations at universities today. This radical change in approach—from the standard sample-playback methodology into a composite toolbox that incorporates extensible procedural, synthesis, and physical modeling techniques—is rapidly evolving within today's game industry toward a future hybrid model of dynamic sound. Due to increasing size, diversity, and complexity inherent in most games, it seems inevitable that sound needs cannot be met with sample-based content alone. I continue to look forward to those creative ways that sound can reinforce the perceived reality of an interactive experience by leveraging the inherent dynamism of simulations.

References

Bonneel, Nicolas, George Drettakis, Nicolas Tsingos, Isabelle Viaud-Delmon, and Doug James. 2008. Fast Modal Sounds with Scalable Frequency-domain Synthesis. *ACM Transactions on Graphics (SIGGRAPH 2008)* 27 (3). http://www-sop.inria.fr/reves/Basilic/2008/BDTVJ08/.

James, Doug L., Jernej Barbič, and Dinesh K. Pai. 2006. Precomputed Acoustic Transfer: Output-sensitive, Accurate Sound Generation for Geometrically Complex Vibration Sources. *ACM Transactions on Graphics (SIGGRAPH 2006)*. http://graphics.cs.cmu.edu/projects/pat/.

Kurachi, Noriko. 2010. Viewpoint. *Computer Graphics World* 33 (9), October. http://www.cgw.com/Publications/CGW/2010/Volume-33-Issue-9-October-2010-/Viewpoint.aspx.

Moss, William, Hengchin Yeh, Jeong-Mo Hong, Ming C. Lin, and Diensh Manocha. 2010. Sounding Liquids: Automatic Sound Synthesis from Fluid Simulation. http://gamma.cs.unc.edu/SoundingLiquids/.

Murch, Walter. 2005. Dense Clarity—Clear Density. *Transom Review* 5 (1), April 1. http://transom.org/?page_id=7006.

Raghuvanshi, Nikunj, John Snyder, Ravish Mehra, Ming Lin, and Naga Govindaraju. 2010. Precomputed Wave Simulation for Real-time Sound Propagation of Dynamic Sources in Complex Scenes. *ACM Transactions on Graphics (SIGGRAPH 2010)* 29 (4), July. http://gamma.cs.unc.edu/PrecompWaveSim/.

Schissler, Carl, and Dinesh Manocha. 2011. *GSound: Interactive Sound Propagation for Games. AES International Conference on Audio for Games*, Paper no. P2–6. New York: Audio Engineering Society

Taylor, Micah T., Anish Chandak, Lakulish Antani, and Dinesh Manocha. 2009. RESound: Interactive Sound Rendering for Dynamic Virtual Environments. *Proceedings of the 17th International ACM International Conference on Multimedia*, Beijing, China. http://gamma.cs.unc.edu/Sound/RESound/.

CHAPTER 30

FOR THE LOVE OF CHIPTUNE

LEONARD J. PAUL

"CHIPTUNE" gets its name from the sound chips in computers that were originally used to balance the processing power of generating sound effects and musical tunes from the CPU. Each computer or videogame machine would have its own sound chip to create audio and it gave a unique sound to each system. The sound of chiptune is the sound of nostalgia filtered through the mono speaker of a blurry television from a videogame console made over thirty years ago. However, we are now seeing a generational shift from those who grew up with videogames and computers from the 1980s to those born into the age of the internet who are listening to chiptune music. There is a certain "retro chic" at play, where the first generation has a nostalgia for the original systems, and the new generation is attracted to the old sound as a way to generate something new, similar to the color photography film digital effects that remind us of the film development from years ago, or the recent trend for Polaroid-esque photos. There is an overlap of the old and the new generations that has created a fascinating blend of styles and approaches to the relatively new label applied to this type of music that has become popularly known as chiptune music.

The synthetic nature of chiptunes allows them to be easily manipulated in real time and most composition systems allow for a degree of live control. Some of the first playback systems for chiptunes (such as the Commodore SID chip) even allowed for live instruments to be modified by the chip's onboard processing. The common format of chiptune creation, being module trackers or "trackers," still finds a home today within modern game-building tools such as Fmod and Unity3D supporting this flexible chiptune format which is explained further in the following sections. A single definition of chiptune is debatable, but one can split the definition of chiptunes into three separate definitions from most restrictive to the most inclusive, relating to hardware, nostalgia, and the chiptune aesthetic.

30.1 HARDWARE

Chiptunes in their strictest definition are songs produced directly from the original audio hardware microchips such as the popular Commodore 64 MOS 6581 SID chip

(Commodore Business Machines 1982) and the Nintendo Entertainment System Ricoh 2a03 sound chip (Taylor 2004). Modifications such as SD card-to-cartridge adapters are allowed to be added to the system, but only at the hardware level and should not pervert the original commonly viewed limitations of the system. The notion of this definition is to preserve the "pure sound" of the chiptune coming directly from the original system. Since the composition software often has an effect on the composition and the sonic capabilities of the song, there can even be a distinction between using an old-school tracker directly on the system, such as SID DUZZ'IT for the Commodore 64 and a new-school tracker such as GoatTracker, which runs on a modern PC but also allows one to export the file such that it can be played on an actual Commodore 64 (see Figure 30.1). Not only does using purely keyboard input of SID DUZZ' IT change one's compositional style, but it also allows for the integration of software samples ("digis"), which GoatTracker does not currently support. Another clear example is Nanoloop for the Gameboy Advance (GBA) that allows FM synthesis, which is too computationally restrictive to use in an actual GBA title but which expands the sonic palette of the GBA via software. In its most refined sense then, an expert chiptune artist can tell the difference not only between chiptunes from different systems but also from different composition programs used to produce the songs.

Chiptunes made for videogames on the original systems are often referred to as VGM (for "videogame music"). The dividing line between the "classic videogame" synthesis sound can be seen as lying between the simple waveform channels of the Nintendo Entertainment System (NES) in 1985 and the sampled sound generation of the SNES in 1990. Accordingly, "chiptunes" can be seen as a modern term for songs composed with these technologies, although not necessarily made for exclusively for videogames.

30.2 Nostalgia

An informed listener will be able to determine which particular sound chip is being emulated, simply by knowing the original specifications of a chip. For this definition, the core constraints of what made the sound of a particular sound chip unique in comparison to other chips is integral in the form and structure of the composition. The overall idea is to capture the "spirit" or "soul" of the chip for the correct feel of the chip but not necessarily all the grit of being constrained by limitations or errors in the original chip design and physical hardware that made compositions on the original systems problematic. A clear example of such a system bug is the Nintendo vibrato bug, where severe clicking and distortion can occur due to a design flaw in the original 2a03 chip (see the Famitracker WIKI: http://famitracker.com/wiki/index.php?title=Sound_hardware). This flaw forces composers to be sensitive to certain vibrato ranges that should not be used. The drawback of

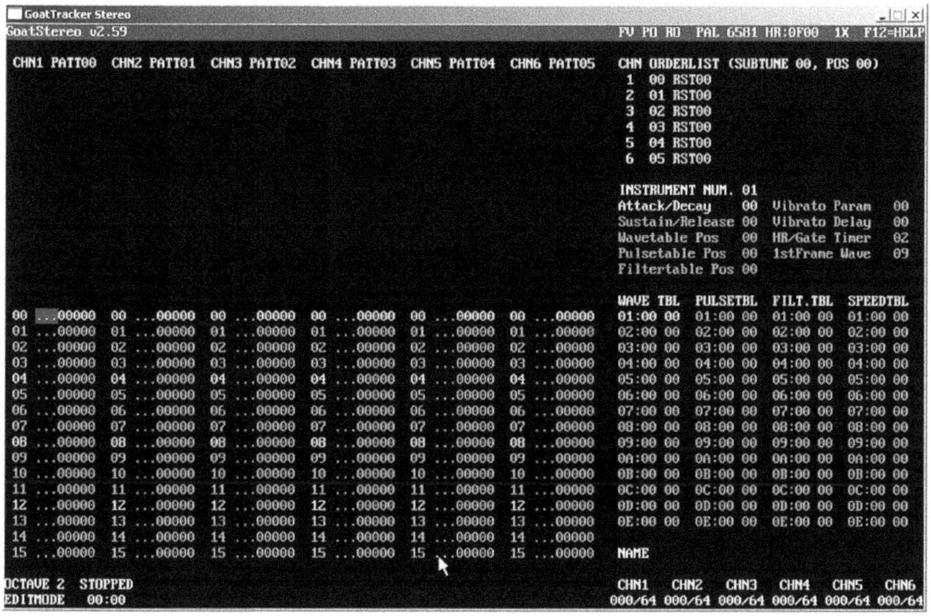

FIGURE 30.1 "GoatTracker—Stereo".

this nostalgia-based definition is how to define a subjective notion of the "core constraints" of a particular sound chip. This nostalgia-based definition covers the work that I completed for the *Retro City Rampage Soundtrack*, in which all of the composers used the open-source PC tracker OpenMPT with a set of samples to emulate the sound of the original NES. The songs would need to be slightly rearranged to play on an actual NES, but with the use of the conversion utility IT2NSF or by hand rearrangement in Famitracker or another NES-native program, the results could be very similar to their originals.

30.3 The Chiptune Aesthetic

The most relaxed definition of chiptune is a subjective one, where the aesthetics of classic sound chips are what establishes a song as being chiptune. This definition usually includes reference to the basic waveforms such as pulse, triangle, sawtooth, and pseudo-random noise with compositional artifacts such as rapid arpeggiation, portamento, and limited polyphony give rise to that "chiptune sound." This broader definition allows for the addition of reverberation, DSP compression effects, delays, and other techniques not originally available on a particular system, to be used to enhance the musicality of the basic waveforms. An example of such a broad definition is composers that use VST ("Virtual Studio Technology") plug-ins such as Plogue

Chipsounds as an element in their compositions. This plug-in has the advantage of using sampled drums and other elements to produce a chiptune. With this definition it can be difficult to define a chiptune song with its closely related electronic genres such as electro, house, or other styles that the song may also reference. In this case, the chiptune sound is likely just another element in a layered mix within a song.

30.4 Genre or Instrument?

With the definitions above it can be seen that the chiptune genre is both a compositional form and an instrument for composition. Chiptune purists tend to enjoy the arcane nature and high level of difficulty of producing chiptunes on original or personally customized hardware, whereas others see the expansion of the chiptune aesthetic into other forms of music as a natural step in an artistic canon. I would argue that the core of chiptune lies in the instrumentation, since basically any style of music can be composed using chiptune methods: If you play a piece of music on a piccolo it doesn't make it "piccolotune."

Making chiptune music can be seen as an orchestration and arrangement exercise, where each channel is its own instrument, and compositions will often need to be rearranged to suit the limited polyphony and waveforms of the target chiptune system. Part of my own attraction to chiptune is the minimalism required to arrange a song for a very limited number of channels. This minimalism can also help the mix of the song. For example, on the Commodore 64, the bass guitar and kick drum are often on the same channel so there is no mix difficulty of their sonic content conflicting with each other as they play at different times. Similar to other chiptune artists, I am forced to be very spartan and focused in my chiptune compositions by limited voices, which boils the song down to its essential elements (Yabsley 2007).

Within the chiptune genre, there are definite stylistic differences between the old-school sound of songs produced in the 1980s and the new-school sounds produced in recent years. Similar to other musical genres and art movements, chiptune music closely refers to the music styles of the times but it is also self-referential as well. "Old-school" European chiptune songs have their roots in the electronic music of pioneers such as Kraftwerk, but they are also self-referential, due to the demoscene producing a distinct composition style that often referred to clear melodic lead lines and repeated musical structures relating to later synthpop music. This contrasts somewhat with the Japanese style exemplified in the Konami-style, which is related to fusion and prog-rock genres with blazing virtuosic lead lines, dense rhythmic patterns, and often alternating odd-meter time signatures throughout the composition. The cross-pollination of genres continues to this day, with artists drawing from existing genres and creating new sounds for chiptunes that refer to dub-step, moombahton, and other emerging music genres.

30.5 Advances in Chiptune Hardware and Software

The creation of chiptune music is a combination of various elements that include the composer, the composition software, and the hardware. The hardware can be broken down further into the sound chip itself and the central chip used to control the sound chip. In the case of the original Nintendo Gameboy, there is no dedicated sound chip, so the sound chip is considered to be the CPU of the system. With the second generation of chip music, the sound chip can be removed entirely from the original system and put under a new system of control. For instance, the HardSID4U (created by Teli Sándor in 2010) is one example of a system where four of the original Commodore 64 SID chips are removed from the control of the original Commodore 64 CPU, and housed in a unit that can respond to a modern PC that sends it messages via a higher-speed USB connection. This addition allows for modulation of the chips at a rate that exceeds the capabilities of the original 1 MHz 6502 and also makes it easier to allow for audio signals input into the SID chip to be processed as well. The primary advantage of the HardSID4U is that it removes the sound chip from electrical interference by other electronic components, and this increases the audio fidelity of the signal from the chip itself via the pro-audio sound components of the HardSID4U. This is similar to "pro-sound mods" that are available for the DMG, NES, and other systems that increase the audio fidelity of the system by modifying the audio signal path of the hardware to reduce interference.

A significant increase in the quality of chiptunes has been from the improved software interface between the composer and the sound chip. In the early days of the Commodore 64, game composers such as Rob Hubbard would be required to learn 6502 assembly language programming in order to make music for the SID chip (McSweeney 1993). These days, there are tracker programs, such as GoatTracker, that have become refined to better suit composers. Over time, this development has allowed for increased efficiency in the use of the mouse, cut and paste functions, and a more user-friendly workflow, as usability is refined over time and incorporates feedback from a variety of users. The response time of a composition system is very important, as it can produce a feedback loop nearing real time such as with traditional instruments. This system is in stark contrast to other systems that required the data to be meticulously hard-coded by hand, compiled and reviewed to allow for updates, such as using MML (music macro language) and other macro or scripting languages to make chiptunes. Although this older type of system produces a different type of music, often even the original composers state that they wouldn't have the patience now to make music in the "old way" (Greening 2010). Sometimes this intensity and amount of focused time required to create chiptunes using the original software and system produced a music that was more minimalist, simply because it required a significant amount of extra time to produce music with a large degree of detail. This minimalism also makes more nuanced chiptunes stand out, such as

Martin Galway spending an entire month coding the guitar solo for the song "Times of Lore" (1988) (Commodore Zone). When the technical challenges were greater, it was more difficult for a composer to be "in the flow" and not struggle against the system in comparison to today's environments. However, even with very limited tools, composers such as Rob Hubbard were able to produce composing environments in which they were able to bring the responsiveness of a system into a workable range. Sometimes, such as with my own composition practice, composers choose to use older software, such as trackers, to produce chiptune music to preserve this compositional environment and limitations to create chiptunes.

Composing a song with assembly language, burning it onto an EPROM (erasable programmable read-only memory) for an NES is very different from making an NES song on the PC with FamiTracker, putting it on a PowerPak via a memory card and playing it on a pro-sound NES system of our "second age" of chiptune. Learning the assembly language of the NES is difficult and time-consuming. As such, the notion of composing a chiptune "from scratch," using just the base system as composers originally using the system would have had access to, is a challenge indeed. Making an instrument from scratch, such as the gAtari by C-trix, is both a technical and artistic feat. One can see that there is a scale of increasing difficulty between creating chiptunes from scratch and simply triggering a chiptune sample from a sample library. Table 30.1 attempts to outline how combinations of hardware and software produce a difference in the general difficulty level of producing a chiptune.

In a live performance setting, the physical and software interface between the performer and the music becomes even more important to allow for a balance between the amount of real-time control parameters and the complexity of the music produced by the system. Traditional instruments such as the violin require significant manual practice to gain an acceptable amount of control over the inputs for a live performance that would be pleasing to a typical audience. With electronic music it is easily possible to replay a very complex song by simply pressing a single button. It can be very difficult for the audience to know what is "live" and what is not. With live chip tune music, most performers appear to try to strike a balance of musical spontaneity and serendipity with repeatability and spectacle of the overall performance. As a performer, I find that performances are less enjoyable for me when I am overwhelmed with performance possibilities, and I prefer to use DJ techniques such as crossfading, song selection, and other, similar practices to allow me to be entirely physically and mentally present in the performance, while having enough flexibility to quickly respond and react to the feel of the audience. If the performer is feeling awkward or not having fun, then it is difficult for the audience to feel otherwise.

30.6 The Demoscene

In the early days of the Commodore 64, a rising number of people began to pirate games and trade them in meetings or via online bulletin board systems (Paul 2005). Pirate groups would often include introductions during the load screen of the cracked software.

Table 30.1 The Chiptune scale

Difficulty	Sound chip	CPU	Composition Software	Examples
Most difficult	ATtiny85	ATtiny85	Custom: coded in assembly language	1-bit Symphony by Tristan Perich (Tristan Perich 2009)
	Original Commodore 64 (C64) MOS 6581	Original C64 MOS 6510 (6502 series)	Original: custom (Rob Hubbard's sound routine and assembly language)	Monty on the Run from a C64 (Hubbard 1985)
	Original C64 MOS 8580	Original C64–MOS 6510 (6502 series)	New: Goattracker	Ride the High Country (Linus 2006)
	Original C64 MOS 6581/8580	New: emulated C64 CPU on PC; HardSID4u	New: Goattracker	Composition using GoatTracker with HardSID4u
	New: Hardware emulation–C64DTV (no filters)	New: CPU emulated by C64DTV	Original: Custom (Rob Hubbard's sound routine and assembly language)	C64DTV is hackable to allow for new songs
	New: Emulator of 6581/8580	New: host system: PC	New: Goattracker	Running GoatTracker with emulated SID
	New: Samples	New: host system	New: OpenMPT	Retro City Rampage Soundtrack (virt, Freaky DNA and Norrin Radd)
	New: Samples and synthesis	New; host system	New: VST plug-ins	Sound examples from Plogue Chipsounds
Easiest	Sampled	New: host system	DAW	Using a chiptune sample library in a song

To show off their skills they would produce increasingly complex graphics and music during these cracking introductions (or "cracktros"). Eventually, these groups began to compete with each other on technical and artistic merits with demos that illustrated their skills. Over time, a lot of great music was produced for these competitions. The core of the demoscene started and remains in Europe, but there are small groups throughout the rest of the world. These competitions (or "compos") continue to this day and are a great way for chiptune artists to prove their skills to their peers and create a stronger community. With its origins in pirating, the demoscene is primarily noncommercial, and this aspect has spread to much of the art created by the community, including its music productions. Although demo parties often play compositions live to their audience, it is focused on the notion of real-time computational functions to generate the productions so human performances were not a focus of these demonstrations or "demos."

30.7 Live Chiptune

The second generation of chiptune is much more focused on live performance than the first generation of chiptune musicians, who were often involved in the demoscene. One of the better-known events for live chiptune began in New York City for the Blip Festival in 2006 and has spread to locations such as Japan and Australia among other countries. The film *Reformat the Planet* provides a good insight into the chiptune music scene that is centered on New York City. The New York scene has more of a DIY culture than a coder culture and has democratized the sound of chiptunes with the growth of relatively easy-to-use cheap and accessible tools, such as Little Sound DJ (LSDJ) on the Gameboy. Chiptunes can be seen as a rebellion against the overproduced sounds from pop music and a return to a sound of electronic music by electronic instruments. This scene is in direct contrast to the existing DJ culture of the time, with its focus on the star power of the DJ rather than the composers who made the music. The chiptune community tends to be humble and accessible in sharing how they create their music, and performances tend to include a large network of peers who have an in-depth understanding of how chiptunes are made.

30.8 The Chiptune Genre

Styles and genres can blur, with bands such as Anamanaguchi, who use the NES as part of their sound to be "chip-punk," or the indie rock band The Advantage, who primarily play cover versions of NES songs. In the case of Anamanaguchi, chiptunes can be viewed as just another element or instrument within the band. For The Advantage, when the synth channels of the NES are orchestrated for guitars, bass, and drums one can hear the influences of the styles of the 1980s music that influenced the original compositions, rather than just hearing the chiptune sound.

Emulation of sound chips is always a touchy topic for chiptune artists. However, when one sees the game system as an instrument similar to a violin then it is easy to see why musicians are so particular about the sound. I believe it comes down to subjectivity, but there is something to be said for the nonlinear sound that comes from the actual hardware instead of an emulation that has a real difficulty with analog circuitry. When listening to the sound of real SID chips in comparison to an emulator, I find that there is a richness to the sound quality of the original chip that is inspiring to work with, in contrast to the sound of an emulator. Each time a song is played, it is slightly different on the real chip, providing it with a certain mystery and life.

The Commodore 64 SID chip is a great example of this phenomenon, as each chip sounds slightly different, due in part to a design issue with the first version of the 6581 chip. Some chiptune artists prefer the original 6581 chip rather than the subsequent 8580 chip that corrected certain flaws but doesn't have the same loudness and noisiness that allows for the common method of generating digitized sound via software. The comparison between various revisions of a sound chip on a particular system doesn't end with the Commodore 64, however, and similar comparisons have been made on the Gameboy and other systems. In the case of the Gameboy, musicians have even gone to great lengths to give an objective argument for their preferences of which Gameboy revision is best (Weixelbaum 2012).

Usually, chiptune music is programmed note-by-note using a tracker instead of being played live. For example, on a Gameboy there are a very limited number of input buttons to interface with the tracker, so it creates a very dense style, similar to trying to put a lot of information within a single text message. To write a novel—or even a few pages—by texting on a phone keypad also breaks up the message, similar to the way tracker music is broken up between patterns of a longer score.

The use of trackers influences the musical composition. Since patterns are often looped or reused in a tracker file, often phrases will not span a pattern and so shorter patterns and melodies tend to dominate over longer solo or melodic lead phrases which are more difficult to compose. The tracker composer will typically loop a single pattern while editing it. Looping a subsection of connected patterns within a song is often difficult with most trackers. It also takes more time to review changes when looping a larger section of the song. For example, if a pattern repeats every four seconds and one element is changed, it just takes four seconds to review the change. However, if a longer solo section of sixteen patterns is being looped and edited, then it requires a maximum of $16 \times 4 = 64$ seconds to hear that change. Conceptually, it is more difficult musically to keep track of the other elements one is attempting to focus on within that sixty-four-second loop, which in turn increases mental complexity and strain during composing, which is less pleasurable in comparison to the nearly instant gratification of hearing a change at most four seconds later.

There are also other user-interface constraints common to the tracker. It is often possible to visually see almost all of a single small pattern on a screen, whereas seeing the details of sixteen patterns becomes problematic. There is simply a cognitive disconnect between seeing the edited notes visually and being able to hear them at one time. This visualization is in direct contrast to traditional musical notation which is very good at visually displaying a long melodic articulated phrase. The tracker composer can watch the notes stream by in "follow mode," but it is often a visual blur with higher tempos. It is possible to drop the tempo during composition and programming and increase the tempo for playback, but often the articulation density at a lower tempo causes strange overmodulated effects when sped up, and the melody can easily become stilted rather than emotive.

It is also very rare for a tracker to allow a level of zoom so that the composer can select a level at which to view the composition besides at the pattern-note level and the pattern-arrangement song level. All that is commonly available to trackers is to change the font size and to hide effects columns to allow the display of more channels simultaneously. This setup is in contrast to most piano-roll style MIDI editors, which allow the composer to easily zoom in and out of a composition on a vertical and horizontal aspect to visualize long note changes over time or easily see a harmonic relation vertically between notes for chords. Here we can see that there are many qualities of tracker music composition which cause a noticeable result in chiptune compositions.

The tracker file format is uniquely constructed to be a near-optimal music format for games. It has the advantage of requiring very little CPU overhead for playback, the playback and creation tools are robust, the tools are typically open-source, and the files themselves are optimized in their use of sample data. A distinct advantage of working with tracker files is that one can easily change the tempo dynamically, and the quality of the playback of the music doesn't become distorted in the way that a linear sampled sound recording does. The instruments can be switched while the song is playing back, and each track can be muted or adjusted in volume in real time, providing many possibilities for adaptive music playback systems for games. The definite drawback of tracker music is that it requires a significant amount of work to achieve a production quality that is as polished as the music from a typical digital audio workstation (DAW). However, the tracker file format is flexible and can include layers of additional information coded in the MIDI format that can be used to send messages to the game in synchronization with the song such as beat or phrase marker information.

As trackers continue to improve in quality and allow for more sophisticated digital signal processing (DSP) effects, the possibilities for using tracker files in videogames continues to grow. With the growth of the mobile market, tracker files continue to provide a good balance of music production quality, file size, and CPU requirements.

30.9 Chiptune Composition

There are many ways to compose chiptunes and the next few sections will investigate a cross-section of various techniques that have been used throughout chiptune history. The simplest form of digital sound generation is 1-bit music which has roots in early computer music history. The basic synthesis capabilities of the NES are then examined to show how the first game console sound chips were originally used and how they are used for composition thirty years later. The Commodore 64 SID chip is then covered to show how its capability of dynamically changing oscillators on its channels, and the possibility for real-time filtering, changes one's chiptune composing style for this system. To conclude, the use of OpenMPT to compose NES-style chiptunes is detailed in reference to the soundtrack for the modern game *Retro City Rampage*.

30.9.1 1-bit Music

One of the attractive elements of chiptune music is its relationship to minimalism. In fact, with PC speaker music or 1-bit music one gets to the fundamental, "atomic" element of sound, in which either the speaker has been "clicked" or not. All one does is organize a series of clicks through time: very minimal indeed. This clicking on and clicking off of the speaker is commonly referred to as "beeper" music.

The history of beeper music dates back to the early days of computers and the origins of computer music itself. Beeper music has seen a small resurgence lately due to the increased availability of trackers for 1-bit music as well as the popularity of the "1-bit Symphony" music release by Tristan Perich that encases an entire hardware music playback system with 8kb for forty minutes of playback. The release is interesting in that each "performance" renders the music each time it is heard and the hardware electronics of the playback system are small enough to be distributed in a regular compact disc case.

In the 1980s, when it was expensive to add a soundchip to an already expensive IBM PC and when costs of the ZX Spectrum were to be kept down, the soundcard was effectively eliminated entirely and the generation of sound was entirely dependent on the CPU for oscillating the series of bits sent to the speaker. Of course, even the most basic sound routines would allow the description of a series of tones over time. The more sophisticated sound routines allow for polyphony and a varied number of effects, such as phase modulation, speech synthesis, volume and envelopes, and different timbres. Continuously supplying the bit-stream for the speaker can be taxing on the CPU and many games would include music only on the title screen.

The sound of 1-bit is determined by the software rather than the hardware. Since the software is changing values so quickly over time, any small change in the manner in which the sound is generated often has a very noticeable effect on the result. Artists such as Mister Beep consider them conceptually almost as different instruments based on their capabilities (Mister Beep, personal communication, 2012). There is an interesting dichotomy between the minimal amount of information required to produce the audio on the hardware and the large amount of digital information required to accurately emulate the sound of the actual system.

30.9.2 The Nintendo Entertainment System (NES)

The next level of complexity for chiptune composers is having access to different waveforms, which can be illustrated by the NES. The NES had four fixed synthesis channels and a 1-bit DPCM sampled sound channel. In terms of compositional elements, the composer has two pulse waves, a triangle wave (fixed volume), noise channel, and DPCM channel, with no effects.

It should be noted that it is possible to affect the volume of the triangle wave (and noise channel) by the information in the DPCM channel. The "frequency" of the noise is the timbral result which makes it sound as though the noise channel is increasing in pitch through sixteen different values, and the type of noise can be switched between two different timbres with one sounding quite buzzy and being infrequently used. The effect of changing the pulse width on the pulse channels is a 2-bit value but since two of the values cannot be perceptually differentiated, there are functionally three different timbres of the pulse channels in practice. Certain NES game cartridges expanded the audio capabilities of the NES, but these games are rare and the nostalgia surrounding the NES sound typically references the stock NES audio hardware.

This brings up an interesting point in that the actual audio hardware of a cartridge-based system is expandable by the chips contained in the cartridge itself. The most popular of the expansion chips is the VRC6, which allows for two more pulse channels and a sawtooth channel. However, in the overall history of the NES, the majority of the games used the basic soundchip, as adding extra soundchips in the cartridge was an additional cost.

A composer, then, really has access only to the timbre of the note and how loud it plays (sometimes) for a total polyphony of five sounds. There is also the issue of sound effects that can play at any time and steal a channel (or more), so sometimes composers would use just the DPCM channel for sound effects and the rest for music. The problem with the DPCM data is that the sample memory ROM was costly and often limited to less than 16 Kb. Adding more ROM to the cartridge for samples increased the cost of the cartridge and therefore you either had voices of your music dropping out when sound effects played or your samples were quite limited in memory size, implying that the sampling rate of the samples was low, the number of samples was low, or both, in an effort to conserve ROM. The DPCM channel was also unable to perform real-time sample rate conversion and the playback rate was limited to sixteen set values and if you wanted to use it for a melodic instrument, the sampled data would have to be included at different pitches.

With these constraints of the channels being fixed and the volume being locked on the triangle wave, often composers make guitar-type melodies on the two pulse waves with an interval relationship between the two, a baseline on the triangle wave and drums on the noise channel. Since there are few possibilities of effects to modulate the sound, NES compositions tend to focus on melody and less so on rhythmic elements. It was also released two years earlier in Japan than the rest of the world, so much of the sound of the early NES soundtracks have a strong Japanese sound to them. The "Konami Style" is a musical style of the Japanese-based game developer Konami's sound team that composed music for many games during the years that the NES was popular. Chiptune composers from today, such as Jake "virt" Kaufman, often reference the Konami style in their contemporary compositions.

In the early 2000s, it was difficult to compose for the NES and so "music macro languages" (MML) and other similar systems were often used. An early NES tracker was Nerdtracker (2003) followed by the release of Famitracker at the end of 2005, which brought in support for expansion chips and other features. In combination with chiptune artists such as Anamanaguchi utilizing the NES for their songs and the ease of making NES songs, it has become a much more popular platform for composing in the last few years.

30.9.3 The Commodore 64 (C64) Sound

In contrast to the NES, the Commodore (C64) has a different set of cultural beginnings and hardware capabilities, which result in a very different sound than the sound of the

NES. The Commodore 64 was released worldwide in July 1982, about a year before the NES was released in Japan. The C64 spread quickly to become one of the best-selling computers in history with affordable cost and high value. Although there were many talented composers from around the globe using the C64, the European composers primarily set the style for the computer. United Kingdom composers Rob Hubbard and Martin Galway arguably set the pace for other C64 composers to follow. The C64 style was much more centered on timbral shifts and rhythm than the NES, simply because the waveforms could be switched on each channel and the filter allowed for subtractive synthesis, which was not possible on the NES. Unfortunately the filter design was unstable in the early version of the SID chip (MOS 6581) and therefore some composers opted not to use it at all (Carr 2001). However, once the SID design was updated to the new MOS 8580 chip, the filter bug was fixed, but unfortunately it also had a side-effect of reducing the effectiveness of the way most "digis" (sampled sound) had been done until then. Since there was a lack of channels, fast arpeggiation was also a trademark of the system. Another trademark of the C64 was its software digitized samples that required additional processing time from the CPU. The 12-bit phase resolution on the pulse wave oscillator also allowed for a dynamic phasor effect not available on the NES. Ring modulation was also an option but not overly used as it reduced the number of effective voices to two, due to one voice being consumed to be used as the modulation signal. Overall, these capabilities meant that the C64 was often best suited to a clear lead line on one voice, arpeggiation for backing chord progression timbre on a second channel, and often the bass and drums were left to alternate on the last channel.

One of the common instrument programming approaches was to compose a sound from multiple oscillators in sequence on a single channel—such as using a quick noise burst followed by a triangle wave to simulate a bass drum. Filtering across multiple channels allowed many different sonic possibilities especially when utilizing different filter types. Changing the resonance and the center frequency of the filter allowed for a great degree of control over the filter. Depending on how tricky one gets, almost any real-time effect is possible on the C64, from time stretching to autotune to vocoding as demonstrated by by Pex "Mahoney" Tufvesson in October 2010 in which eleven real-time effects ran on an unmodified C64 (Tufvesson 2010).

30.9.4 NES Chiptunes Using OpenMPT

Much of my own interest in the NES has been driven by being one of the composers and sound designer for the NES-style indie game *Retro City Rampage*. When I was asked to develop the sound for the game, I looked into various options, and since the game wasn't running on an actual NES, the strongest toolset for making mod music at the time was Open ModPlug Tracker (OpenMPT). In addition, since we wanted composer Jake Kaufman to be involved, we decided that his format of making NES-style tunes using the Impulse Tracker format in Open MPT would be a great choice. Since the format was open source, it was easy to integrate the playback routine into the game code and have it quickly running on several different systems. The game uses the Impulse Tracker mod file as a source but renders it to an audio stream during run-time, making it easy to port the game to various systems and to keep the CPU requirement to a minimum. I wanted the audio playback to occur at real time to generate the noise waveform dynamically but all of the synthesis waveforms are actually just short digitized samples. To port the music to an actual NES tracker such as Famitracker or to use a conversion program such as IT2NES would take some time but the quality of the songs would remain quite similar to their sources.

Recently, I have begun composing in Famitracker, but I find that the interface for OpenMPT is much better aligned to the way my mind works during the compositional process. It is very important for the composer to be able to stay "in the zone" or "in the flow" during the artistic musical process or elements can become confused and not as focused as one might like. Jake Kaufman has switched entirely over to Famitracker for his latest release of FX4 and continues to highlight his talents as a leading chiptune composer. For the release of the soundtrack, the songs were mastered with a minimal amount of equalization but no reverb or other elements were added in the goal of keeping them sounding as NES authentic as possible.

30.9.5 Tracker Tricks

Although the NES does not have any capability for real-time effects in hardware, there are possibilities of simulating certain effects using certain tracker "tricks." Likely, the easiest audio effect to duplicate in the tracker is a delay effect which is simply done by duplicating the notes of a particular channel at a quieter volume delayed in time. For feedback, one just continues to produce time delayed notes with decreasing volume, commonly around 50 percent and eventually reaching silence after three or more echoes. This is most easily accomplished by using a second channel for the echoing notes such as the following which can be seen with the Figure 30.2 and 30.3 examples.

The most obvious difficulty that we can see here (see Figure 30.2) is that if we attempt to collapse the delayed notes into a single channel to use fewer channels then we will lose our extra delayed note at C-4 at position 8 due to the new D♯6 note. The easiest way to rectify this situation is to change the delay time interleaved with the tempo of the notes such that they do not occur at exactly the same time. An example follows as #1. Here we can see that simply increasing the delay time by two rows (or one-tenth of a second) we can now represent the echo effect on one channel (see Figure 30.3). If we had built this delay effect in Famitracker and defined the echo effect in the instrument, then the delayed versions of the C-4 note would have been cut off at the first D♯6 note.

One of the exciting elements of working with trackers is developing sets of tricks that add technical dexterity to your compositions. Effects include flanging, in which a note is duplicated on the pulse channels but a vibrato is applied to one of the channels. Chorusing can be done very similarly except that the frequency of one note is bent slightly and held slightly out of tune with the other note. Using a very fast vibrato can

#1	1: Pulse1 –	2: Pulse2 –	
0	C-4 01.64- ---	--- -- -- ---	-
1	== --- -- -- ---	--- -- -- ---	-
2	--- -- -- ---	--- -- -- ---	-
3	--- -- -- ---	--- -- -- ---	-
4	--- -- -- ---	--- -- -- ---	-
5	--- -- -- ---	--- -- -- ---	-
6	C-4 01.32- ---	--- -- -- ---	-
7	--- -- -- ---	--- -- -- ---	-
8	D#6 01.64- ---	--- -- -- ---	-
9	--- -- -- ---	--- -- -- ---	-
10	--- -- -- ---	--- -- -- ---	-
11	--- -- -- ---	--- -- -- ---	-
12	C-4 01.16- ---	--- -- -- ---	-
13	--- -- -- ---	--- -- -- ---	-
14	D#6 01.32- ---	--- -- -- ---	-
15	--- -- -- ---	--- -- -- ---	-
16	--- -- -- ---	--- -- -- ---	-
17	--- -- -- ---	--- -- -- ---	-
18	C-4 01.08- ---	--- -- -- ---	-
19	--- -- -- ---	--- -- -- ---	-
20	D#6 01.16- ---	--- -- -- ---	-
21	--- -- -- ---	--- -- -- ---	-
22	--- -- -- ---	--- -- -- ---	-
23	--- -- -- ---	--- -- -- ---	-
24	--- -- -- ---	--- -- -- ---	-
25	--- -- -- ---	--- -- -- ---	-
26	D#6 01.08- ---	--- -- -- ---	-
27	--- -- -- ---	--- -- -- ---	-

FIGURE 30.2 One-channel delay.

#0	1: Pulse1 –	2: Pulse2 –	
0	C-4 01.64 ---	--- -- -- ---	-
1	== -- ---	--- -- -- ---	-
2	--- -- -- ---	--- -- -- ---	-
3	--- -- -- ---	--- -- -- ---	-
4	--- -- -- ---	--- -- -- ---	-
5	--- -- -- ---	C-4 01.32 ---	-
6	--- -- -- ---	== -- ---	-
7	--- -- -- ---	--- -- -- ---	-
8	D#6 01.64 ---	C-4 01.16 ---	-
9	== -- ---	== -- ---	-
10	--- -- -- ---	--- -- -- ---	-
11	--- -- -- ---	--- -- -- ---	-
12	--- -- -- ---	D#6 01.32 ---	-
13	--- -- -- ---	== -- ---	-
14	--- -- -- ---	--- -- -- ---	-
15	--- -- -- ---	--- -- -- ---	-
16	--- -- -- ---	D#6 01.16 ---	-
17	--- -- -- ---	== -- ---	-
18			

FIGURE 30.3 Two-channel delay.

change the timbre of the pulse and triangle waves to make it sound "thicker." Rapidly changing the pulse width can give a feeling of filtering, as the number of overtones for each duty cycle changes with each pulse width but keeps the pitch of the fundamental the same. To simulate reverb, which is basically a set of quiet random delays, one can

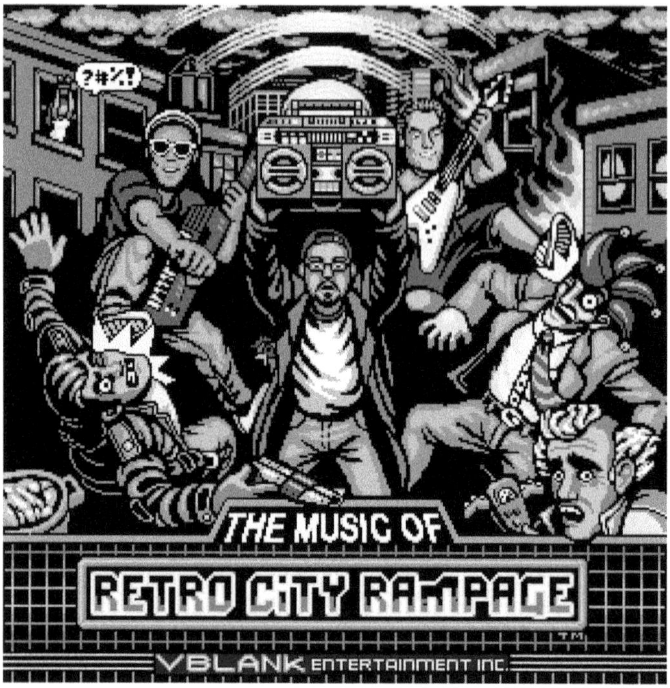

FIGURE 30.4 *Retro City Rampage* jacket cover.

have an initial delayed note for the first reflection and then a sequence of somewhat random decreasing volume modulations which fall to silence. There are also instrument effects which one can apply, such as including a short attack of the note that is an octave higher to simulate a plucked sound similar to a plucked string on a guitar. Depending on the phrasing, one can modulate the bends and note leaps of a lead line on a pulse or triangle wave to sound like an electric guitar. A more complex timbre can be created by triggering sounds on multiple channels, such as the noise channel and DPCM channel being triggered together to give an accent on sampled drum hits with the higher frequencies of the noise channel. The possibilities are endless, just as with any other form of composition, and being able to code your own tracker allows for an even greater level of control which is often prized by chiptune composers.

The composition program used by all composers for *Retro City Rampage* was the free and open source OpenMPT for the Windows OS. This tracker has advanced GUI capabilities over older DOS-based trackers. Using the impulse tracker (.it) format allowed the soundtrack for *Retro City Rampage* (which is currently approximately two hours and twenty minutes long) to occupy just 4MB of game data. A typical MP3 compression scheme would normally require well over 100MB of storage.

30.9.5.1 An In-depth View of a Tracker Song

This blazing rock theme opens with a heavy power chord riff on the two pulse wave channels and continues to build with the addition of a bassline on the triangle wave and drums on the noise channel and DPCM 1-bit lo-fi sampled sound channel. Detailed articulations in the volumes and portamentos give a classic Kaufman (virt) feeling right from the outset. A drumroll at the end of the first pattern kicks off the main drum pattern. The riff breaks into a climbing line with an octave to possibly simulate a pinch harmonic or feedback in the guitar line. A dueling solo follows between the two pulse waves, punctuated by interval hits and slow bends that compliment the detailed bass line and tom rolls. At a detailed level, relentless use of alternating duty cycles, volume gating, echoes, and subtle frame delays punctuate virt's mastery of the NES 80s guitar fusion rock form. When the riff returns at 0:32, it plays a background role with a complimentary riff to layer another lead onto the theme. Almost hidden in the riff at the end of one of the patterns is a clever use of octave arpeggios which serve to change the timbre of the pulse wave without adding any confusion regarding the melodic intervals of the song. At 0:48 heavy power-chord break hits on all instruments and weaves back into another riff. When this rhythm riff repeats, the notes are punctuated even higher as it continues to grow with finger-tapping frenetic parallel lead lines. When the main riff returns at 1:04, the lead grows to astronomical seven-string Steve Vai guitar heights and never misses a step. At 1:20, paired power chord slides push even higher with more drum rolls and searing lead lines. By 1:36 we are dropped into a triplet power chord thrash fest. By the time we arrive at 1:52 to recap the main theme, we realize that we are just getting prepped for the satisfying loop to the beginning!

The screen grab of the score in OpenMPT from 0:53 to 0:57 at 120 bpm (Figure 30.5) provides a good indication of the amount of detail that is present in every second of this section of the song. It begins in the pulse wave channel with a deep vibrato using the H8F command from position 78 to 83. The DPCM samples show how the noise channels are often paired to add more high-frequency transients to each hit, which was commonly used by the Konami sound team. The kick drum is instrument number 22 at C-5 and the snare is D-5 in the noise channel. If this were an actual NES, they would need to be defined as separate instrument samples as there is no sample rate conversion on the DPCM channel. The tom drum sample is instrument 23 in the DPCM channel. Once again, on the NES, each tom drum sample with a different playback rate would need to have its own sample. The combinations of commands SD1 and SD2 are used to slightly delay the onset of the note by a number of ticks show rapid triplet patterns (for example, in the pulse 1 channel from positions 96 to 102) which help break up the quantization often found in less accomplished tracker songs (or electronic songs in general for that matter). Each row that appears in this pattern is three ticks (one-sixtieth of a second) so SD1 is a one-sixtieth-second delay and SD2 is a two-sixtieth-second delay. A slide down in the pulse 1 wave is predicated by a strong vibrato from 116 to 122 and falls from positions 123 to 127. While all of this is happening, a small quiet unison riff with the bassline on the triangle channel is occurring in the complimentary pulse wave from 112 to 127, with quick octave trills using the JC0 command repeated using J00 to change the timbre to a "plucked effect." Using the octave makes the note stand out and provides more sonic energy to the note similar to when a guitar string is plucked and there is an emphasis on

#15	1: Pulse1	2: Pulse2	3: Triangle	4: Noise	5: DPCM
66	F-5 01 -- ---	C-3 01ᴗ44 ---	F-4 15 -- ---	--- -- -- ---	C-5 22 -- SD2
67	--- -- -- ---	== -- -- ---	--- -- -- S00	--- -- -- ---	--- -- -- ---
68	F#5 01 -- ---	C-5 01ᴗ44 ---	F-5 15 -- ---	A-5 24ᴗ40 ---	D-5 22 -- ---
69	G-5 01 -- SD1	--- -- -- ---	--- -- -- S00	--- -- -- ---	--- -- -- ---
70	A#5 01 -- SD2	A#4 01ᴗ44 ---	D#5 15 -- ---	--- -- -- ---	C-5 22 -- ---
71	--- -- -- ---	== -- -- ---	--- -- -- S00	--- -- -- ---	--- -- -- ---
72	C-6 01 -- ---	F#4 01ᴗ44 ---	B-4 15 -- ---	G-5 24ᴗ20 ---	C-5 22 -- ---
73	--- -- -- ---	--- -- -- ---	--- -- -- S00	--- -- -- ---	--- -- -- ---
74	A#5 01 -- ---	C-4 01ᴗ44 ---	F-4 15 -- ---	--- -- -- ---	C-5 22 -- ---
75	== -- -- ---	== -- -- ---	--- -- -- S00	--- -- -- ---	--- -- -- ---
76	C-6 01 -- ---	C#4 01ᴗ44 ---	F#4 15 -- ---	A-5 24ᴗ40 ---	D-5 22 -- ---
77	--- -- -- ---	--- -- -- ---	--- -- -- S00	--- -- -- ---	--- -- -- ---
78	--- -- -- H8F	F#3 01ᴗ44 ---	B-3 15 -- ---	--- -- -- ---	C-5 22 -- ---
79	--- -- -- H00	== -- -- ---	--- -- -- S00	--- -- -- ---	--- -- -- ---
80	--- -- -- H00	C#4 01ᴗ44 ---	F#4 15 -- ---	G-5 24ᴗ20 ---	C-5 22 -- ---
81	== -- -- H00	--- -- -- ---	--- -- -- S00	--- -- -- ---	--- -- -- ---
82	--- -- -- H00	C-4 01ᴗ44 ---	F-4 15 -- ---	--- -- -- ---	C-5 22 -- ---
83	--- -- -- H00	--- -- -- ---	--- -- -- S00	--- -- -- ---	--- -- -- ---
84	A#5 01 -- ---	A#3 01ᴗ44 ---	D#4 15 -- ---	A-5 24ᴗ40 ---	D-5 22 -- ---
85	--- -- -- ---	--- -- -- ---	--- -- -- S00	--- -- -- ---	--- -- -- ---
86	C-6 01 -- ---	C#4 01ᴗ44 ---	F#4 15 -- ---	--- -- -- ---	C-5 22 -- ---
87	== -- -- ---	--- -- -- ---	--- -- -- S00	--- -- -- ---	--- -- -- ---
88	F#5 01 -- ---	C-4 01ᴗ44 ---	F-4 15 -- ---	G-5 24ᴗ20 ---	C-5 22 -- ---
89	--- -- -- ---	--- -- -- ---	--- -- -- S00	--- -- -- ---	--- -- -- ---
90	F-5 01 -- ---	A#3 01ᴗ44 ---	D#4 15 -- ---	--- -- -- ---	C-5 22 -- ---
91	== -- -- ---	--- -- -- ---	--- -- -- S00	--- -- -- ---	--- -- -- ---
92	D#5 01 -- ---	G-3 01ᴗ44 ---	C-4 15 -- ---	A-5 24ᴗ47 ---	D-5 22 -- ---
93	== -- -- ---	--- -- -- ---	--- -- -- S00	--- -- -- ---	A-5 23 -- SD1
94	C-5 01 -- ---	F-3 01ᴗ44 ---	A#3 15 -- ---	--- -- -- ---	E-5 23 -- SD2
95	== -- -- ---	--- -- -- ---	--- -- -- S00	--- -- -- ---	--- -- -- ---
96	G-5 01 -- ---	G-3 01ᴗ44 ---	C-4 16 -- ---	C-5 24ᴗ40 ---	C-5 22 -- ---
97	--- -- -- ---	F-3 01ᴗ44 SD1	A#3 16 -- SD1	--- -- -- ---	--- -- -- ---
98	D#5 01 -- SD2	G-3 01ᴗ44 SD2	C-4 16 -- SD2	--- -- -- ---	C-5 22 -- SD1
99	--- -- -- ---	== -- -- ---	--- -- -- SC2	--- -- -- ---	--- -- -- ---
100	--- -- -- ---	A#3 01ᴗ44 ---	D#4 15 -- ---	A-5 24ᴗ40 ---	D-5 22 -- ---
101	C-5 01 -- SD1	--- -- -- ---	--- -- -- ---	--- -- -- ---	--- -- -- ---
102	--- -- -- ---	--- -- -- ---	--- -- -- ---	--- -- -- ---	C-5 22 -- SD1
103	--- -- -- ---	--- -- -- ---	--- -- -- ---	--- -- -- ---	--- -- -- ---
104	F-5 01 -- ---	C-4 01ᴗ44 ---	C-4 16 -- ---	C-5 24ᴗ40 ---	C-5 22 -- ---
105	--- -- -- ---	A#3 01ᴗ44 SD1	A#3 16 -- SD1	--- -- -- ---	C-5 22 -- SD1
106	C-5 01 -- ---	C-4 01ᴗ44 SD2	C-4 16 -- SD2	--- -- -- ---	C-5 22 -- SD2
107	--- -- -- ---	== -- -- ---	--- -- -- SC2	--- -- -- ---	--- -- -- ---
108	D#5 01 -- ---	A#3 01ᴗ44 ---	D#4 15 -- ---	A-5 24ᴗ40 ---	D-5 22 -- ---
109	--- -- -- ---	--- -- -- ---	--- -- -- ---	--- -- -- ---	--- -- -- ---
110	--- -- -- F10	--- -- -- ---	--- -- -- ---	--- -- -- ---	--- -- -- ---
111	--- -- -- ---	--- -- -- ---	--- -- -- ---	--- -- -- ---	--- -- -- ---
112	D#5 01 -- ---	C-5 03ᴗ30 JC0	C-4 15 -- ---	G-5 24ᴗ40 ---	D-5 22 -- ---
113	--- -- -- ---	== -- -- ---	--- -- -- S00	--- -- -- ---	D-5 22 -- SD1
114	--- -- -- ---	C-6 03ᴗ30 J00	C-5 15 -- ---	G-5 24ᴗ30 SD2	C-5 22 -- SD2
115	--- -- -- ---	== -- -- ---	--- -- -- S00	--- -- -- ---	--- -- -- ---
116	--- -- -- H8F	C-5 03ᴗ30 J00	C-4 15 -- ---	--- -- -- ---	C-5 22 -- ---
117	--- -- -- H00	== -- -- ---	--- -- -- S00	G-5 24ᴗ40 SD1	C-6 23 -- SD1
118	--- -- -- H00	A#5 03ᴗ30 J00	A#4 15 -- ---	--- -- -- ---	C-6 23 -- SD2
119	--- -- -- H00	== -- -- ---	--- -- -- S00	--- -- -- ---	--- -- -- ---
120	--- -- -- H00	C-5 03ᴗ30 J00	C-4 15 -- ---	G-5 24ᴗ30 ---	F-5 23 -- ---
121	--- -- -- H00	== -- -- ---	--- -- -- S00	--- -- -- ---	F-5 23 -- SD1
122	--- -- -- H00	A-5 03ᴗ30 J00	A-4 15 -- ---	A-5 24ᴗ40 SD2	A-5 23 -- SD2
123	--- -- -- E10	== -- -- ---	--- -- -- S00	--- -- -- ---	--- -- -- ---
124	--- -- -- E00	C-4 03ᴗ30 J00	C-4 15 -- ---	--- -- -- ---	A-5 23 -- ---
125	--- -- -- E00	== -- -- ---	--- -- -- S00	G-5 24ᴗ40 SD1	D-5 23 -- SD1
126	--- -- -- E00	A#5 03ᴗ30 J00	A#4 15 -- ---	--- -- -- ---	D-5 23 -- SD2
127	--- -- -- E00	== -- -- ---	--- -- -- S00	--- -- -- ---	--- -- -- ---

FIGURE 30.5 In-depth song analysis: score from 0:53 to 0:57.

the upper harmonics at the initial pluck of the note. The repetition of the C-5 in the pulse 2 channel can be seen as an open string where notes are articulated on the same string.

As one can clearly see with a small detailed look at approximately four seconds within a 120-second composition, the amount of detail and technical dexterity demonstrated is staggering, especially when one considers that the entire soundtrack for the game spans over two hours of total music.

30.10 Conclusions

We live in an increasingly digital world and the sound of chiptune helps us have a conversation with the machines that construct our surroundings when they used to have their own unique voices. There is a personality to working with older digital systems that has been largely lost in the internet age but chiptune is doing its best to reclaim a personal human to machine connection.

If you are at all interested in chiptunes, then the best thing to do is to listen a lot to existing chiptunes and to start making your own. The appendix provides a good starting point to listening to good examples of chiptunes on many different systems from an old-school perspective of the first time around, when songs were made on the hardware, and the new-school approach of utilizing the system within modern contexts.

I feel that chiptunes are a type of folk-art that anyone can enjoy by downloading free software for their computer or buying an older game console from the pawn shop and experimenting on their own. Chiptunes are a look into our own kinder, gentler past with machines and give a new language for the future that will continue to evolve with each generation.

A Listener's Guide to Chiptunes

This is not meant to be an exhaustive list but just an overview of the more common chiptune platforms. There are also many chiptunes using the Roland Sound Canvas, the Gravis Ultrasound, Commodore Vic 20, Amstrad CPC, the Atari Lynx, and many more. Chiptunes tend not to be made on older arcade console hardware due to their often unique set of chips and the need to burn custom EPROMs to change the audio data.

4mat. 1990. *Phantasmagoria* (Amiga). http://www.scenemusic.net/demovibes/song/11243/.

Animal style. 2010. *Trenchvent* (Sega Genesis). http://www.8bitpeoples.com/discography/8BP106.

Baldwin, Neil. 2012. *Duty Cycle Generator* (NES). http://dutycyclegenerator.com/.

System	Signature sound	Old-school examples	New-school examples
One-bit "Beeper" music	Very buzzy sounding monophonic digital sound	Raw Recruit (Tim Follin 1988), The Secret of Monkey Island (Michael Land 1990)	One-bit Symphony (Tristan Perich 2009), A Thousand Furious Bees (Mister Beep 2011)
Amiga	Aliasing noise (upsampling artifacts), limited polyphony of four hard-panned two left and two right, samples from the ST-01 and higher sample set, eight-bit crunchy sound	Phantasmagoria (4mat 1990), Substance (Moby and ATZ 1991)	Acid Chunks (Manley 2012), Acidjazzed Evening (Suni 2000)
Atari 2600	Two simple sounds, characteristic noise sound, notes are often very much out of tune	Desert Falcon (Bob Polaro 1988), Pitfall II (David Crane 1982)	Disasteradio—Fax Of Life (Disasteradio 2009), Booty Plus Plus (cTrix 2012)
Atari ST	Simple waves often combined with eight-bit samples	Astaroth: The Angel of Death (Hippel 1989), Crapman (de Man 1993)	Ancient Art of Chiptune (gwEm 2008), Dynamite (stu 2003)
Commodore 64	Noise channel quality, "gloopy" filter, sometimes a very distorted sample	Arkanoid (Galway 1987), Monty on the Run (Hubbard 1985)	I'll be a Pimp in Cabrini Green (Randall 2005), Edge of Disgrace (Dane 2008)
Gameboy	Two pulse waves, wave channel (usually triangle) and noise channel	Tetris (Hirokazu Tanaka 1989), Kirby's Dream Land 2 (Hirokazu Ando and Tadashi Ikegami 1995)	Masada Gestalt (Zen Albatross 2011), Love Bit (Toriena 2012)
Gameboy Advance	Two pulse waves, wave channel, noise channel, and two sampled sound channels (FM sounds with Nanoloop)	Sonic Advance 3 (Sonic Team 2004), Mega Man Zero (Ippo Yamada 2002)	Out-House #12 (Henry Homesweet 2011), Live @ Blip Tokyo (10/23/11) (Je Deviens DJ en 3 Jours 2011)
MSX	Three pulse waves, noise channel	Metal Gear 2: Solid Snake (Theme of) (Masatosh Ikariko 1987), Space Manbow (Konami Kukeiha Club 1989)	SYNCtified (Inverse Phase 2012), Lucky Star (Otobeya 2008)
NES	Fixed pulse, triangle and noise channels, common intervals between pulse channels, triangle as bass and noise as kick, snare and high hat	Super Mario Bros. (Kondo 1985), Shatterhand (Mizutani and Yamanishi 1991)	First Flight (virt 2012), Rokko Chan Soundtracks—Boss Rush—eight-bit Arranged (Freaky DNA 2012)

(*Continued*)

(Continued)

System	Signature sound	Old-school examples	New-school examples
Sega Genesis	Sampled sounds (usually for drums) combined with chimes, filtered bass, wood hits and bell sounds on the FM channels plus sometimes some simple waves	Adventures of Batman and Robin (Jesper Kyd 1994), NBA Live 95 (Traz Damji 1994)	Trenchvent (Animal style 2010), I am Still Breathing (Little Scale 2010)
SNES	Eight channels of compressed ADPCM with reverb	Panel de Pon—Sea Stage (Masaya Kuzume 1995), Earthbound (Keiichi Suzuki and Hirokazu Tanaka 1994)	Super Anthropomorphic Space Turismo (Velathnos 2012), Hometown (sleepytimejesse 2011)
ZX Spectrum	Typically 1–5 channels of very buzzy sounds	And so it is (Yerzmyey 2000), Astro Marine Corps. (Jose Antonio Martin Tello 1989)	Restless Fighter (Shiru 2010), Latrax Minus (Mellona and Rob F. 2010)

Brimble, Allister. 1992. *Project X* (Amiga). http://www.exotica.org.uk/wiki/Project-X#Game.2FBrimble_Allister.2FProject-X.lha.

Bucky. 2012. Retro Game Audio. http://retrogameaudio.tumblr.com/.

C64 DTV Hacking Wiki: http://wikiapiary.com/wiki/C64_DTV_Hacking_Wiki.

Crane, David. 1982. *Pitfall II*. Atari 2600 [special sound hardware in cartridge]. http://vgmdaily.wordpress.com/2010/06/29/8-bit-mondays-pitfall-ii-lost-caverns-theme-david-crane/.

cTrix. 2012. *Booty Plus Plus* (Aatri 2600). http://ctrix.bandcamp.com/track/booty-plus-plus-atari-2600.

Damji, Traz. 1994. *NBA Live 95* (SNES). http://project2612.org/details.php?id=501.

Dane. 2008. *Edge of Disgrace* (Commodore 64). http://noname.c64.org/csdb/release/?id=72550.

de Man, Joris. 1993. *Crapman* (Atari ST). http://low.fi/~visy/ym-archive/Scavenger%20%28Joris%20de%20Man%29/.

Disasteradio. 2009. *Fax of Life* (Atari 2600 Cartridge). http://www.youtube.com/watch?v=Y80BzcS5bKw.

Follin, Tim. 1988. *Raw Recruit* (ZX Spectrum 48K). http://www.worldofspectrum.org/infoseekid.cgi?id=0004043.

Freaky DNA. 2012. *Rokko Chan Soundtracks*—Boss Rush—8-bit Arranged ("No Boss, No Rush") (NES) http://rokkochan.bandcamp.com/track/boss-rush-chipped-by-freaky-dna.

gwEm. 2008. *Ancient Art of Chiptune* (Atari ST). http://gwem.bandcamp.com/track/ancient-art-of-chiptune.

Hippel, Jochen. 1989. *Astaroth : The Angel of Death* (Atari ST). http://www.atarimania.com/game-atari-st-astaroth-the-angel-of-death_11109.html.

Hirokazu Ando, and Tadashi Ikegami. 1995. *Kirby's Dream Land 2* (Gameboy). http://kirby.wikia.com/wiki/Kirby%27s_Dream_Land_2/Music.

Hoffman. 2011. *8-Bit Jungle* (Amiga). http://www.scenemusic.net/demovibes/compilation/216/.

Homesweet, Henry. 2011. *Out-House #12* (Gameboy Advance). http://henryhomesweet.bandcamp.com/album/out-house.

Hubbard, Rob. 1985. *Monty on the Run*. Commodore 64. http://noname.c64.org/csdb/sid/?id=14328.

Ikariko, Masatosh. 1987. *Metal Gear 2: Solid Snake (Theme of)* (MSX PSG + SCC). http://www.bgmsx.com/index.php?contenido_servicio_tabla=show_themes&contenido_cat_hist=,7.

Inverse Phase. 2012. *SYNCtified* (MSX PSG+SCC). http://inversephase.bandcamp.com/track/synctified.

Je Deviens DJ en 3 Jours. 2011. *Live @ Blip Tokyo (10/23/11)* (Gameboy Advance). http://soundcloud.com/jddj3j/live-blip-tokyo-10-23-11.

Konami Kukeiha Club. 1989. *Space Manbow* (MSX). http://www.bgmsx.com/index.php?contenido_servicio_tabla=show_themes&contenido_cat_hist=,2.

Kondo, Koji. 1985. *Super Mario Bros* (NES). http://www.zophar.net/music/nsf/super-mario-bros-1.html.

Kuzume, Masaya. 1995. *Panel De Pon: Sea Stage* (SNES). http://www.vgmpf.com/Wiki/index.php?title=Tetris_Attack_%28SNES%29.

Kydd, Jesper. 1994. *Adventures of Batman and Robin*. (Sega Genesis). http://project2612.org/details.php?id=80.

Land, Michael. 1990. *The Secret of Monkey Island* (PC). http://www.youtube.com/watch?v=1IOL4q5tDDQ.

Linus. 2006. *Ride the High Country*. SID file. http://noname.c64.org/csdb/release/?id=33932.

Little Scale. 2010. *I Am Still Breathing* (Sega Genesis). http://www.iimusic.net/catalog/2010/12/little-scale-i-am-still-breathing.

Manley, Joss. 2012. *Acid Chunks* (Amiga). http://www.kittenrock.co.uk/?p=422.

Mellona, and Rob F. 2010. *Latrax Minus* (ZX Spectrum). http://www.youtube.com/watch?v=DIjuTn5sUuQ.

Mister Beep. 2011. *A Thousand Furious Bees* (ZX Spectrum). http://freemusicarchive.org/music/Mister_Beep/.

Mister Beep 2012. Z80 (Liner Notes by Mister Beep). http://www.kittenrock.co.uk/?p=384.

Mizutani, Iku, and Kouichi Yamanishi. 1991. *Shatterhand* (NES). http://www.zophar.net/music/nsf/shatterhand.html.

Moby and ATZ. 1991. *Substance* (Amiga). http://ada.untergrund.net/?p=demo&i=137.
Old School Game and Demo Misc. *Kohina Internet Radio Playlist*: http://www.kohina.com/playlist.php.
Otobeya. 2008. *Lucky Star* (MSX PSG + SCC). http://park19.wakwak.com/~cherry/oto/.
Owens, Paul. 2008. *Blip Festival: Reformat the Planet*. DVD. Documentary. http://www.imdb.com/title/tt1183662/.
Perich, Tristan. 2009. *1-bit Symphony*. Electronics in CD Case. http://www.1bitsymphony.com/.
Provinciano, Brian. 2012. *Retro City Rampage* (PC). Vblank Entertainment. http://retrocityrampage.com/.
Randall. 2005. *I'll Be a Pimp in Cabrini Green* (Commodore 64). http://noname.c64.org/csdb/release/?id=20574.
Shiru. 2010. *Restless Fighter* (ZX Spectrum). http://battleofthebits.org/arena/Entry/Restless+Fighter/4548/.
sleepytimejesse. 2011. *Hometown* (SNES). http://www.reverbnation.com/artist/artist_songs/1293140?song_id=8012357.
Sonic Team. 2004. *Sonic Advance 3* (Gameboy Advance). http://www.sonicstyle.org/music/GBA/SonicAdvance3/.
Stagediver. 2011. *Imaginary Dystopian Future Wars EP* (Amiga).
stu. 2003. *Dynamite* (Atari STo. http://www.8bitpeoples.com/discography/by/stu.
Suni, Janne. 2000. *Acidjazzed Evening* (Amiga). http://www.scene.org/file.php?file=%2Fparties%2F2000%2Fassembly00%2Foldskool%2Fmusic%2Facid_jaz.zip&fileinfo.
Suzuki, Keiichi and Hirokazu Tanaka. 1994. *Earthbound* (SNES). http://www.zophar.net/music/spc/earthbound.html.
Tanaka, Hirokazu. 1989. *Tetris* (Gameboy). http://www.zophar.net/music/gbs/tetris.html.
Tello, Jose Antonio Martin. 1989. *Astro Marine Corps* (ZX Spectrum). http://www.worldofspectrum.org/infoseekid.cgi?id=0000295.
Toriena. 2012. *Love Bit* (Gameboy). http://soundcloud.com/toriena/love-bit.
Velathnos. 2012. *Super Anthropomorphic Space Turismo* (SNES). http://battleofthebits.org/arena/Entry/Super+Anthropomorphic+Space+Turismo/7003/.
virt, Freaky DNA, and Norrin Radd. *Retro City Rampage Soundtrack*. http://retrocity-rampage.bandcamp.com/.
virt. 2012. *First Flight* (NES FX4). http://virt.bandcamp.com/track/first-flight.
Whittaker, David. 1989. *Xenon 2—Megablast* (Amiga). http://www.exotica.org.uk/wiki/Xenon_2_-_Megablast.
Yamada, Ippo. 2002. *Mega Man Zero* (Gameboy Advance).
Yerzmyey. 2000. *And so It Is* (ZX Spectrum). http://yerzmyey.i-demo.pl/Ye_Olde_Square/05-Yerzmyey-And_so_it_is_2000.mp3.
Zen Albatross. 2011. *Masada Gestalt*. (Gameboy). http://www.scenemusic.net/demovibes/song/11243/.

References

Anderson, Janus and Dallas Snell. 2007. "Is It Fun Yet?" In *GDC Lyon 2007*.

Carr, Neil. 2001. An Interview with Stefan Hartwig. http://www.remix64.com/interview_stefan_hartwig.html.

Commodore Business Machines. 1982. *Commodore 64 Programmer's Reference Guide*.

Commodore Zone. [n.d.] Interviews with Martin Galway. http://trondal.com/c64sid/mgalway.html; accessed October 19, 2013.

Gilmore, Thomas. 2012. A Brief History of Chipmusic. *The Music Network*. http://www.themusicnetwork.com/music-features/industry/2012/02/27/a-brief-history-of-chipmusic/.

Greening, Chris. 2010. Game Music : Interview with the *Mega Man* 1 & 2 Sound Team: Reunited 20 Years On (November 2010). http://www.squareenixmusic.com/features/interviews/manamimatsumae.shtml.

McSweeney, Anthony. 1993. Rob Hubbard's Music: Disassembled, Commented and Explained. http://www.1xn.org/text/C64/rob_hubbards_music.txt.

Nash, Chris. 2011. Supporting Virtuosity and Flow in Computer Music. http://revisit.info/files/PhD_Thesis_600dpi.pdf.

Paul, Leonard. 2005. Electronic Folk Art: Commodore 64. http://www.videogameaudio.com/ElecFolkArt-Feb2005/ElecFArt.d64.

Taylor, Brad. 2004. 2A03 Technical Reference. http://nesdev.com/2A03%20technical%20reference.txt.

Tufvesson, Pex. 2010. Pex "Mahoney" Tufvesson's Homepage! http://www.livet.se/mahoney/cubase64.

Weixelbaum, Herbert. 2012. Game Boy Sound Comparison. http://www.herbertweixelbaum.com/comparison.htm.

Yabsley, Alex. 2007. The Sound of Playing: A Study into the Music and Culture of Chiptunes. Bachelor's thesis, Queensland Conservatorium, Griffith University, Australia. http://www.woolyss.com/chipmusic/chipmusic-discovery/A_study_into_the_music_and_culture_of_chiptunes.pdf.

CHAPTER 31

PROCEDURAL AUDIO THEORY AND PRACTICE

ANDY FARNELL

It is worth revisiting the distinction between traditional and new media. In the former, content such as images or sound is acquired by measurement for subsequent playback, a process of data sampling (Nyquist 1928), and so we refer to such media as being data-driven. At various stages the sampled data may be transformed, or edited. It may, for example, be filtered to change its tone and color, or sliced, and the pieces rearranged in time. Sections may be faded in or out, such that they overlap smoothly. They may be stretched or shortened in duration, or effects added to create false sensations of space and atmosphere. All of these aspects are features of nonlinear editing and are the staple functionality of modern digital software for producing video and music.

New interactive media, of which computer games are an important class, do not predefine the timing, order, or quality of visual and audio events. Their final form is deferred until run-time. Choices made by the player through interaction as well as random elements, shape the flow of the experience, limiting, expanding, and selecting possible paths of play. Even the supremacy of the event itself may be challenged, and a model based on a continuum of experience substituted.

The director of ordinary media obtains almost total control over the experience of the user. This control includes fine details such as the color mix of the projection and the acoustic properties of the theater. Issues of cultural subjectivity aside, crafting this experience is precisely the acknowledged role of the film director, music producer, or other person exercising artistic control in a project having definitive form. Similar to the designer of a rollercoaster ride, the rider is a captive audience, a point passing along a precise and carefully planned trajectory.

Contrasting this user experience with the experiences of a computer game player, we see these vary from guided or semilinear to completely free. The semilinear design is like that of a maze, where the player makes path choices at prescribed junctures. The latter form is analogous to an adventure playground or sandbox, in which the player roams freely, interacting with objects at will. Contrasting some artistic demands; for the director of a film, definitive form is a blessing, as it limits scope. In film, everything appears uniquely for the first time, framed perfectly only at its precise narrative moment, and in the form that the artist intends (Tarkovsky 1989). But for virtual reality to obtain a significant scope, a game designer's concern is contingency, hoping to depict every rock, every flower, every cloud and blade of grass, such that it can be seen from every possible angle and in every narrative context. Data representing the prior manifestations of objects are clearly no longer adequate: instead, we need object models that capture the myriad behaviors. In procedural media, an analysis of physical and perceptual qualities of natural objects or processes yields algorithms run by objects within a virtual world. These are designed for interactivity with a player, and with each other. Examples of procedural media include: Plants and animals (Prusinkiewicz and Lindenmayer 1990); Landscapes (Chen et al. 2008) (Prusinkiewicz and Hammel 1993); Mazes (Roden and Parberry 2005b); Cities (Greuter et al. 2003; Wonka et al. 2003); Textures like fire and water (Ebert 1994; Roden and Parberry 2005a); Behavioral characters (Perlin 1995); Sounds (Cook 2002; Farnell 2010a); and Music (Lerdahl and Jackendoff 1983; Winkler 1998). Much more is left out, since these examples are selected as apropos to computer games and omit obvious areas such as artificial intelligence for nonplayable character (NPC) entities, soundscape, and atmosphere compositions by such artists as Brian Eno and Karlheinz Essl, procedural story generation and other dynamic media.

31.1 Palette of Content

However, within interactive media there is a further distinction to be made. An historical method of constructing interactive media may be termed "real-time selective". In selective systems, sampled data (although still essentially static) is presented by a control system in such a way as to provide the impression of a freely navigable virtual reality (Krueger 1977). These pieces of data, variously called clips, fragments, slices, or grains, are combined with techniques to randomize and interleave them to create visible textures or sound atmospheres. With sound, such granular techniques (Jones and Parks 1988; Roads 1978, 2004; Truax 1988) can do much to make sampled data go further in terms of the ability to introduce variability. Image quilting (Efros and Freeman 2001), clever tiling arrangements, or stochastic analysis and resynthesis methods (Heeger and Bergen 1995) are commonly used in procedural texture generation. In advanced use, systematic organizations of data fragments, or extracted basis functions that can reconstruct them, can generate a dictionary, perhaps a multidimensional data structure with complex relationships based on features. The inspiration here is from techniques used to identify chemical compounds in forensic diagnostic analysis, by breaking them up and using a spectrometer. Applied to images and sound, this technique facilitates data compression and sparse resynthesis methods (see for example Davies and Daudet 2006). For audio, interesting side effects and applications are blind stream separation for MIR (Plumbley et al. 2010), applications, hybridization and cross-synthesis, for example by

Collins and Sturm (Sturm et al. 2009). The limiting condition of this concept is that selective methods differ only in scale and the resolution of time or function selection, and are ultimately limited to variations on the analysis material supplied.

By contrast now, consider procedural methods that start with no a priori data set, but produce signals by purely computational means, ex nihilo. Examples are rooted in the history of computer music (see Cook 2002; Moore 1990; Roads and Strawn 1987), where well-known vibrational equations are taken and turned to computable routines (Benson 2007). Extraordinary compactness and efficiency are two immediate advantages. Several classes of purely computational synthesis may be observed (for some overviews see Roads 1996; Tolonen, Välimäki, and Karjalainen 1998). Direct signal synthesis by summation (Horner and Beauchamp 1996; Risset and Mathews 1969), by table-based nonlinear transfer function (Le Brun 1979), or by modulation (Chowning 1973) are long-established and well-documented methods. These differ conceptually from so-called "physical models" of the lumped mass (Cook 2002; Florens and Cadoz 1991) or waveguides (Smith 1987), being pure abstractions or phenomenal analogies, rather than simplifications of physics. There is ground between these approaches, such as simplification to modal behaviors (Adrien 1991), direct implementation of equations in differential variables (Bilbao 2009), or recursive transformation (Trautmann and Rabenstein 2003), that offer a compromise between pure signal methods and the complexity of explicit modeling.

A key issue here, from the viewpoint of creative utility, is the parametric access provided by respective methods (Farnell 2010b). Indeed, it is meaningful responsive, interactive behaviors that form the most central theme in Krueger's virtual-reality foundations, not necessarily the modeling process itself. Variability and interactivity risk the price of either using great quantities of data and having great complexity of run-time selection, or high complexity of control variables for signal models. A considered design approach is therefore necessary to navigate the middle ground and create optimal sounding objects.

31.2 Procedural Concepts and Philosophies Revisited

31.2.1 Genesis and Transformation

The philosophical thrust of procedural thinking generally, that distinguishes it from sampling, is that its practice is epistemological rather than genealogical. It lies at the intersection of physics, computational modeling, and psychology. The development of predictive, naturalistic knowledge replaces the earlier activity of the hunter and curator. In their application area, new interactive media are constructed from simple precepts and delivered at run-time by a computational process. In the visual and tangible realms, the apprehension of conditions and process is required, but it is usually the results of the process, not the process itself, that are of interest. For example, Prusinkiewicz and Lindenmayer (1990) highlight Turing's work on morphogenesis (1952) in the development of L-System methods, as well as Chomsky's interpretation of generative grammar as structurally innate (1956). A central concept is that of rewriting rules that govern the evolution of a system from one state to the next. Koch (1905), Mandelbrot (1982), Conway (analyzed in Gardner 1970), and Wolfram (1985) have all demonstrated seminal examples, and analogous sonic manifestations have been realized by Chareyron (1990), Miranda (1995), and more recently by Serequera and Miranda (2010). In the sonic realm, however, we are more interested in tracing the trajectory of the process itself, of which the outcome is almost always an increase in entropy. In contrast, models based on life processes tend toward a decrease in entropy (Farnell 2010a) and the emergence of order, or dynamic equilibrium, as a desired feature. A continuum of concepts can be noted. We can see recursion, for example, in waveguide methods of Karplus and Strong (1983), or infinite impulse type digital filter and waveguides more generally (Smith 1987), as a rewriting technique in which the transforming filter coefficients take the role of a substitution rule at a point in the buffer, which assumes the role of the rewritten expression, and a wavetable oscillator as a case of one-dimensional data tiling. Overall, simple production rules and iterated rewriting lead to high complexity of form.

31.2.2 Form and Structure

Sonic procedural thinking might be understood at the highest level though a mixture of Platonic idealism and reductionism. Interactions and complex gestures may be idealized as combinations of prototypical processes that are adapted to special cases. The sonic manifestation of any target object may be reduced to simpler basis actions. For example, bouncing may be seen in terms of impact energy decay due to loss (Warren and Verbrugge 1984). These are identified as a relatively small set of phenomena; possibly fewer than ten are useful basic abstractions, such as collision, modal resonance, damped contact, stick-slip interaction, and turbulent excitation, among others. All are further reducible to transient propagation, simple harmonic, or chaotic phenomena. For example, a squeaking door is a stick-slip friction model combined with a square laminar radiator with the modal absorption properties of wood. A particular snapping twig may be derived as a case of an ideal cylindrical fragmentation model, given certain parametric characteristics, length, diameter, material strength, and so forth. All real instances are of course imperfect and uniquely parameterized. Grammatically, there is a shift from proper noun to an active gerund verb exemplar, such that process supplants static quality. Central to this is the notion of signature process. Signature processes pertain to the build up and decay of energy in a system that may uniquely identify it or carry semantics. Musically, these are outlined by Schaeffer (1977). Furthering this musical agenda Tanaka (2007) examines the gesture and motor affect through modern transducers (accelerometer, gyrosensor, image tracking), particularly in a mobile computing context. Fundamentally, physical interaction processes have been identified and explored variously by Hertz, Helmholtz, and others, and explored computationally in more recent times by researchers in Sonic Interaction (Avanzini and Rocchesso 2004; Delle Monache et al. 2009; Rath 2003; Vicario, Rocchesso, and Fernström 2001).

31.2.3 Phenomenological Concerns

An important conceptual overlap into phenomenology identifies human interaction as a vital part of procedural audio. Synthetic models may be quite weak in terms of flexible usage, or difficult to design for all but a narrow range of parameters unless perceptual factors, interactive cognitive ergonomics, and human understanding of acoustic ecology is brought to the design of sounding objects. Here Hug, Franinović, and Visell (2007), Lai and colleagues (2011), Erkut, Delle Monache, and Hug (2010), Ekman and colleagues (2005), and others have contributed essential work. In terms of choosing methods, a proposed partition into phenomenal and essential methods may be useful (Farnell 2010b). In particular the contribution of the Sounding Object and CLOSED projects (Vicario, Rocchesso, and Fernström 2001) are highly relevant to procedural thinking for computer game design.

31.3 Principles and Practice of Procedural Audio

The following section summarizes some key concepts important to the design of effective sounding objects using a physically informed approach, for computer games or tangible sonic interaction. These are anecdotal, since they lack a formal experimental history; they are offered to raise possibilities for future work as well as to guide other practitioners.

Centrality of process: Above all, it is important that we remain mindful of sound as a process of transforming energy. If we think about conservation and boundary conditions, questions like, "What are the sources and sinks of energy?" and "What detail can be removed from the model without affecting the sound?" become essential considerations.

Separation of design: A layered, decoupled approach that treats the model, method, and implementation as separate concerns is valuable. If a model seems to depend on a particular synthesis method, there are changes in computational process that would need to be considered, for instance a shift from additive to granular. Similarly, if that computational process breaks down, it is necessary to consider what it is really part of a properly abstracted model. Designers need to practice implementing designs in different languages (to learn a lot about the limitations and strengths of an implementation): see

Dan Stowell's (2009) redesign of some Pure Data examples for how such an implementation might work.

Informed deconstruction: Thorough research of the object or process and its physical mechanisms should be undertaken. Analytical recording, service manuals, anatomical diagrams, and even a screwdriver may aid the study. The most extensive sample library may not help as much as actually picking up and holding the object under study.

Identification of signature process: One aspect of the model tends to dominate all others. It may be causally prior to others, or an energetic choke point that modulates other vibrational mechanisms. Identifying this will help with top-down design of control structure.

Interactive object behavior: Whether a procedural sounding object is acted on directly by a person or mediated by more complex control systems, its interactive nature should never be ignored. Considered as an object exposing a parametric interface, the range, resolution, scaling curves, volatility, and interaction of all input values must be explored. Ideally these are orthogonal, a minimal set offering maximal flexibility. The object should be unconditionally stable for all settings.

Compactness: Quite simply, the least amount of code required to obtain a given sonic behavior. Compact code can be useful for purely software engineering reasons, being easy to understand, extend, and maintain.

Efficiency: Not to be confused with compactness, a model that is lighter on CPU usage will always be preferred, since most procedural sounding objects are intended for real-time or embedded applications.

Sparse design goal: This objective is an aesthetic quality of the sound but is inextricably linked to the design. You will tend to find that the first 10 or 20 percent of the code covers most of the perceptual requirements, the next 40 percent moves the result from just adequate to high-quality realism, and the last efforts are sometimes unnoticeable or actually worsen the result for some listeners. Sparse design opens the door for dynamic Level Of Audio Detail (LOAD) in properly designed digital signal processing (DSP) engines.

"Good enough" design: Although similar to the principle of sparse design, this is a reminder that sounding objects are made mainly for the purposes of arts and entertainments. It is easy to get carried away with elaborate models based on accurate physical models. Think carefully about the precision and resolution of computed values throughout your code and strive for a good balance between cost metrics and perceived aesthetic quality. Even in critical sonification tasks, high precision may not actually have perceptual significance.

Plasticity of tools: A good design, based on a well-researched model that is properly decoupled, will not pivot on the implementation framework, or any specific feature of it. This process allows the skill of procedural audio object design to develop in isolation of any tool or workflow. Tools will tend to fall into the categories of: low-level languages; open-ended DSP-specific languages; toolkits; and prepared sounding objects. With the level of skill required decreasing through the list, and the level of flexibility decreasing. Prepared sounding objects are becoming available through middleware audio engine developers such as Audiogaming (Audiogaming 2009).

Contextual design: Take account of how and where your models will be used within a game or interactive application. Will other sounds mask or distract from it? Will the sound be focal or incidental within its context? Will it be moving and modified by Doppler shift, or confined within an acoustically colored space? Consider possible interactions with other software or subsystems performing occlusion and spatial modeling. Avoid duplicated work and respect proper design boundaries by considering whether any feature is really part of your model, or part of the environment.

Psychoacoustic and subjective dimensions: Be aware of common psychoacoustic phenomena and leverage these, or avoid them, to enhance your design. Be aware of subjective effects and try to conduct at least some informal listening tests with ears other than your own.

Acoustic viability: Create models that make sense physically. Avoid impossible transitions, paradoxical couplings of parameters, and unnatural energy decays. Remember that the hyperreal is a often only a slight extrapolation of reality that maintains physical plausibility.

31.4 Conclusions

Procedural audio is a different way of thinking about sound. It has a rich genealogy, with roots in simulation, generative mathematics, and algorithms. Its current main areas of application are computer games, animated film, and embedded interactive sound in products. Although numerous examples have been demonstrated in the last thirty years, it has not enjoyed much uptake throughout these decades because the simple replay of samples has been effective and cheap.

With increasing demand for flexible, interactive audio and ever-increasing CPU power, the motivation and means for procedural audio are now overwhelming. We continue to pursue a coherent philosophy that combines elements of interaction, modeling, physics, perception, synthesis, and analysis. Procedural audio is a wonderfully cross-disciplinary subject, with future research directions encompassing technical and humanistic investigations into the meaning and emotion of sounds, as well as effective production methods. Research outcomes will be tools, techniques, and media examples of new virtual interactive sounds.

References

Adrien, Jean-Marie. 1991. The Missing Link: Modal Synthesis. In *Representations of Music Signals*, ed. Giovanni De Poli, Aldo Piccialli, and Curtis Roads, 269–298. Cambridge, MA: MIT Press.

Audiogaming. 2009. http://www.audiogaming.net/.

Avanzini, Federico, and Davide Rocchesso. 2004. Physical Modeling of Impacts: Theory and Experiments on Contact Time and Spectral Centroid. *Proceedings of the Conference on Sound and Music Computing 2004*, 287–293. http://smcnetwork.org/files/proceedings/2004/P46.pdf.

Benson, David J. 2007. *Music: A Mathematical Offering*. Cambridge, UK: Cambridge University Press.

Bilbao, Stefan. 2009. *Numerical Sound Synthesis: Finite Difference Schemes and Simulation in Musical Acoustics*. Chichester, UK: Wiley.

Chareyron, J. 1990. Digital Synthesis of Self-modifying Waveforms by Means of Linear Autonoma. *Compute Music Journal* 14: 25–41.

Chen, Guoning, Gregory Esch, Peter Wonka, Pascal Müller, and Eugene Zhang. 2008. Interactive procedural street modeling. *Proceedings of ACM SIGGRAPH Three-dimensional Graphics and Realism*, 27. http://www.sci.utah.edu/~chengu/street_sig08/street_sig08.pdf.

Chomsky, Noam. 1956. Three Models for the Description of Language. *Transactions on Information Theory*, 2: 113–124.

Chowning, John. 1973. The Synthesis of Complex Audio Spectra by Means of Frequency Modulation. *Journal of the Audio Engineering Society* 21: 526–534.

Cook, Perry. 2002. *Real Sound Synthesis for Interactive Applications*. Natick, MA: A. K. Peters.

Davies, M. E., and L. Daudet. 2006. Sparse Audio Representations Using the MCLT. *Signal Processing* 86: 457–470.

Delle Monache, Stefano, Delphine Devallez, Carlo Drioli, Federico Fontana, Stefano Papetti, Pietro Polotti, and Davide Rocchesso. 2009. *Sound Design Toolkit: Closing the Loop of Sound Evaluation and Design. CLOSED Project. University of Verona. Deliverable* 2.3, 1–62. http://closed.ircam.fr/uploads/media/CLOSED_D2.3.pdf.

Ebert, David S. 1994. *Texture and Modeling: A Procedural Approach*. New York: Academic Press.

Efros, Alexei A., and William T. Freeman. 2001. Image Quilting for Texture Synthesis and Transfer. *Proceedings of SIGGRAPH '01*, 341–346. New York: ACM. doi: 10.1145/383259.383296.

Ekman, Inger, Laura Ermi, Jussi Lahti, Jani Nummela, Petri Lankoski, and Frans Mäyrä. 2005. Designing Sound for a Pervasive Mobile Game. *Advances in Computer Entertainment Technology* 110–116. doi: 10.1145/1178477.1178492.

Erkut, Cumhur, Stefano Delle Monache, and Daniel Hug. 2010. Basic Exploration of Narration and Performativity for Sounding Interactive Commodities. *Proceedings of the 5th International Conference on Haptic and Audio Interaction Design*: 65–74. Berlin: Springer-Verlag.

Farnell, Andy J. 2010a. *Designing Sound*. Cambridge, MA: MIT Press.

——. 2010b. Behaviour, Structure and Causality in Procedural Audio. In *Game Sound Technology and Player Interaction*, ed. Mark Grimshaw, 313–339. Hershey, PA: Information Science Reference.

Florens, J. and C. Cadoz. 1991. The Physical Model. In *Representations of Musical Signals*, ed. Giovanni De Poli, Aldo Piccialli, and Curtis Roads. 19-29. Cambridge, MA: MIT Press.

Gardner, Martin. 1970. Mathematical Games: The Fantastic Combinations of John Conway's New Solitaire Game "Life." *Scientific American* 223:120–123.

Greuter, Stefan, Jeremy Parker, Nigel Stewart, and Geoff Leach. 2003. Real-time Procedural Generation of "Pseudo Infinite" Cities. *Proceedings of the 1st International Conference on Computer Graphics and Interactive Techniques in Australia and South East Asia*, 87–94.

Heeger, David J., and James R. Bergen. 1995. Pyramid Based Texture Analysis/Synthesis. *Computer Graphics Proceedings* 229–238. http://people.csail.mit.edu/torralba/courses/6.869/lectures/lecture5/heegerbergen.pdf.

Horner, Andrew, and James Beauchamp. 1996. Piecewise-linear Approximation of Additive Synthesis Envelopes: A Comparison of Various Methods. *Computer Music* 20 (2): 72–95.

Hug, Daniel, Karmen Franinović, and Yon Visell. 2007. Sound Embodied: Explorations of Sonic Interaction Design for Everyday Objects in a Workshop Setting. *Proceedings of the 13th International conference on Auditory Display*, 334–338. http://www.icad.org/Proceedings/2007/FraninovicHug2007.pdf.

Jones, Douglas, and Thomas W. Parks. 1988. Generation and Combinations of Grains for Music Synthesis. *Computer Music Journal* 12 (2): 27–33.

Karplus, Kevin, and Alex Strong. 1983. Digital synthesis of plucked-string and drum timbres. *Computer Music Journal* 7 (2): 43–55.

Koch, Helfe von. 1905. Une méthode géométrique élémentaire pour l'étude de certaines questions de la théorie des courbes planes. *Acta Mathematica* 30: 145–174.

Krueger, Myron. 1977. Responsive environments. *Proceedings AFIPS 46 National Computer Conference*, 423–433. Anaheim, California.

Lai, Chi-Hsai, Matti Niinimäki, Koray Tahiroglu, Johan Kildal, and Teemu Ahmaniemi. 2011. Perceived physicality in audio-enhanced force input. *ICMI: Proceedings of the 13th International Conference on Multimodal Interfaces*, 287–294. New York: ACM.

Le Brun, Marc. 1979. Digital Waveshaping Synthesis. *Journal of the Audio Engineering Society* 27 (4): 250–266.

Lerdahl, Fred, and Ray Jackendoff. 1983. *A Generative Theory of Tonal Music*. Cambridge, MA: MIT Press.

Mandelbrot, Benoit. B. 1982. *The Fractal Geometry of Nature*. New York and London: W. H. Freeman.

Miranda, Eduardo Reck. 1995. Granular Synthesis of Sounds by Means of a Cellular Automaton. *Leonardo* 28 (4): 297–300.

Moore, F. Richard. 1990. *Elements of Computer Music*. New York: Prentice Hall.

Nyquist, H. 1928. Certain Topics in Telegraph Transmission Theory. *Transactions of the AIEE*, 47 (2): 617–644.

Perlin, K. 1995. Real Time Responsive Animation with Personality. *IEEE Transactions on Visualization and Computer Graphics* 1 (1): 5–15.

Plumbley, M., T. Blumensath, L. Daudet, R. Gribonval, and M. E. Davies. 2010. Sparse Representations in Audio and Music: From Coding to Source Separation. *Proceedings of the IEEE* 98 (6): 995–1005.

Prusinkiewicz, Przemyslaw, and Mark Hammel. 1993. Fractal Model of Mountains with Rivers. In *Proceeding of Graphics Interface '93*, 174–180. http://algorithmicbotany.org/papers/mountains.gi93.pdf.

Prusinkiewicz, Przemyslaw, and Astrid Lindenmayer. 1990. *The Algorithmic Beauty of Plants*. New York: Springer-Verlag.

Rath, Matthias. 2003. An Expressive Real-time Sound Model of Rolling. *Proceedings of the 6th International Conference on Digital Audio Effects*. http://www.elec.qmul.ac.uk/dafx03/proceedings/pdfs/dafx40.pdf.

Risset, J. C., and M. V. Mathews. 1969. Analysis of musical-instrument tones. *Physics Today*, 22: 23–30.

Roads, Curtis. 1978. Automated Granular Synthesis of Sound. *Computer Music Journal* 2 (2): 61–62.

——. 1996. *The Computer Music Tutorial*. Cambridge, MA: MIT Press.

——. 2004. *Microsound*. Cambridge, MA: MIT Press.

Roads, Curtis, and John Strawn. 1987. *Foundations of Computer Music*. Cambridge, MA: MIT Press.

Roden, Timothy, and Ian Parberry. 2005a. Clouds and Stars: Efficient Real-time Procedural Sky Rendering Using 3d Hardware. *Proceedings of ACM SIGCHI International Conference on Advances in Computer Entertainment Technology ACE 2005*, 434–437. New York: ACM.

——. 2005b. Procedural Level Generation. In *Game Programming Gems* 5: 579–588. Charles River Media. https://larc.unt.edu/ian/research/content/.

Schaeffer, Pierre. 1977. *Traité des objets musicaux*. Paris: Éditions du Seuil.

Serequera, Jaime, and Eduardo Miranda. 2010. Cellular Autonoma Sound Synthesis with an Extended Version of Multitype Voter Model. *AES Convention Paper 8029*, 128.

Smith, Julius O. 1987. Music Applications of Digital Waveguides. *Technical Report STAN-M-39*, Center for Computer Research in Music and Acoustics. https://ccrma.stanford.edu/files/papers/stanm39.pdf.

Stowell, Daniel. 2009. *Designing Sound in SuperCollider*. http://en.wikibooks.org/wiki/Designing_Sound_in_SuperCollider/Print_version.

Sturm, Bob L., Curtis Roads, Aaron McLeran, and John J. Shynk. 2009. Analysis, Visualization, and Transformation of Audio Signals Using Dictionary-based Methods. *Journal of New Music Research* 38: 325–341.

Tanaka, Atau. 2007. Interaction, Experience, and the Future of Music. In *Consuming Music Together: Social and Collaborative Aspects of Music Consumption Technologies*, ed. Kenton O'Hara and Barry Brown, 267–288. Dordrecht, Netherlands: Springer.

Tarkovsky, Andrey. 1989. *Sculpting in Time: Reflections on the Cinema*. Austin: University of Texas Press.

Tolonen, Tero, Vesa Välimäki, and Matti Karjalainen. 1998. *Evaluation of Modern Sound Synthesis Methods*. Helsinki: Helsinki University Technology. http://www.acoustics.hut.fi/publications/reports/sound_synth_report.pdf.

Trautmann, Lutz, and Rudolf Rabenstein. 2003. *Digital Sound Synthesis by Physical Modeling Using the Functional Transformation Method*. New York: Kluwer Academic.

Truax, Barry. 1988. Real-time Granular Synthesis with a Digital Signal Processor. *Computer Music Journal* 12 (2): 14–26.

Turing, Alan. 1952. The Chemical Basis of Morphogenesis. *Philosophical Transactions of the Royal Society*, Series B, *Biological Sciences* 237 (641): 37–72.

Vicario, Giovanni Bruno, Davide Rocchesso, and Mikael Fernström. 2001. *The Sounding Object*. Project jointly by University of Udine Dipartimento di Scienze Filosofiche e Storico-Sociali, University of Verona Dipartimento di Informatica, University of Limerick, Speech, Music and Hearing centre KTH—Stockholm. www.soundobject.org/.

Warren, W. H. and Verbrugge, R. R. 1984. Auditory Perception of Breaking and Bouncing Events: A Case Study in Ecological Acoustics. *Journal of Experimental Psychology* 10 (5): 704–712.

Winkler, Todd. 1998. *Composing Interactive Music*. Cambridge, MA: MIT Press.

Wolfram, Stephen. 1985. Some Recent Results and Questions about Cellular Automata. In *Dynamical Systems and Cellular Automata*, ed. Jacques Demongeot, Eric Goles, and Maurice Tchuente, 153–167. London: Academic Press.

Wonka, Peter, Michael Wimmer, François Sillion, and William Ribarsky 2003. Instant Architecture. *ACM Transactions on Graphics* 22 (3): 669–677.

CHAPTER 32

LIVE ELECTRONIC PREPARATION

Interactive Timbral Practice

RAFAŁ ZAPAŁA

As I was tuning my surrounding octophonic sound installation once, a colleague of mine, a double bass player, suddenly appeared in the concert hall. He glanced at me and my attempts, only to sum up what he witnessed as "You and those toys of yours." What he was referring to was the typical collection of electronic equipment that I had: a laptop, sound card, mixer, and set of speakers. Such a disdainful approach, streaked with a particular distrust toward electronic instruments, is still a rather common attitude among some circles of musicians. Many traditionally educated musicians playing acoustic instruments still approach electronics with a certain timidity. Dozens of years after the development of digital electronic instruments, this timidity is inexplicable. It is all the more bizarre considering that other computer-based skills are commonly used in everyday life. A computer-to-microphone combination does not particularly differ from a computer-to-printer pairing, and handling the basic sequencer functions is now as intuitive as handling the menu of a mobile phone. Through rehearsal recording, the metronome, synchronization, tuning, or working with tape, the laptop appears to me an obvious element of the musician's experience, side by side with acoustic instruments. It is still, however, a rare sight in the practice room. Clearly, there exists a dissonance in today's art of sound creation. On the one hand, musicians who are fascinated with tradition let us rediscover virtuosity, the technical richness developed through centuries. However, on a number of occasions, they underestimate the technical and aesthetic changes introduced by electronic music. They tend to treat electronic music like a fashionable fad or like a toy.

On the other hand, an autonomous area of art—*ars electronica*—continues to advance in electronic studios, academic think-tanks, and underground innovatory circles. An experimental, scholarly, interdisciplinary approach dominates among these artists. The disadvantages of the approach, such as issues of liveness, one-time effects, and concentration on the technology itself rather than on its artistic application, are covered in other chapters of this Handbook. Both of the above-mentioned attitudes have advantages and disadvantages, and they complement each other. Naturally, there are performers and ensembles who effectively integrate both approaches. Nonetheless, I would argue that such integration is required in concert institutions and systems of education.

"You and those toys of yours": I would never think of referring this way to the double bass, the bow, rosin, or the foldable music stand. I would, rather, envisage a set of masterfully crafted objects, a fascinating product of human culture, an unusual artifact whose construction secrets have been mastered over centuries. I would also hear the potential richness of profound, noble sounds in numerous varieties, sounding different in different musicians' hands, sounding different with different strings, bows, or spaces where sound is performed. I would instantly remember the whole available performing artistry, techniques shaped over centuries, virtuosity forged through hours of practice. As a composer, I would also realize the interpretation potential inherent in the performer's unique musicality built on his or her personal virtues and vices, reflections and experiences. Eventually, I would recollect the merits of the traditional concert, which can hardly be substituted by an *acousmatic* event. Pierre Schaeffer, who coined the term "acousmatic," defines it as "referring to a sound that one hears without seeing the causes behind it" (Schaeffer 1966, 91). Driven by the composer's imagination, I would probably automatically immerse myself in the images of beautiful double bass sounds. But imagination leads me straight to developing a timbral thought leading toward the granular timbral cloud, toward spectral evolutions or glitch techniques of the click'n'cut area, or the vast richness of possibilities that the studio-based electronic composition has endowed us with today.

Naturally, I can record the sound of the double bass in the old-fashioned way, transpose it at will in the studio, and then present it to the audience via speakers. With reference to the analog times of electronic music, we can still come across the term "tape music," although the hard disk has supplanted tape, and the label "acousmatic music" is more suitable today. Although the term "fixed sounds" denotes an almost total control of one's composition, it is, nonetheless, a pity to part with the traditional stage drama, the potential richness of interpretations. Perhaps the acoustic musician should be juxtaposed with the tape, which is known in the French tradition as "mixed music." A number of such compositions have been created, which make it possible to take advantage of the best possibilities afforded by acoustic and electronic audio. The basic problem is combining these two areas, in particular the question of synchronization: "Analogue (and fixed-digital) playback systems rarely gave the performer control of one of the most precious part of expressive musical potential: control over *time,* the speed or tempo of performance... In fact, any "fixed sound track" works especially poorly with the more strictly notated traditions of Western music... The necessary synchronisation between the two forces could only be enforced one way: the objective clock time of the prerecorded part was the inflexible conductor of the live part" (Emmerson 2009).

Synchronization problems are diminished when techniques of dynamic electronics control are introduced during the concert. Although the idea of an electronic instrument was important from the very onset of electronic music, for a long time what was possible to achieve in the studio was impossible to achieve in real time. Today, "live electronics" techniques appear closest to the ideal of the new electronic instrument. The laptop is an instrument in its own right, although it is still not treated like one (see below). A number of potential strategies and techniques have been proposed to make these two separate worlds—acoustic and electronic—meet on more or less equal terms. However, in my opinion it is not a meeting of two worlds that would be the ideal solution, but their integral fusion, an abolition of the borderline between them, a borderline that is basically artificial and unnecessary. Both of these areas of musical art need and complement each other.

The key to full fusion of the acoustic and electronic audio is their *live* interaction. In this chapter I would like to draw attention to a specific composition and performance technique, the electronic treatment of the acoustic source. It is a variety of live electronics in this particular form in which interactivity binds the two. Besides discussing the technique, I will also attempt to indicate a number of aspects of electronic music in the context of a traditional *live* concert, which should be developed or redefined. Three basic elements make this technique the most promising and powerful in creating new and original sound. At the same time, however, the technique is one that fully uses centuries-old tradition of performing technique, the wearing of acoustic instruments, the maturity of electronic music, and the processing power of computers, which makes *live* interaction possible.

32.1 Evolution of the Acoustic Audio

Real-time electronic treatment of acoustic sound is a logical and natural consequence of the development of key aesthetic ideas of the twentieth century. A brief review of evolution of acoustic and electronic music will facilitate understanding of the context of the proposed technique. The twentieth century brought increased interest in inharmonic timbres. The Italian Futurists for instance, notably Luigi Russolo, suggested construction of new instruments: "We must break out of this limited circle of sounds and conquer the infinite variety of noise-sounds" Russolo stated in 1913 (Russolo 2002). The range of instruments was extended with everyday objects: tins, bottles, gramophones, radio waves, typewriters, pistols, anvils, steel sheets, and other objects were used. As a consequence, the sound universe was extended to every possible timbre, from a single micrograin of sound to whole acoustic spaces.

New timbres are represented by orchestral textural concepts understood as a complex sound. We come across this type of intuition by the 1920s, in the concepts of Edgard Varèse: "When new instruments will allow me to write music as I conceive it, the movement of sound-masses, of shifting planes, will be clearly perceived in my work, taking the place of the linear counterpoint. When these sound-masses collide, the phenomena of penetration or repulsion will seem to occur" (Varèse 2002, 18). Varèse brought his concepts to life through emancipating the percussion timbres, the natural source of noise timbres for a traditional orchestra. A group of techniques elaborated upon by his successors expanded the composer's vision. Here, one can include certain compositions by Stockhausen, the micropolyphonic textures of Ligeti, or the stochastic "sound clouds" of Iannis Xenakis. Certain compositions of the so-called Polish school of composers also constituted an original phenomenon within the vein. Labeled in Poland "sonoristic," this technique was, above all, associated with the works of Penderecki, Serocki, and Kotoński. Other proposals concerning the use of traditional ensembles for new purposes included the search for new systems of organization of all twelve pitches (Lutosławski) and microtonal scales (Partch). A detailed understanding of harmonic sound's inner structure exerted a profound influence upon orchestral textures. Spectral music is today an influential vein associated with the works of Gerard Grisey, Jonathan Harvey, Tristan Murail, and earlier with the concepts of Giacinto Scelsi.

Extended techniques of playing instruments, initiated by Henry Cowell's "cluster technique," likewise constituted a proposal intended to expand their timbre. Cage's preparation compositions were particularly significant, for eventually they led to "stepping outside" a given instrument, in which Cage enriched music with a "noninstrument" entity. The introduction of objects and nonmusical sounds brings, in my opinion, a fundamental change in the treatment of the instrument, which is only partially responsible for the timbre, and only becomes the starting point for the target timbre. Sometimes, with very radical preparations, it changes its character altogether. In other words, almost the entire timbral potential of acoustic material has now been explored. A talented and creative performer has always been able to find an individual language of artistic expression with any instrument, but since the twentieth century, traditional orchestral sound has been viewed by some as a problem to solve, rather than a source of inspiration.

32.2 Evolution of the Electronic Audio

It was electronic music that exerted the greatest influence upon the evolution of twentieth-century music. The impact of the technology was so profound that it basically led to what can be argued to be a separate, independent branch of music, *electronica*. Just like the evolution of electronic technology in general, new phenomena in electronic music emerged, which brought in a permanent influx of new, exciting artistic concepts. For example, the ability to record sound became a milestone in the development of the art of music, and resulted in an avalanche of new practices. Pierre Shaeffer's *musique concrète* and *objét sonores* broke the barrier between so-called "musical" and "nonmusical" timbres, and became part of the aforementioned contexts of noise emancipation and extension of the timbral palette.

Sound recording technologies opened up new possibilities for so-called studio-based composition. Whole new methods of processing and editing, and the evolution of material in the domain of time and frequency became possible. In fact, it can be argued that, in its present shape, acousmatic music is a direct result of the emergence of the sound recording phenomenon. With electronic technology, recorded music could separate the sound from the context in which it originated, in what R. Murray Schafer has referred to as a "schizophonic" quality (Schafer 1994). Phenomena such as ambient music, soundscapes, field recordings and site-specific art are only some of the consequences of the emergence of sound recording techniques.

The emergence of the spectral analysis of sounds, in which sounds can be broken down into their constituent frequencies, put particular emphasis upon the microcosm of sound's inner structure. New techniques of sound synthesis that draw on spectral analysis still form one of the most dynamically developing fields of the art of sound. Furthermore, structural models of sound have shaped the approach to musical texture. Work in the studio erased boundaries between the micro level of the composition (sound design) and the macro level of the piece's texture. In this way, "The new structure of a work is thus born from the elementary structure of sound material" (Gisele Brelet in Fubini 1997).

The growth of electronics has also resulted to some extent in the treatment of timbre as the dominant musical factor. Timbre's evolution, spread between a single tone, through the timbre of harmonic proportions, all the way to white noise, marks the new concepts of consonance and dissonance. Today, electronic music is a mature form of art. It has already developed a canon, as well as a vast number of movements and trends, or aesthetic paradigms. The significance of the first ideas of electronic music appears obvious and indisputable. In simple, polarizing terms, it could be said that acoustic music is grounded in tradition, and electronic music is focused on experimenting: what is indispensable is a third way, a fusion of both of these branching areas. The twentieth-century ideas discussed above influenced both acoustic and electronic compositions. Noise, sound-masses, microtonality, spectralism, the role of chance, aleatoric music, montage, and glitch, among other techniques, now interpenetrate and overlap.

32.3 Interactive Audio

The fusion of acoustic and electronic music is particularly valuable in the performing practice, in musicians' awareness, but also in systems of education, methods of musical notation, and the updated definition of the composer. Digital technology has developed so much that almost all aspects of electronic analysis and synthesis can be used and processed in real time. Software tools such as Max/MSP, Pd, Supercollider, and so on make it possible to implement the scientific developments related to sound structure and processing. Here, I will present, codify, and define the techniques (both performing and composing), whose interactive character obscures the borderline between acoustic and electronic audio. In line with such logical assumptions, I would like to name this specific technique the "live electronic preparation" (LEP) technique.

As described above, the practice of processing acoustic instruments is not a new one. A number of its aspects are worth clarifying, so that a specific coherent technique could be created. It concerns both the composer's approach and the performer's practice, providing a standardized tool and a platform for communication that defines a specific standard of work, of interactively combining electronics and acoustics. Of course, such attempts can be numerous. What is critical is that works should possess the quality of objectivity and establish a certain type of standardized approach that might be adopted by performers. Such a tool should combine the tradition of performing art and the timbral developments of acoustic instruments with the present and future electronic processing techniques. Below, I examine particular elements of this phenomenon, trying to define its starting, standardized form.

32.4 Live Electronic Preparation: Source

In the LEP technique, the source (the starting point) is always the acoustic timbre. The term "acoustic" (sound, instrument, music) here refers to the opposite of "electric," that is, the type of timbre which is generated without electric current. Sometimes, instruments that require muscular strength to produce sound are labeled "acoustic," yet such an understanding is too restrictive (Emmerson 2009, 168). A musical instrument is an object with which to perform music, even if originally it was used for a different purpose. It is an artifact developed by generations of experts. All acoustic instruments, whether orchestral, ethnic, early, or modern, can thus become sources for the LEP technique.

The instrument itself is one thing, but the playing technique perfected alongside its technical development is just as fascinating. Performing skills have matured with social changes and changing aesthetic needs. Virtuosity is where a peculiar "liveliness" is contained: in precision, intuition, and interpretation, but also in inaccuracies, flaws, in creating *live*, in taking micro decisions at each moment of playing. The abundance of methods and playing techniques can become the starting point for electronic processing. Naturally, electronic processing also concerns instruments prepared in the traditional way. For example, a piano can be prepared with acoustic objects such as screws and pieces of wood or rubber (acoustic preparation) that can then become a source for electronic transformations (electronic preparation). We arrive at an interesting situation of a seemingly double-extension of an instrument's timbral possibilities: acoustic *and* electronic preparation.

These are also groups of instruments that can be treated as a single acoustic source. Nothing stands in the way of a gamelan ensemble, a jazz combo, or a whole symphony orchestra constituting one large source transformed *live*. Thus, the composer obtains a huge textural richness developed by existing musical cultures. Here, the starting point could be historical, classical styles, ethnic music textures, or contemporary techniques of sound material organization. Moreover, not only a single instrument or group of instruments, but every object that generates sound unplugged can become an LEP source. Even the entire timbral environment, natural or man-made soundscapes, can be treated this way.

Acoustic sound, as a rule, has very rich inner structure, and very dynamically evolves over time. Furthermore, different performing techniques, playing dynamics, or performers' interpretations change this structure. Such richness is still remarkably difficult to achieve with purely electronic timbres. Paradoxically, this potential is rooted in the imperfections: the unique contents of inharmonic elements, possible differences in construction (each instrument is unique), and in the playing technique. It is also a question of interpretation, which arises from the combination of individual qualities of the musician and individual qualities of the instrument in the unique conditions of a given concert. Acoustic sound contains a certain obliqueness, an acceptance of tiny timbral differences. This creates a situation in which a performance is never the same. A performance is an interaction with instruments, with the audience, with the acoustic environment, and with oneself, which is to say, it is the traditional method of presenting music. It would be regretable to renounce so many possibilities. The LEP technique makes it possible to fully retain them. Each nuance of acoustic sound can be captured by means of a microphone, and with that capture, the whole richness of acoustic musical culture becomes the source for the LEP technique.

32.5 Live Electronic Preparation: Capturing

Acoustic signals are captured in real time using microphones. Equipment setups, and their methods of use (e.g., techniques of microphone placement) can differ, just as acoustic instruments differ. As a composer, I understand these conditioning factors and accept the differences. Naturally, the choice of microphone, its placement, as well as the choice of speakers influences the final timbral effect. Selection of these devices becomes a process of creative, artistic performance decisions, and in this way, a new form of interpretation. In this way, the role of the person responsible for the technical aspects grows; he or she becomes a performer, an interpreter in his or her own right. Amplification itself is a form of transformation and can become a creative composer's toolset (e.g., in the form of strong amplification of very quiet, "noisy" timbres inside the instrument, which may be brought up to the level of our perception). The influence of the electronic musician upon the final shape of the composition is evident.

32.6 Live Electronic Preparation: Electronic Processing

The signal of an acoustic instrument is changed by electronic means. Acoustic timbre, the intricate and evolving spectral-temporal complex, constitutes dynamic material for processing with a computer program. It is organic live matter, permanently evolving, dynamically and artistically shaped by a more or less sensitive performer. Similarly, processing algorithms may also react to changes in the source sound, in a flexible way following the performer's interpretation. An important element of the LEP technique is the clarity of idea: avoidance of extra, purely electronic sources, electronic sounds, samples, prerecorded material. The source is always the dynamic acoustic stream.

The processing power of modern-day computers makes it possible to transfer most of the strategies typically employed in the studio to the stage. Thus, the vast palette of means, the potential revealed by the developments of the history of electronic music can be employed and implemented *live*. Here, incorporation of digital signal processing (DSP) studio techniques is only natural. Traditionally, these are divided into working techniques in the time and frequency domains, and their combinations. In his *Computer Music Tutorial* (1997), Curtis Roads lists the following as the basic transformations of sound: dynamic range processing (envelope shaping, noise reduction, different types of compressing), digital filtering, convolution, fixed and variable time delay processing, time/pitch changing, reverberation, and sound spatialization.

Although countless techniques of sound synthesis have already become associated with purely electronic music, ways of understanding sound and knowledge of its structure let one consciously process it. Were the LEP technique to carry out the idea of common structure of material and musical texture, full awareness of all internal sound processes, as well as of its transformation would be indispensable. Micro-level composing, knowledge of analysis, synthesis, and sound design are exceedingly useful.

32.7 Live Electronic Preparation: Interaction

In the proposed concept, the key element, the essence of the LEP technique, is interaction. It concerns the composing method: the composer must be aware of both the acoustic level and its planned transformation; they must be in control of the interactive evolution of both of these levels and of the final mixed timbre. The process also reveals an interactive character on the level of the performing technique. Above all, the musician engages both media simultaneously. The performer plays an acoustic sound and they also control the electronics, monitoring both levels aurally and reacting to them in a dynamic, natural way. Basically, what is at their disposal is simply a different instrument, and the fusion of timbres is an automatic process. The combined timbre is subject to all interpretation processes. Here, interaction takes place in contact with technology, program, and interface.

However, a different type of interaction is also possible and desirable: a situation in which transformations are controlled by a different musician, an LEP musician. The stake of their mutual interaction is still one common timbre and its evolution as planned by the composer. Such performance of music may be profound, an almost intimate interaction, in which each phrase is the result of mutual creation. It is a deeper interaction than for example, playing in a duo; it is joint creation, joint interpretation, in which mutual understanding assumes paramount importance. Such types of interactive performance are needed.

32.8 Live Electronic Preparation: Projection

The ultimate objective of live electronic preparation is a complex timbre, a fusion of acoustic sound and its transformation in the speaker. Above all, one has to realize that transformation (electronic timbre in the speaker) is always an added element. Acoustic sound is always fully audible, and when it is filtered digitally, its natural volume is not decreased. An important, LEP technique-specific feature, is that electronic timbre always adds something to acoustic timbre, that is, it is always an additive strategy. Subtraction strategies are popular in electronic music, yet in such cases the timbre that constitutes the source would have to be hidden, performed outside the concert hall, or, say, inside soundproof rooms. Rather, what the LEP technique aims to achieve is to extend the natural timbre of the acoustic instrument by means of electronic preparation.

One of my basic assumptions is an equal level of volume between both constituents. The effect is to remain as natural and chamber-like as possible. The underlying assumption is the greatest consistency, the best level of fusion possible. For this reason, the question of speaker arrangement becomes vital. Here, I suggest a departure from the traditional arrangement of stereo loudspeakers on the edge of the stage. Speakers should be located as close to the instrument as possible. Here, naturally, avoidance of feedback becomes a key issue; it is very difficult, particularly in case of radical transformations, to place the speaker too close to the microphone. In the case of small instruments, and

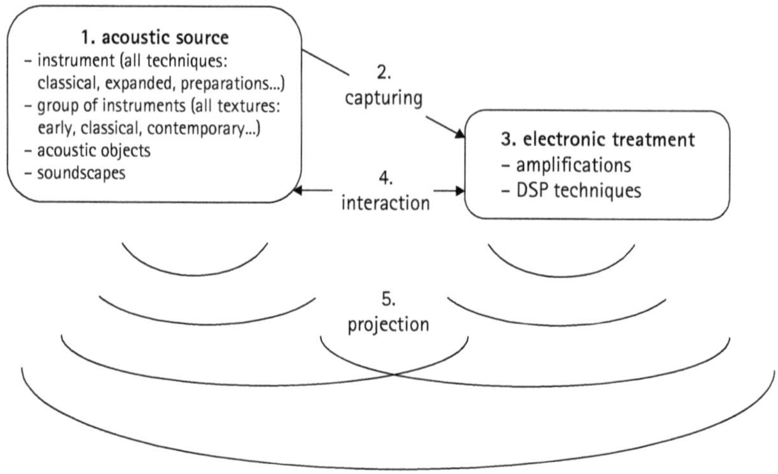

FIGURE 32.1 LEP technique diagram.

in situations when transformations do not reveal spatial character, one can use a single speaker. What is vital besides the level of timbre fusion is the natural spatial perception of the created timbre. For instance, in an extended ensemble, the transformed instrument should have a clearly defined placement in space, and should not be blurred by the transformation. In other words, the sound should come from the place where the acoustic instrument is located. Sounds thus give the use of speakers a chamber-like quality. In the case of larger instruments, ensembles and orchestras, amplification should be adequate to the size and loudness of the transformed source. With a technique presented this way, we can talk about organic relation, full mutual interdependence, full interaction. Acoustic and electronic timbres are fully united, and are subject to interpretation; they are dynamically created *live*.

It is well known that electronic sound processing can be a powerful tool. The timbre can be radically altered, and the original obscured. For this reason, I would like to draw attention to a certain ecology of the employment of processing, a respect for the source. With electronics, it is easy to obscure, to kill the characteristics of natural sound. What I believe is important is to treat transformation as a means of underlining the natural qualities of acoustic timbre. A transformation that in some way results from the natural temporal-spectral structure of the acoustic source would be an ideal solution. Figure 32.1 illustrates a technique diagram for the LEP.

32.9 Composition

What we need is a redefinition and update of such key notions such as composer, performer, and score. I discuss them here in the context of the LEP technique, but

I find the conclusions relevant in the context of the whole of musical art of our time. A number of these changes are simply taking place; on a number of occasions they are already established, although they do not always appear evident. Above all, it is the notion of composer that requires extension. The area of operation of a composer who works with the LEP technique increases. Both command of the acoustic and electronic media, as well as their integration, are indispensable elements of the composer's technique today. Naturally, all previously held skills related to the traditional art of instrumentation and the command of traditional and modern performing techniques remain vital. However, the ability to use the electronic studio becomes essential. It is required if only to understand changes in musical paradigms, which often take place in close relation to electronics. However, in addition to a command of basic musical programs, such as DAWs and the LEP technique, competence should cover at least basic programming skills (e.g., in the environments that slowly become standard: Max/MSP, Pd, or SuperCollider).

Work on a LEP composition has an interactive character and takes place on a number of connected and interrelated levels:

1. The concept: the mental idea of a composition, aural imagination, or idea of a piece.
2. Acoustic score: Writing the score encompasses traditional composing for an instrument with consideration of the planned transformations. The result is the score, frequently only its draft, an outline, or general sketch of the form. Collaboration with the instrumentalist, which is very helpful at this level, becomes necessary in the following stage.
3. Recording: Recording of the instrumental part is necessary to work on electronics. Even if it is not the final version, it is material for precise composition of the computer program.
4. Electronic score: Work in the studio, writing the program (such as a patch dedicated to the composition) based on the recorded material. Writing the program is thus a form of composing.
5. Interaction: Comparing all stages which dynamically evolve and influence each other.

Particular stages of work frequently overlap and are developed simultaneously.

Thus, it becomes clear that the result of the LEP composer's work is two scores: acoustic (i.e., a set of commands, notes, or graphic instructions for the acoustic instrument), and electronic (i.e., a functional computer program, a patch written for a particular composition and integrated with the acoustic score). I want to clearly state my opinion on the above: I find it self-evident that writing a patch in a musical program environment is a new form of composing, and the program is a new form of the score. The composer's education should include this new aspect as a necessary component. Figure 32.2 illustrates the LEP composition strategy.

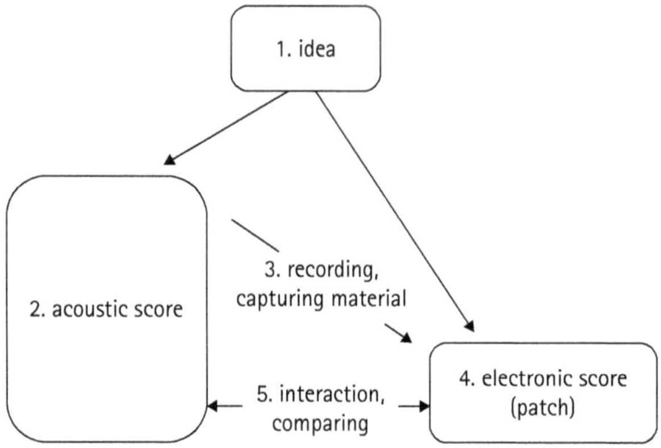

FIGURE 32.2 LEP composition strategy.

32.10 Performance

There is a need among musicians for greater awareness of electronic tools. As for composers, repeated contact with electronics means better understanding of electronic music aesthetics, and understanding of new paradigms of contemporary music, both of which are often associated with the electronic medium. Today, electronics supports almost every area of human activity. A laptop with a soundcard can be very helpful in practice (recording of rehearsals, tuning, advanced metronome, and documentation). With the LEP technique it is not only helpful but simply essential; it becomes an extension of the instrument. As such, it requires experience, a new kind of proficiency (e.g., familiarity with MIDI knobs and sliders), a new way of listening, and an awareness of both media.

A musician should be prepared to perform electronic parts of a piece. The ideal studio should feature a basic electronic setup that would find a permanent place in the rehearsal room side by side with the acoustic instrument. Practice indicates what standard elements such a set should comprise: a laptop with soundcard and music software; MIDI controller, MIDI pedal, or other interfaces; a microphone suitable for acoustic instrument; and a set of speakers or headphones.

The large number of brands and types of recording equipment available on the market lets one build a setup to meet one's needs, preferences, comfort requirements, and financial capacity. For the purposes of this article, I am going to refer to such a standardized setup as the "LEP setup," although, obviously, it is a setup typical of many types of live electronics (see Figure 32.3). It is with such a standardized setup that performers should practice. It is for such a standardized setup that LEP pieces should be composed. I believe that acceptance of the standard would be valuable for a number of reasons, including the fact that what discourages practitioners from performing electronic music is often the specific and different technology employed in every work. A standardized

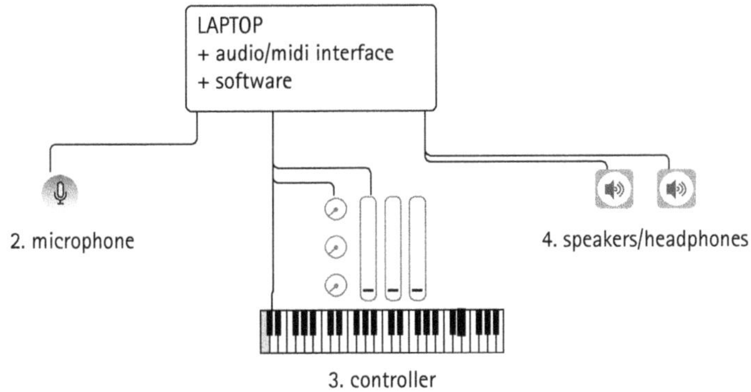

FIGURE 32.3 Standardized LEP setup.

setup stands a chance of becoming commonly recognized as an instrument that is worth practicing, that is worth having as an extension to the acoustic instrument.

In a situation in which the performer also controls the electronic part, it is natural that this control must be simplified and limited to activating processes (for instance by means of pressing a computer key). Despite such a somewhat-automated situation, it is still *live* processing of *live* sound, with its idiosyncrasies and richness: it is still a different phenomenon than playing with tape. A good practice (already relatively common, popularized, among others, by IRCAM and CNMAT), is the use of the MIDI foot pedal by the performer whose hands are busy delivering the acoustic part. Performances may also concern the conductor (the performer playing a huge acoustic instrument, i.e., the orchestra), who can be equipped with the LEP set and thus control electronic transformations of the entire ensemble.

32.11 Live Electronic Preparation: Performer

Finally, the objective of the LEP technique is to arrive at a situation in which electronics, like acoustic instruments, are flexible and intricate, are subject to interpretation, and interactively extend the tradition of acoustic music. LEP transformation controlled by the performer while playing an acoustic instrument is a simpler solution: the value of the electronics is contained in the program rather than elsewhere. But the essence of live electronic preparation is the interaction between two performers in order to jointly shape sound, phrase, and texture. The final timbre is intended to result from subtle, intricate interaction between two musicians: an acoustic instrument virtuoso and an electronic interface virtuoso. Each musical object, the dynamics of each phrase, and each tone would be a fusion of acoustic sound and its transformation, jointly created in concert. Every timbre should be handmade. Leaving all occurrences in the performer's hands is intended to stress the element of interpretation and emotionality in the performance. It is no longer the handling of the interface; it becomes playing music.

Here arises a need to define the electronic musician who becomes a real performer. I can see this new type of musician, the electronic performer, as a result of evolution: from acoustician through musician dealing with sound projection (on a number of occasions the composer himself played this role, such as Karlheinz Stockhausen). Mastering a traditional instrument to a professional level takes several years. Today, it is difficult to imagine a similar proficiency in playing the electronic interfaces in addition to the professional training required by musicians. We find such competence only in major contemporary music studios, research centers, sometimes among electronic music DJs, and sound artists who specialize in particular types of interfaces. As a matter of fact, we still lack a good name for a "live electronics musician." Within the framework of the discussed technique, the name "LEP performer" appears justified.

Further attempts to standardize notation and define the tools are necessary to make it possible for composers to write in a manner similar to the traditional methods, to write for *this* instrument, and for various artists to perform it (just like the Max/MSP environment appears to play the role of the standard in musical programming, which is as difficult to standardize). The question of the electronic instrument is of substance here. The previously mentioned LEP set could become the basic, standard set. However, such an advanced version of the LEP technique entails that each element of the set can be extended. In compositions that assume the inclusion of an LEP performer, one can plan complex, experimental microphone techniques, significantly more advanced interaction in playing transformations, as well as unusual, advanced, multichannel spatialization of the transformed sound, and so on.

The controller is the element that lets one manage the program (electronic score), to *play* in the traditional meaning of the word. New unusual interfaces emerge all the time. Development of this fascinating area of research (particularly popular in NIME conference circles), will continue, and its further influence upon the musical art remains an open issue. Unprecedented development of electronic tools does not, however, find its reflection in performing practice. Despite their potential, live electronic interfaces still do not operate as a "legitimate" instrument. One of the reasons for this lack of legitimization is the prevailing paradigm of electronic composing as having to do with creating an instrument for oneself. A number of new electronic instruments have been developed, as have new interfaces controlled by gesture, breath, dance, and so on. Emphasis is placed on creating new interfaces, new ways of controlling electronics. Unquestionably, it is a very interesting area, yet despite their creators' efforts, such devices typically stand no chance of becoming legitimate instruments. In most cases, these tools are unique, used only once by a single composer or performer. Composers and performers do not employ them in a categorized way. On a number of occasions, it is the tool that defines the piece.

I would argue that we don't need new interfaces, we need a new virtuoso. I suggest adopting the assumption that the choice of the interface remains responsibility of the electronic performer. The performer selects the instrument interface that suits them best, with which they master their skills. Such a decision would simply be an element of the musician's skill, similar to choosing a particular microphone technique, or sound projection, or, for example, selecting the type of bow. As with any other instrument, it is a question of his preferences, abilities, but also of his financial capacity.

32.12 Live Electronic Preparation: Notation

A number of LEP technique-related questions require further reflection and refinement. The score in particular demands detailed consideration: a standardized notation is indispensable to make it possible for different artists performing in different environments and using individualized LEP sets to recreate compositions. Traditional notation has to be augmented with controller notations, bearing in mind their variety, potential configurations, and individual preferences. Electronic component notation should precisely determine the actions to be performed, as well as the required operation of the program (the electronic score). The score and the program should be prepared in such a way as to make the performance of a piece possible by an LEP artist equipped with both a traditional set of knobs and sliders, and, say, a set of movement detectors. It is not an easy task, for a number of reasons, including the different types of music data, and controller ranges, MIDI or Open Sound Control messages, effect values, ranges of frequency, volume, duration, and so on. The varied nature of musical data does not make the task easy.

With reference to an electronic score (or program), a desirable solution should be a standard for patch preparation. The *concert patch* presented during Max/MSP courses at IRCAM, or the diagram of the patch to work with a CNMAT MIDI pedal are such attempts. Furthermore, patches should retain open structure to allow the LEP performer to configure the program control method.

32.13 Conclusions

Electronics in the service of music, including the art of *live electronics*, has been developing very rapidly. When the frontline of progress sees the emergence of still newer devices, it is perhaps worthwhile to take a step back and look in retrospect at the pros and cons. What do we gain from the use of electronics? What can we lose? How do we combine traditional values with the potential of today? It does not concern timbre alone, but also the technique of composing and performing music. It is also worthwhile to look for universal potential, for repeatable techniques, which can be reproduced in various circumstances and with various equipment.

Live electronics preparation (LEP) is one such attempt (naturally, by no means the only one) to standardize some of the phenomena from the milieu of live electronics. LEP is a bridge between the worlds of acoustic and electronic music. It lets one maintain all the potential (current and future) of acoustic instruments, acoustic textures, and techniques of sound organization, as well as expanding it by adding the achievements of electronics. It occurs in an organic merger, an interactive process that takes place both during composing and during concert performance. I suggested that electronic transformation be treated as a form of instrumental preparation, hence the proposed name for the technique ("live electronic preparation"). And just like "plugged" music is chronologically the successor to acoustic music, in the LEP technique, electronic transformation is only derivative of the acoustic source.

The ability to use electronic equipment, the "LEP setup," including the choice of microphones and loudspeakers, command of musical programs, virtuosity in handling controllers—all this belongs to the realm of the new type of musician (the LEP performer). The LEP part should find its reflection in standard notation, whereas perfecting these skills should enter curricula as a separate form of performing specialization.

References

Fubini, Enrico. 1997. *Historia Estetyki Muzycznej (film)*. Krakow: Musica Iagellonica.

Emmerson, Simon. 2009. Combining the Acoustic and the Digital: Music for Instruments and Computers or Pre-recorded Sound. In *The Oxford Handbook of Computer Music*, ed. Roger T. Dean, 167–188. Oxford: Oxford University Press.

Roads, Curtis. 1997. *The Computer Music Tutorial*. Cambridge: MIT Press.

Russolo, Luigi. 2002. The Art of Noises: Futurist Manifesto. In *Audio Culture: Readings in Modern Music*, ed. Christoph Cox and Daniel Warner, 10–14. New York: Continuum.

Schaeffer, Pierre. 1966. *Traité des objets musicaux*. Paris: Éditions du Seuil.

Schafer, R. Murray. 1994. *The Soundscape: Our Sonic Environment and the Tuning of the World*. Rochester, VT: Destiny.

Varèse, Edgard. 2002. The Liberation of Sound. In *Audio Culture: Readings in Modern Music*, ed. Christoph Cox and Daniel Warner, 17–21. New York: Continuum.

CHAPTER 33

NEW TOOLS FOR INTERACTIVE AUDIO, AND WHAT GOOD THEY DO

TIM VAN GEELEN

INTERACTIVE audio has its roots in linear audio. Consequently, many audio production tools and techniques have been tailored to linear audio, including music, and sound for film and broadcast, for instance. These tools have become much more powerful over the years. Their main function, however, has hardly changed since the early days of recording and producing professional sound: to provide maximum control and power to the creator, and deliver a fixed, finite, and final product for the consumer to use. Emphasis is on the word "use" here because the amount of use the consumer, or user, can really achieve ends with the mere observing of the product. At most, the user can play along, or dance to the final product. Both of these activities are hardly interactive.

The process of creating interactive sound, in any form or fashion, requires tools focused not on maximizing the control and power of the creator, but on maximizing the usage, and interaction potential, of the final product. In practice, the vast majority of tools available for producing linear audio are largely unsuited for creating interactive audio. What remains is a limited but increasing set of tools and techniques. This chapter explores interactive audio tools in the form of hardware (including but not limited to instruments, computer interfaces, and sensory devices), software, techniques (for composition, production, or system design), and other useful tools. The chapter concludes with a statement on what the advancements in tools provide us with as creative human beings.

33.1 AN OVERVIEW OF TOOLS

Every tool available can be classified into one of two functional groups: Creation tools, and Control tools (Paine and Drummond 2009). Creation tools are used to generate sound or music, and they include synthesizer algorithms, pattern generators, stringed or percussive devices, and so on. Control tools, on the other hand, do not generate any sound themselves, but serve as processors for preproduced musical elements, like loops, recorded phrases, and so on. In addition, Control tools allow for the modification of preexisting musical elements in the form of sound effects (digital signal processors), acoustic modifiers, mixers, time- and pitch-changing algorithms, and so on. Creation tools are always also Control tools to some degree, because a generated sound that is controlled no further than its initial state would be considered noise. An example of a Creation tool is a piano or a guitar: they do not modify any existing sound, but generate sound according to user input. An example of a Control tool is a pitch-shifter, which can change the pitch of a sound emanating from a Creation tool, but which does not *generate* the sound itself.

As mentioned above, the line between these two types of tools is not fixed, since most Creation tools also offer at least some amount of Control. The acoustic piano for example does not just generate sound from the strings, but the lid also functions as a high-frequency filter. The opposite can also be true. For instance, control of an otherwise unmusical sound can be perceived as musical, by rhythmical modulation, or by resonant filtering to create a tone out of something initially toneless. The reason why I use these labels suggested by Paine and Drummond (2009), instead of the industry-standard labels "instruments" and "signal processors," is that the latter labels refer to actual tangible or digital devices, while Creation and Control tools can also encompass a composition strategy (Creation) or a software engine that uses smart sample selection based on tension within a computer game (Control). The strategies or techniques in question are not generally associated with instruments or signal processors, since they do not actually create or alter anything audible, but do influence the production and use of an audio product. Theoretically, even something like the material of the buttons on a coffee maker could be seen as a Control tool, since they alter the way the machine sounds when operated (and indeed this knowledge is used to some extent in product design), but for the sake of brevity, I will not be discussing devices not clearly intended for musical or other sonic production.

Traditional human–computer interfaces generally do not feel natural at first use: to some extent, they have a learning curve. The intended use of these interfaces is usually clear: a keyboard is meant to input letters, numbers, and symbols; a mouse is intended to input cursor coordinates, and so on. While not necessarily limiting their use to such means (one can easily trigger musical notes with a standard textual keyboard), the design of traditional human–computer interfaces does not encourage ways of exploration beyond their intended use. While Control devices such as MIDI controllers aimed at DJs and producers have become more user-friendly, they are still based on traditional interfaces. Other design philosophies do not take the machine as a starting point, but the natural communication methods of humans instead. Voice recognition is one such approach, gestural control is another.

33.2 Gestural Control of Musical Sound

The practice of analyzing gestures that are natural to the user, instead of interaction through system-dictated means, is gaining more interest in both academic and commercials fields. From Apple's iPad, with its multitouch screen, and Nintendo's Wiimote to more professional tools like the Karlax and pen-style tablets, the use of gestural input is increasingly common. When trying to classify gestural systems as either Creation or Control tools, we can clearly see that they lean heavily on the Control side. The Theremin, one of the first electronic instruments, is a perfect example of a Creation tool, shaping the pitch and volume of a sine-wave purely by moving one's hands over and in front of the device. It proved to be very difficult to master, since no haptic feedback was present while playing the device. Gestural musical devices in the digital era have since mostly been based on Control, with most devices lacking any form of haptic or tactile feedback common in Creation tools.

Robert Hatten (2004, xxi) defines musical gestures in terms of interactive sound as "emergent gestalts that convey affective motion, emotion, and agency by fusing otherwise separate elements into continuities of shape and force." When we compare this statement to traditional human–computer interaction (or most any human–device interaction that does not use gestures), it becomes clear that that traditional human–computer interaction is not very useful for creating interactive sound. The singular-event basis of traditional interaction is hardly ideal for creating combinations of continuous input values. While gestural input promises great control over interactivity, it is also a very different way of developing a system, since according to Hatten's theory, singular events, however complex, do not effectively convey motion or emotion. Gesture requires a new approach from traditional system and interaction designers. The difficulty in incorporating gesture is increased by the way in which most computer systems work, because one input value can mean a great many different things, depending on multiple other input values, special and chronological context, level of familiarity with the system, and so on. Gesturally, natural connections between action and perception have to be found. While some musical instruments have very clear inherent gestures, like those of a drummer's arms, most aspects of music, and controlled sound in general, have much less obvious gestures. With the possibilities of creating sounds not playable by any traditional instruments, new gestures have to be found, since no natural association with any kind of gestural movement exists. Many performers that use these possibilities to any significant degree, like electronic music DJs, use very nonmusical ways of input, doing no more than moving knobs, faders, and trackpads, all of which feature only very minimal bodily involvement and provide a generally unattractive visual performance. The latter is something that seems to come naturally to a lot of performers that use traditional instruments.

There are, however, examples of successful gestural control over nontraditional sound-generating tools, such as the work of Chris Vik, known by the artist name of *synaecide*. One of Vik's performances at a 2011 Microsoft show (a conference on new computer technology for designers and developers) made gestural use of the Kinect system, with various audio generation software components. While most early uses of musical gestures were merely controlling a very limited number of musical elements, like starting loops and filtering out high frequencies, this performance used only a limited amount of preproduced sound, with both gestural control and gestural creation filling the biggest part of the performance.

One could argue that tools used for creating music, in whatever form, are all in some way newer interpretations of previous musical instruments, and gestural control over such devices is no more than translating the original way of controlling an instrument to other domains, aided by technology. For example, drumpads are sometimes built into keyboards to control percussive sounds. There have, however, been artists that use little to no traditional control over an instrument, yet create new and inspiring music with these nontraditional creation and control tools. For instance, Amon Tobin's album *Foley Room* was created using only manipulated field recordings. The songs were, however, completely manufactured with precision editing of many parameters, and without performance in mind. While Tobin is not known to use gestural input to any great extent, his music does sound like it was conceived in a very natural and fluent way, and has been featured in many high-budget films, games, and commercials.

Many gestural input devices are used in game consoles, like the Nintendo Wiimote and the Microsoft Kinect. These devices have already been used for gesturally controlling nonmusical parameters, and working with them has become commonplace in many design studios. With games like *Guitar Hero* and *Rock Band*, gestural input for controlling and creating music has been implemented as a main feature in games. But interactive music does not always have to be controlled so obviously by the user. As long as the user becomes aware of their direct or indirect influence on the music (in any form or way), interactivity or rather *re*activity is achieved, and this can become nearly as intricate and interesting as when the main goal is to directly control the music. For example, a more indirect form of interactive music is the RUMR system by Electronic Arts, as used in *SSX* (Conley 2012). RUMR allows for interactivity with user-chosen music, linked to actions within the game. It functions roughly as follows: First, a beat-detection system determines the tempo of a predetermined track (this can be any track the user provides, such as music ripped from a CD). Then the system makes sure it has a couple of bars in the buffer at all times, allowing for any modulation. User actions, in this case snowboard tricks, trigger different real-time effects, such as looping half of a bar or triplets, and so on. Every one of these effects starts at a music interval, and ends at one, creating a fluid experience. Longer user actions, such as jumps and falls, do not trigger singular modulation events, but use a run-time parameter linked to a parameter of one or multiple audio effects, such as filters and reverb. These effects are somewhat related to the user actions, as short actions cause short, intense effects like quick pitch shifting and triplet looping, while big actions, like doing tricks in the air, cause a thin sound, with high and low frequencies filtered out, and a big reverb. All in all, this approach uses only a single audio track coupled with real-time, musically timed audio effects, yet it feels very natural.

33.3 Creation and Control Hardware

In the 1960s, 70s, and 80s, hardware for music and audio generation was focused on creating new or improved versions of traditional instruments. In the 1980s, synthesizers became more and more powerful tools for creating new and previously unknown sounds. Apart from MIDI, almost all hardware falls into the Creation tools category, with built-in control elements. Very few new control elements were present at that time. From the 1990s and onward, control of sound sources became more important, and subsequently technology allowed for new ways of interacting with sound, through MIDI controllers with drumpads, sliders, knobs, and so on. Now, audio controllers using new communication protocols and vastly different interaction styles are available. Due to the fast growth in this field, it is difficult to index all new developments, but a very good attempt has been in the Taxonomy of real-time Interfaces for Electronic Music performance, or TIEM (Drummond and Paine 2008).

Sheizaf Rafaeli (1988) defines interactivity as a message with multiple relations to previous messages and to the relationship between them. This would imply that many tools that we see as interactive are actually not interactive, but rather reactive, because they have only one relation to a message (e.g., you turn a knob, the volume changes). Even in multisensory tools such as the iPad, every sensory input is related to one message, together combining a possibly complex set of messages, but without influences between these messages themselves. This could imply that true interactivity is less apparent than reactivity, because there are relationships between messages that may at first hand not be obvious or even recognizable at all. In other words, the user knows that he or she is directly influencing a message, but does not know exactly how. A good example of this is the *Inception* app for the iOS platform. In this app, "dreams" are created by combining various musical elements, based on many sensory inputs on the user's phone. It reads inputs like volume level over time (through the microphone), GPS location, movement, direction and speed, incoming light, and so on. Combining these values, new dreams are revealed. This represents a truly interactive system, in which the outcome is to some extent unpredictable. This doesn't have to be the case, because the user can have enough knowledge of the input values to predict the output of the combined values, thus directly controlling messages in an interactive way. Though this unpredictable nature is not necessarily native to interactive systems or tools, it does stimulate interactivity rather than reactivity.

With the rapid increase in smartphone use worldwide since the start of the decade, new chances for interactivity have arisen. Taking on the view that interactivity needs multiple relations between messages, the introduction of sensors in smartphones and other smart devices, such as tablets, gives rise to a new level of interactivity. With the number of sensors in each device growing per generation, and the data coming from them becoming freely accessible, options abound.

33.4 From Reactive to Interactive

If we do a basic usage analysis of a musical instrument, for instance the bass guitar, we can determine fairly easily if its use is reactive or interactive. The control that a bass player has over the instrument is in the dimensions of loudness, tone, pitch, and timing. This control is issued through, respectively, pressure on the strings, placement of the right-hand fingers and articulation techniques, placement of the left hand fingers and the timing of finger movement. All of these techniques are directly related to one, sometimes two dimensions of the sound. When playing music with others, the musician decides what to play according to the musical choices of the other players and the style and form of the given piece. As such, the actual sonic production of the instrument could be classified as reactive, and the choice of each message as interactive, since a musician can remember what has been played before, what other musicians have been playing, and so on. Because the aspects of the sound that can be controlled are limited, the musician can focus on creative choices rather than on how to control his instrument. When the number of input values grows (as could be the case in multisensory electronic devices such as smartphones and tablets), this control of the tool will become increasingly indirect and more difficult to predict, taking away from the focus on creative choices. Concluding from this, one could argue that with an increase in sensory inputs of a given tool, its use becomes decreasingly *consciously* creative. I say "consciously" because the definition of creativity as a completely conscious process is debatable. One can assume, however, that with full focus on what choices to make and complete control of the tool at hand (for instance, through muscle memory), creativity can take place more consciously without being distracted by the workings of the tool.

In the early 2000s, a strong rise in music- and sound-making hardware has taken place. Many previously very expensive devices, like digital pad controllers, knob and fader boxes, and MIDI-enabled instruments other than keyboards, became affordable for the prosumer and consumer markets. While powerful, these devices are mostly still on a one-sensor one-message basis; that is to say, every input value represents one sonic parameter. With the coming of more powerful digital devices, with more versatile interfaces, such as tablets, touch tables and smartphones, even more power became available for the prosumer and consumer markets. In terms of expandability, these devices often offer very little. The next step offers a combination of flexibility, powerful interfacing, and more modular, expandable, and customizable design, without the need for advanced engineering or programming skills. A good example of this is the rise of Open Source Hardware, offering a fairly simple, standardized protocol for interoperable hardware, which is mostly programmable using simplified programming languages. Furthermore, all Open Source Hardware's design needs to be publicly available so that anyone can study, alter, and distribute it. This in essence enables nonprofessionals to create their own digital devices, including many possible implementations for sound creation and control. The Arduino is a good example of a powerful, affordable, Open Source Hardware platform. While most of the above-mentioned hardware tools offer some form of feedback, or the option for the user to define a type of feedback, they initially invite the user to react, instead of interact. When viewing interactivity as a complex combination of inputs and outputs, combined with the community-driven, public availability of open-source project, multiuser systems seem a likely next step. The Reactable is one such example, where users can place autonomously functioning "sound blocks" on a touchtable that function as a sound source or modifier. Others interact with these by adding blocks, changing positions, reconnecting blocks to others, and so on.

33.5 Software

In the beginning of the digital audio era, most if not all audio creation, recording, and control software was heavily based on an analog counterpart. With growing digital power came a focus on usability. As such, audio and music software gained functionality especially useful for situations in which digital devices were previously available only in form dedicated to a single use, such as on stage or while traveling. The trend of flexible, customizable tools continues in the software realm, with Open Source software such as Pure Data, modular graphical programming languages such as Max/MSP and Reaktor, and very precise communication between software and hardware through the OSC protocol. While such programs are traditionally developed for, and only available on, desktop and laptop computers, they are increasingly being converted to, and developed for, mobile devices. This started with the ability to control desktop software through dedicated software interfaces on mobile devices such as tablets and smartphones, and expanded to full standalone and cloud-based software. The increase in digital processing power and storage is also showing a rise in physical-modeling software. Previously, there has been a focus on reproduction and modification of recorded sound, of which physical modeling can be seen as the opposite approach.

When looking at human–human interaction, rather than human–computer interaction, we can also see a rise in software that facilitates non-"face-to-face" musical interaction. An example of this is Ohm Studio, "the first real-time collaborative digital audio workstation" (Force n.d.). Besides standard digital audio workstation features, it allows musicians to directly invite others into their project. Furthermore, we can see that the previously solid line between creator and consumer is blurring more and more. In a world where "everyone has access to everything" (Curtis 2011), recorded music is becoming more and more like a fast-food consumable, and often part of a multitasking experience. Listener-to-music interaction seems like a logical next step. We already see this in art installations and game music that reacts to the user, though these are often still more reactive than truly interactive.

33.6 Techniques and Resources

One of the fields in which interactive audio has always found a place to grow is that of electronic gaming. Looking at interactive audio in games, there are a few techniques that are commonly used and referred to as interactive but that are really very simple switching or mixing tools controlled by very basic logic, and are thus more reactive tools (Rafaeli 1988). An example of this switching or mixing is *layering*, in which a musical piece is mixed together in real time from usually two to five tracks, called *stems*. These stems are mixed in when certain locations or situations arise, such as an enemy approaching, or the player is entering a building (Geelen 2008). While this can be viewed as somewhat interactive, I find the behavior of this mixing to be so basic and straightforward that it becomes predictable and unconvincing very quickly. *Branching* is another technique somewhat similar to layering (Buffoni 2012), but the stems have a different feel, such as a more dissonant harmony or a more active rhythm track. While this behavior is already more directly related to user choices, it is still very predictable and repetitive, since the number of unique combinations remains small, unless a very large number of pieces of precomposed music are available in the form of stems. Furthermore, every stem has to complement the other stems in some way, so as to not create undesirable combinations. In practice, for games, this usually results in only a handful of different stems for a very small number of tracks and situations. A third technique that is often used is that of *transitions*. This system relies not on the combination of tracks or stems, but on individually composed pieces for many different situations within a game, and it tries to tie together all possible transitions using short pieces of music specially composed to fluidly move from one piece to another. This approach allows for more dynamic and faster movement between pieces, creating a greater involvement with the rest of the game. It is, however, very labor intensive, since a transition piece has to be created for every possible jump from one piece to the next. Even if using very simple transitions like long chords or cross-fades, these transitions have to be auditioned, preferably together with the other (visual) media involved, to check if the desired effect is achieved. This technique can be very successful provided that enough music is made available to the system. It works in games where the general setting of the situation is changed very little (such as in racing games) and dynamic gameplay is achieved through many different short events (such as gaining on a car, finishing a lap, crashing, etc.). If the music is beat-driven and tempo-steady, transitions can be short and will sonically fit easier.

In film, music has to be scored very synchronously to picture, in order to support the situation of any particular moment in the story. To do this effectively, composers have come up with and applied multiple techniques to link emotions, characters, locations, and so forth, to musical elements. While in film music the director predetermines all these situations and the composer follows, in an interactive environment this is not so simple. The composer is no longer able to control the music while the game or interactive event progresses. So even if a composer for an interactive product like a video game uses these techniques to link onscreen happenings to musical elements, these don't automatically piece together well, since we don't know when, if at all, these things will happen. With growing interaction, the linear-based methods mentioned earlier are quickly becoming a much too time- and cost-intensive option, and they often have heavy constraints of investment. New options have to be found, and these options have to do the following:

(1) Provide ways to very quickly and accurately have music interact with very specific human input.
(2) Allow for fluid musical movement between different musical segments.

Any linear composed or rendered music will lack ability (1) to some extent, but may allow for very natural movement between pieces. Any prerecorded or rendered music will have trouble with ability (2). Purely generative music, where all the elements that make up a musical piece are generated and controlled in real time, will allow for both. However, this poses another set of problems:

(1) Generative music is controlled not by human creativity, but by computational interpretations of creativity. This has proven to be a very difficult thing to achieve in a broad stylistic context, since the computer has to be "taught" exactly how to interpret input in many different musical ways.
(2) Generative music requires extensive resources, since by far most music contains multiple sonic sources and an extremely wide timbral palette. Therefore, the generative system must have a great number of different ways to create and control musical sound.

While there have been successful generative musical systems in certain styles, such as David Cope's generative Bach chorales and Robert M. Keller's Jazz Impro-Visor, other styles have proven to be very difficult to create generative systems for. Part of this is because the timbral palette of both of these examples is fairly straightforward, and can therefore be described, or taught to a computer with create precision. Most other styles cannot be characterized as easily. Software such as Band-in-a-Box, can generate patterns in many styles, but a large number of these patterns are preprogrammed and the actual generative part of the patterns is limited to small, style-independent factors.

Despite the difficulty of teaching a computer how to make fitting music, certain basic mathematical models have been successfully applied to a wide range of musical aspects and styles. What follows is a comprehensive list of useful techniques and models, with a short description of each. Also included are techniques that stem from film and pop music composing, and are not mathematical models as such, but they are useful compositional approaches, and effective for interactive music.

33.7 The Jazz Approach

Jazz is a style that is very unique in some aspects. In jazz, musicians adhere to some basic common agreements on form, chords, tempo, and basic melody. Everything else is freely interpretable for every musician. However, structure and cohesion remain because the players are "at the service" of the musician currently taking the lead, and they can and will follow her down the musical route she wishes to take. For instance, if the leading musician plays a very soft and slow melody, with only small melodic changes, the harmony instruments would back this up with long, mellow chords. When the leading musician decides to increase his speed and intensity of playing, the others can follow immediately. This leading role can be switched, or handed over, from musician to musician in a very fluent and quick manner. By constantly listening to what the other musicians are doing, a jazz musician can determine his current role in the piece. If every musician is openly listening to the others, clashes in terms of harmony, rhythm, or accents can be avoided. This lead and backing approach can theoretically be applied to any musical style, and could also work well in interactive environments. Because the basic structure of the piece is the same for everyone, a precomposed piece can quickly be rearranged to another lead and backing formation. If there is a lead and backing formation for every instrument in the piece, at any moment the piece can move between the different versions, without strange and unnatural changes in the music, and without transitions. One could allow the leading part of a piece to be entirely real-time generated and highly reactive to the user, while the backing remains preproduced. This does negate the interactive nature of a jazz performance, unless the generated leading part can be interpreted by the system in some way, with the backing altered or chosen accordingly. Using Markov chains is a common technique in generative music. In a Markov-chains system, a probability matrix is used to randomly decide on a new music element: a rhythmical pattern, a melodic progression, a chord progression, and so on. This would greatly improve the experience, but also is much more labor intensive, since a new backing version for each possible change in the leading part has to be on hand at all times. The real power of this approach lies not in the interactivity, which still depends on common techniques like Markov chains, but in the adaptive nature of everything around the leading part. Science also supports the concept of one leading sound among many other sounds (Beck 2012), although this effect is more apparent when trying to block out all the other sounds. Nevertheless, a leading part is most intensely experienced when it is supported, and does not have to fight for attention. In fact, our auditory cortex actually boosts a chosen signal if other signals seem unrelated, in order to give our higher brain a clearer signal to interpret.

33.8 Nonmusical Interactive Audio Techniques

For interactive audio other than music, specific techniques have also been developed. Real-time processing of basic audio properties like volume, panning, and filtering has become commonplace in interactive art and games. More advanced processing like reverberation, complex filtering, and time and pitch manipulation are now also possible in both web-based and client- or console-based software, sometimes even hardware-accelerated (Brandon 2008). Signal processing alone is already a powerful thing, but the real advance in this field is in generative audio. For instance, by modeling various natural phenomena, many sounds found in nature but also in human-made objects can be generated with great precision. A short analysis of one of these models, used in the Xoxos plug-in Water, for instance, would follow a series of stages. First, generate a sine wave directly related to water droplet size. Then, duplicate and randomize slight adjustments in sine-wave pitch and timing. Then, modulate the sine-wave pitch upwards very quickly, spread the sine-waves across the stereo image, and finally set the volume and low-pass filters according to impact surface type. This process allows for a very precise control over liquid-like sounds, even though it is a fairly simple algorithm. Anything from a tiny stream to a giant downpour can be instantly created. While recordings of these kinds of natural sound abound, they require great amounts of memory and storage and are virtually nonadaptable. These types of generative processes can be a bit more processor intensive, but they offer great benefits. This particular example requires only some basic waveform generation and processing, but more complex models like those found in the Fauna plug-in by Xoxos can be more strenuous on the system. Fauna models the vocal tract and organs of animals and can create realistic and fictional vocal sounds.

Blurring the line between Control and Creation are techniques such as granular (re)synthesis and FFT analysis. While both require some form of sonic input, and can thus be classified as Control tools, they allow a very deep level of alteration to any sound, so much that the end result is often indistinguishable from the original. As such, these techniques are very powerful examples of interacting with sound to create new sound.

33.9 Conclusions

With the business models of software sales through apps, digital distribution, and various other technological advances, creative tools are reaching many more people than before. Eventually, nearly all creative tools will be available to the masses. Meanwhile, developers are increasingly focusing on usability and smart-functionality. At the rate advances in music technology are going, we can safely assume that sometime in the near future, the digital power of a device is no longer going to be an issue for anyone looking to create music. With such a diverse palette of easily accessible tools available to us, the focus of the creative process will shift further away from the tools and toward control over our own internal and physical creative processes. The playing of an instrument is slowly losing its value in terms of creative freedom, while the physical aspect of it remains a beautiful thing to most humans. At the same time, advances in communication technology have given us all the power to take on creative roles that would previously have been unthinkable. In this context, an individual can be a composer, remixer, critic, publisher, end user, marketer, and musician, all using just the power of current musical and communications technology available to the Western world. While we still see these professions as largely independent from each other, the lines between them are fading. As such, our need for functionality in musical creation tools (and other tools as well) will no longer be divided between Creation and Control, but rather will focus on a bigger goal. This goal is ever-moving. For example, take audio recording tools: at first, the goal of these devices was no more than capturing a basic representation of sound. Then the goal became the capturing of multiple sounds at the same time, in order to combine them. Further on, the adaptation of sound became an integral part of recording devices, beginning with simple volume control, all the way to complex analysis and resynthesis of the sound. With the coming of open-media hosting platforms, and social media being widely accepted and used, publication and distribution have become a part of many recording tools. On a more philosophical level, the question what a tool should and shouldn't do is becoming more pressing. With massive amounts of music of widely varying quality being created, modified, published, and distributed every day, new problems arise as quickly as previous problems have been overcome by advancing music creation and communication technology. To find the answer to this question, we have to find out what a creative process, and in this context, a musically creative process really is. After all, a tool that does not need any human input is not a tool at all. Many people would argue that writing a beautiful song on a piano requires a creative process. If we are able to generate the exact same outcome, without any actual interaction in the conventional sense with the piano, will we still argue that this process is equally creative? Some form of user input is considered by most people necessary for a musical piece to be of some quality, but I believe that what form this should take is the great debate that will keep us searching for a better understanding of beauty.

References

Beck, Melinda. 2012. What Cocktail Parties Teach us. *Wall Street Journal*, April 23. http://online.wsj.com/article/SB10001424052702303459004577361850069498164.html?mod=WSJ_hps_editorsPicks_1.

Brandon, Alexander. 2008. Next-Gen Audio Square-Off: PlayStation 3 vs. Xbox 360. *Gamastura*. http://www.gamasutra.com/view/feature/132208/nextgen_audio_squareoff_.php?print=1.

Buffoni, Louis-Xavier. 2012. Making Music Interactive: Elaboration on the Feature Set in Wwise. *Games Sound Design*. http://www.gamesounddesign.com/making-interactive-music-for-games-part-one.html.

Conley, Ted. 2012. Demonstration of SSX's RUMR system. *Designing Sound*. http://designingsound.org/2012/03/demonstration-of-ssxs-rumr-system/.

Drummond, John, and Garth Paine. 2008. *TIEM*. 2008. http://vipre.uws.edu.au/tiem/.

Force, Ohm. *Ohm Studio*. n.d. http://www.ohmstudio.com/ (accessed April 11, 2013).

Geelen, Tim van. 2008. Realizing Groundbreaking Adaptive Music. In *From Pac-Man to Pop Music: Interactive Audio in Games and New Media*, ed. Karen Collins, 93–102. Aldershot, UK: Ashgate.

Hatten, Robert. 2004. *Interpreting Musical Gestures, Topics, and Tropes: Mozart, Beethoven, Schubert*. Bloomington: Indiana University Press.

Paine, Garth, and Jon Drummond. 2009. Developing an Ontology of New Interfaces for Realtime. *EMS*. 2009. http://www.ems-network.org/ems09/papers/paine.pdf.

PressPausePlay. Directed by Victor Köhler David Dworsky. Performed by Hillman Curtis. 2011.

Rafaeli, Sheizaf. 1988. *From New Media to Communication*. Beverly Hills, CA: Sage.

Tobin, Amon. 2007. *Foley Room* [CD]. Ninja Tune.

Vik, Chris. *Live Looping with Ableton and Xbox Kinect*. http://chrisvik.wordpress.com/2011/06/18/live-looping-with-ableton-and-xbox-kinect/.

Index

Page numbers followed by *t* and *f* indicate tables and figures, respectively. Numbers followed by "n" indicate notes.

www.ingramcontent.com/pod-product-compliance
Ingram Content Group UK Ltd.
Pitfield, Milton Keynes, MK11 3LW, UK
UKHW050220160926
475595UK00015B/329